METALLIC MATERIALS

CORROSION TECHNOLOGY

Editor
Philip A. Schweitzer, P.E.
Consultant
York, Pennsylvania

1. *Corrosion and Corrosion Protection Handbook: Second Edition, Revised and Expanded*, edited by Philip A. Schweitzer
2. *Corrosion Resistant Coatings Technology*, Ichiro Suzuki
3. *Corrosion Resistance of Elastomers*, Philip A. Schweitzer
4. *Corrosion Resistance Tables: Metals, Nonmetals, Coatings, Mortars, Plastics, Elastomers and Linings, and Fabrics: Third Edition, Revised and Expanded (Parts A and B)*, Philip A. Schweitzer
5. *Corrosion-Resistant Piping Systems*, Philip A. Schweitzer
6. *Corrosion Resistance of Zinc and Zinc Alloys*, Frank C. Porter
7. *Corrosion of Ceramics*, Ronald A. McCauley
8. *Corrosion Mechanisms in Theory and Practice*, edited by P. Marcus and J. Oudar
9. *Corrosion Resistance of Stainless Steels*, C. P. Dillon
10. *Corrosion Resistance Tables: Metals, Nonmetals, Coatings, Mortars, Plastics, Elastomers and Linings, and Fabrics: Fourth Edition, Revised and Expanded (Parts A, B, and C)*, Philip A. Schweitzer
11. *Corrosion Engineering Handbook*, edited by Philip A. Schweitzer
12. *Atmospheric Degradation and Corrosion Control*, Philip A. Schweitzer
13. *Mechanical and Corrosion-Resistant Properties of Plastics and Elastomers*, Philip A. Schweitzer
14. *Environmental Degradation of Metals*, U. K. Chatterjee, S. K. Bose, and S. K. Roy
15. *Environmental Effects on Engineered Materials*, edited by Russell H. Jones
16. *Corrosion-Resistant Linings and Coatings*, Philip A. Schweitzer
17. *Corrosion Mechanisms in Theory and Practice: Second Edition, Revised and Expanded*, edited by Philippe Marcus
18. *Electrochemical Techniques in Corrosion Science and Engineering*, Robert G. Kelly, John R. Scully, David W. Shoesmith, and Rudolph G. Buchheit
19. *Metallic Materials: Physical, Mechanical, and Corrosion Properties*, Philip A. Schweitzer

ADDITIONAL VOLUMES IN PREPARATION

METALLIC MATERIALS
Physical, Mechanical, and Corrosion Properties

Philip A. Schweitzer, P. E.
Consultant
York, Pennsylvania, U.S.A.

CRC Press
Taylor & Francis Group
Boca Raton London New York

CRC Press is an imprint of the
Taylor & Francis Group, an **informa** business

CRC Press
Taylor & Francis Group
6000 Broken Sound Parkway NW, Suite 300
Boca Raton, FL 33487-2742

© 2003 by Taylor & Francis Group, LLC
CRC Press is an imprint of Taylor & Francis Group, an Informa business

First issued in paperback 2019

No claim to original U.S. Government works

ISBN 13: 978-0-367-44688-8 (pbk)
ISBN 13: 978-0-8247-0878-8 (hbk)

**Visit the Taylor & Francis Web site at
http://www.taylorandfrancis.com**

**and the CRC Press Web site at
http://www.crcpress.com**

Preface

Selecting the most appropriate material of construction for an application involves the making of numerous important decisions. This is true whether it be for the construction of a bridge, a household appliance, a piece of chemical processing equipment, or the decorative facing of a building. Factors such as physical and mechanical properties, corrosion resistance, workability, and cost must all be taken into consideration.

With the introduction of new metallic alloys and advances in the production of the so-called exotic metals, what was the best choice several years ago may no longer be so. Over the years, improvements have been made to specific properties of various alloys. These improvements include methods to increase mechanical, physical, and corrosion resistance properties. Alternatives in composition have also been formulated to improve the workability of many alloys.

In order to conduct a meaningful evaluation of a design, all the data needed to select the most appropriate material must be available. It is the purpose of this book to supply as much of this information as possible for commercially available metallic materials.

Chapter 1 provides background relating to the physical and mechanical properties of metals and defines the terminology. Chapter 2 provides a brief description of the various forms of corrosion to which metals may be susceptible.

Chapters 3 through 14 cover the wrought ferrous metals and alloys, providing physical, mechanical, and corrosion-resistance properties. Typical applications are also included for each metal or alloy. Similarly, Chapter 15 covers wrought nickel and high nickel alloys.

Chapter 16 provides a table of comparative corrosion resistance of wrought stainless steel and high nickel alloys.

Many applications require castings. The properties of casting will vary somewhat from the properties of the same wrought material. Chapter 17 covers the cast ferrous, nickel, and high nickel alloys.

Chapters 18 through 26 provide information on wrought and cast non-ferrous metals and their alloys, covering the same areas as in the previous chapters.

It is hoped that this book will provide invaluable insight to assist the designer in the selection of the most appropriate material for a specific application.

Philip A. Schweitzer

Contents

1
Physical and Mechanical Properties

I. INTRODUCTION

Metals have been widely used for thousands of years, commencing with the Bronze Age which took place approximately 3000 to 100 years BC. The Iron Age, which we are experiencing today, presumably replaced the Bronze Age. Although we still use considerable amounts of bronze, our steel use is many times greater.

Traditionally metals have been classified as ferrous and nonferrous. The ferrous category refers to base metals of iron, while the nonferrous metals are iron free. Ferrous alloys are used in quantities which exceed all other metals combined.

At the present time there are available for use in excess of 45,000 different metallic alloys. Although the steels and cast irons make up the largest use on a weight basis, the number of different nonferrous alloys exceed the number of ferrous alloys. The primary nonferrous alloys are those in which the base metal consists of either aluminum, copper, nickel, magnesium, titanium, or zinc.

The engineer or designer is faced with the problem of material selection for his or her project. A decision must be based on information that will permit selection of a material that will possess the necessary physical, mechanical, and corrosion resistance properties in addition to cost considerations. Cost is not only the raw material cost, but rather the finished manufactured cost in conjunction with estimated life of the finished product. The raw material with the lowest cost is not necessarily the most economical choice.

Part of the selection process necessitates the examination of the phys-

1

ical and mechanical properties. Physical behavior deals with electrical, optical, magnetic, and thermal properties. Mechanical behavior deals with the reaction of the body to a load or force. Corrosion resistance must also be taken into account. This applies whether the exposure is to the natural atmosphere, to a more aggressive atmosphere, or to physical contact with a corrodent. The specific application will determine which of the properties will be of greatest importance.

Physical and mechanical properties will be discussed in this chapter, while corrosion will be discussed in Chapter 2. We will consider the properties of

1. Modulus of elasticity
2. Tensile strength
3. Yield strength
4. Elongation
5. Hardness
6. Density
7. Specific gravity
8. Specific heat
9. Thermal conductivity
10. Thermal expansion coefficient
11. Impact strength

A. Modulus of Elasticity

The modulus of elasticity is a measure of a metal's stiffness or rigidity, which is a ratio of stress to strain of a material in the elastic region. Figure 1.1 illustrates how this property is determined; the slope of the line represents the elastic portion of the stress–strain graph (i.e., it is the stress required to produce unit strain). It is a good indication of the atom bond strength in crystalline materials. The uniaxial modulus of elasticity is often referred to as Young's modulus and is represented by E. Table 1.1 lists the moduli of some common materials.

Since the atom bond strength decreases with increasing temperature the moduli also decrease as temperature increases. Refer to Figure 1.2. Modulus has the same dimensions as stress, psi.

B. Tensile Strength

Tensile strength, also referred to as ultimate tensile strength, is the maximum resistance of a material to deformation in a tensile test carried to rupture. As stress is continuously applied to a body, a point will be reached where stress and strain are no longer related in a linear manner. In addition, if the

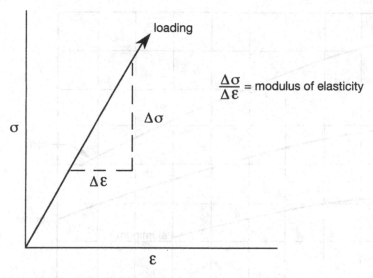

FIGURE 1.1 Determination of modulus of elasticity.

force is released, the bar will not return to its original length or shape since some permanent deformation has taken place. Figure 1.3 shows schematically an engineering stress–strain curve for an easily deformed metal. The elastic limit extends from the point of origin to the proportional limit, where departure from a linear relationship between stress and strain occurs. Since this point of departure is difficult to measure, a scheme was developed whereby a line is constructed parallel to the elastic line but is offset by a strain of 0.2% on the strain axis. The American Society for Testing and Materials (ASTM) specification E8-90 defines this offset. The point where the constructed line intersects the actual stress–strain curve is called the yield stress. The proportional limit is almost never used to define the yield stress. As the stress is increased, the slope of the curve depends on the plastic behavior of the metal being tested. For most metals the stress to maintain

TABLE 1.1 Typical Moduli at 73°F/20°C

Material	$E \times 10^6$ (psi)
Aluminum alloys	10.3
Plain carbon steel	29
Copper	16
Titanium	17

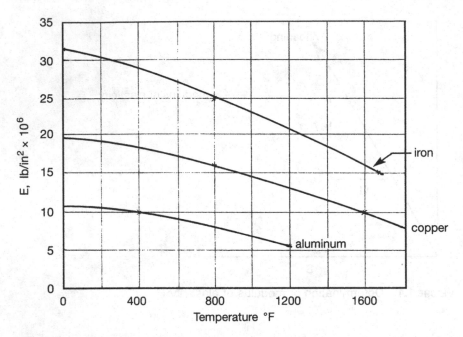

FIGURE 1.2 Young's modulus versus temperature for some common metals.

plastic flow increases due to strain hardening. Therefore the stress must increase with increasing plastic strain and the curve rises to its maximum value—the ultimate tensile strength, or tensile strength.

C. Yield Strength

Yield strength is the stress at which plastic deformation is fully developed in some portion of the material. Strain-aging types of metallic materials, such as annealed or normalized low-carbon steels, show a sudden transition from elastic to plastic behavior as the applied stress reaches a critical value. This gives a true yield point, an observable physical phenomenon from which the stress can be determined quite accurately.

In other metals the transition is gradual. For these materials the yield strength is defined as the stress required to cause a predetermined amount of plastic strain, called the offset, as described under tensile strength and illustrated in Figure 1.3.

In brittle materials, yielding does not really take place. On occasion values of yield strength are listed for brittle materials like cast iron. That is the stress at some arbitrary amount of strain. This is a special case since the material has essentially no linear portion of the curve and fracture occurs at

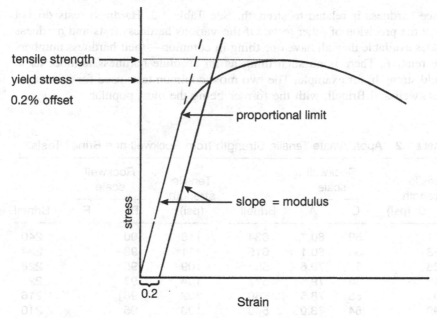

FIGURE 1.3   Typical stress–strain curve for a ductile metal.

very small strains. It does not yield in the conventially accepted meaning of the term.

D. Elongation

Elongation is a measure of ductility, as measured by the percentage of elongation. Increasing the gauge length of a specimen will decrease the percent of elongation to fracture. This is because after the neck forms all subsequent deformation takes place in the vicinity of the neck. The behavior around the neck is the same regardless of the length of the specimen. Therefore in shorter gauge lengths, a larger fraction of the specimen's length is deforming during the test. In longer specimens, the portion away from the neck is not continuing to deform after the onset of necking, therefore a smaller percentage of the specimen's length is contributing to the total deformation. Because of this it is necessary to compare percents of elongation of various metals with the same gauge length when comparing ductility.

E. Hardness

The hardness test is the most utilized mechanical property test of all methods available. These tests do not require much time and are very informative since hardness is related to strength. See Table 1.2. Hardness tests do not have the precision of other tests. Of the various hardness tests and hardness scales available they all have one thing in common—their hardness numbers are relative. There is no such thing as an absolute hardness number as in yield strength, for example. The two most common tests used for metals are Rockwell and Brinell, with the former being the most popular.

TABLE 1.2 Approximate Tensile Strength from Rockwell and Brinell Tests

Tensile strength × 10^3 (psi)	Rockwell scale C	A	Brinell	Tensile strength (psi)	Rockwell scale B	F	Brinell
351	59	80.7	634	146	100		240
358	58	80.1	615	114	99		234
325	57	79.6	595	109	95		228
313	56	79	577	104	97		222
301	55	78.5	560	102	96		216
292	54	78.0	543	100	95		210
293	53	77.4	525	98	94		205
273	52	76.8	512	94	93		200
264	51	76.3	496	92	92		195
255	50	75.9	481	90	91		190
246	49	75.2	469	89	90		185
238	48	74.5	451	88	89		180
229	47	74.1	442	86	88	75.1	176
221	46	73.6	432	84	87		172
215	45	73.1	421	83	86		169
208	44	72.5	409	82	85		165
194	42	71.5	390	81	84		162
182	40	70.4	371	80	83		159
171	38	69.4	353	77	82		156
161	36	68.4	336	73	81		153
152	34	67.4	319	72	80		150
146	32	66.3	301	70	79		147
138	30	65.3	286	69	78		144
131	28	64.3	271	68	77		141
125	26	63.3	258	67	76		139
119	24	62.4	247	66	75	99.6	137
115	22	61.5	237	65	74	99.1	135
110	20	60.5	226	63	72	98.0	130
				61	70	96.8	125
				59	68	95.6	121

The basic Rockwell tester has a number of scales which consist of various indenters used in combination with a variety of loads. The most common scales employ either a 1/16-in. diameter steel ball, a 1/8-in. diameter ball, or a diamond indenter. Each indenter can be used with a load of 60, 100, or 150 kg, giving a total of nine common scales. The Rockwell tester measures the depth of indentation which is automatically converted to a hardness number. The scale is arbitrary and very nonlinear.

The Rockwell C scale, 150 kg with the diamond pyramid indenter, and the Rockwell A scale, 60 kg with the diamond indenter, are normally used for steel and similarly hard alloys. Aluminum alloys are usually measured on the B scale, 100 kg with a 1/16-in. diamond sphere, while some copper alloys are measured on the K scale.

The Brinell hardness test measures the diameter of the indentation microscopically resulting from the penetration of a hardened steel or tungsten carbide ball that has been pressed into the surface under a specified load, usually 500 or 3000 kg.

Other hardness tests that are used include the Mohs scale, the Vickers or diamond pyramid hardness (DPH), and the Knoop hardness test. As mentioned previously the various hardness scales are not linear, however the DPH and Knoop scales are more linear than any other scale. For example a DPH value of 300 is probably close to three times the hardness of a material measured with a 100 DPH value.

F. Density

The density of a metal is its mass per unit volume. The customary unit is pounds per cubic foot.

G. Specific Gravity

Specific gravity is the ratio of the density of the metal to the density of a reference material, usually water at a specified temperature and pressure. Specific gravity is unitless.

H. Thermal Conductivity

Thermal conductivity is the quantity of heat flow under steady state conditions through unit area per unit temperature gradient in the direction perpendicular to the area.

The heat flow through a wall per unit of area is called the thermal flux, J, which is proportional to the thermal gradient, the proportionality constant being the thermal conductivity. The equation is

$$J = K \frac{\Delta T}{\Delta X}$$

where $\Delta T/\Delta X$ is the thermal gradient (i.e., temperature change) per unit thickness X and K is the thermal conductivity. The thermal conductivity is expressed as Btu ft/hr ft^2 °F in English units and as kcal m/s m^2 K in the metric system. These values are used when calculating the heat transfer through a metal wall.

I. Thermal Expansion Coefficient

The thermal expansion coefficient represents a change in dimension per unit temperature change. Values are usually given as in./in.°F or cm/cm°C. Thermal expansion can be expressed as change in either volume, area, or length, with the last being the most frequently used.

Thermal expansion data are usually reported to three significant figures and can be measured accurately by a number of means.

J. Impact

Charpy and Izod impact tests determine the amount of energy absorbed in deforming and fracturing a standard specimen by impact loading with a hammer.

The Charpy V-notch test is more commonly used for metals in this country, while the Izod impact test is used for plastics and metals in Europe. A schematic of the impact testing machine is shown in Figure 1.4 with the Charpy specimen shown above the machine.

The specimen is placed on the anvil and the pendulum hammer is released from its starting position, impacting the specimen and thereby causing a fast fracture to take place. The fracture may be of either a brittle or ductile nature or a combination of both. A pointer attached to the hammer will show on a calibrated scale the foot-pounds of energy absorbed. A brittle material will absorb very little energy and the pointer will point to the right part of the scale near the zero mark. A reading below 25 ft-lb for steels is considered unacceptable and would not be used at the temperature of the test.

These tests provide data to compare the relative ability of materials to resist brittle failure as the service temperature decreases. Body-centered steels, such as ferritic steels, tend to become embrittled as the operating temperature decreases. The impact tests can detect the transition temperature.

Defining the transition temperature poses somewhat of a problem since the transition in behavior is not sharp. Several proposals for defining the

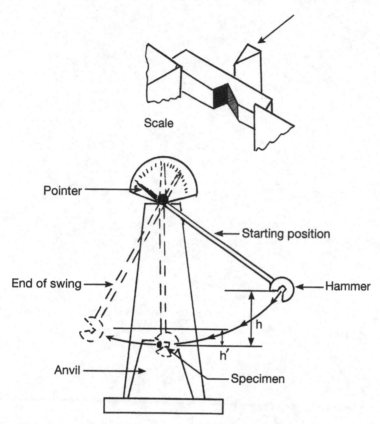

FIGURE 1.4 Schematic diagram of Charpy impact testing apparatus.

transition temperature have been made. Measures have been proposed or used which define the temperature as that at which

1. An arbitrarily chosen energy level of, say, 30 ft-lb occurs.
2. The impact energy is the average of the impact energy at high temperature and the impact energy at low temperature.
3. The fracture surface contains a certain percentage of cleavage (brittle) and fibrous (ductile) appearance, approximately 50:50.
4. Some arbitrarily predetermined amount of strain occurs in the lateral direction in the vicinity of the notch.

These tests are an indication of a metal's toughness or ability to resist crack propagation. The impact test indicates at what temperature it is safe to use the particular material being tested. It does not provide data that can be used for design purposes.

Figure 1. Schematic diagram of a hard impact testing machine

2

Corrosion of Metallic Materials

A wide variety of metals and alloys is available; these include a family of ferrous alloys and alloys of nonferrous materials as well as numerous "pure" metals. Over the years these materials have been produced as needs arose in industry for materials to handle specific corrodents and/or to operate at elevated temperatures in the presence of corrodents.

As may be expected, no one material is completely corrosion resistant, although some come close. However, as the overall corrosion resistance improves, material cost increases. Because of this, the most corrosion-resistant material is not always selected for the application. Compromises must be made. This can be done effectively by making judicious decisions.

Although other forms of attack must be considered in special circumstances, uniform attack is one form most common confronting the user of metals and alloys. The rate of uniform attack is reported in various units. In the United States it is generally reported in inches penetration per year (ipy) and milligrams per square decimeter per day (mdd). Multiply the ipy value by 1000 to convert from ipy to mpy (i.e., 0.1 in. $\times$ 1000 = 100 mpy). Conversion of ipy to mdd or vice versa requires knowledge of the metal density. Conversion factors are given in Table 2.1. The subject of uniform corrosion will be discussed later.

Corrosion is the destructive attack of a metal by a chemical or electrochemical reaction. Deterioration by physical causes is not called corrosion, but is described as erosion, galling, or wear. In some instances corrosion may accompany physical deterioration and is described by such terms as erosion corrosion, corrosive wear, or fretting corrosion.

TABLE 2.1 Conversion Factors from Inches per Year (ipy) to Milligrams per Square Decimeter per day (mdd)

Metal	Density (g/cc)	0.00144 density ($\times 10^{-3}$)	696 $\times$ density
Aluminum	2.72	0.529	1890
Brass (red)	8.75	0.164	6100
Brass (yellow)	8.47	0.170	5880
Cadmium	8.65	0.167	6020
Columbium	8.4	0.171	5850
Copper	8.92	0.161	6210
Copper-nickel (70/30)	8.95	0.161	6210
Iron	7.87	0.183	5480
Duriron	7.0	0.205	4870
Lead (chemical)	11.35	0.127	7900
Magnesium	1.74	0.826	1210
Nickel	8.89	0.162	6180
Monel	8.84	0.163	6140
Silver	10.50	0.137	7300
Tantalum	16.6	0.0868	11550
Titanium	4.54	0.317	3160
Tin	7.29	0.198	5070
Zinc	7.14	0.202	4970
Zirconium	6.45	0.223	4490

Note: Multiply ipy by (696 $\times$ density) to obtain mdd. Multiply mdd by (0.00144/density) to obtain ipy.

Direct chemical corrosion is limited to unusual conditions involving highly aggressive environments or high temperature or both. Examples are metals in contact with strong acids or alkalies.

Electrochemical reaction is the result of electrical energy passing from a negative area to a positive area through an electrolyte medium. With iron or steel in aerated water, the negative electrodes are portions of the iron surface itself, covered by porous rust (iron oxides), and positive electrodes are areas exposed to oxygen. The positive and negative electrode areas interchange and shift from place to place as the corrosion reaction proceeds. The term *Rusting* applies to the corrosion of iron-based alloys with the formation of corrosion products consisting largely of hydrous ferric oxides. Nonferrous metals and alloys corrode, but do not rust.

All structural metals corrode to some extent in material environments. Bronzes, brasses, stainless steels, zinc, and aluminum corrode so slowly under the service conditions in which they are placed that they are expected to survive for long periods without protection. When these same metals are placed into contact with more aggressive corrodents, they suffer attack and are degraded. Corrosion of structural grades of iron and steel, however, proceeds rapidly unless the metal is amply protected.

Ordinarily iron and steel corrode in the presence of both oxygen and water. If either of these ingredients are absent, corrosion will not take place. Rapid corrosion may take place in water, the rate of corrosion being accelerated by the velocity or the acidity of the water, by the motion of the metal, by an increase in temperature, by the presence of certain bacteria, or by other factors. Conversely, corrosion is retarded by protective layers (films) consisting of corrosion products or absorbed oxygen. High alkalinity of the water also retards the rate of corrosion on steel surfaces.

There are nine basic forms of corrosion that metallic materials may be subject to:

1. Uniform corrosion
2. Intergranular corrosion
3. Galvanic corrosion
4. Crevice corrosion
5. Pitting
6. Erosion corrosion
7. Stress corrosion cracking
8. Biological corrosion
9. Selective leaching

In addition there are other forms that specific metals or alloys are subject to. Prevention or control of corrosion can usually be achieved by use of a suitable material of construction, use of proper design and installation techniques, and by following specific in-plant procedures, or a combination of these.

I. UNIFORM CORROSION

A metal resists corrosion by forming a passive film on the surface. This film is formed naturally when the metal is exposed to air for a period of time. It can also be formed more quickly by a chemical treatment. For example, nitric acid if applied to an austenitic stainless steel will form this protective film. Such a film is actually a form of corrosion, but once formed it prevents future degradation of the metal, as long as the film remains intact. It does not provide an overall resistance to corrosion, since it may be subject to chemical attack. The immunity of the film to attack is a function of the film composition, the temperature, and the aggressiveness of the chemical. Examples of such films are the patina formed on copper, the rusting of iron, the tarnishing of silver, the fogging of nickel, and the high temperature oxidation of metals.

A. Passive Films

There are two theories regarding the formation of these films. The first theory states that the film formed is a metal oxide or other reaction compound. This is known as the *oxide film theory*. The second theory states that oxygen is adsorbed on the surface forming a chemisorbed film. However, all chemisorbed films react over a period of time with the underlying metal to form metal oxides. Oxide films are formed at room temperature. Metal oxides can be classified as network formers, intermediates, or modifiers. This division can be related to thin oxide films on metals. The metals that fall into network-forming or intermediate classes tend to grow protective oxides that support anion or mixed anion/cation movement. The network formers are noncrystalline, while the intermediates tend to be microcrystalline at low temperatures.

B. Passive Film on Iron

Iron in iron oxides can assume a valence of two or three. The former acts as a modifier and the latter as a network former. The iron is protected from the corrosion environment by a thin oxide film 1–4 mm in thickness with a composition of $\sqrt{Fe_2O_3}/Fe_3O_4$. This is the same type of film formed by the reaction of clean iron with oxygen or dry air. The $\sqrt{Fe_2O_3}$ layer is responsible for the passivity, while the Fe_3O_4 provides the basis for the formation of a higher oxidizing state. Iron is more difficult to passivate than nickel, because with iron it is not possible to go directly to the passivation species $\sqrt{Fe_2O_3}$. Instead, a lower oxidation state of Fe_3O_4 is required, and this film is highly susceptible to chemical dissolution. The $\sqrt{Fe_2O_3}$ layer will not form until the Fe_3O_4 phase has existed on the surface for a reasonable period of time. During this time the Fe_3O_4 layer continues to form.

C. Passive Film on Nickel

The passive film on nickel can be achieved quite readily in contrast to the formation of the passive film on iron. Differences in the nature of the oxide film on iron and nickel are responsible for this phenomenon. The film thickness on nickel is between 0.9 and 1.2 mm, while the iron oxide film is between 1.5 and 4.5 mm. There are two theories as to exactly what the passive film on nickel is. It is either entirely NiO with a small amount of nonstoichiometry giving rise to Ni^{3+} and cation vacancies, or it consists on an inner layer of NiO and an outer layer of anhydrous Ni(OH)$_2$. The passive oxide film on nickel once formed cannot be easily removed by either cathodic treatment or chemical dissolution.

D. Passive Film on Austenitic Stainless Steel

The passive film formed on stainless steel is duplex in nature, consisting of an inner barrier oxide film and an outer deposit hydroxide or salt film. Passivation takes place by the rapid formation of surface-absorbed hydrated complexes of metals, which are sufficiently stable on the alloy surface that further reaction with water enables the formation of a hydroxide phase that rapidly deprotonates to form an insoluble surface oxide film. The three most commonly used austenite stabilizers, nickel, manganese, and nitrogen, all contribute to the passivity. Chromium, a major alloying ingredient, is in itself very corrosion resistant and is found in greater abundance in the passive film than iron, which is the majority element in the alloy.

E. Passive Film on Copper

When exposed to the atmosphere over long periods of time, copper will form a coloration on the surface known as patina, which in reality is a corrosion product that acts as a protective film against further corrosion. When first formed the patina has a dark color that gradually turns green. The length of time required to form the patina depends on the atmosphere, because the coloration is given by copper hydroxide compounds. In a marine atmosphere, the compound is a mixture of copper/hydroxide/chloride and in urban or industrial atmospheres copper/hydroxide/sulfate. These compounds will form in approximately 7 years. When exposed in a clean rural atmosphere, tens or hundreds of years may be required to form patina.

F. Passive Film on Aluminum

Aluminum forms a thin, compact, and adherent oxide film on the surface which limits further corrosion. When formed in air at atmospheric temperatures it is approximately 5 mm thick. If formed at elevated temperatures or in the presence of water or water vapor it will be thicker. This oxide film is stable in the pH range of 4–9. With a few exceptions the film will dissolve at lower or higher pH ranges. Exceptions are concentrated nitric acid (pH 1) and concentrated ammonium hydroxide (pH 13). In both cases the oxide film is stable.

The oxide film is not homogeneous and contains weak points. Breakdown of the oxide film at weak points leads to localized corrosion. With increasing alloying content and on heat-treatable alloys the oxide film becomes more nonhomogeneous.

G. Passive Film on Nickel

The passive film formed on nickel will not protect the nickel from corrosive attack in oxidizing environments, such as nitric acid. When alloyed with chromium a much improved stable film results, producing a greater corrosion resistance to a variety of oxidizing media. However, these alloys are subject to attack in environments containing chloride or other halides, especially if oxidizing agents are present. Corrosion will be in the form of pitting. The addition of molybdenum or tungsten will improve the corrosion resistance.

H. Passive Film on Titanium

Titanium forms a stable, protective, strongly adherent oxide film. This film forms instantly when a fresh surface is exposed to air or moisture. Addition of alloying elements to titanium affect the corrosion resistance because these elements alter the composition of the oxide film.

   The oxide film of titanium is very thin and is attacked by only a few substances, most notable of which is hydrofluoric acid. Because of its strong affinity for oxygen, titanium is capable of healing ruptures in this film almost instantly in any environment when a trace of moisture or oxygen is present.

I. Passive Film on Tantalum

When exposed to oxidizing or slightly anodic conditions tantalum forms a thin, impervious layer of tantalum oxide. This passivating oxide has the broadest range of stability with regard to chemical attack or thermal breakdown compared to other metallic films. Chemicals or conditions which attack tantalum, such as hydrofluoric acid, are those which penetrate or dissolve the oxide film.

J. Uniform Corrosion Rates

When exposed to a corrosion medium, metals tend to enter into a chemical union with the elements of the corrosion medium, forming stable compounds similar to those found in nature. When metal loss occurs in this manner, the compound formed is referred to as the corrosion product and the metal surface is referred to as being corroded. An example of such an attack is that of halogens, particularly chlorides. They will react with and penetrate the film on stainless steel, resulting in general corrosion. Corrosion tables are developed to indicate the interaction between a chemical and a metal. This type of attack is termed uniform corrosion. It is one of the most easily measured and predictable forms of corrosion. Many references exist which report average or typical rates of corrosion for various metals in common media. One such reference is Schweitzer, Philip A. (Ed.) *Corrosion Resistance Tables*, Fourth Edition, Vols. 1–3 (Marcel Dekker, New York, 1995).

Since corrosion is so uniform, corrosion rates for materials are often expressed in terms of metal thickness lost per unit of time. One common expression is *mils per year* (mpy); sometimes *millimeters per year* is used. Because of its predictability, low rates of corrosion are often tolerated and catastrophic failures are rare if planned inspection and monitoring is implemented. For most chemical process equipment and structures, general corrosion rates of less than 3 mpy are considered acceptable. Rates between 2 and 20 mpy are routinely considered useful engineering materials for the given environment. In severe environments, materials exhibiting high general corrosion rates of between 20 and 50 mpy might be considered economically justifiable. Materials which exhibit rates of general corrosion beyond this are usually unacceptable. It should be remembered that not only does the metal loss need to be considered, but where the metal is going must also be considered. Contamination of product, even at low concentrations, can be more costly than the replacement of the corroded component.

Uniform corrosion is generally thought of in terms of metal loss due to chemical attack or dissolution of the metallic component onto metallic ions. In high temperature situations, uniform loss is more commonly preceded by its combination with another element rather than its oxidation to a metallic ion. Combination with oxygen to form metallic oxide, or scale, results in the loss of the material in its useful engineering form as it ultimately flakes off to return to nature.

To determine the corrosion rate a prepared specimen is exposed to the test environment for a period of time and then removed to determine how much metal has been lost. The exposure time, weight loss, surface area exposed, and density of the metal are used to calculate the corrosion rate of the metal using the formula

$$\text{mpy} = \frac{22.273 \; WL}{D \; A \; T}$$

where

WL = weight loss, g
D = density, g/cm^3
A = area, in^2
T = time, days

The corrosion rates calculated from the formula or taken from the tables will assist in determining how much corrosion allowance should be included in the design based on the expected lifetime of the equipment.

II. INTERGRANULAR CORROSION

Intergranular corrosion is a localized form of corrosion taking place at the grain boundaries of a metal with little or no attack on the grain boundaries themselves. This results in a loss of strength and ductility. The attack is often rapid, penetrating deeply into the metal and causing failure.

In the case of austenitic stainless steels the attack is the result of carbide precipitation during welding operations. Carbide precipitation can be prevented by using alloys containing less than 0.03% carbon, by using alloys that have been stabilized with columbium or titanium, or by specifying solution heat treatment followed by a rapid quench that will keep carbides in solution. The most practical approach is to use either a low carbon content or stabilized austenitic stainless steel.

Nickel-based alloys can also be subjected to carbide precipitation and precipitation of intermetallic phases when exposed to temperatures lower than their annealing temperatures. As with austenitic stainless steels, low carbon content alloys are recommended to delay precipitation of carbides. In some alloys, such as alloy 625, niobium, tantalum, or titanium is added to stabilize the alloy against precipitation of chromium or molybdenum carbides. These elements combine with carbon instead of the chromium or molybdenum.

III. GALVANIC CORROSION

This form of corrosion is sometimes referred to as dissimilar metal corrosion and is found in the most unusual places, often causing professionals the most headaches. Galvanic corrosion is also often experienced in older homes where modern copper water tubing is connected to the older existing carbon steel water lines. The coupling of the copper to the carbon steel causes the carbon steel to corrode. The galvanic series of metals provides details of how galvanic current will flow between two metals and which metal will corrode when they are in contact or near each other and an electrolyte is present (e.g., water). Table 2.2 lists the galvanic series.

When two different metallic materials are electrically connected and placed in a conductive solution (electrolyte), an electric potential exists. This potential difference will provide a stronger driving force for the dissolution of the less noble (more electrically negative) material. It will also reduce the tendency for the more noble material to dissolve. Notice in Table 2.2 that the precious metals of gold and platinum are at the higher potential (more noble, or cathodic) end of the series (protected end), while zinc and magnesium are at the lower potential (less noble, or anodic) end. It is this principle that forms the scientific basis for using such materials as zinc to sacrificially protect a stainless steel drive shaft on a pleasure boat.

TABLE 2.2 Galvanic Series of Metals and Alloys

Corroded end (anodic)	
Magnesium	Muntz metal
Magnesium alloys	Naval bronze
Zinc	Nickel (active)
Galvanized steel	Inconel (active)
Aluminum 6053	Hastelloy C (active)
Aluminum 3003	Yellow brass
Aluminum 2024	Admiralty brass
Aluminum	Aluminum bronze
Alclad	Red brass
Cadmium	Copper
Mild steel	Silicon bronze
Wrought iron	70/30 Cupro-nickel
Cast iron	Nickel (passive)
Ni-resist	Iconel (passive)
13% chromium stainless steel	Monel
(active)	18-8 Stainless steel type 304
50/50 lead tin solder	(passive)
Ferretic stainless steel 400 series	18-8-3 stainless steel type 316
18-8 stainless steel type 304	(passive)
(active)	Silver
18-8-3 Stainless steel type 316	Graphite
(active)	Gold
Lead	Platinum
Tin	Protected end (cathodic)

You will note that several materials are shown in two places in the galvanic series, being indicated as either active or passive. This is the result of the tendency of some metals and alloys to form surface films, especially in oxidizing environments. These films shift the measured potential in the noble direction. In this state the material is said to be passive.

The particular way in which metals will react can be predicted from the relative positions of the materials in the galvanic series. When it is necessary to use dissimilar metals, two materials should be selected which are relatively close in the galvanic series. The further apart the metals are in the galvanic series, the greater the rate of corrosion.

The rate of corrosion is also affected by the relative areas between the anode and the cathode. Since the flow of current is from the anode to the cathode, the combination of a large cathodic area and a small anodic area is undesirable. Corrosion of the anode can be 100–1000 times greater than if the two areas were equal. Ideally the anode area should be larger than the cathode area.

The passivity of stainless steel is the result of the presence of a corrosion-resistant oxide film on the surface. In most material environments it will remain in the passive state and tend to be cathodic to ordinary iron or steel. When chloride concentrations are high, such as in seawater or in reducing solutions, a change to the active state will usually take place. Oxygen starvation also causes a change to the active state. This occurs when there is no free access to oxygen, such as in crevices and beneath contamination of partially fouled surfaces.

Differences in soil concentrations, such as moisture content and resistivity, can be responsible for creating anodic and cathodic areas. Where there is a difference in concentrations of oxygen in the water or in moist soils in contact with metal at different areas, cathodes will develop at relatively high oxygen concentrations, and anodes at points of low concentrations. Stained portions of metals tend to be anodic and unstrained portions cathodic.

When joining two dissimilar metals together, galvanic corrosion can be prevented by insulating the two metals from each other. For example, when bolting flanges of dissimilar metals together, plastic washers can be used to separate the two metals.

IV. CREVICE CORROSION

Crevice corrosion is a localized type of corrosion occurring within or adjacent to narrow gaps or openings formed by metal-to-metal or metal-to-nonmetal contact. It results from local differences in oxygen concentrations, associated deposits on the metal surface, gaskets, lap joints, or crevices under a bolt or around rivet heads where small amounts of liquid can collect and become stagnant.

The material responsible for the formation of the crevice need not be metallic. Wood, plastics, rubber, glass, concrete, asbestos, wax, and living organisms have been reported to cause crevice corrosion. Once the attack begins within the crevice, its progress is very rapid. It is frequently more intense in chloride environments.

Prevention can be accomplished by proper design and operating procedures. Nonabsorbant gasketting material should be used at flanged joints, while fully penetrated butt welded joints are preferred to threaded joints. In the design of tankage, butt welded joints are preferable to lap joints. If lap joints are used, the laps should be filled with fillet welding or a suitable caulking compound designed to prevent crevice corrosion.

The critical crevice corrosion temperature of an alloy is that temperature at which crevice corrosion is first observed when immersed in a ferric chloride solution. Table 2.3 lists the critical crevice corrosion temperature of several alloys in 10% ferric chloride solution.

V. PITTING

Pitting is a form of localized corrosion that is primarily responsible for the failure of iron and steel hydraulic structures. Pitting may result in the perforation of water pipe, making it unusable even though a relatively small percentage of the total metal has been lost due to rusting. Pitting can also cause structural failure from localized weakening effects even though there is considerable sound material remaining.

The initiation of a pit is associated with the breakdown of the protective film on the surface. The main factor that causes and accelerates pitting is electrical contact between dissimilar metals, or between what are termed concentration cells (areas of the same metal where oxygen or conductive salt concentrations in water differ). These couples cause a difference of potential that results in an electric current flowing through the water or across moist steel, from the metallic anode to a nearby cathode. The cathode may be brass or copper, mill scale, or any other portion of the metal surface that is cathodic to the more active metal areas. However, when the anodic area is relatively large compared with the cathodic area, the damage is spread out and usually negligible. When the anodic area is relatively small, the metal loss is concentrated and may be serious. For example, it can be expected when large areas of the surface are generally covered by mill scale, applied coatings, or deposits of various kinds, but breaks exist in the con-

TABLE 2.3 Critical Crevice Corrosion Temperatures in 10% Ferric Chloride Solution

Alloy	Temperature (°F/°C)
Type 316	27/−3
Alloy 825	27/−3
Type 317	36/2
Alloy 904L	59/15
Alloy 220S	68/20
E-Brite	70/21
Alloy G	86/30
Alloy 625	100/38
AL-6XN	100/38
Alloy 276	130/55

tinuity of the protective material. Pitting may also develop on bare clean metal surfaces because of irregularities in the physical or chemical structure of the metal. Localized dissimilar soil conditions at the surface of steel can also create conditions that promote pitting. Figure 2.1 shows diagrammatically how a pit forms when a break in mill scale occurs.

If an appreciable attack is confined to a small area of metal acting as an anode, the developed pits are described as deep. If the area of attack is relatively large, the pits are called shallow. The ratio of deepest metal penetration to average metal penetration, as determined by weight loss of the specimen, is known as the pitting factor. A pitting factor of 1 represents uniform corrosion.

Performance in the area of pitting and crevice corrosion is often measured using critical pitting temperature (CPT), critical crevice temperature (CCT), and pitting resistance equivalent number (PREN). As a general rule, the higher the PREN, the better the resistance. The PREN will be discussed further in Chapter 7, which deals with the corrosion resistance of stainless steel since it is determined by the chromium, molybdenum, and nitrogen contents.

Prevention can be accomplished by proper materials selection, followed by a design that prevents stagnation of material and alternate wetting and drying of the surface. Also, if coatings are to be applied, care should be taken that they are continuous, without "holidays."

VI. EROSION CORROSION

Erosion corrosion results from the movement of a corrodent over the surface of a metal. The movement is associated with the mechanical wear. The increase in localized corrosion resulting from the erosion process is usually

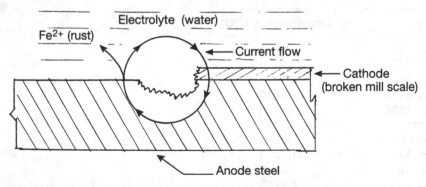

FIGURE 2.1   Formation of pit from break in mill scale.

related to the removal or damage of the protective film. The mechanism is usually identified by localized corrosion which exhibits a pattern that follows the flow of the corrodent.

This type of corrosion is also referred to as impingement attack and is caused by contact with high velocity liquids resulting in a pitting type of corrosion. It is most prevalent in condenser tubes and pipe fittings such as elbows and tees. Prevention can be accomplished by one or more means:

1. Reduce the velocity.
2. Select a harder material.
3. Properly design the piping system or the condensers.

Erosion corrosion also occurs when abrasive materials destroy or damage the passive film on the surface of the metal, thereby permitting corrosion to take place.

An additional subset of erosion corrosion is the case of cavitation, prevalent in pump impellers. This form of attack is caused by the formation and collapse of tiny vapor bubbles near a metallic surface in the presence of a corrodent. The protective film is again damaged, in this case by the high pressures caused by the collapse of the bubbles.

When two metal surfaces are in contact and experience a very slight relative motion causing damage to one or both surfaces, fretting corrosion, a special form of erosion corrosion, takes place. The movement causes mechanical damage to the protective film which can lead to erosion corrosion when a corrodent is present. This corrosion usually takes the form of a pitting type attack.

VII. STRESS CORROSION CRACKING

Certain alloys (or alloy systems) in specific environments may be subject to stress corrosion cracking (SCC). Stress corrosion cracking occurs at points of stress. Usually the metal or alloy is virtually free of corrosion over most of its surface, yet fine cracks penetrate through the surface at the points of stress. Depending on the alloy system and corrodent combination, the cracking can be intergranular or transgranular. The rate of propagation can vary greatly and is affected by stress levels, temperature, and concentration of the corrodent. This type of attack takes place in certain media. All metals are potentially subject to SCC. The conditions necessary for stress corrosion cracking are

1. Suitable environment
2. Tensile stress
3. Sensitive metal
4. Appropriate temperature and pH values

An ammonia-containing environment can induce SCC in copper-containing alloys, while with low alloy austenitic stainless steels a chloride-containing environment is necessary. It is not necessary to have a high concentration of corrodent to cause SCC. A solution containing only a few parts per million of the critical ion is all that is necessary. Temperature and pH are also factors. There is usually a threshold temperature below which SCC will not take place and a maximum or minimum pH value before cracking will start.

Normally stress corrosion cracking will not occur if the part is in compression. Failure is triggered by a tensile stress that must approach the yield stress of the metal. The stresses may be induced by faulty installation or they may be residual stresses from welding, straightening, bending, or accidental denting of the component. Pits, which act as stress concentration sites, will often initiate SCC.

Alloy content of stainless steels, particularly nickel, determine the sensitivity of the metal to SCC. Ferritic stainless steels, which are nickel free, and the high nickel alloys are not subject to stress corrosion cracking. An alloy with a nickel content greater than 30% is immune to SCC. The most common grades of stainless steel (304, 304L, 316, 316L, 321, 347, 303, and 301) have nickel contents in the range of 7–10% and are the most susceptible to stress corrosion cracking.

Examples of stress corrosion cracking include the cracking of austenitic stainless steels in the presence of chlorides, caustic embrittlement cracking of steel in caustic solutions, cracking of cold-formed brass in ammonia environments, and cracking of monel in hydrofluorosilicic acid. Table 2.4 is a partial listing of alloy systems subject to stress corrosion cracking.

In severe combinations, such as type 304 stainless steel in a boiling magnesium chloride solution, extensive cracking can be generated in a matter of hours.

Fortunately in most industrial applications the progress of SCC is much slower. However, because of the nature of the cracking it is difficult to detect until extensive corrosion has developed, which can lead to unexpected failure.

Tensile stresses can assist in other corrosion processes, such as the simple mechanical fatigue process. Corrosion fatigue is difficult to differentiate from simple mechanical fatigue, but is recognized as a factor when the environment is believed to have accelerated the normal fatigue process. Such systems can also have the effect of lowering the endurance limit such that fatigue will occur at a stress level below which it would be normally expected.

It is important that any stresses which may have been induced during the fabrication be removed by an appropriate stress-relief operation. The

TABLE 2.4   Alloy Systems Subject to Stress Corrosion Cracking

Alloy	Environment
Carbon steel	Anhydrous liquid ammonia, HCN, ammonium nitrate, sodium nitrite, sodium hydroxide
Aluminum base	Air, seawater, salt and chemical combination
Magnesium base	Nitric acid, caustic, HF solutions, salts, coastal atmospheres
Copper base	Primarily ammonia and ammonium hydroxide, amines, mercury
Martensitic and precipitation hardening stainless steels	Seawater, chlorides, H_2S solutions
Austenitic stainless steels	Chlorides (organic and inorganic), caustic solutions, sulfurous and polythiuric acids
Nickel base	Caustic above 600°F (315°C) fused caustic, hydrofluoric acid
Titanium	Seawater, salt atmosphere, fused salt
Zirconium	$FeCl_3$ or $CuCl_2$ solutions

design should also avoid stagnant areas that could lead to pitting and the initiation of stress concentration sites.

VIII.   BIOLOGICAL CORROSION

Corrosive conditions can be developed by living microorganisms as a result of their influence on anodic and cathodic reactions. This metabolic activity can directly or indirectly cause deterioration of a metal by the corrosion process. This activity can

1.   Produce a corrosive environment
2.   Create electrolytic concentration cells on the metal surface
3.   Alter the resistance of surface films
4.   Have an influence on the rate of anodic or cathodic reaction
5.   Alter the environmental composition

Because this form of corrosion gives the appearance of pitting it is first necessary to diagnose the presence of bacteria. This is also referred to as microbial corrosion.

The term *microorganism* covers a wide variety of life forms including bacteria, blue-green cyanobacteria, algae, lichens, fungi, and protozoa. All microorganisms may be involved in the biodeterioration of materials. Pure cultures never occur under natural conditions; rather mixed cultures prevail. Of the mixed cultures only a few actually become actively involved in the process of corrosion. The other organisms support the active ones by adjusting the environmental conditions in such a manner as to support their growth. For example, in the case of metal corrosion caused by sulfate reducing bacteria (SRB), the accompanying organisms remove oxygen and produce simple carbon compounds like acetic acid and/or lactic acid as nutrients for SRB.

Bacteria are the smallest living organisms on this planet. Some can only live with and others without oxygen. Some can adapt to changing conditions and live either aerobically or anaerobically. There is a wide diversity with regard to their metabolisms. They are classified as to their source of metabolic energy as follows:

Energy source	Classification
Light	Phototrophs
Chemical reactors	Chemotrophs
Inorganic hydrogen donators	Lithotrophs
Organic hydrogen donators	Organotrophs
Carbon dioxide (cell source)	Autotrophs
Organic molecules (cell source)	Heterotrophs

These six terms may be combined to describe easily the nutritional requirements of a bacterium. For example, if energy is derived from inorganic hydrogen donators and biomass is derived from organic molecules, they are called mirotrophs (chemolithoorganotrophs).

An important feature of microbial life is the ability to degrade any naturally occurring compound. Exceptions to this rule are a few manmade materials like highly polymerized and halogenated compounds.

In addition to energy and carbon sources, nitrogen, phosphorus, and trace elements are needed by microorganisms. Nitrogen compounds may be inorganic ammonium nitrate as well as organically bound nitrogen (e.g., amino acids, nucleotides). With the help of an enzyme called nitrogenuse, bacteria are able to fix nitrogen from atmospheric nitrogen, producing ammonia, which is incorporated in cell constituents.

Phosphorus is taken up as inorganic phosphate or as organically bound phosphoroxylated compounds, such as phosphorus-containing sugars and lipids. Phosphorus in the form of adenosine triphosphate (ATP) is the main energy storing compound.

For many of the metabolic purposes trace elements are needed. Cobalt aids in the transfer of methyl groups from/to organic or inorganic molecules (vitamin B_{12}, cobalamine, is involved in the methylation of heavy metals such as mercury). Iron as Fe^{2+} or Fe^{3+} is required for the electron transport system, where it acts as an oxidizable/reducible central atom in cytochromes or in nonheme iron sulfur proteins. Magnesium acts in a similar manner in the chlorophyll molecule. Copper is an essential part of a cytachrome which, at the terminal end of the electron transport system, is responsible for the reduction of oxygen to water.

Since life cannot exist without water, water is an essential requirement for microbial life and growth. Different microorganisms have different requirements as to the amount of water needed. A solid material is surrounded by three types of water: hygroscopic, pellicular, and gravitational. Only pellicular and gravitational water are biologically available and can be used by microorganisms. The biologically available water is usually measured as the water activity, a_w, of a sample:

$$a_w = \frac{V_s}{P_w}$$

where V_s is the vapor pressure of the solution and P_w is the vapor pressure of pure water, at the same temperature. Most bacteria require an a_w value in excess of 0.90.

Hydrogen ion concentration is another important factor affecting growth. Microorganisms are classified as to their ability to grow under acidic, neutral, or alkaline conditions, being given such titles as acidophiles, neutrophiles, or alkalopholes. Most microorganisms thrive in the neutral pH range of 6–8.

Microbial growth is also affected by redox potential. Under standard conditions hydrogen is assumed to have a redox potential of −421 mV, and oxygen has a redox potential of 820 mV. Metabolism can take place within that range.

Available oxygen is another factor that influences microbial growth. Microbial growth is possible under aerated as well as under totally oxygen free conditions. Those organisms living with the amount of oxygen contained in the air are called aerobes, while those that perform their metabolism without any free oxygen are called anaerobes. These latter are able to use bound oxygen (sulfate, carbon dioxide) or to ferment organic compounds.

Temperature is another important factor affecting microbial growth. Microbial life is possible within the range of −5 to 110°C. Microorganisms are also classified as to the temperature range in which they thrive, as in the following table:

Microorganism	Temperature range
Psychrophiles	−5°C to 20°C
Psychrotrophes	5°C to 30°C
Mesophiles	20°C to 45°C
Moderate thermophiles	40°C to 55°C
Thermophiles	55°C to 85°C
Extreme thermophiles	up to 110°C

Most of the organisms live in the mesophilic range of 20 to 45°C, which corresponds to the usual temperature on the surface of the earth.

A. Corrosion of Specific Materials

Microbially induced corrosion (MIC) may occur for metallic materials in many industrial applications. MIC has been reported in the following industrial applications:

Industry	Location of MIC
Chemical processing	Pipelines, stainless steel tanks, flanged joints, welded areas, after hydrotesting with natural river or well water
Nuclear power generating	Copper-nickel, stainless steel, brass, and aluminum-bronze cooling water pipes, carbon and stainless steel piping, and tanks
Underground pipeline	Water-saturated clay type soils of near neutral pH with decaying organic matter and a source of sulfate reducing bacteria
Metalworking	Increased wear from breakdown of machinery oils and emulsions
Onshore and offshore oil and gas	Mothballed and flooded systems, oil and gas handling systems, particularly in environments soured by sulfate reducing bacteria–produced sulfides
Water treatment	Heat exchangers and piping
Sewage handling and treatment	Concrete and concrete reinforced structures
Highway maintenance	Culvert piping
Aviation	Aluminum integral wiring, tanks including fuel storage tanks

MIC of metallic materials is not a new form of corrosion. The methods by which microorganisms increase the rate of corrosion of metals and/or the susceptibility to localized corrosion in an aqueous environment are

1. *Production of metabolites.* Bacteria may produce inorganic acids, organic acids, sulfides, and ammonia, all of which may be corrosive to metallic materials.
2. *Destruction of protective layers.* Organic coatings may be attacked by various microorganisms leading to the corrosion of the underlying metal.
3. *Hydrogen embrittlement.* By acting as a source of hydrogen and/or through the production of hydrogen sulfide, microorganisms may influence hydrogen embrittlement of metals.
4. *Formation of concentration cells at the metal surface and in particular oxygen concentration cells.* A concentration cell may be formed when a biofilm or bacterial growth develops heterogeneously on the metal surface. Some bacteria may tend to trap heavy metals such as copper and cadmium within their extracellular polymeric substance, causing the formation of ionic concentration cells. These lead to localized corrosion.
5. *Modification of corrosion inhibitors.* Certain bacteria may convert nitrite corrosion inhibitors used to protect mild steel to nitrate, while other bacteria may convert nitrate inhibitors used to protect aluminum and aluminum alloys to nitrite and ammonia.
6. *Stimulation of electrochemical reactors.* An example of this type is the evolution of cathodic hydrogen from microbially produced hydrogen sulfide.

MIC can result from

1. Production of sulfuric acid by bacteria of the genus thiobacillus through the oxidation of various inorganic sulfur compounds. The concentration of the sulfuric acid may be as high as 10–12%.
2. Production of hydrogen sulfide by sulfate reducing bacteria.
3. Production of organic acids.
4. Production of nitric acid.
5. Production of ammonia.
6. Production of hydrogen sulfide.

Prevention

There are many approaches that may be used to prevent or minimize MIC. Among the choices are

1. Material change or modification
2. Environment or process parameter modification
3. Use of organic coatings
4. Cathodic protection
5. Use of biocides
6. Microbiological methods
7. Physical methods

The approach to follow depends upon the type of bacteria present. A technique that has gained importance in addition to the preventative methods is that of "simulation of biogenic attack." By simulation of the biogenic attack, a quick-motion effect can be produced that will allow materials to be tested for their compatibility for a specific application. In order to conduct the simulation properly it is necessary that a thorough knowledge of all the processes and participating microorganisms be known. The situation may be modeled under conditions that will be optimal for the microorganisms resulting in a reduced time span for the corrosion to become detectable.

IX. SELECTIVE LEACHING

When one element in a solid alloy is removed by corrosion, the process is known as selective leaching, dealloying, or dezincification. The most common example is the removal of zinc from brass alloys. When the zinc corrodes preferentially, a porous residue of copper and corrosion products remain. The corroded part often retains its original shape and may appear undamaged except for surface tarnish. However, its tensile strength and particularly its ductility have been seriously reduced.

Dezincification of brasses takes place in either localized areas on the metal surface, called plug type, or uniformly over the surface, called layer type. A plug of dezincified brass may blow out leaving a hole, while a water pipe having layer type dezincification may split open.

Conditions which favor dezincification are

1. High temperature
2. Stagnant solutions, especially if acidic
3. Porous inorganic scale formation

Brasses which contain 15% or less of zinc are usually immune. Dezincification can also be suppressed by alloying additions of tin, aluminum, arsenic, or phosphorus.

Other alloy systems are also susceptible to this form of corrosion, including the selective loss of aluminum in aluminum-copper alloys and the loss of iron in cast iron–carbon steels.

X. CORROSION MECHANISMS

Most of the commonly used metals are unstable in the atmosphere. These unstable metals are produced by reducing ores artificially; therefore they tend to return to their original state or to similar metallic compounds when exposed to the atmosphere. Exceptions to this are gold and platinum, which are already in their metal state.

Corrosion by its simplest definition is the process of a metal returning to the material's thermodynamic state. For most materials this means the formation of the oxides or sulfides from which they originally started when they were taken from the earth before being refined into useful engineering materials. Most corrosion processes are electrochemical in nature, consisting of two or more electrode reactions: the oxidation of a metal (anodic partial reaction) and the reduction of an oxidizing agent (cathodic partial reaction). The study of electrochemical thermodynamics and electrochemical kinetics is necessary in order to understand corrosion reactions. For example, the corrosion of zinc in an acidic medium proceeds according to the overall reaction

$$Zn + 2H^+ \rightarrow Zn^{2+} + H_2 \tag{1}$$

This breaks down into the anodic partial reaction

$$Zn \rightarrow Zn^{2+} + 2e \tag{2}$$

and the cathodic partial reaction

$$2H^+ + 2e \rightarrow H_2 \tag{3}$$

The corrosion rate depends on the electrode kinetics of both partial reactions. If all of the electrochemical parameters of the anodic and cathodic partial reactions are known, in principle the rate may be predicted. According to Faraday's law a linear relationship exists between the metal dissolution rate at any potential V_M and the partial anodic current density for metal dissolution i_{aM}:

$$V_M = \frac{i_{aM}}{nF} \tag{4}$$

where n is the charge number (dimensionless) which indicates the number of electrons exchanged in the dissolution reaction and F is the Faraday constant (F = 96,485 C/mol). In the absence of an external polarization, a metal in contact with an oxidizing electrolytic environment acquires spontaneously a certain potential, called the corrosion potential, E_{corr}. The partial anodic current density at the corrosion potential is equal to the corrosion current density i_{corr}. Equation (4) then becomes

$$V_{corr} = \frac{i_{corr}}{nF} \qquad (5)$$

The corrosion potential lies between the equilibrium potentials of the anodic and cathodic partial reactions.

The equilibrium potential of the partial reactions is predicted by electrochemical thermodynamics. The overall stoichiometry of any chemical reaction can be expressed by

$$0 = \varepsilon v_i \beta_i \qquad (6)$$

where β designates the reactants and the products. The stoichiometric coefficients v_i of the products are positive and of the reactants negative. The free enthalpy of reaction ΔG is

$$\Delta G = \varepsilon v_i \mu_i \qquad (7)$$

where μ_i is the chemical potential of the participating species. If Reaction (6) is conducted in an electrochemical cell, the corresponding equilibrium potential E_{rev} is given by

$$\Delta G = -nFE_{rev} \qquad (8)$$

Under standard conditions (all activities equal to one),

$$\Delta G^0 = -nFE^0 \qquad (9)$$

where ΔG^0 represents the standard free enthalpy and E^0 represents the standard potential of the reaction.

Electrode reactions are commonly written in the form

$$\varepsilon v_{ox,i} \beta_{ox,i} + ne = \varepsilon v_{red,i} \beta_{red,i} \qquad (10)$$

where $v_{ox,i}$ represents the stoichiometric coefficient of the "oxidized" species, $\beta_{ox,i}$ appearing on the left side of the equality sign together with the free electrons, and $v_{red,i}$ indicates the stoichiometric coefficients of the reducing species, $\beta_{red,i}$ appearing on the right side of the equality sign, opposite to the electrons. Equation (10) corresponds to a partial reduction reaction and the stoichiometric coefficients $v_{ox,i}$ and $v_{red,i}$ are both positive.

By setting the standard chemical potential of the solvated proton and of the molecular hydrogen equal to zero, $\mu_{H^+}^0 = 0$; $\mu_{H_2}^0 = 0$, it is possible to define the standard potential of the partial reduction reaction of Eq. (10) with respect to the standard hydrogen electrode. The standard potential of an electrode reaction that corresponds to the overall reaction

$$\varepsilon v_{ox,i} \beta_{ox,i} + \frac{n}{2} H_{2(PH_2 = 1\,bar)} = \varepsilon v_{red,i} \beta_{red,i} + nH^+_{(aH^+ = 1)} \qquad (11)$$

Table 2.5 indicates the standard potential of selected electrode reactions.

For a given reaction to take place there must be a negative free energy change, as calculated from the equation

$$\Delta G = -nFE \tag{12}$$

For this to occur the cell potential must be positive. The cell potential is taken as the difference between two half-cell reactions, the one at the cathode minus the one at the anode.

If we place pure iron in hydrochloric acid, the chemical reaction can be expressed as

$$Fe + 2HCl \rightarrow FeCl_2 + H_2\uparrow \tag{13}$$

On the electrochemical side we have

$$Fe + 2H^+ + 2Cl^{2-} \rightarrow Fe^{2+} + Cl^{2-} + H_2\uparrow \tag{14}$$

The cell potential is calculated to be

TABLE 2.5 Standard Potentials of Electrode Reactions at 25°C

Electrode	E°/V
Li⁺ + e = Li	-3.045
Mg²⁺ + 2e = Mg	-2.34
Al³⁺ + 3e = Al	-1.67
Ti²⁺ + 2e = Ti	-1.63
Cr²⁺ + 2e = Cr	-0.90
Zn²⁺ + 2e = Zn	-0.76
Cr³⁺ + 3e = Cr	-0.74
Fe²⁺ + 2e = Fe	-0.44
Ni²⁺ + 2e = Ni	-0.257
Pb²⁺ + 2e = Pb	-0.126
2H⁺ + 2e = H₂	0
Cu²⁺ + 2e = Cu	0.34
O₂ + 2H₂O + 4e = 4OH	0.401
Fe³⁺ + e = Fe²⁺	0.771
Ag⁺ + e = Ag	0.799
Pt²⁺ + 2e = Pt	1.2
O₂ + 4H⁺ + 4e = 2H₂O	1.229
Au³⁺ + 3e = Au	1.52

E = Cathode half-cell − Anode half-cell

$$E = E\left(\frac{H^+}{H_2}\right) - E\left(\frac{Fe}{Fe^{2+}}\right)$$

$$E = 0 - (-0.440)$$

$$E = +0.440$$

Since the cell is positive the reaction can take place. The larger this potential difference, the greater the driving force of the reaction. Other factors will determine whether or not corrosion does take place and if so at what rate. For corrosion to take place there must be a current flow and a completed circuit, which is then governed by Ohm's law ($I = E/R$). The cell potential calculated here represents the peak value for the case of two independent reactions. If the resistance were infinite, the cell potential would remain as calculated but there would be no corrosion. If the resistance of the circuit were zero, the potentials of each half-cell would approach each other while the rate of corrosion would be infinite.

At an intermediate resistance in the circuit, some current begins to flow and the potentials of both half-cells move slightly toward each other. This change in potential is called polarization. The resistance in the circuit is dependent on various factors, including the resistivity of the media, surface films, and the metal itself. Figure 2.2 shows the relationship between the polarization reactions at each half-cell. The intersection of the two polarization curves closely approximate the corrosion current and the combined cell potentials of the freely corroding situation.

The corrosion density can be calculated by determining the surface area once the corrosion current is determined. A corrosion rate in terms of metal loss per unit time can be determined using Faraday's laws.

In addition to estimating corrosion rates, the extent of the polarization can help predict the type and severity of corrosion. As polarization increases, corrosion decreases. Understanding the influence of environmental changes on polarization can aid in controlling corrosion. For example, in the iron–hydrochloric acid example, hydrogen gas formation at the cathode can actually slow the reaction by blocking access of hydrogen ions to the cathode site, thereby increasing circuit resistance, resulting in cathodic polarization and lowering the current flows and corrosion rate. If the hydrogen is removed by bubbling oxygen through the solution, which combines with the hydrogen to form water, the corrosion rate will increase significantly.

There are three basic causes of polarization: concentration, activation, and potential drop. Concentration polarization is the effect resulting from the excess of a species which impedes the corrosion process (as in the previous hydrogen illustration), or with the depletion of a species critical to the corrosion process.

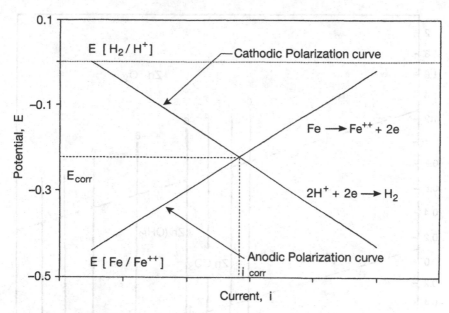

FIGURE 2.2   Polarization of iron in acid.

Activation polarization is the result of a rate-controlling step within the corrosion reaction. In the H^+/H_2 conversion reaction, the first step of the process,

$$2H^+ + 2e \rightarrow 2H$$

proceeds rapidly, while the second step,

$$2H \rightarrow H_2$$

takes place more slowly and can become a rate-controlling factor.

Potential drop is the change in voltage associated with effects of the environment and the current circuit between the anode and cathode sites. Included are the effects of surface films, corrosion products, resistivity of the media, etc.

Other factors affecting corrosion include temperature, relative velocities between the metal and the media, surface finish, grain orientation, stresses, and time.

Since corrosion is an electrochemical reaction and reaction rates increase with increasing temperature, it is logical that corrosion rates will increase with increasing temperature.

In some instances increasing the velocity of the corrodent over the surface of the metal will increase the corrosion rates when concentration

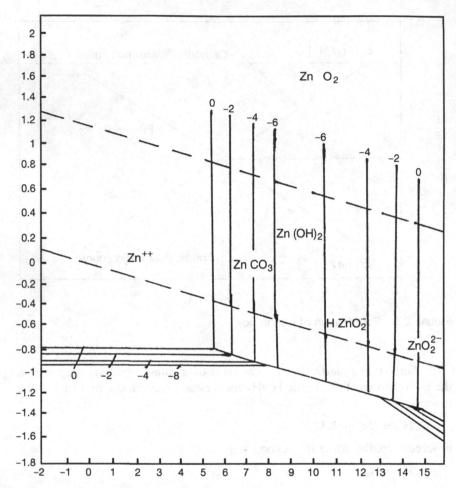

FIGURE 2.3   Potential/pH diagram for the system $Zn-CO_2-H_2O$ at 77°F (25°C).

polarization occurs. However, with passive metals, increasing the velocity can actually result in lower concentration rates since the increased velocity shifts the cathodic polarization curve so that it no longer intersects the anodic polarization curve in the active corrosion region.

Rough surfaces or tight crevices can promote the formation of concentration cells. Surface cleanliness is also a factor since deposits or films can act as initiation sites. Biological growths can behave as deposits or change the underlying surface chemistry to promote corrosion.

Variations within the metal surface on a microscopic level can influence the corrosion process. Microstructural differences such as second phases or grain orientation will affect the manner in which the corrosion process will take. The grain size of the material plays an important role in determining how rapidly the material's properties will deteriorate when the grain boundaries are attacked by corrosive environments.

Stress is a requirement for stress corrosion cracking or fatigue, but can also influence the rate of general corrosion. The severity of corrosion is affected by time. Corrosion rates are expressed as a factor of time. Some corrosion rates are rapid and violent, while most are slow and almost imperceptible on a day-to-day basis.

Potential/pH diagrams (Pourbaix diagrams) represent graphically the stability of a metal and its corrosion products as a function of the potential and pH of an aqueous solution. The pH is shown on the horizontal axis and the potential on the vertical axis. Pourbaix diagrams are widely used in corrosion because they easily permit the identification of the predominant species at equilibrium for a given potential and pH. However, being based on thermodynamic data, they provide no information on the rate of possible corrosion reactions.

In order to trace such a diagram the concentration of the dissolved material must be fixed. Figure 2.3 shows a simplified Pourbaix diagram for zinc. The numbers indicate the H_2CO_3 concentration in the moisture film, for example, 10^{-2}, 10^{-4} mol/L. The diagram shown takes into account the formation of zinc hydroxide, of Zn^{2+}, and of the zincate ions $HZnO_2^-$ and ZnO_2^{2-}. At high potentials ZnO_2 may possibly be formed, but because the corresponding thermodynamic data are uncertain they are not presented in the diagram. The broken lines indicate the domain of thermodynamic stability of water.

REFERENCES

1. PA Schweitzer. Encyclopedia of Corrosion Technology. New York: Marcel Dekker, 1998.
2. P Marcus, J Oudar. Corrosion Mechanisms in Theory and Practice. New York: Marcel Dekker, 1995.
3. F Mansfield. Corrosion Mechanisms. New York: Marcel Dekker, 1987.
4. PA Schweitzer. Corrosion Engineering Handbook. New York: Marcel Dekker, 1996.
5. HH Uhlig. Corrosion and Corrosion Control. New York: John Wiley and Sons, 1963.
6. PA Schweitzer. Corrosion and Corrosion Protection Handbook, 2nd ed. New York: Marcel Dekker, 1989.
7. GT Murray. Handbook of Materials Selection. New York: Marcel Dekker, 1997.

3

Carbon Steel

I. INTRODUCTION

Smelting of iron to extract it from its ore is believed to have started around 1300 BC in Palestine. Tools of iron appeared about this time and an iron furnace has been found. Steel is basically an alloy of iron and carbon, with the carbon content up to approximately 2 wt%.

Steel, because of its strength, formability, abundance, and low cost, is the primary metal used for structural applications. As the term "plain carbon steel" implies, these are alloys of iron and carbon. These steels were the first developed, are the least expensive, and have the widest range of applications.

In the Society of Automotive Engineers (SAE) system plain carbon steels are identified by a four digit number, the first two digits are 1 and 0 and the second two digits indicate the weight percent of carbon content. For example, a 1045 steel is a plain carbon steel alloy containing 0.45 wt% carbon. The equivalent in the Unified Numbering System (UNS) is G10450. The American Iron and Steel Institute (AISI) numbering system for plain carbon steels also includes 11XX, 12XX, and 15XX. A brief description of each type of alloy is as follows:

10XX Plain carbon, Mn 1.00% max.
11XX Plain carbon, resulfurized
12XX Plain carbon, resulfurized and rephosphorized
15XX Plain carbon, Mn range 1.00–1.65% max.

Steel is produced from pig iron by the removal of impurities in an open-hearth furnace, a basic oxygen furnace, a Bessemer convertor, or an electric furnace. In the United States over 80% of the steel is produced in the basic open-hearth furnace.

As a result of the method of production, the following elements are always present in steel: carbon, manganese, phosphorus, rubber, silicon, and traces of nitrogen, oxygen, and aluminum. The most important of these is carbon, and it is necessary to understand the effects of carbon on the internal structure of the steel to understand the heat treatment of carbon steel.

There are numerous terms which describe the phases, microstructures, and microstructural constituents. It should be kept in mind that a crystalline solid phase is a specific arrangement of atoms (crystal structure, e.g., body-centered cubic structure, face-centered cubic structure), which can only be seen by x-rays, while a microstructure is a particular arrangement of a phase or phases, which may be seen with the aid of a microscope. Figure 3.1 is an abbreviated iron–iron carbide phase diagram to 6.7% carbon which will be referred to in the following discussion of terms.

1. *Ferrite.* This is called the α phase since it is on the extreme left of the diagram. It is an interstitial solid solution of carbon in BCC iron. This is a relatively weak iron since it only contains a maximum of 0.022 wt% carbon at 727°C (eutectoid temperature).

2. *Austenite.* This is the γ phase, which consists of an interstitial solid solution of carbon in FCC iron, where the carbon atom resides in the largest interstitial sites in the center of the cube. The FCC latice has a larger interstitial site than does the BCC latice and therefore can contain more carbon, up to 2.11 wt% at 2084°F (1148°C) and up to 0.8 wt% carbon at 1338°F (727°C). Under equilibrium conditions this phase cannot exist below 1338°F (727°C).

3. *Cementite.* This phase is Fe_3C, the iron carbide compound. It has an orthorhombic crystal structure and is hard and brittle.

4. *Martensite.* This is a metastable phase that results from the rapid cooling (quenching) of a steel from the γ region to room temperature. It cannot be found on the equilibrium diagram. Martensite is considered to be supersaturated with carbon because it would be more stable if the carbon atoms were not present. This metastable phase is very hard and brittle, and most often is not present in the finished product. It plays the role of an intermediate step in the heat treatment process, where the objective is to obtain a more desirable microstructure.

5. *Pearlite.* This is a microstructure consisting of parallel platelets of α and Fe_3C. This product results when austenite of 0.8% carbon is rather slowly cooled. It has a higher hardness than the ferrite phase but much less than the hard martensite phase.

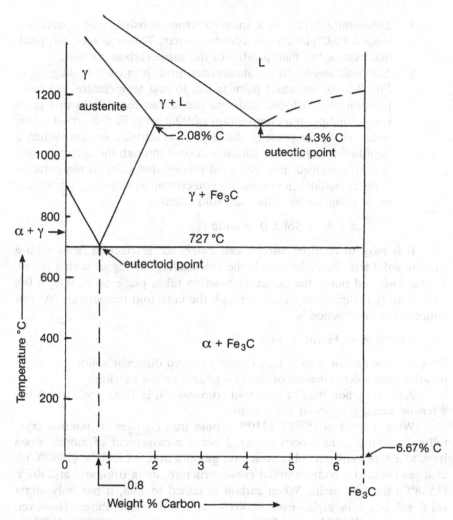

FIGURE 3.1 Abbreviated iron–iron carbide diagram to 6.7% carbon.

6. *Bainite*. Bainite is a microstructure consisting of small particles of Fe_3C in ferrite matrix. It is obtained by cooling the austenite phase at moderately slow rates. This microstructure has very desirable mechanical properties. Its hardness is between that of pearlite and the hard brittle martensite, yet it has good ductility and toughness.

7. *Spheroidite.* This is a microstructure consisting of spherically shaped Fe_3C particles in a ferrite matrix. This is a very soft product, even softer than pearlite of the same carbon content.

8. *Eutectoid steel.* In the abbreviated iron–iron carbide diagram of Fig. 3.1 the eutectoid point is the lowest temperature and composition at which the austenite phase can exist. It is that point corresponding to a composition of 99.2 wt% Fe 0.8 wt% C. This is the eutectoid point of the iron–iron carbide system. When a liquid of eutectic composition is cooled through the eutectic point, it is transformed into two solid phases that exist in the parallel-plate formation known as the eutectic microstructure. At the eutectic point we have the eutectoid reaction

solid A → Solid B + solid C

It is easy to confuse *eutectic* and *eutectoid.* Euctectoid deals with a solid-to-solid transformation, while the eutectic region begins with a liquid. At the eutectoid point, the eutectoid reaction takes place on cooling a 0.8 wt% C alloy composition slowly through the eutectoid temperature. At this temperature the reaction is

Austenite → Ferrite + Fe_cC

This is a reaction of a solid transforming to two different solids forming a parallel-plate microstructure of the two phases called pearlite.

A plain carbon steel of eutectoid composition is 1080 steel, which has a tensile strength of about 112,000 psi.

When heated to 1670°F (910°C), pure iron changes its internal crystalline structure from a body-centered cubic arrangement of atoms, alpha iron, to a face-centered cubic structure, gamma iron. At 2535°F (1390°C) it changes back to a body-centered cubic structure, delta iron, and at 2302°F (1538°C) the iron melts. When carbon is added to iron, it has only slight solid solubility in alpha iron (0.001% at room temperature). However, gamma iron will hold up to 2.0% carbon in solution at 2069°F (1130°C). The alpha iron–containing carbon, or any other element in solid solution, is called ferrite, and the gamma iron–containing element in solid solution is called austenite. Usually when not in solution in the iron, the carbon forms the compound Fe_3C (iron carbide), which is extremely hard and brittle and is known as cementite.

Carbon steel in equilibrium at room temperatures will contain both ferrite and cementite. The physical properties of the ferrite are approximately those of pure iron and are characteristic of the metal. The presence of cementite does not in itself cause steel to be hard, rather it is the shape and distribution of the carbides in the iron that determine the hardness of the steel. The fact that the carbides can be dissolved in austenite is the basis of the heat treatment of steel, because the steel can be heated above the critical temperature to dissolve all the carbides, and then suitable cooling through the cooling range will produce the desired size and distribution of carbides in the ferrite.

If austenite that contains 0.80% carbon (eutectoid composition) is slowly cooled through the critical temperature, ferrite and cementite are rejected simultaneously, forming alternate plates of lamellae. This microstructure is called pearlite because when polished and etched it has a pearly luster. If the austenite contains less than 0.80% carbon (hypoeutectoid composition), free ferrite will first be rejected on slow cooling through the critical temperature until the composition of the remaining austenite reaches 0.80% carbon, when the simultaneous rejection of both ferrite and carbide will again occur, producing pearlite. So a hypoeutectoid steel at room temperature will be composed of areas of free ferrite and areas of pearlite; the higher the carbon percentage, the more pearlite present in the steel. A 1040 steel is typical of a hypoeutectoid steel. It will be somewhat softer than a 1080 steel and have a tensile strength of about 75,000 psi.

When austenite that contains more than 0.80% carbon (hypereutectoid composition) is slowly cooled, cementite is thrown out at the austenite grain boundaries, forming a cementite network, until the austenite again contains 0.80% carbon, at which time pearlite is again formed. Thus a hypereutectoid steel, when slowly cooled, will have areas of pearlite surrounded by a thin carbide network.

As the cooling rate is increased, the spacing between the pearlite lamellae becomes smaller, with resulting greater dispersion of carbide preventing slip in the iron crystals; the steel becomes harder. Also, with an increase in the rate of cooling, there is less time for the separation of excess ferrite or cementite, and the equilibrium amount of these constituents will not be precipitated before the austenite transforms to pearlite. Thus, with a fast rate of cooling, pearlite may contain more or less carbon than that given by the eutectoid composition. When the cooling rate becomes very rapid (as obtained by quenching), the carbon does not have sufficient time to separate out in the form of carbide, and the austenite transforms to a highly stressed structure supersaturated with carbon, called martensite. This structure is exceedingly hard but brittle, and requires tempering to increase the ductility. Tempering consists of heating martensite to a temperature below critical, causing the carbide to precipitate in the form of small spheroids. The higher the tempering temperature, the larger the carbide particle size, the greater the ductility of the steel, and the lower the hardness.

In a carbon steel it is possible to have a structure consisting either of parallel plates of carbide in a ferrite matrix, the distance between the plates depending upon the rate of cooling, or of carbide spheroids in a ferrite matrix, the size of the spheroids depending on the temperature to which the hardened steel was heated.

II. HEAT TREATING OF STEEL

The following definitions of terms have been adopted by the American Society of Metals (ASM), the Society of Automotive Engineers, and the American Society for Testing and Materials (ASTM) in essentially identical form:

Heat treatment.   An operation or combination of operations involving the heating and cooling of a metal or alloy in the solid state for the purpose of obtaining certain desirable conditions or properties.

Quenching.   Rapid cooling by immersion in liquids or gases or by contact with metal.

Hardening.   Heating and quenching certain iron-based alloys from a temperature either within or above the critical range to produce a hardness superior to that obtained when the alloy is not quenched. Usually restricted to the formation of martensite.

Annealing.   Annealing is a heating and cooling operation that usually implies relatively slow cooling. The purpose of such a heat treatment may be (1) to remove stresses, (2) to induce softness, (3) to alter ductility, toughness, or electrical, magnetic, or other physical properties, (4) to refine the crystalline structure, (5) to remove gases, or (6) to produce a definite microstructure. The temperature of the operation and the rate of cooling depend upon the material being treated and the purpose of the treatment. Certain specific heat treatments coming under the comprehensive term annealing are as follows:

Process Annealing.   Heating iron-based alloys to a temperature below or close to the lower limit of the critical temperatures, generally 1000 to 1300°F (540 to 725°C).

Normalizing.   Heating iron-based alloys to approximately 1000°F (500°C) above the critical temperature range, followed by cooling to below that range in still air at ordinary temperature.

Patenting.   Heating iron-based alloys above the critical temperature range, followed by cooling below that range in air, molten lead, or a molten mixture of nitrates or nitrites maintained at a temperature usually between 800 and 1050°F (425 and 555°C), depending on the carbon content of the steel and the properties required in the finished product. This treatment is applied in the wire industry to medium or high carbon steel as a treatment to precede further wire drawing.

Spherodizing.   Any process of heating and cooling steel that produces a rounded or globular form of carbide. The following spherodizing methods are used: (1) prolonged heating at a temperature just below the lower critical temperature, usually followed by relatively slow cooling; (2) for small objects of high carbon steels, the spherodizing result is achieved more rapidly by prolonged heating to temperatures alternately within and slightly below the critical range; (3) tool steel is generally spherodized by heating to a temperature of 1380 to 1480°F (750 to 805°C) for carbon steels and higher for many alloy tool steels, holding at heat 1 to 4 hr and cooling slowly in the furnace.

Tempering (drawing).   Reheating hardened steel to a temperature below the lower critical temperature, followed by any desired rate of cooling. Although the terms tempering and drawing are practically synonymous as used in common practice, the term tempering is preferred.

III. PHYSICAL AND MECHANICAL PROPERTIES

The physical and mechanical properties of carbon steel are determined by the carbon content, the finishing operation, and any heat treatments performed. Table 3.1 lists the chemical compositions of selected carbon steels. Cold finishing is frequently used to improve the surface finish by machining and/or cold drawing. A 12% reduction in area by cold drawing can increase the yield strength by as much as 50%. Cold-finished bars fall into four classifications:

1. Cold drawn bars
2. Turned and polished bars
3. Cold drawn ground and polished bars
4. Turned, ground, and polished bars

The most commonly available steels for cold-finished carbon steel bars are:

1018	1137
1020	1141
1045	1212
1050	1213
1117	1215

Cold-finished carbon steel bars are used for bolts, nuts, motor and transmission power shafting, piston pins, bushings, oil-pump shafts and gears, etc.

TABLE 3.1   Chemical Composition of Selected Carbon Steels

AISI no.	Chemical composition ranges and limits (%)			
	C	Mn	P	S
1010	0.08–0.13	0.30–0.60	0.04 max.	0.05 max.
1020	0.18–0.23	0.30–0.60	0.04 max.	0.05 max.
1030	0.28–0.34	0.60–0.90	0.04 max.	0.05 max.
1035	0.32–0.38	0.60–0.90	0.04 max.	0.05 max.
1040	0.37–0.44	0.60–0.90	0.04 max.	0.05 max.
1045	0.43–0.50	0.60–0.90	0.04 max.	0.05 max.
1050	0.48–0.55	0.60–0.90	0.04 max.	0.05 max.
1055	0.50–0.60	0.06–0.90	0.04 max.	0.05 max.
1060	0.55–0.65	0.60–0.90	0.04 max.	0.05 max.
1070	0.65–0.75	0.60–0.90	0.04 max.	0.05 max.
1080	0.75–0.85	0.60–0.90	0.04 max.	0.05 max.
1090	0.85–0.98	0.60–0.90	0.04 max.	0.05 max.
1095	0.90–1.03	0.30–0.50	0.04 max.	0.05 max.
1137	0.32–0.37	1.35–1.65	0.04 max.	0.08–0.13
1144	0.40–0.44	0.70–1.00	0.04 max.	0.08–0.13

Besides improved mechanical properties, cold-finished steel has better machining properties than hot-rolled products. The surface finish and dimensional accuracy are also greatly improved by cold finishing. Table 3.2 gives average mechanical properties of cold-drawn steel. Table 3.3 lists the mechanical properties of hot-worked 1020 carbon steel.

The properties of plain carbon steels can be quite varied depending upon the type of heat treatment to which they are subjected. Table 3.4 provides the mechanical properties of selected carbon steels in the quenched and tempered condition, while Table 3.5 does likewise in the normalized and annealed condition.

When a carbon steel component is subjected to alternating stresses, the fatigue resistance of the steel must be considered. The fatigue limit of steels is strongly dependent on the type of microstructure, with the microstructures that give higher strengths also giving higher fatigue limits. In smooth specimens fatigue limit is approximately 50% of the ultimate tensile strength up to strength levels of about 87,000 psi, decreasing somewhat at higher stress levels. These fatigue limits can be increased by surface treatments such as carburization, nitriding, and shot peening. Some fatigue data that have been averaged from numerous values are shown in Table 3.6.

TABLE 3.2 Average Mechanical Properties of Cold-Drawn Steel

AISI no.	Tensile strength (psi)	Yield strength (psi)	Elongation in 2 in. (%)	Brinell hardness	Reduction of area (%)
1010	67,000	55,000	25.0	137	57
1015	71,000	60,300	22.0	149	55
1020	75,000	63,700	20.0	156	52
1025	80,000	68,000	18.5	163	50
1030	87,000	73,900	17.5	179	48
1035	92,000	78,200	17.0	187	45
1040	97,000	82,400	16.0	197	40
1045	102,000	86,200	15.0	207	35
1120	78,000	66,300	19.5	159	49
1137	105,000	89,200	16.0	217	35

IV. CORROSION RESISTANCE

Carbon steels are primarily affected by overall or general corrosion. Iron occurs naturally in the form of various oxides, the ores of which are refined to produce steel. Therefore, in atmospheric service they tend to return to their oxide form by a process known as rusting. The corrosion of steel is very complex, having over a dozen variables determining the rate of corrosion. Water is the most common corrosive solvent in everything from dilute solutions to concentrated acids and salt solutions. Some organic systems are also capable of causing severe corrosion.

Atmospheric corrosion of steel is a function of location. In country air the products of corrosion are either oxides or carbonates. In industrial at-

TABLE 3.3 Mechanical and Physical Properties of AISI 1020 Steel (Hot Worked)

Modulus of elasticity × 10^6 (psi)	29–30
Tensile strength × 10^3 (psi)	75
Yield strength × 10^3 (psi)	65.7
Elongation in 2 in. (%)	36
Brinell hardness	156
Density (g/cm³)	7.85
Thermal conductivity (Btu/hr ft² °F)	27
Specific heat (Btu/lb °F)	0.1
Thermal expansion × 10^{-6} (in./in. °F)	8.4

TABLE 3.4   Mechanical Properties of Selected Carbon Steels in
the Quenched and Tempered Condition

AISI no.	Tempering temperature		Yield strength (psi)	Elongation (%)	Hardness (RB)
	°C	°F			
1040	315	600	86,000	20	255
	540	1000	71,000	26	212
1060	315	600	113,000	13	321
	540	1000	97,000	17	277
1080	315	600	142,000	12	388
	540	1000	117,000	16	321
1095	315	600	118,000	10	375
	540	1000	98,000	15	321
1137	315	600	122,000	10	285
	540	1000	88,000	24	229

TABLE 3.5   Mechanical Properties of Selected Carbon Steels in
Normalized and Annealed Condition

AISI no.	Treatment[a] (°C/°F)		Yield strength (psi)	Elongation (%)	Hardness (RB)	Izod impact (ft–lb)
1020	N	870/1600	50,000	35.8	131	87
	A	870/1600	43,000	36.5	111	91
1040	N	900/1650	54,000	28.0	170	48
	A	790/1450	51,000	30.2	149	33
1060	N	900/1650	61,000	18.0	229	10
	A	790/1450	54,000	22.5	179	8
1080	N	900/1650	76,000	11.0	293	5
	A	790/1450	56,000	24.7	174	4.5
1095	N	900/1650	73,000	9.5	293	4
	A	790/1450	55,000	13.0	192	2
1137	N	900/1650	58,000	22.5	197	47
	A	790/1450	50,000	26.8	174	37

[a]N = normalized; A = annealed; temperature is that to which the piece was
heated.

TABLE 3.6 Typical Fatigue Limits of Selected
Carbon Steels

AISI no.	Condition	Ultimate tensile stress (ksi)	Fatigue limit (ksi)
1020	Normalized	57–63	27–28
1030	Normalized	65–74	33–35
1033	Quenched and tempered		31–33
1045	Pearlitic	105–109	47–48
1055	Quenched and tempered	103–152	42–44
1144	Pearlitic	100–104	48–49

mospheres sulfuric acid is present, and near the ocean some salt is in the air. Corrosion is more rapid in industrial areas because of the presence of the acid, and it is higher both near cities and near the ocean because of the higher electrical conductivity of the rain and the tendency to form soluble chlorides or sulfates, which cause the removal of protective scale.

When steel is exposed to a clean, dry atmosphere, the surface is covered with a 20- to 50-Å-thick oxide film consisting of an inner layer of Fe_2O_3. This film practically prevents further corrosion. If small amounts of water are present, FeOOH may also form.

In noncontaminated atmospheres the initiation of corrosion on a clean metal surface is a very slow process, even if the atmosphere is saturated with water vapor. Under these conditions initiation of corrosion may occur at surface inclusions of MnS, which dissolve when the surface becomes wet. However, the presence of solid particles on the surface is a more important factor for the start of corrosion. Airborne dust that has settled on the steel surface may prompt corrosion by absorbing SO_2 and water vapor from the atmosphere. Of greater importance are particles of hygroscopic salts, such as sulfates or chloride, which form a corrosive electrolyte on the surface. Rusting is rapidly initiated in SO_2^--polluted atmosphere.

Sulfur dioxide may be adsorbed on steel surfaces under atmospheric conditions. The rate of adsorption on rusty or polished steel depends on the relative humidity; high concentrations of SO_2 (>10 ppm) create a low pH in the surface film. The sulfur dioxide is oxidized on moist particles or in droplets of water to sulfuric acid. At concentrations below 90% sulfuric acid is corrosive to steel.

In general carbon steel should not be used in contact with dilute acids. At concentrations between 90 and 95% sulfuric acid, steel can be used up to the boiling point, between 80 and 90% it is serviceable at room temper-

TABLE 3.7 Compatibility of Carbon Steel with Selected Corrodents

Chemical	°F/°C	Chemical	°F/°C
Acetaldehyde	130/54	Lard oil	X
Acetic acid, all conc.	X	Linoleic acid	X
Acetic acid vapors	X	Linseed oil	90/32
Acetone	300/149	Magnesium chloride, 30%	80/27
Aluminum chloride, dry	X	Mercuric chloride	X
Aluminum fluoride	X	Mercuric nitrate	100/38
Ammonium chloride	X	Methyl alcohol	200/93
Ammonium hydroxide, 25%	210/99	Methyl ethyl ketone	200/93
Aqua regia, 3:1	X	Methylene chloride	100/38
Benzene	140/60	Mineral oil	100/38
Boric acid	X	Nitric acid	X
Bromine gas, dry	X	Oil vegetable	160/71
Bromine gas, moist	X	Oleum	80/27
Calcium chloride	140/60	Oxalic acid, all conc.	X
Calcium hydroxide, all conc.	X	Perchloric acid	X
Citric acid, all conc.	X	Petrolatum	X
Diesel fuels	200/93	Phenol	210/99
Ethanol	240/116	Phosphoric acid	X
Ferric chloride	X	Potassium chloride, 30%	210/99
Formaldehyde, to 50%	X	Potassium hydroxide, 50%	X
Formic acid	X	Propylene glycol	210/99
Glucose	170/77	Sodium chloride, 30%	150/66
Green liquor	400/204	Sodium hydroxide, to 30%	210/99
Hydrobromic acid	X	Sulfur dioxide, wet	X
Hydrochloric acid, dil.	X	Sulfuric acid, to 90%	X
Hydrochloric acid, 20%	X	Water, demineralized	X
Hydrofluoric acid, dil	X	Water, distilled	X
Hydrofluoric acid, 30%	X	Water, salt	X
Hydrofluoric acid, vapors	X	Water, sea	X
Hydrogen sulfide, dry	90/32	Water, sewage	90/32
Hydrogen sulfide, wet[a]	450/232	White liquor	X
Iodine	X	Wines	X
Lactic acid	X	Xylene	200/93

[a]Hydrogen embrittlement may occur depending upon conditions.
Note: The chemicals listed are in the pure state or in a saturated solution unless otherwise indicated. Compatibility is shown to the maximum allowable temperature for which data are available. Incompatibility is shown by an X.
Source: Ref. 1.

ature. Carbon steel is not normally used with hydrochloric, phosphoric, or nitric acids.

If iron contamination is permissible, steel can be used to handle caustic soda up to approximately 75% and 212°F (100°C). Stress relieving should be employed to reduce caustic embrittlement.

Carbon steel is susceptible to stress corrosion cracking in the presence of hydroxides, gaseous hydrogen, gaseous chlorine, hydrogen chloride, hydrogen bromide, hydrogen sulfide gas and aqueous nitrite solutions, even in low concentrations.

In general there are four types of organic compounds that can be corrosive to carbon steel:

1. Organic acids, such as acetic or formic acid.
2. Compounds that hydrolyze to produce acids. This includes chlorinated hydrocarbons (e.g., carbon tetrachloride), which react with water to produce hydrochloric acid.
3. Chelating agents, which take up or combine with transition elements.
4. Inorganic corrosives dissolved and dissociated in organic solvents. For example, hydrochloric acid dissolved in methanol.

Table 3.7 provides the compatibility of carbon steel with selected corrodents. A more comprehensive list can be found in Ref. 1.

REFERENCES

1. PA Schweitzer. Corrosion Resistance Tables. 4th ed. Vols. 1–3. New York: Marcel Dekker, 1995.
2. GT Murray. Introduction to Engineering Materials. New York: Marcel Dekker, 1993.
3. PA Schweitzer. Carbon steel and low-alloy steel. In: PA Schweitzer, ed. Corrosion and Corrosion Protection Handbook. 2nd ed. New York: Marcel Dekker, 1988.
4. DM Berger. Fundamentals and prevention of corrosion. In: PA Schweitzer, ed. Corrosion and Corrosion Protection Handbook. 2nd ed. New York: Marcel Dekker, 1988.
5. PA Schweitzer. Atmospheric Degradation and Corrosion Control. New York: Marcel Dekker, 1999.
6. GM Kirby. The corrosion of carbon and low-alloy steels. In: PA Schweitzer, ed. Corrosion Engineering Handbook. New York: Marcel Dekker, 1996.
7. PA Schweitzer. Encyclopedia of Corrosion Technology. New York: Marcel Dekker, 1998.
8. GT Murray. Handbook of Materials Selection for Engineering Applications. New York: Marcel Dekker, 1997.
9. HH Uhlig. Corrosion and Corrosion Control. New York: John Wiley & Sons, 1963.

4
Low-Alloy Carbon Steels

I. MAKEUP OF LOW-ALLOY STEELS

Low-alloy carbon steels are those steels containing about 2–8% total alloying element content and manganese, silicon, and copper content greater than that for the carbon steels, which are 1.65% Mn, 0.6% Si, and 0.6% Cu. It is possible to subdivide these hardenable steels as follows:

1. Low-carbon quenched and tempered steels having yield strengths in the 50–150 ksi range and total Ni-Cr-Mo content to approximately 4%.
2. Medium-carbon ultrahigh-strength steels having strengths in the range of 200 ksi. AISI 4130 and 4340 steels are typical examples.
3. Bearing steels, such as 8620 and 2100.
4. Chrome-molybdenum heat-resisting steels containing up to 1% Mo and 9% Cr in the AISI 9XXX series.

In addition a series of "weathering steels," which contain small additions of copper, chromium, and nickel to form a more adherent oxide during atmospheric exposure, are also produced. An example is U.S. Steel's Cor-Ten steel.

When relatively large amounts of alloying elements are added to steel, the characteristic behavior of carbon steel is obliterated. Most alloy steel is medium- or high-carbon steel to which various elements have been added to modify its properties to an appreciable extent, but it still owes its distinctive characteristics to the carbon that it contains. The percentage of alloy element required for a given purpose ranges from a few hundredths of 1% to possibly as high as 5%.

When ready for service these steels will usually contain only two constituents, ferrite and carbide. The only way that any alloying element can affect the properties of steel is to change the dispersion of carbide in the ferrite or change the properties of the carbide. The effect on the distribution of carbide is the most important factor. In large sections when carbon steels fail to harden throughout the section, even under a water quench, the hardenability of the steel can be increased by the addition of any alloying element (except possibly cobalt). The elements most effective in increasing the hardenability of steel are manganese, silicon, and chromium.

Elements such as molybdenum, tungsten, and vanadium are effective in increasing the hardenability when dissolved in the austenite, but they are usually present in the austenite in the form of carbides. The main advantage of these carbide-forming elements is that they prevent the agglomeration of carbides in tempered martensite. Tempering relieves the internal stresses in the hardened steel and causes spheroidization of the carbide particles, with resultant loss in hardness and strength. With these stable carbide-forming elements present, higher tempering temperatures may be employed without sacrificing strength. This permits these alloy steels to have a greater ductility for a given strength, or a greater strength for a given ductility, than plain carbon steels.

The third factor that contributes to the strength of alloy steel is the presence of the alloying element in the ferrite. Any element present in solid solution in a metal will increase the strength of the metal. The elements most effective in increasing the strength of the ferrite are phosphorus, silicon, manganese, nickel, molybdenum, tungsten, and chromium.

A final effect of alloying elements is their influence on the austenitic grain size. Martensite, when formed from coarse-grained austenite, has considerably less resistance to shock than that formed from a fine-grained austenite. Aluminum is the most effective element for fine-grained growth inhibitors.

Table 4.1 provides a summary of the effects of various alloying elements. This table indicates only the trends of the elements. The fact that one element has an influence on one factor does not prevent it from exerting an influence on another factor.

The high strength low-alloy (HSLA) structural steels are sometimes referred to as microalloyed steels. These are plain carbon steels with less than 3% carbon and 0.7% manganese but with microalloying additions (fractions of a percent) of other elements. These steels may be quenched and tempered but are seldom used in this condition. They have yield strengths in excess of 40 ksi with some varieties as high as 100 ksi. These steels are extremely weldable, and are available as hot-rolled sheets, strips, bars,

TABLE 4.1 Trends of Influence of Alloying Elements

Element	As dissolved in		As dissolved carbide in austenite: fine-grain toughness	As dispersed carbide in tempering: high-temperature strength and toughness	As fine nonmetallic dispersion: fine-grain toughness
	Ferrite: strength	Austenite: hardenability			
Al	Moderate	Mild	None	None	Very strong
Cr	Mild	Strong	Strong	Moderate	Slight
Co	Strong	Negative	None	None	None
Cb	Little	Strong	Strong	Strong	None
Cu	Strong	Moderate	None	None	None
Mn	Strong	Moderate	Mild	Mild	Slight
Mo	Moderate	Strong	Strong	Strong	None
Ni	Mild	Mild	None	None	None
P	Strong	Mild	None	None	None
Si	Moderate	Moderate	None	None	Moderate
Ta	Moderate	Strong	Strong	Strong	None
Ti	Strong	Strong	Very strong	Little	Moderate
W	Moderate	Strong	Strong	Strong	None
V	Mild	Very strong	Very strong	Very strong	Moderate

plates, and structured sections such as channels, I-beams, wide flanged beams, and special shapes. Cold-rolled sheets and forgings are also available. These steels can be divided into the following groups:

1. As-rolled pearlitic structural steels which contain carbon and manganese and have minimum yield strength of 40–50 ksi
2. Ferrite-pearlite steels with micro amounts of alloying ingredients
3. Weathering steels having small amounts of copper and/or microalloying elements
4. Acicular ferrite steels
5. Dual-phase steels
6. Inclusion-shaped controlled steels
7. Hydrogen-induced cracking-resistant steels with low carbon and sulfur

Table 4.2 lists the various steels along with their alloying elements. In addition to the AISI-SAE designations of the steel alloys, the American Society for Testing and Materials (ASTM) has also developed a series of specifications that apply to carbon and low-alloy steels. Table 4.3 presents selected ASTM specifications for carbon and low-alloy steels.

The addition of such alloying elements as nickel, chromium, molybdenum, and to a lesser degree vanadium and tungsten provides the carbon steel with good hardenability. Consequently these alloys exhibit a high quenched hardness and strength throughout thick sections (up to a few inches). They find application as shafts, thick discs, load supporting rods, etc. As a result of the addition of the alloying elements and the heat treating operation the cost is increased.

When steels are welded, microstructural changes occur in the heat-affected zone (HAZ). Since microstructure and properties are related, the properties in the HAZ will differ from those of the bulk material. Low-carbon steels are generally less affected by welding since there is less chance for the brittle martensite to form in the HAZ.

The alloy content (hardenability) also affects the weldability. The higher the hardness of a steel, the poorer the weldability. Usually steels with a high hardness often contain a higher percentage of martensite, which is extremely sensitive to cracking. The carbon equivalent (CE) equation has been developed to express the weldability:

$$CE = C + \frac{Mn}{6} + \frac{Ni}{15} + \frac{Cu}{15} + \frac{Cr}{5} + \frac{Mo}{5} + \frac{V}{5}$$

where the concentration of the alloying element is expressed in weight percent. The higher the concentration of the alloying elements, the higher the

TABLE 4.2 Chemical Composition of Low-Alloy Carbon Steels

	Alloy no.[a]	Composition (%)
Manganese steel	13XX	Mn 1.75
Nickel steels	23XX	Ni 3.5
	25XX	Ni 5.00
Nickel-chromium steels	31XX	Ni 1.25; Cr 0.65, 0.08
	32XX	Ni 1.75; Cr 1.07
	33XX	Ni 3.50; Cr 1.50, 1.57
	34XX	Ni 3.00; Cr 0.77
Molybdenum steels	40XX	Mo 0.20, 0.25
	44XX	Mo 0.40, 0.52
Chromium-molybdenum steel	41XX	Cr 0.50, 0.80, 0.95; Mo 0.12, 0.20, 0.25, 0.30
Nickel-chromium-molybdenum steels	43XX	Ni 1.82; Cr 0.50, 0.80; Mo 0.25
	43BVXX	Ni 1.82; Cr 0.50; Mo 0.12, 0.25; V 0.03 min.
	47XX	Ni 1.05; Cr 0.45; Mo 0.20, 0.35
	81XX	Ni 0.30; Cr 0.40; Mo 0.12
	86XX	Ni 0.55; Cr 0.50; Mo 0.20
	87XX	Ni 0.55; Cr 0.50; Mo 0.25
	88XX	Ni 0.55; Cr 0.50; Mo 0.35
	93XX	Ni 3.25; Cr 1.20; Mo 0.12
	94XX	Ni 0.45; Cr 0.40; Mo 0.12
	97XX	Ni 1.00; Cr 0.20; Mo 0.20
	98XX	Ni 1.00; Cr 0.80; Mo 0.25
Nickel-molybdenum steels	46XX	Ni 0.85, 1.82; Mo 0.20, 0.25
	48XX	Ni 3.50; Mo 0.25
Chromium steels	50XX	Cr 0.27, 0.40, 0.50, 0.65
	51XX	Cr 0.80, 0.87, 0.92, 0.95, 1.00, 1.05
	50XXX	Cr 0.50; C 1.00 min.
	51XXX	Cr 1.02; C 1.00 min.
	52XXX	Cr 1.45; C 1.00 min.
Chromium-vanadium steel	61XX	Cr 0.60, 0.80, 0.95; V 0.10, 0.15
Tungsten-chromium steel	72XX	W 1.75; Cr 0.75
Silicon-manganese steel	92XX	Si 1.40, 2.00; Mn 0.65, 0.82, 0.85; Cr 0.00, 0.65

[a]XX or XXX indicates the carbon content in hundredths of a percent.

TABLE 4.3   Selected ASTM Specifications for Carbon and Low-Alloy Steels

ASTM no.	Description
A6	Rolled steel structural plate, shapes, sheets, and bars
A20	Plate for pressure vessels
A29	Carbon and alloy bars, hot rolled and cold finished
A36	Carbon steel plate, bars, and shapes
A108	Standard quality cold-finished bars
A131	Carbon and HSLA steel plates, bars, and shapes
A238	Carbon steel plates of low or intermediate strength
A242	HSLA steel plates, bars, and shapes
A248	Carbon-Si plates for machine parts and general construction
A304	Alloy bars having hardenability requirements
A322	Hot-rolled bars
A331	Cold-finished bars
A434	Hot-finished or cold-finished quenched and tempered bars
A440	Carbon steel plates, bars, and shapes of high strength
A441	Mn-V HSLA steel plates, bars, and shapes
A505	Hot-rolled and cold-rolled alloy steel sheet and strip
A506	Regular quality hot- and cold-rolled alloy sheet and strip
A507	Drawing quality hot- and cold-rolled alloy sheet and strip
A534	Carburizing steels for antifriction bearings
A535	Special quality ball and roller bearing steel
A510	Carbon wire rods and course round wire
A514	Quenched and tempered alloy steel plates of high strength
A545	Cold heading quality medium carbon wire for machine screws
A546	Cold heading quality medium carbon for hexagonal bolts
A547	Cold heading quality alloy wire for hexgonal head bolts
A548	Cold heading quality alloy wire for tapping or sheet screws
A549	Cold heading carbon steel for wood screws
A573	Carbon steel plates requiring toughness at ambient temperature
A575	Merchant quality hot-rolled steel bars
A576	Special quality hot-rolled steel bars
A588	HSLA steel plates with 50 ksi yield strength
A633	HSLA normalized steel plates, bars, and shapes
A656	V-Al-N and Ti-Al HSLA steel plates
A659	Commercial quality hot-rolled carbon sheet and strip

hardenability and the higher the CE. Of all the elements carbon has the greatest effect. Steels with a low CE usually have good weldability.

The thickness of the section affects the cooling rate and hence weldability, since a faster cooling rate favors martensite formation. A compensated carbon equivalent (CCE) relating to thickness is as follows:

$$CCE = CE + 0.0025t$$

where t is the thickness in millimeters. Table 4.4 lists several steel families, their CE, and relative weldabilities. Part thicknesses have not been taken into account and strength values are averaged from selected steels in each group.

Pre– or post–heat treatments are not normally required for plain carbon steels having a carbon content less than 0.22% and with manganese content to 1.6% in order to be welded. A low-alloy filler metal should be used. It is recommended that carbon steels containing more than 0.5% carbon be given pre– and post–heat treatments, while higher carbon content steels *must* be given such treatments.

As the strength of HSLA steels increase the weldability decreases. HSLA steels having the same carbon content as plain carbon steels have about the same weldability. Microalloy amounts of a few tenths of a percent do not affect the weldability. Quenched and tempered structural steels with carbon contents less than 0.22% have good weldability. Heat treatments are not required for heat-treated low-alloy steels having a CE less than 0.45 in order to be welded. Above this value heat treatments are required. These CE values include common steels such as 4130, 4140, 4340, 5140, 8640, and 300M, which should be welded in the annealed state followed by a postweld heat treatment.

In order to avoid hydrogen-induced cracking the hydrogen content of all steels and filler metals must be maintained at an extremely low level (parts per million). Vacuum-melted steels are preferred.

TABLE 4.4   Weldability of Several Families of Steel

Steel type	Strength level (ksi)	Carbon equivalent	Weldability
C-Mn	47–71	0.2–0.4	Good
C-Mn + Cr + Ni + V	84–87	0.50	Fair to good
C-Mn + Cr + Ni + Mo + Si	87–109	0.60	Fair with preheating
C-Mn + Cr + Mo	114–126	0.70	Poor

Table 4.5 provides room temperature mechanical properties along with hardness numbers of selected low-alloy steels in the normalized and annealed conditions, while Table 4.6 provides the same information in the quenched and tempered condition. In each table two tempering temperatures have been included. It is possible to extrapolate how the mechanical properties may change at other temperatures. Fatigue limits of selected low-alloy carbon steels are shown in Table 4.7, and thermal properties are given in Table 4.8.

Plain carbon steel has little or no resistance against atmospheric corrosion. It is not capable of developing its own protective coating; therefore

TABLE 4.5 Mechanical Properties of Selected Low-Alloy Carbon Steels in Normalized and Annealed Condition

AISI no.	Treatment[a] (°C/°F)	Yield strength (ksi)	Elongation (%)	Hardness (RB)	Izod impact (ft-lb)
1340 N	870/1600	81	22.0	248	68
A	800/1475	63	22.5	207	52
3140 N	870/1600	87	19.7	262	40
A	815/1500	61	24.5	197	34
4130 N	870/1600	63	25.5	197	64
A	865/1585	52	28.2	156	46
4140 N	870/1600	95	17.7	307	17
A	815/1500	61	25.5	197	40
4340 N	870/1600	125	12.2	363	12
A	840/1490	69	22.0	217	38
5140 N	870/1600	69	22.7	229	28
A	830/1575	43	28.6	167	30
5150 N	870/1600	77	20.7	255	23
A	825/1520	52	22.0	197	19
6150 N	870/1600	89	21.8	269	26
A	815/1500	60	23.0	197	20
8620 N	915/1675	52	26.3	183	74
A	870/1600	56	31.3	149	83
8650 N	870/1600	100	14.0	302	10
A	795/1465	56	22.5	212	22
9255 N	900/1650	84	19.7	269	10
A	845/1550	71	21.7	119	6.5
9310 N	890/1630	83	18.8	269	88
A	845/1550	64	63.8	241	58

[a]N = normalized; A = annealed; temperature was that to which piece was heated.

TABLE 4.6 Mechanical Properties of Selected Low-Alloy Carbon Steels in Quenched and Tempered Conditions

AISI no.	Tempering temperature		Yield strength (ksi)	Elongation (%)	R.A. (%)	Hardness (RB)
	°C	°F				
1340	315	600	206	12	43	453
	540	1000	120	17	58	295
4130	315	600	200	11	43	435
	540	1000	132	17	57	315
4140	315	600	208	9	43	465
	540	1000	121	18	58	285
4340	315	600	230	10	40	486
	540	1000	156	13	51	360
5130	315	600	204	10	46	440
	540	1000	136	15	56	305
5150	315	600	230	6	40	475
	540	1000	150	15	54	340
6150	315	600	228	8	39	483
	540	1000	155	13	50	395
8630	315	600	202	10	42	430
	540	1000	130	17	54	310
8740	315	600	225	11	46	495
	540	1000	165	15	55	363
9255	315	600	260	4	10	578
	540	1000	160	15	32	352

TABLE 4.7 Typical Fatigue Limits of Selected Low-Alloy Carbon Steels

AISI no.	Condition	Ultimate tensile stress (ksi)	Fatigue limit (ksi)
1340	Quenched	158–163	71–74
4140	Quenched and tempered		55–63
4340	Quenched and tempered	159	72–79
4340	Quenched and tempered	203	84–88
4340	Quenched and tempered	246	91–94
4340	Normalized		73
8640	Quenched and tempered		78

TABLE 4.8 Thermal Properties of Selected Low-Alloy Carbon Steels

AISI no.	Condition	Thermal conductivity at 200°C (W/m·K)	Coefficient of thermal expansion (10^{-6}/°C)	
			20–100°C	20–600°C
4130	Quenched and tempered		12.2	14.6
4140	Quenched and tempered	42.3	12.3	14.5
4340			12.3	14.5
5140	Quenched and tempered	43.6		14.6
8622			11.1	

unalloyed steel used in outdoor applications is usually protected with a coating of antirust paint, or zinc, or aluminum. The low-alloy steels have a somewhat better corrosion resistance but still should have some protective coating.

Phosphate coatings are often applied to carbon steel and low-alloy carbon steel. This is also known as "bonderizing" or "parkerizing." These coatings are secondary barriers that are formed by immersion, spraying, or other application techniques with a phosphate solution.

Phosphate coatings by themselves are not normally used to provide corrosion protection. Their main advantage is in providing an excellent surface for the bonding of paint films. If used alone, the presence of pinholes in the phosphate film would result in rapid rusting under severe corrosive conditions.

Water and oxygen diffuse through paint film at a relatively high rate. The protective ability of the painted phosphate film is influenced by the properties of the phosphate film because the film is in contact with the diffused water and oxygen. Film properties of concern are solubility, uniformity, and porosity.

The solubility of phosphate films varies with pH, the type of phosphate, and its morphology. In the alkaline range of pH >10 the solubility of phosphate film is directly related to the adhesion of the paint film. Formation of corrosion cells under the paint film causes an increase in pH at the cathodic area since the cathodic reactions produced OH^- ions, and pH at the interface between the steel and paint film reaches 13.5. During the cathodic electrodeposition process for primers, pH at the surfaces reaches a value of 10–12, whereas in the anodic electrodeposition process pH at the surface is 3–4. Solubility of a phosphate film is low in the pH range of 6–8 and increases in lower or higher pH ranges. Therefore the solubility of a phosphate film at pH of approximately 10 affects paint adhesion.

Solubility also varies depending on the coating process, whether by spray or immersion. The denser the phosphate film, the lower is its solubility. A looser film is formed by spraying and has a greater solubility than one formed by immersion.

The specific phosphate compound also determines the solubility. For example, phosphated steel provides better adhesion for paint films than does phosphated zinc.

The uniformity of phosphate films is influenced by the surface condition of the base metal and the nucleation of the phosphates. Carbonaceous residues on steel surfaces resulting from rolling lubricants during the annealing of steels have an effect on the corrosion performance of painted phosphate-coated steels. The porosity of a phosphate film is increased as the result of carbon contamination on the steel surface, which promotes the formation of defects in the film.

Weathering steels were developed to provide protection from atmospheric corrosion. They are produced by alloying carbon steel with copper, chromium, nickel, phosphorus, silicon, and manganese. One such example is U.S. Steel's Cor-Ten steel.

These low-alloy steels have improved corrosion resistance in outdoor atmospheres, in rural areas, or in areas having relatively low pollution levels. The protective action of copper and other alloying elements is due to a resistant form of oxide that forms a protective coating under atmospheric conditions but has little or no favorable effect when immersed continuously in water or when exposed to severe industrial corrosive conditions.

In an industrial atmosphere, steel with 0.32% copper will corrode only half as much after 5 years as steel with 0.05% copper. A high strength low-alloy steel having the following composition (percentage) will corrode only half as much as steel having 0.32% copper:

C	0.12 max.
Mn	0.20–0.50
P	0.07–0.16
S	0.05 max.
Si	0.75 max.
Cu	0.30–0.50
Cr	0.50–1.25
Ni	0.55 max.

It will be noted that, in addition to copper, this high strength alloy also contains notable amounts of chromium and nickel, both of which are helpful in increasing strength and adding resistance to corrosion. Phosphorus, which it also contains, is another element that aids in providing protection against atmospheric corrosion.

In general, the presence of oxygen or of acidic conditions promotes the corrosion of carbon steel. Alkaline conditions inhibit corrosion. Factors that affect the corrosion resistance of these steels are

Climatic conditions
Pollution levels
Degree of sheltering from the atmosphere
Specific composition of the steel

Exposure to most atmospheres results in a corrosion rate that becomes stabilized in 3–5 years. Over this period a protective film or patina is formed that is dark brown to violet. The patina is a tightly adhering rust formation on the surface of the steel that cannot be wiped off. Since the formation of this film is dependent on pollution in the air, in rural areas where there is little or no pollution a longer time may be required to form this film. In areas that have a high pollution level of SO_2 loose particles are formed with a much higher corrosion rate. This film of loose particles offers little or no protection against continued corrosion.

When chlorides are present, such as in a marine environment, the protective film will not be formed. Under these conditions corrosion rates of the weathering steels are equivalent to those of unalloyed carbon steel.

In order to form a patina a series of wet and dry periods is required. If the steel is installed in such a manner as to be sheltered from the rain, the dark patina does not form. Instead, a rust, lighter in color, forms that provides the same resistance. The corrosion rate of the weathering steels will be the same as the corrosion rate of unalloyed steel when it is continuously exposed to wetness, such as in soil or water.

Since the patina formed has a pleasant appearance the weathering steels can be used without the application of any protective coating of antirust paint, zinc, or aluminum.

In order to receive the maximum benefit from the weathering steels consideration must be given to the design. The design should eliminate all possible areas where dirt, dust, water, and corrosion products can accumulate. When pockets are present, the time of wetness increases, which leads to the development of corrosive conditions. The design should make maximum use of exposure to the weather. Sheltering from the rain should be avoided.

While the protective film is forming, rusting will proceed at a relatively high rate, during which time rusty water is produced. This rusty water may stain masonry, pavements, and the like. Consequently, steps should be taken to prevent detrimental staining effects, such as coloring masonry brown, so that staining effects will not be obvious. The ground area exposed to staining can be covered with an easily exchangeable material such as gravel.

The corrosion mechanism for weathering steels is similar to that of unalloyed carbon steels. The rust forms a more dense and compact layer on the weathering steels than on unalloyed carbon steels. The rust layer more effectively screens the steel surface from the corrosive environments of the atmosphere. The corrosion process may be affected in several ways by this rust layer. The cathodic reaction may be affected by the low diffusion rate of oxygen, whereas the anodic reaction may be retarded by limiting the supply of water and corrosion-stimulating ions that can reach the surface of the steel. In addition, the increased electrolyte resistance may also decrease the corrosion rate.

Each of the alloying ingredients reacts in a different manner to improve the resistance of weathering steels to atmospheric corrosion.

Copper has the most pronounced effect of any of the individual elements in decreasing the corrosion rate. An increase in the copper content from 0.01 to 0.4% decreases the corrosion rate by up to 70%. Only a slight improvement in the corrosion resistance results from an increase of copper in the range of 0.2–0.5%. Several theories have been proposed regarding the mechanism by which copper improves the corrosion resistance. One theory is that the beneficial effect is due to the formation of a surface coating of metallic copper, which acts either as protection on itself or promotes anodic passivation by supporting the cathodic reaction. Another theory is that copper ions dissolved from the base metal are able to precipitate sulfide ions originating either from sulfide inclusions in the steel or from the atmospheric pollution, and thus eliminate their detrimental effect. The most probable theory is that copper forms basic sulfates with low solubility which precipitate within the pores of the rust layer, thereby decreasing their porosity. Weathering steels usually contain 0.2–0.5% copper.

When added in combination with copper, chromium and nickel further increase the corrosion resistance of weathering steels. Chromium is usually added to a content of 0.4–1%; whereas nickel is usually added up to 0.65%. Chromium appears to be more effective than nickel. The mechanical properties of the steel are improved by both elements.

Chromium is enriched in the inner rust layer together with copper and phosphorus. They promote the formation of a dense layer of amorphorus FeOOH next to the steel surface. This layer acts as a barrier to the transport of water, oxygen, and pollutants. Nickel is supposed to act by forming insoluble basic sulfates on pores of the rust layer.

Phosphorus also helps to improve the corrosion resistance of weathering steels. By increasing the phosphorus content from less than 0.01 to 0.1% a 20–30% improvement in the corrosion resistance of copper-bearing steels is realized. Phosphorus may form layers of insoluble phosphates in the rust, acting as transportation barriers in the same manner as the basic sulfates previously mentioned. A comparison of the corrosion rates of carbon, steel, a copper-phosphorus low-alloy steel, and a chromium-vanadium-copper low-alloy steel is shown in Table 4.9.

As indicated previously, weathering steels possess no particular advantage of corrosion protection in heavily polluted industrial atmospheres or in direct contact with corrodents. They will suffer the same types of corrosion as other low-alloy carbon steels.

Carbon steel and low-alloy carbon steels can be subjected to a number of types of localized corrosion, including stress corrosion cracking (SCC), sulfide stress cracking (SSC), pitting, hydrogen effects, and corrosion fatigue.

II. TYPES OF CORROSION

A. Stress Corrosion Cracking

Stress corrosion cracking occurs at points of stress. Usually the metal or alloy is virtually free of corrosion over most of its surface, yet fine cracks penetrate through the surface at the points of stress. The conditions necessary for stress corrosion cracking are

TABLE 4.9 Atmospheric Corrosion of Various Steels in Different Atmospheric Types

Atmospheric type	Exposure time (years)	Carbon steel	A242 (K11510) Cu-P Steel	A558 (K11430) Cr-V-Cu Steel
		Average reduction in thickness (mil)		
Urban	3.5	3.3	1.3	1.8
Industrial	7.5	4.1	1.5	2.1
Rural	3.5	2.0	1.1	1.4
	7.5	3.0	1.3	1.5
Severe marine	0.5	7.2	2.2	3.8
80 ft from ocean	2.0	36.0	3.3	12.2
	3.5	57.0		28.7
	5.0	Destroyed	19.4	38.8

1. A suitable environment. Chemicals capable of causing SCC in carbon steel and low-alloy carbon steels include anhydrous liquid ammonia, HCN, ammonium nitrate, sodium nitrite, sodium hydroxide, hydrogen sulfide, and hydrogen gas.
2. A tensile stress, either residual or operational.
3. Appropriate temperature and pH values.

One advantage of carbon steel is that SCC can be prevented by stress relieving after fabrication.

B. Sulfide Stress Cracking

Many corrosion processes produce hydrogen ions, which combine with electrons from the base metal to form hydrogen atoms. Two such formed hydrogen atoms may combine to form a hydrogen molecule. The majority of such molecules will form hydrogen bubbles and float away harmlessly. However, a percentage of the hydrogen atoms will diffuse into the base metal and embrittle the crystalline structure. Sulfide stress cracking will occur when a critical concentration of hydrogen is reached while a tensile stress exceeding a threshold level is present. Although H_2S does not actively participate in the SSC reaction, sulfides act to promote the entry of the hydrogen atoms into the base metal.

The susceptibility of carbon steels to SSC is directly related to their strength or hardness levels. As carbon or low-alloy carbon steel is heat treated to progressively higher levels of hardness, the time to failure decreases rapidly for a given stress level.

Temperature is also a factor. The most severe sulfide stress cracking takes place in the temperature range of 20 to 120°F (-7 to 49°C). Above 120°F (49°C) the diffusion rate of hydrogen is so fast that the hydrogen passes through the material so quickly that the critical concentration is not reached. Below 20°F (-7°C) the diffusion rate is so slow that the crucial concentration is not reached.

By carefully monitoring the processing of carbon and low-alloy carbon steels, and keeping the hardness level below 22HRC, these steels will have acceptable resistance to SSC.

C. Pitting

Carbon and low-alloy carbon steels may pit under low flow or stagnant conditions. The pits are generally shallow. In seawater the pitting rate is 5 to 45 mpy, while the overall corrosion rate in the aerated splash zone is as high as 17 mpy.

D. Corrosion Fatigue

Fatigue failures are the results of prolonged cyclic loading. If corrodents are present, the fatigue problem is worse, sometimes involving corrosion deposits accumulating in the cracks to concentrate the cycling stresses.

E. Uniform Corrosion

Carbon and low-alloy carbon steels are primarily affected by uniform, or general, corrosion. The most common current corrosive solvent is water, in everything from dilute solutions to concentrated acids and salt solutions. Some organic systems are also capable of causing serious corrosion.

F. Microbiologically Influenced Corrosion

Stagnant hydrotest water is frequently the cause of this type of corrosion. In water systems there are many bacteria that accumulate and grow on metal surfaces in colonies, changing the localized chemistry to highly corrosive conditions.

REFERENCES

1.   PA Schweitzer. Corrosion and Corrosion Protection Handbook. New York: Marcel Dekker, 1989, pp. 53–68.
2.   GN Kirby. The corrosion of carbon and low-alloy steels. In: PA Schweitzer, ed. Corrosion Engineering Handbook, New York: Marcel Dekker, 1996, pp. 35–52.
3.   GT Murray. Carbon and low-alloy steels. In: GT Murray, ed. Handbook of Materials Selection for Engineering Applications, New York: Marcel Dekker, 1997, pp. 125–148.
4.   PA Schweitzer. Atmospheric Degradation and Corrosion Control. New York: Marcel Dekker, 1999, pp. 31–42.
5.   V Kucera, E Mattsson. Atmospheric corrosion. In: M Florian, ed. Corrosion Mechanisms. New York: Marcel Dekker, 1986, pp. 249–255.

5

Cast Iron and Cast Steel

At the present time there are three designation systems commonly used for cast materials. They are (1) the Unified Numbering System (UNS), (2) Alloy Casting Institute (ACI), and (3) the American Society for Testing and Materials (ASTM).

The Unified Numbering System is the most widely accepted. In this system the metals and alloys are divided into 18 series. The UNS designations start with a single letter, which is followed by five numerical digits. Where possible the letter is suggestive of the family of metals (e.g., A is for aluminum alloys, C for copper alloys, N for nickel alloys, etc.). Table 5.1 provides the complete listing of the letter designations. If possible common designations are used within the five numerical digits for user convenience. Examples include

A 92024 Aluminum alloy 2024
C 36000 Copper alloy 360
S 31600 Type 316 stainless steel
N 10276 Nickel alloy C276

A complete listing of all UNS numbers assigned to date can be found in the publication *Metals and Alloys in the Unified Numbering System*.

Designations in the ACI system begin with two letters, which are followed by two or three numerical digits. Some also end with additional letters and/or numerical digits. Usually they begin with either a C (for corrosion-resistant materials) or an H (for heat-resistant materials). The second letter ranges from A to Z, depending upon the nickel content, and to a lesser degree upon the chromium content. For example, an alloy containing 12% chro-

69

TABLE 5.1   Letter Prefixes Used in UNS

Prefix	Alloy series
A	Aluminum and aluminum alloys
C	Copper and copper alloys
D	Steels with special mechanical properties
E	Rare earths and rate earth–like metals and alloys
F	Cast irons
G	Carbon and alloy steels
H	AISI H-steels (hardenability controlled)
J	Cast steels, except tool steels
K	Miscellaneous steels and ferrous alloys
L	Low melting temperature metals and alloys
M	Miscellaneous nonferrous metals and alloys
N	Nickel and nickel alloys
P	Precious metals and alloys
R	Reactive and refractory metals and alloys
S	Heat- and corrosion-resistant (stainless) steels
T	Tool steels
W	Welding filler metals
Z	Zinc and zinc alloys

mium and no nickel would begin CA, while a material with 100% nickel would begin CZ. Alloys in between begin with intermediate letters. The maximum content is indicated by the numerical digits (percent ×100). The presence of other alloying ingredients is indicated by additional letters. Examples are shown in Table 5.2.

Nickel-copper alloys do not follow the scheme shown in Table 5.2. They use M as the first letter (examples are M35-1 and M25.5). Nickel-molybdenum alloys use the letter N as the beginning letter, such as N7Mn and N12MV.

Since ACI no longer exists, ASTM has adopted the system and assigns new designations as other alloys are developed.

ASTM also has their own system of designations for many special carbon and alloy steel products as well as for cast iron. Some designations indicate the material type such as WCA, WCB, and WCC for welded carbon steel grades A, B, and C. Some grades are numbered in sequence as added to a specification, and others indicate a property, such as strength. The UNS numbers have not been adopted for these materials since they have no relation with any common designation. Refer to Table 5.3 for examples.

TABLE 5.2 Examples of ACI Designations

| | Alloying elements (%) | | | |
Designation	Chromium, nominal	Nickel, nominal	Carbon, max.	Other alloying elements, nominal
CA15	12	—	0.15	—
CD4MCu	25	6	0.04	Mo 3; Cu 3
CF8M	19	10	0.08	Mo 2.5
CF3M	19	10	0.03	Mo 2.5
CN7M	21	29	0.07	Mo 2.5
CW2M	16	68	0.02	Mo 1.6
CZ100	0	100	1.0	
HK40	25	20	0.40	

I. CAST IRONS

The general term *cast iron* is inclusive of a number of alloys of iron, carbon, and silicon. Typically these alloys have carbon contents of approximately 1.8 to 4% and silicon contents of 0.5 to 3%. This composition range describes all grades of cast irons with properties ranging from highly wear-resistant hard materials to ductile energy-absorbing alloys suitable for applications involving high stress and shock loads. The carbon content of the alloy can be present in several different forms: graphite flakes, irregular graphite nodules, graphite spheres, iron carbides, or cementite and combinations of these. The basic types of cast irons are gray iron, ductile (nodular) iron, compacted graphite iron, white iron, malleable iron, and high-alloy cast irons.

TABLE 5.3 Examples of ASTM and UNS Designations

ASTM designation	UNS designation
Grade WCC: welded carbon steel casting, grade C	JO2503
Grade LCB: low temperature welded carbon steel casting, grade B	JO3
Class 3 cast iron	F12802
Grade 135-125: 135 min. tensile, 125 min. yield strength; steel casting	None

A. Gray Iron

This is the most common cast iron. When the material fractures it has a gray appearance, thus the name gray iron. Gray iron contains 1.7 to 4.5% carbon and 1 to 3% silicon. It is the least expensive of all cast metals and, because of its properties, it has become the most widely used cast material on a weight basis.

Various ASTM standards deal with gray iron castings, but the one referred to most commonly is A48. This standard classifies the alloy on the basis of tensile strength with values ranging from 20,000 psi to 60,000 psi. Standard A247 is often used in conjunction with A48 to describe the graphite structure. Standards A126, A278, and A319 also deal with gray iron.

The gray irons have good fluidity at pouring temperatures, which makes them ideally suited for casting intricate shapes in all sizes. Under normal conditions little or no shrinkage takes place during solidification, which simplifies the job of pattern making as compared to other alloys.

In gray iron the carbon is in the form of graphite flakes. Silicon additions assist in making the Fe_3C unstable. As the metal slowly cools in the mold the Fe_3C will decompose to graphite. Gray iron has relatively poor toughness because of the stress concentration effect of the graphite flake tips. The mechanical properties vary with cooling rate and are measured from separately cast bars poured from the same metal as the casting. These bars are designated A, B, C, and S. The A, B, and C bars are listed in increasing diameter and are selected for use by the relative size of the casting and its cooling rate. The S designation is a special bar agreed upon by the manufacturer and customer. These bars are used in the three-point bending fixture (per ASTM A438) to determine traverse breaking load and deflection.

In general gray iron castings are not recommended for applications where impact resistance is required. Gray iron castings are normally used in neutral or compressive applications because the graphite flake form acts as an internal stress raiser. The graphite flake form provides advantages in machining, sound damping, and heat transfer applications.

Tensile strength specifications for gray iron are written as a class of iron. A class 30 gray iron would have a minimum of 30 ksi tensile strength in a separately cast A size bar. Typical designations run for class 20 to class 60. For automotive applications (SAE specifica-J431) a tensile strength–to–Brinell hardness ratio is used to classify the gray irons. The term *class* is a common designation and the term *grade* as used in the SAE specification refers to a tensile/hardness ratio. Refer to Table 5.4 for grades of gray iron.

Gray irons possess considerably greater strength in compression than in tension. Refer to Table 5.5 for a comparison of compressive strength to tensile strength for various grades of gray iron.

TABLE 5.4 Grades of Gray Iron from SAE J431

Grade	Brinell hardness	Test bar (t/h)[a]	Description
G1800	120 to 187	135	Ferritic-pearlitic
G2500	170 to 229	135	Pearlitic-ferritic
G3000	187 to 241	150	Pearlitic
G3500	207 to 255	165	Pearlitic
G4000	217 to 269	175	Pearlitic

[a]t/h = tensile strength ÷ hardness.

As mentioned previously gray iron possesses an excellent damping capacity. A typical comparison of the damping capacity of gray iron with other metals is shown in Table 5.6.

Graphite is essentially an inert material and is cathodic to iron, consequently the iron will suffer rapid attack in even mildly corrosive atmospheres. Gray iron is subject to a form of corrosion known as graphitization, which involves the selective leaching of the iron matrix leaving only a graphite network. Even though no apparent dimensional change has taken place there can be sufficient loss of section and strength to lead to failure. In general gray iron is used in the same environment as carbon steel and low-alloy steels, although the corrosion resistance of gray iron is somewhat better than that of carbon steel. Corrosion rates in rural, industrial, and seacoast environments are generally acceptable. The advantage of gray iron over carbon steel in certain environments is the result of a porous graphite-iron corrosion product film that forms on the surface. This film provides a

TABLE 5.5 Compressive Strength/Tensile Strength Comparison of Gray Iron

Grade	Tensile strength (ksi)	Compressive strength (ksi)	Compressive ÷ tensile strength	Tensile modulus $E \times 10^6$ (psi)
20	22	33	3.7	10
30	31	109	3.5	14
40	57	140	2.45	18
60	62.5	187.5	3.0	21

TABLE 5.6   Damping Capacity of Gray Iron

Type of metal	Relative decrease in amplitude of vibrations per cycle
Carbon steel	1.0–2.0
Malleable iron	3.3–6.3
Ductile iron	3.0–9.4
Hypoeutectoid gray iron	
3.2% C; 2.0% Si	40
3.7% C; 1.8% Si	126

particular advantage under velocity conditions, such as in pipe lines. This is the reason for the widespread use of underground gray iron water pipes.

Gray iron is not resistant to corrosion in acid except for concentrated acids where a protective film is formed. It is not suitable for use with oleum because it has been known to rupture in this service with explosive violence.

Gray iron exhibits good resistance to alkaline solutions such as sodium hydroxide and molten caustic soda. Likewise it exhibits good resistance to alkaline salt solutions such as cyanides, silicates, carbonates, and sulfides. Acid and oxidizing salts rapidly attack gray iron. Gray iron will contain sulfur at temperatures of 350–400°F (149–205°C). Molten sulfur must be air free and solid sulfur must be water free.

Gray iron finds application in flue gas handling such as in wood- and coal-fired furnaces and heat exchangers. Large quantities are also used to produce piping which is buried. Normally gray iron pipe will outlast carbon steel pipe depending upon soil type, drainage, and other factors.

B. Ductile (Nodular) Iron

Ductile iron has basically the same chemical composition as gray iron with a small chemical modification. Just prior to pouring the molten iron an appropriate inoculant such as magnesium is added. This alters the structure of iron to produce a microstructure in which the graphite form produced during the solidification process is spheroidal instead of the flake form. The flake form has better machinability, but the spheroidal form yields much higher strength and ductility. The matrix can be ferritic, pearlitic, or martensitic depending on the heat treatment process. Graphite nodules surrounded by white ferrite, all in a pearlite matrix, is the most common. Other elements that can be used to produce the nodular graphite form include yttrium, calcium, and cerium.

ASTM A536 (SAE J434) covers the property specifications. 60-40-18 is an example of such a specification. This specification calls for a minimum of 60 ksi tensile strength, 40 ksi minimum yield strength, and a minimum elongation in 2 in. of 18%. These properties are usually obtained from a Y, or keel, block, which is a separately cast test casting. Table 5.7 lists properties of ductile iron.

The corrosion resistance of ductile iron is comparable to that of gray iron with one exception. Under velocity conditions the resistance of ductile iron may be slightly less than that of gray iron since it does not form the same type of film that is present on gray iron.

C. Austenitic Gray Cast Iron

Austenite iron is also referred to as Ni-resist alloy. This group consists of high nickel austenitic cast irons used primarily for their corrosion resistance. Table 5.8 lists the chemical composition of the Ni-resist alloys. These alloys have improved toughness over unalloyed gray iron but with relatively low tensile strengths, ranging from 20,000 to 30,000 psi. ASTM A571 covers ductile Ni-resist alloys which have an improved low temperature impact toughness. Table 5.9 presents the mechanical properties of Ni-resist alloys.

The corrosion resistance lies between that of gray iron and the 300 series stainless steels. It finds wide application in hydrogen sulfide–containing oil field applications. Excessive attack is prevented by the formation of a protective film. It is superior to gray iron in atmospheric exposure, sea-

TABLE 5.7 Properties of Ductile Iron from ASTM A536

Grade and heat treatment	Tensile strength, min. (psi)	Yield strength, min. (psi)	Elongation in 2 in. (%)	Brinell hardness	Microstructure
60-40-18[a]	60,000	40,000	18	149–187	Ferrite
65-45-12	65,000	45,000	12	—	—
80-55-06[b]	80,000	55,000	6	187–255	Pearlite and ferrite
100-70-03	100,000	70,000	3	—	—
120-90-02[c]	120,000	90,000	2	240–300	Tempered martensite

[a]May be annealed after casting.
[b]As-cast grade with a higher manganese content.
[c]Oil quenched and tempered to desired hardness.

TABLE 5.8 Composition of Ni-Resist Alloys

ASTM spec.	Cast alloy designation	Percent							Other elements (%)
		C	Mn	Si	S^a	Cr	Ni	Fe	
A436	Type 1	3.00	0.5/1.5	1.00/2.80	0.12	1.5–2.5	13.5–17.5	Balance	5.5–7.5 Cu
A436	Type 1b	3.00	0.5/1.5	1.00/2.80	0.12	2.5–3.5	13.5–17.5	Balance	5.5–7.5 Cu
A436	Type 2	3.00	0.5/1.5	1.00/2.80	0.12	1.5–2.5	16.0–22.0	Balance	0.05 Cu^a
A436	Type 2b	3.00	0.15/1.5	1.00/2.80	0.12	3.0–6.0	18.0–22.0	Balance	0.05 Cu^a
A436	Type 3	2.60	0.15/1.5	1.00/2.00	0.12	2.5–3.5	28.0–32.0	Balance	0.5 Cu^a
A436	Type 4	2.60	0.5/1.5	5.00/6.00	0.12	4.5–5.5	29.0–32.0	Balance	0.5 Cu^a
A436	Type 5	2.40	0.5/1.5	1.00/2.00	0.12	0.10^a	34.0–36.0	Balance	0.5 Cu^a
A436	Type 6	3.00	0.5/1.5	1.50/2.50	0.12	1.0–2.0	18.0–22.0	Balance	3.5–5.5 Cu; 1.00 Mo^a
A571		2.2/2.7	3.75–4.5	1.50/2.5	—	0.20^a	21.0–24.0	Balance	0.08 P^a

^{a} Maximum.

TABLE 5.9 Properties of Ni-Resist Alloys

ASTM spec.	Cast alloy designation	Tensile strength (ksi)[a]	Yield strength (ksi)[a]	Elongation in 2 in. (%)	Brinell hardness	Charpy impact test (ft-lb)[b]	Charpy impact test Test temp.[c] (°F/°C)
A436	Type 1	25	—	—	131–183	100	RT
A436	Type 1b	30	—	—	149–212	80	RT
A436	Type 2	25	—	—	118–174	100	RT
A436	Type 2b	30	—	—	171–248	60	RT
A436	Type 3	25	—	—	118–159	150	RT
A436	Type 4	25	—	—	149–212	80	RT
A436	Type 5	20	—	—	94–124	150	RT
A436	Type 6	25	—	—	124–174	—	—
A571[d]	—	65	30	30	121–171	15	−195/−320

[a]Minimum.
[b]Typical unnotched.
[c]RT = room temperature.
[d]Annealed.

water, caustic soda or sodium hydroxide, and dilute and concentrated (unaerated) sulfuric acid. The copper in type 1 provides the best resistance to sulfuric acid.

D.   Austenitic Ductile Cast Irons

These alloys are commonly called ductile Ni-resist. They are similar to the austenitic gray irons except that magnesium is added just prior to pouring to produce a nodular graphite structure. As a result of this nodular structure higher strengths and greater ductility are produced as compared to the flake graphite structure. Although several different grades are produced, type 2D is the most commonly used grade. Table 5.10 shows the properties of austenitic ductile cast irons.

The corrosion resistance is similar to that of austenitic gray iron, although those containing 2% or more chromium are superior. Table 5.11 shows the compatibility of Ni-resist with selected corrodents.

E.   White Iron

This alloy is also referred to as Ni-hard. The carbon in these alloys is essentially all in solution and the fracture surface appears white. These alloys contain nickel in the range of 4–5% and chromium in the range of 1.5–3.5%.

White iron solidifies with a "chilled" structure. Instead of forming free graphite, the carbon forms hard, abrasion-resistant, iron-chromium carbides. These alloys are used primarily for abrasive applications. After machining, the material is generally heat treated to form a martensitic matrix for maximum hardness and wear resistance. The only property required by ASTM is the Brinell hardness, with 600 being typical, although some of the alloys are capable of having Brinell hardnesses in excess of 700.

TABLE 5.10   Properties of Austenitic Ductile Cast Iron

Grade	Tensile strength (ksi)	Yield strength (ksi)	Elongation (%)
D-2	58	30	8
D-2C	58	30	20
D-5	55	30	10
D-5S	65	30	10

TABLE 5.11 Compatibility of Ni-Resist Alloy with Selected Corrodents

Chemical	Maximum temp. °F	°C	Chemical	Maximum temp. °F	°C
Acetic anhydride	X		Ethanol amine	200	93
Acetone	140	60	Ethyl acetate	90	32
Acetylene	90	32	Ethyl chloride, dry	90	32
Alum	100	38	Ethylene glycol	460	238
Aluminum hydroxide, 10%	470	243	Ethylene oxide	X	
Aluminum potassium sulfate	100	38	Ferric sulfate	460	238
Ammonia, anhydrous	460	238	Ferrous sulfate	X	
Ammonium carbonate, 1%	90	32	Fuel oil	X	
Ammonium chloride	210	99	Furfural, 25%	210	99
Ammonium hydroxide	90	32	Gallic acid	90	32
Ammonium nitrate, 60%	120	49	Gas, natural	90	32
Ammonium persulfate, 60%	120	49	Gasoline, leaded	400	204
Ammonium phosphate	X		Gasoline, unleaded	400	204
Ammonium sulfate	130	54	Glycerine	320	160
Amyl acetate	300	149	Hydrochloric acid	X	
Aniline	100	38	Hydrogen chloride gas, dry	X	
Arsenic acid	X		Hydrogen sulfide, dry	460	238
Barium carbonate	X		Hydrogen sulfide, wet	460	238
Barium chloride	X		Isooctane	90	32
Barium hydroxide	X		Magnesium hydroxide	X	
Barium sulfate	X		Magnesium sulfate	150	66
Barium sulfide	X		Methyl alcohol	160	71
Benzene	400	204	Methyl chloride	X	
Black liquor	90	32	Phosphoric acid	X	
Boric acid	X		Sodium borate	90	32
Bromine gas	X		Sodium hydroxide, to 70%	170	77
Butyl acetate	X		Sodium nitrate	90	32
Calcium carbonate	460	238	Sodium nitrite	90	32
Calcium hydroxide	90	32	Sodium peroxide, 10%	90	32
Calcium nitrate	210	99	Sodium silicate	90	32
Calcium sulfate	440	227	Sodium sulfate	X	
Carbon dioxide, dry	300	149	Sodium sulfide	X	
Carbon dioxide, wet	X		Steam, low pressure	350	177
Carbon monoxide	300	149	Sulfate liquors	100	38
Carbon tetrachloride	170	77	Sulfur	100	38
Carbonic acid	460	238	Sulfur dioxide, dry	90	32
Chlorine gas, dry	90	32	Tartaric acid	100	38
Chromic acid	X		Tomato juice	120	49
Cyclohexane	90	32	Vinegar	230	110
Diethylene glycol	300	149	Water, acid mine	210	99
Diphenyl	210	99	White liquor	90	32

Note: The chemicals listed are in the pure state or in a saturated solution unless otherwise indicated. Compatibility is shown to the maximum allowable temperature for which data are available. Incompatibility is shown by an X. When compatible, corrosion rate is less than 20 mpy.
Source: Ref. 1.

The Ni-hard alloys are very brittle. Elongation in the hardened condition is typically 2%. There is essentially no difference in the corrosion resistance between gray iron and white iron.

F. High Silicon Cast Irons

High silicon cast irons are sold under the trade name of Duriron and Durichlor 51, which are tradenames of the Duriron Company. These alloys contain 12–18% silicon, with 14.5% being nominal (14.2% minimum is required for good corrosion resistance), 1% carbon, and the balance iron.

These alloys are particularly susceptible to thermal and mechanical shock. They can not be subjected to sudden fluctuations in temperature nor can they withstand any substantial stressing or impact. The high silicon irons are extremely brittle and difficult to machine. The mechanical and physical properties are shown in Table 5.12.

When high silicon cast irons are first exposed to a corrosive environment, surface iron is removed, leaving behind a silicon oxide layer which is very adherent and corrosion resistant. These alloys are extremely corrosion resistant. One of the main uses is in the handling of sulfuric acid. It is resistant to all concentrations of sulfuric acid up to and including the normal boiling point. Nitric acid above 30% concentration can be handled to the boiling point. Below 30% the temperature is limited to 180°F (82°C).

When 4.5% chromium is added to the alloy it becomes resistant to severe chloride-containing solutions and other strongly oxidizing environments. The chromium-bearing grade (Durichlor) will handle hydrochloric acid up to 80°F (27°C). Table 5.13 lists the compatibility of high silicon iron with selected corrodents.

TABLE 5.12 Mechanical and Physical Properties of High Silicon Iron

Property	Duriron	Durichlor
Modulus of elasticity $\times 10^6$ (psi)	23	23
Tensile strength $\times 10^3$ (psi)	16	16
Elongation in 2 in. (%)	Nil	Nil
Brinell hardness	520	520
Density (lb/in.^{3})	0.255	0.255
Specific gravity	7.0	7.0
Specific heat at 32–212°F (Btu/lb °F)	0.13	0.13
Coefficient of thermal expansion $\times 10^{-6}$ (Btu/ft^2 hr/°F/in.)		
at 32–212°F		7.2
at 68–392°F	7.4	

TABLE 5.13 Compatibility of High Silicon Iron[a] with Selected Corrodents

Chemical	Maximum temp. °F	Maximum temp. °C	Chemical	Maximum temp. °F	Maximum temp. °C
Acetaldehyde	90	32	Barium carbonate	80	27
Acetamide			Barium chloride	80	27
Acetic acid, 10%	200	93	Barium hydroxide		
Acetic acid, 50%	200	93	Barium sulfate	80	27
Acetic acid, 80%	260	127	Barium sulfide	80	27
Acetic acid, glacial	230	110	Benzaldehyde	120	49
Acetic anhydride	120	49	Benzene	210	99
Acetone	80	27	Benzene sulfonic acid, 10%	90	32
Acetyl chloride	80	27	Benzoic acid	90	32
Acrylic acid			Benzyl alcohol	80	27
Acrylonitrile	80	27	Benzyl chloride	90	32
Adipic acid	80	27	Borax	90	32
Allyl alcohol	80	27	Boric acid	80	27
Allyl chloride	90	32	Bromine gas, dry	X	
Alum	240	116	Bromine gas, moist	80	27
Aluminum acetate	200	93	Bromine, liquid		
Aluminum chloride, aqueous			Butadiene		
Aluminum chloride, dry			Butyl acetate		
Aluminum fluoride	X		Butyl alcohol	80	27
Aluminum hydroxide	80	27	n-Butylamine		
Aluminum nitrate	80	27	Butyl phthalate	80	27
Aluminum oxychloride			Butyric acid	80	27
Aluminum sulfate	80	27	Calcium bisulfide		
Ammonia gas			Calcium bisulfite	X	
Ammonium bifluoride	X		Calcium carbonate	90	32
Ammonium carbonate	200	93	Calcium chlorate	80	27
Ammonium chloride, 10%			Calcium chloride	210	99
Ammonium chloride, 50%	200	93	Calcium hydroxide, 10%		
Ammonium chloride, sat.			Calcium hydroxide, sat.	200	93
Ammonium fluoride, 10%	X		Calcium hypochlorite	80	27
Ammonium fluoride, 25%	X		Calcium nitrate		
Ammonium hydroxide, 25%	210	99	Calcium oxide		
Ammonium hydroxide, sat.			Calcium sulfate	80	27
Ammonium nitrate	90	32	Caprylic acid	90	32
Ammonium persulfate	80	27	Carbon bisulfide	210	99
Ammonium phosphate	90	32	Carbon dioxide, dry	570	299
Ammonium sulfate, 10–40%	80	27	Carbon dioxide, wet	80	27
Ammonium sulfide			Carbon disulfide		
Ammonium sulfite			Carbon monoxide		
Amyl acetate	90	32	Carbon tetrachloride	210	99
Amyl alcohol	90	32	Carbonic acid	80	27
Amyl chloride	90	32	Cellosolve	90	32
Aniline	250	121	Chloracetic acid, 50% water	80	27
Antimony trichloride	80	27	Chloracetic acid	90	32
Aqua regia, 3:1	X		Chlorine gas, dry		

TABLE 5.13   Continued

Chemical	Maximum temp. °F	Maximum temp. °C	Chemical	Maximum temp. °F	Maximum temp. °C
Chlorine gas, wet			Magnesium chloride, 30%	250	121
Chlorine, liquid			Malic acid	90	32
Chlorobenzene	80	27	Manganese chloride		
Chloroform	90	32	Methyl chloride		
Chlorosulfonic acid, dry			Methyl ethyl ketone	80	27
Chromic acid, 10%	200	93	Methyl isobutyl ketone	80	27
Chromic acid, 50%	200	93	Muriatic acid		
Chromyl chloride	210	99	Nitric acid, 5%	180	82
Citric acid, 15%			Nitric acid, 20%	180	82
Citric acid, conc.	200	93	Nitric acid, 70%	186	86
Copper acetate			Nitric acid, anhydrous	150	66
Copper carbonate			Nitrous acid, conc.	80	27
Copper chloride	X		Oleum	X	
Copper cyanide	80	27	Perchloric acid, 10%	80	27
Copper sulfate	100	38	Perchloric acid, 70%	80	27
Cresol			Phenol	100	38
Cupric chloride, 5%			Phosphoric acid, 50–80%	210	99
Cupric chloride, 50%			Picric acid	80	27
Cyclohexane	80	27	Potassium bromide, 30%	100	38
Cyclohexanol	80	27	Salicylic acid	80	27
Dichloroacetic acid			Silver bromide, 10%		
Dichloroethane (ethylene dichloride	80	27	Sodium carbonate		
			Sodium chloride, to 30%	150	66
Ethylene glycol	210	99	Sodium hydroxide, 10%	170	77
Ferric chloride	X		Sodium hydroxide, 50%	X	
Ferric chloride, 50% in water			Sodium hydroxide, conc.	X	
Ferric nitrate, 10–50%	90	32	Sodium hypochlorite, 20%	60	16
Ferrous chloride	100	38	Sodium hypochlorite, conc.		
Ferrous nitrate			Sodium sulfide, to 50%	90	32
Fluorine gas, dry	X		Stannic chloride	X	
Fluorine gas, moist			Stannous chloride	X	
Hydrobromic acid, dilute	X		Sulfuric acid, 10%	212	100
Hydrobromic acid, 20%			Sulfuric acid, 50%	295	146
Hydrobromic acid, 50%	X		Sulfuric acid, 70%	386	197
Hydrochloric acid, 20%[b]	80	27	Sulfuric acid, 90%	485	252
Hydrochloric acid, 38%			Sulfuric acid, 98%	538	281
Hydrocyanic acid, 10%	X		Sulfuric acid, 100%	644	340
Hydrofluoric acid, 30%	X		Sulfuric acid, fuming		
Hydrofluoric acid, 70%	X		Sulfurous acid	X	
Hydrofluoric acid, 100%	X		Thionyl chloride		
Hypochlorous acid			Toluene		
Iodine solution, 10%			Trichloroacetic acid	80	27
Ketones, general	90	32	White liquor		
Lactic acid, 25%	90	32	Zinc chloride		
Lactic acid, conc.	90	32			

Footnotes to table on next page.

TABLE 5.14 Properties of Malleable Iron

ASTM spec.	Alloy grade	Tensile strength (ksi)	Yield strength (ksi)	Elongation in 2 in. (%)	Brinell hardness
A47	22010	49	52	10	156 max.
A47	24018	52	35	18	156 max.
A220	40010	60	40	10	147–197
A220	45008	65	45	8	156–197
A220	45006	65	45	6	156–197
A220	50005	70	50	5	170–229
A220	60004	80	60	4	197–241
A220	70003	85	70	3	217–269
A220	80002	95	80	2	241–285
A220	90001	105	90	1	269–321

G. Malleable Iron

Malleable iron and ductile iron are very similar, but malleable iron is declining in use because of economic reasons. Malleable iron contains a carbon form referred to as *temper carbon* graphite. This carbon form is generated by a heat treatment of the as-cast product after solidification. The casting is annealed by holding at a temperature of 1500 to 1780°F (800–970°C) for up to 20 hr. During solidification after casting, an unstable form of iron carbide is produced. The heat treatment breaks down the iron carbide to form a graphitic nodule, called temper carbon, and austenite. Upon cooling the austenite transforms to the various products of ferrite, pearlite, or martensite, depending on the cooling rate. It is the cost of this heat treatment operation that has been the reason for the decline in usage.

ASTM A47 and 220 are the most widely used specifications. Table 5.14 lists the properties of the malleable irons.

In general there is little difference in corrosion resistance between gray iron and malleable iron. Malleable iron may be inferior to gray iron in flowing conditions since there are no graphite flakes to hold the corrosion products in place; therefore the attack continues at a constant rate rather than declining with time.

H. Compacted Graphite Iron

The structure of compacted graphite iron is between that of gray iron and ductile iron. The graphite shape is between a true nodular and the flake shape of gray iron. The graphite form is termed *vernucular* and is a short blunt flake. Production is similar to ductile iron with an additional alloying element like titanium.

Compacted graphite iron retains many of the desirable properties of gray iron but has improved strength and ductility. It is noted for its high fluidity and a solidification shrinkage rate between gray iron and the higher shrinkage rate of nodular iron. The true advantage is the combination of the fluidity, which allows pouring a more intricate casting with thinner sections, and its mechanical properties. Typical values for a compacted iron with ferritic matrix are tensile strengths in the 35–55 ksi range with a yield strength of 25–40 ksi and 6% elongation.

Commercial applications are somewhat limited because of the very careful control of the molten metal chemistry required and solidification parameters. There is little difference in the corrosion resistance between compacted graphite iron and gray iron.

II. CAST STEEL

Cast carbon and low-alloy steels are widely used because of their low cost, versatile properties, and the wide range of available grades. The carbon steels are alloys of iron and carbon with manganese (<1.65%), silicon, sulfur, phosphorus, and other elements in small quantities. The latter elements are present either for their desirable effects or because of the difficulty of re-moving them. Steel castings are generally grouped into four categories:

1. Low carbon castings with less than 0.20% carbon
2. Medium carbon castings with 0.20–0.50% carbon
3. High carbon castings with more than 0.5% carbon
4. Low-alloy castings with alloy content less than 8%

The strengths of ferritic steel castings, rolled, forged, and welded metal are virtually identical. Cast steels of the same hardness as forged, rolled, or welded steels have essentially the same ductility. The impact values of wrought steel are generally given in the longitudinal direction and are usu-ally higher than those for cast steel. However, since the impact value for wrought steels in the traverse direction are typically 50 to 75% of those in the longitudinal direction, the impact values of cast steels normally lie somewhere between the traverse and longitudinal impact properties of the same carbon steel. Table 5.15 lists the mechanical properties of selected cast steels.

Carbon and low-alloy steels are used for water, steam, air, and other mild services, providing that the moisture content is below the saturation point. Carbon steels can be used to handle carbon dioxide, carbon monoxide, hydrogen cyanide, sulfur dioxide, chlorine, hydrogen chloride, hydrogen fluoride, and nitrogen. It is important that these gases be dry. Contamination from air and humidity will cause excessive attack and/or stress corrosion cracking.

Protective surface films may be formed in specific corrosive environments. In the presence of concentrated hydrofluoric acid a fluoride film is formed, while in the presence of concentrated sulfuric acid a ferrous sulfate film protects the steel. Care must be taken not to damage the film since extremely high corrosion rates will result. Conditions such as high velocities, condensing water (humidity from the air), and hydrogen bubbles floating across a surface will damage the protective film.

Steel may be used for sodium hydroxide and other alkaline compounds providing the temperature remains below 150°F (66°C). Above this temperature SCC and excessive corrosion may develop. Neutral salts, brines, and organics are normally noncorrosive to steel.

TABLE 5.15 Mechanical Properties of Selected Cast Steels

Steel	Tensile strength (ksi)	Endurance limit (ksi)	
		Unnotched	Notched
Normalized and tempered			
1040	94	38	28
1330	93	38	32
4135	113	51	33
4335	126	63	35
8630	111	54	33
Quenched and tempered			
1330	122	59	37
4135	146	61	41
4335	168	77	48
8630	137	65	39
Annealed			
1040	84	33	26

Cast carbon and low-alloy steels are often used for hydrogen service. However, hydrogen attack can occur as the temperature and hydrogen partial pressure increase. In hydrogen attack, atomic hydrogen diffuses into the steel, combines with carbon, and forms methane. Structural integrity is lost as, over time, high pressure methane pockets develop.

Anhydrous ammonia can be handled in steel, provided that small amounts of water are added to prevent SCC. Stress corrosion cracking of carbon low-alloy steel can also be caused by high temperature hydroxides, nitrates, carbonates, moist gas mixtures of carbon dioxide and carbon monoxide, hydrogen cyanide solutions, amine solutions, and hydrogen sulfide. Stress corrosion cracking may be minimized by postweld heat treatment.

Cast carbon and low-alloy steels are usually protected from atmospheric corrosion by painting and/or coating systems. Although the weathering wrought steels develop a protective film, this is not necessarily true of comparable cast steels. The natural segregation in cast steels may produce different results.

There is little difference in the corrosion rate between plain carbon steel and low-alloy carbon steel.

REFERENCE

1. PA Schweitzer. Corrosion Resistance Tables, 4th ed. Vols. 1–3. New York: Marcel Dekker, 1995.

6

Introduction to Stainless Steels

In all probability the most widely known and most commonly used material of construction for corrosion resistance is stainless steel. Stainless steels are iron-based alloys containing 10.5% or more chromium. There are currently over 70 types of stainless steels.

In the United States annual stainless steel consumption is approaching 2 million metric tons. In addition to being an important factor in industrial process equipment it also finds application in a wide variety of household items.

Worldwide production of stainless steel exceeds 12.5 million metric tons. The first trials of adding chromium to mild steel took place in the early 1900s. This apparently was the result of observing that chromium-plated steel parts were highly corrosion resistant. This experiment resulted in the production of the ferritic family of stainless steels. Documentation of this class of steel began to appear in the 1920s. In 1935 the first Americal Society for Testing and Materials (ASTM) specifications for stainless steels were published.

Stainless steel is not a singular material, as its name might imply, but rather a broad group of alloys, each of which exhibits its own physical, mechanical, and corrosion-resistant properties.

These steels are produced both as cast alloys [Alloy Casting Institute (ACI) types] and wrought forms [American Iron and Steel Institute (AISI) types]. Generally, all are iron based with 12 to 30% chromium, 0 to 22% nickel, and minor amounts of carbon, columbium, copper, molybdenum, selenium, tantalum, and titanium. They are corrosion resistant and heat resistant, noncontaminating, and easily fabricated into complex shapes.

I. STAINLESS STEEL CLASSIFICATION

There are three general classification systems used to identify stainless steels. The first relates to metallurgical structure and places a particular stainless steel into a family of stainless steels. The other two, namely, the AISI numbering system and the Unified Numbering System, which were developed by ASTM and SAE to apply to all commercial metals and alloys, define specific alloy compositions. Tables 6.1, 6.2, and 6.3 provide a comparison between AISI and UNS designations for selected stainless steels.

The various stainless steel alloys can be divided into seven basic families:

1. Ferritic
2. Martensitic
3. Austenitic
4. Precipitation hardenable
5. Superferritic
6. Duplex (ferritic-austenitic)
7. Super austenitic

TABLE 6.1   Austenitic Stainless Steels

AISI type	UNS designation	AISI type	UNS designation
201	S20100	308	S30800
202	S20200	309	S30900
204	S20400	309S	S30908
204L	S20403	310	S31000
205	S20500	310S	S31008
209	S20900	314	S31400
22-18-S	S20910	316	S31600
18-8.8 plus	S20220	316L	S31603
301	S30100	316F	S31620
302	S30200	316N	S31651
302B	S30215	317	S31700
303	S30300	317L	S31703
303Se	S30323	321	S32100
304	S30400	329	S32900
304L	S30403	330	N08330
	S30430	347	S34700
304N	S30451	348	S34800
305	S30500	384	S38400

TABLE 6.2   Ferritic Stainless Steels

AISI type	UNS designation	AISI type	UNS designation
405	S40500	446	S44600
409	S40900	439	S43035
429	S42900	444	S44400
430	S43000	26-1	S44627
430F	S43020	26-3-3	S44660
430FSe	S43023	29-4	S44700
434	S43400	29-4C	S44735
436	S43600	29-4-2	S44800
442	S44200		

A. Ferritic Family

The name is derived from the analogous ferrite phase, or relatively pure iron component, of carbon steels, cooled slowly from the austenite region. The ferrite phase for pure iron is the stable phase existing below 1670°F (910°C). For low carbon Cr-Fe alloys the high temperature austenite phase exists only up to 12% Cr; immediately beyond this composition the alloys are ferritic at all temperatures up to the melting point.

Chromium readily forms an oxide which is transparent and happens to be extremely resistant to further degradation. It is less noble than iron and when alloyed with steel tends to form its oxide first. Increasing the chromium content in steel gradually above the 2% level improves mild atmospheric corrosion resistance steadily up to approximately 12%, where cor-

TABLE 6.3   Martensitic Stainless Steels

AISI type	UNS designation	AISI type	UNS designation
403	S40300	420F	S42020
410	S41000	422	S42200
414	S41400	431	S43100
416	S41600	440A	S44002
416Se	S41623	440B	S44003
420	S42000	440	S44004

rosion is essentially stopped. For exposure to mild wet environments the addition of approximately 11% chromium is sufficient to prevent rusting of steel, hence the term stainless.

Ferritic stainless steels are magnetic, have body-centered cubic atomic structures and possess mechanical properties similar to these of carbon steel, though are less ductile. They can be hardened moderately by cold working but not by heat treatment.

Continued additions of chromium will improve corrosion resistance in more severe environments, particularly in terms of resistance in oxidizing environments, at both moderate and elevated temperatures. Chromium contents in the ferritic stainless steels are limited to approximately 28%. These alloys are known as 400 series stainless since they were identified with numbers beginning with 400 when AISI had the authority to designate alloy compositions.

Specific members of the ferritic families will be covered in Chapter 10.

B. Martensitic Family

The name is derived from the analagous martensite phase in carbon steels. Martensite is produced by a shear type phase transformation on cooling a steel rapidly (quenching) from the austenite region of the phase diagram. These alloys are hardenable because of the phase transformation from body-centered cubic to body-centered tetragonal. As with the alloy steels this transformation is thermally controlled. The martensitic stainless steels are nominally 11–13% chromium and are ferromagnetic.

Since the corrosion resistance of these stainless steels is dependent upon the chromium content, and since the carbon contents are generally higher than the ferritic alloys, it is logical that they are less corrosion resistant. However, their useful corrosion resistance in mild environments coupled with their high strengths make members of the martensitic family useful for certain stainless steel applications. Details of the specific family members are covered in Chapter 13.

C. Austenitic Family

This third group is named after the austenite phase, which for pure iron exists as a stable structure between 1670 and 2552°F (910 and 1400°C). It is the major or only phase of austenitic stainless steel at room temperature, existing as a stable or metastable structure depending upon composition. By virtue of their austenite-forming alloy additions, notably nickel and manganese, these stainless steels have the face-centered austenite structure from far below 32°F (0°F) up to near melting temperatures.

This family of stainless steel accounts for the widest usage of all the stainless steels. These materials are nonmagnetic, are not hardenable by heat treatment—but can be strain-hardened by cold work—have face-centered cubic structures, and possess mechanical properties similar to those of mild steels, but with better formability. The strain hardening from cold work induces a small amount of ferromagnetism.

It has been established that certain elements, specifically chromium, molybdenum, and silicon, are ferrite formers. Aluminum and niobium may also act as ferrite formers depending upon the alloy system. Other elements such as nickel, manganese, carbon, and nitrogen tend to promote the formation of austenite.

Once the corrosion resistance plateau of 18% chromium is reached, the addition of approximately 8% nickel is required to cause a transition from ferritic to austenitic. This alloy is added primarily to form the austenitic structure, which is very tough, formable, and weldable. An additional benefit is the increased corrosion resistance to mild corrodents. This includes adequate resistance to most foods, a wide range of organic chemicals, mild inorganic chemicals, and most natural environmental corrosion.

The corrosion resistance of the austenitic stainless steels is further improved by the addition of molybdenum, titanium, and other elements. Physical, mechanical, and corrosion-resistant properties of individual members of the austenitic stainless steels is covered in Chapter 8.

D. Precipitation Hardenable Stainless Steels

A thermal treatment is utilized to intentionally precipitate phases causing a strengthening of the alloy. An alloy addition of one or more of titanium, niobium, molybdenum, copper, or aluminum generates the precipitating phase. The final alloy can be solution treated since all alloying elements are in solid solution and the material is in its softest or annealed state. In this condition the material can be formed, machined, and welded. After fabrication the unit is exposed to an elevated temperature cycle (aging) which precipitates the desired phases to cause an increase in mechanical properties.

There are three types of precipitation hardenable (PH) stainless steels: martensitic, austenitic, and semiaustenitic. The relationship between these alloys is shown in Fig. 6.1. The semiaustenitic steels are supplied as an unstable austenite, which is the workable condition and must be transformed

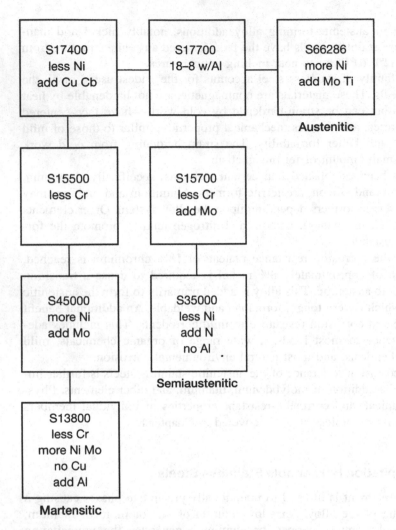

FIGURE 6.1 Precipitation hardening stainless steels.

to martensite before aging. The martensitic and austenitic PH stainless steels are directly hardened by thermal treatment.

These alloys possess high mechanical properties, but not as high as the low-alloy martensitic steels, in conjunction with useful corrosion resistant properties. In general their corrosion resistance is below that of type 304 stainless steel although certain specific alloys approach the corrosion resistance of type 316 stainless steel.

The physical, mechanical, and corrosion-resistance properties of the individual family members are discussed in Chapter 12.

E. Superferritic Stainless Steels

During the 1970s developmental efforts were directed at producing ferritic materials that could exhibit a high level of general and localized pitting resistance. The first commercially significant alloy that could meet these specifications was an alloy containing 26% chromium and 19% molybdenum. In order to obtain the desired corrosion resistance and acceptable fabrication characteristics, the material was electron beam refined under a vacuum and introduced as E-Brite alloy. Carbon plus nitrogen contents were maintained at levels below 0.020%. Other alloys were developed.

The superferritic alloys exhibit excellent localized corrosion resistance. The superferritic materials alloyed with nickel exhibit improved mechanical toughness and are less sensitive to contamination from interstitial elements. However, their availability is still limited to thicknesses less than approximately 0.20 in. This is related to the formation of embrittling phases during cooling from annealing temperatures. Greater thicknesses cannot be cooled quickly enough to avoid a loss of toughness.

Individual family members will be discussed in Chapter 11 where physical, mechanical, and corrosion-resistance properties will be presented.

F. Duplex Stainless Steels

The duplex stainless steels contain roughly 50% austenite and 50% ferrite, which provides improved corrosion resistance. These alloys contain relatively high amounts of chromium with only enough nickel and austenitizers to develop 50% austenite.

The duplex stainless steels contain molybdenum as an alloying ingredient which is responsible for the improved corrosion resistance in chlorendic environments. Molybdenum also reduces the susceptibility to chloride pitting, crevice corrosion, and stress corrosion cracking. The general corrosion resistance of the duplex stainless steels is slightly greater than that of 316 stainless steels in most media. These alloys also offer higher strengths than those typically found with austenitic steels.

Care must be taken when selecting these alloys since the boundary between acceptable and poor performance is very sharp. They should not be used under conditions that operate close to the limits of their acceptability.

The duplex stainless steels are not as ductile as the austenitic family of stainless steels. Welding requires more care than with the austenitic alloys due to a greater tendency to compositional segregation and sensitivity to weld heat input.

Physical, mechanical, and corrosion-resistance properties of each family member will be covered in Chapter 14.

G. Superaustenitic Stainless Steels

The superaustenitic stainless steels were developed to provide alloys with better resistance to localized corrosion. Included in this family of stainless steels are those that have improved pitting resistance, those that have improved crevice corrosion resistance, and those that have good general corrosion resistance to strong acids.

Physical, mechanical, and corrosion-resistance properties of each family member will be covered in Chapter 9.

II. FABRICATING STAINLESS STEEL

All stainless steels can be fabricated by conventional manufacturing methods.

A. Hot Forming

Stainless steels are readily formed by hot operations such as rolling, extrusion, and forging—the last method results in finished or semifinished parts.

Hot rolling is generally a steel mill operation for producing standard mill forms and special shapes.

Extrusion is usually associated with softer, nonferrous metals. In extrusion a shaped piece is made by forcing a bar or billet through a die, the exiting cross-section conforming to the die opening. Several companies produce hot extrusions in stainless steel. Relatively small quantities are both feasible and economical. Virtually any shape whose cross-section will fit into a 6.5-in. (165-mm) circle can be extruded. The maximum product weight for hot extrusion is about 30 lb/ft (44.91 kg/m). The minimum allowable cross-section area is 0.28 in.^{2} (18.07 mm^2), and the minimum web thickness is 0.125 in. (3.175 mm). Hollow shapes as well as solids can be produced.

Forging is used extensively for all types of stainless steel and in sizes from a few ounces to thousands of pounds and parts smaller than one inch to many feet long. Special operations such as drawing, piercing, and coining further enhance forging capabilities.

A unique feature of forgings is that the continuous grain follows the contour of the part, as shown in Fig. 6.2. In comparison is the random structure of a cast part and the straight-line orientation of grain in a machined part. This difference in grain structure is responsible for secondary advantages in forged stainless steel as follows:

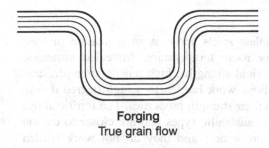

Forging
True grain flow

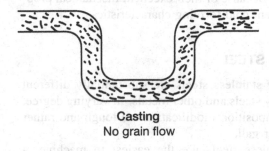

Casting
No grain flow

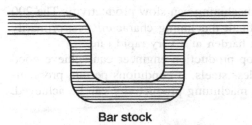

Bar stock
Grain flow broken by machining

FIGURE 6.2 Grain flow in forgings, castings, and machined parts.

1. Through grain refinement and flow, forging puts the strength where it's needed most.
2. A higher strength-to-weight ratio permits the use of thinner, lighter weight sections without sacrificing safety.
3. Forging develops the full impact resistance, fatigue resistance, ductility, creep-rupture life, and other mechanical properties.
4. Tolerances of a few thousandths of an inch are routinely maintained from part to part, simplifying machining requirements.
5. Forgings are solid, nonporous, and uniform in metallurgical structure.

B. Cold Forming

The mechanical properties of stainless steels serve as an indication of their relative formability at ambient or room temperature. Annealed austenitic grades are typified as having low yield strengths, high tensile strengths, and high elongation. Some of these alloys work harden to a high degree during cold work, which further increases their strength properties. The ferritic alloys have much lower ductility than the austenitic types and are closer to carbon steel with respect to mechanical properties; and they do not work harden significantly during cold forming. Because of their excellent mechanical properties, stainless steels have excellent cold-forming characteristics.

III. MACHINING STAINLESS STEEL

The machining characteristics of stainless steels are substantially different from those of carbon steel or alloy steels and other metals. In varying degree, most stainless steels without composition modification are tough and rather gummy, and they tend to seize or gall.

While the 400 series stainless steels are the easiest to machine, a stringy chip produced during the machining can slow productivity. The 200 and 300 series have the most difficult machining characteristics, primarily because of their tendency to work harden at a very rapid rate.

An experienced machine shop production engineer can achieve good productivity with any of the stainless steels. If conditions permit, problems can be minimized, and maximum machining productivity can be achieved. Following are three suggestions:

1. Specify a free machining stainless steel.
2. Use a special analysis stainless steel that is more suited for machining.
3. Specify stainless steel bar for machining that is in a slightly hardened condition.

When sulfur, selenium, lead, copper, aluminum, or phosphorus, either separately or in combination, is present in sufficient quantity, the machining characteristics of stainless steel are improved. The effect of these alloying ingredients is to reduce the friction between the workpiece and the tool, thereby reducing the tendency of the chip to weld to the tool. In addition sulfur and selenium form inclusions that reduce the friction forces and traverse ductility of the chips, causing them to bread off more readily.

Special analysis stainless steels are used when the end use of a free machining stainless steel does not meet the requirements of the application. For example type 303 is a free machining stainless steel which for a particular application may not be suitable to use in place of type 304. In this circumstance it would be possible to produce the type 304 in a special analysis which would improve its machining characteristics.

When conditions require maximum resistance to corrosion in the alloy selected, and there is no room for compromise in the composition of the stainless steel, bar stock should be ordered in a slightly hardened condition. This may result in a small improvement in machinability.

IV. WELDING STAINLESS STEEL

The most common method of joining stainless steel is by welding. Nearly all of the stainless steels can be welded by most methods employed in industry today. The methods employed include shielded metal arc (SMAW: stick electrodes), gas tungsten arc with inert gas shielding (TIG), and gas metal arc also with inert gas shielding (MIG). In submerged arc welding the bare wire and the arc take place in a pool of molten flux.

Hydrogen pickup from moisture (e.g., in the covered electrode flux coating) is a potential problem mainly with martensitic grades. With austenitic grades, moisture pickup is less critical, but it still is good practice to keep the SMAW electrodes in drying ovens and to warm the metal to at least 71°F (21°C) before welding.

Because of the difference between the stainless alloys and carbon and low-alloy steels there are variations in welding techniques, for example:

1. It is important that procedures be followed to preserve corrosion resistance in the weld area immediately adjacent to the weld, referred to as the heat-affected zone (HAZ).
2. It is desirable to maintain optimum mechanical properties in the joint.
3. Certain steps are necessary to minimize problems of heat distortion.

The alloy content of stainless steel is the primary difference between stainless steel and other steel types. It is the alloy content which provides corrosion resistance. In welding it is necessary to select a weld rod that provides weld filler metal having corrosion-resistance properties as nearly identical to the base metal as possible or better. Table 6.4 provides recommendations of filler and material to be used for welding various grades of stainless steel.

Proper weld rod selection not only insures preservation of the corrosion-resistance properties, but it also is important in achieving optimum mechanical properties.

TABLE 6.4   Filler Metals Suggested for Welding
Stainless Steel

Type of stainless to be welded	Electrode or filler rod material to be used
430	308
444	In-82, 316L
XM-27	XM-27
304	308
304L	347 or 308L
316	310
316L	316Cb or 316L
317	317
317L	317Cb
321	347
347	347
20Cb-3	ER 320LR
904L	Alloy 625
20Mo-6	Alloy 625
A1-6XN	Alloy 625

The thermal conductibility of stainless steel is about half that of carbon or low-alloy steels. Hence, heat is not dissipated as quickly. There are four methods to overcome this situation: lower weld current settings, skip-weld techniques to minimize heat concentration, use of backup chill bars or other cooling techniques to dissipate heat, and proper joint design. It is important that the heat generated through welding be dissipated, otherwise corrosion problems may result. One such potential problem is carbide precipitation (sensitization) that can lead to intergranular corrosion in corrosive environments. The lack of proper heat dissipation can also lead to heat distortion of the finished product.

The use of beveled joints in thinner gauges permits the use of several light passes, thus avoiding the high temperature that would be reached in a single heavy pass.

Cleaning of the edges to be welded is also important. Contamination from grease or oil can lead to carburization in the weld area with subsequent reduction of corrosion resistance. Postweld cleanup is also important. Carbon steel cleaning tools (files, brushes, etc.) or grinding wheels that have been used on carbon steel should not be used for this purpose. Such tools can leave fine particles embedded in the stainless steel surface that will later rust and stain if not removed by chemical cleaning.

A. Martensitic Grades

Because of the phase change from austenite to martensite that takes place during cooling after welding (and in the HAZ during welding) there is a change in volume, increased hardness, and loss of ductility. Special care must be taken to prevent cracking. These grades should also be properly tempered after welding to prevent them from being subject to hydrogen-assisted cracking (HAC) and sulfide stress cracking (SSC).

Filler metal for welds can be identical to the base metal or it can be an austenitic stainless steel composition.

B. Ferritic Grades

The ferrite structure in the entire temperature range below the melting point makes these alloys easier to weld than the martensitic stainless steels. However, there are three major difficulties associated with welding ferritic stainless steels:

1. Excessive grain growth
2. Sensitization
3. Lack of ductility

Some of these problems can be minimized by heat treating after welding or use of one of the ferritic alloys with low carbon and nitrogen contents.

Filler metal can be either a similar composition or an austenitic composition (types 308, 309, 316L, or 310).

C. Austenitic Grades

The 200 and 300 series are the most weldable of the stainless steels. The primary problem arising is that of chromium carbide precipitation (sensitization), which leads to intergranular corrosion. This problem can be minimized by using a low carbon or stabilized grade of stainless steel.

Preheating is not required, while postheating is necessary to redissolve precipitated carbides and to stress relieve components that are to be used in environments that may lead to stress corrosion cracking (SCC).

D. Precipitation Hardenable Grades

These grades are suitable for welding with little need for pre- or post-heat treatments except to restore or improve mechanical properties.

They may be welded with a rod of similar composition or one of austenitic composition.

E. Duplex Grades

The duplex grades are readily weldable with a filler metal of similar but
enhanced composition. Although modern grades are of low carbon content,
corrosion problems may arise from conversion of ferrite to sigma phase in
the heat-affected zone.

V. POSTWELD HEAT TREATMENT

Postweld heat treatment (PWHT) consists of solution annealing and tempering or stress relieving, whichever is required.

The martensitic grades of stainless steel will have been welded in the
annealed condition because in the hardened condition the hard brittle martensitic structure does not have sufficient ductility to withstand the stress
created by the thermal change. Prior to welding they must be fully annealed;
otherwise severe weld cracking will occur. Having been preheated at 400–
600°F (205–316°C) and welded with a controlled interpass temperature, they
are given PWHT high enough to form austenite and subsequently quenched
to produce the hardened structure desired.

If less than 0.5-in. (6.35-mm) thick the ferrite grades of stainless need
not be preheated. Thicker sections require preheating at approximately 300°F
(150°C) to reduce shrinkage stresses and yield strength. Postweld heat treatment is used to transform any residual martensite to ferrite and to relieve
stresses.

No preheating is required for the austenitic grades. The main concern
is sensitization in the 800–1550°F (425–850°C) range. For the regular carbon grades a PWHT consists of a solution anneal at about 1830°F (1000°C)
followed by a water quench to avoid sensitization. This treatment is required
only if the material is to be exposed in environments which are conducive
to intergranular attack (IGA). For non-IGA exposures a thermal stress relief
at about 1600–1750°F (870–950°C) for 2 hr/in. of thickness or 2 hr minimum will reduce residual stresses and prevent stress corrosion cracking. Low
carbon or stabilized grades do not require a solution anneal. If SCC is potentially a problem, thermal stress relief as described above would be
required.

The martensitic precipitation hardening grades are austenitic above
1830°C (1000°C) but undergo the martensitic transformation on cooling to
room temperature. They harden by heat treatment in the 900–1100°F (480–
590°C) range due to the influence of such alloying elements as copper, molybdenum, aluminum, titanium, and columbium.

The semiaustenitic PH grades do not undergo the martensitic transformation. Postweld heat treatment at 1200–1600°F (650–815°C) causes pre-

cipitation of austenite-forming elements, permitting a martensite transformation upon cooling to room temperature. They are then precipitation hardened at 800–1100°F (425–595°C).

The austenitic PH grades remain austenitic at all temperatures. They are annealed at 2000–2050°F (1090–1120°C) then hardened at 1200–1400°F (650–760°C).

The duplex alloys may be solution annealed after welding if there is concern over sigma phase transformation in the heat-affected zone or to assure the desired ferrite/austenite balance and impact resistance.

VI. SOLDERING

Relatively few problems arise from temperature when soldering stainless steel. However, aggressive fluxes are necessary to prepare the surface for soldering. Because of this phosphoric acid type fluxes are recommended since they are not corrosive at room temperatures.

VII. BRAZING

All stainless steels can be brazed. However, since brazing alloys are usually composed of copper, silver, and zinc, substantially high temperatures are required. This can lead to such high temperature problems as carbide precipitation and a reduction of the corrosion resistance.

VIII. PASSIVATION

The corrosion resistance of the stainless steels is the result of the passive oxide film which forms on the exposed surfaces. Under normal circumstances this film will form immediately upon exposure to oxygen. Some fabrication processes can impede the formation of this film. To guarantee the formation of this protective layer, stainless steels are subjected to passivation treatments.

The most common passivation treatments involve exposing the metal to an oxidizing acid. Nitric and nitric/hydrochloric acid mixtures find the widest usage. The nitric-hydrochloric acid mixtures are more aggressive and are used to remove the oxide scales formed during thermal treatment. This process provides two benefits. It removes the oxide scale and passivates the underlying metal. Second, the passivation process will remove any chromium-depleted layer that may have formed as a result of scale formation.

For passivation treatments other than for scale removal, less aggressive acid solutions are used. The purpose of these treatments is to remove any contaminents that may be on the component's surface that could prevent the formation of the oxide layer locally. The most common contaminent is embedded or free iron particles from forming or maching tools. A 10% nitric acid solution is effective in removing free iron. For martensitic, ferritic, and precipitation hardening grades a nitric acid solution inhibited with sodium dichromate is used so as not to attack the stainless steel too aggressively. A 1% phosphoric acid solution and 20% nitric acid solution are used for the more resistant stainless alloys.

IX. SANITIZING

When stainless steel is to be used in food service it requires treatment to remove bacteria or other microorganisms. It is quite common to use chlorine water or hypochlorite solution for this purpose. These solutions should be made up using demineralized water. This process can be successful provided that the solution is properly drained and flushed. A conductivity test may be used on the rinse water to ensure that the discharge is substantially equivalent to the demineralized water used in formulating the sanitizing solutions. If not thoroughly rinsed, chloride pitting, crevice corrosion, or stress corrosion cracking may occur.

Other safer alternative oxidizing solutions such as ammonium persulfate, hydrogen peroxide, dilute peracetic acid, or a citric/nitrate solution should be considered. Another possible approach is the use of a nonoxidizing biocide such as hexamethylene biguanide or other environmentally safe biocides. These are free of the hazards associated with chlorine, hypochlorite, chlorine dioxide, or other halogenated agents.

X. PREPARING FOR SERVICE

Once fabrication is complete and the material is ready to be placed in service it is essential that steps are taken to preserve the protective film of chromium oxide. The most common causes of problems are

   Iron contamination
   Organic contamination
   Welding contamination

A. Iron Contamination

As mentioned previously embedded iron can be removed by pickling. This is primarily an operation required on fabricated vessels. However, care must be exercised in the storage and handling of stainless steel sheet or plate to prevent the surface from becoming contaminated with embedded iron. If cleanliness on the surface is extremely important, such as in pharmaceutical or food environments where product contamination would be detrimental, the sheet or plate can be ordered with a protective adhesive paper on the surface. Leaving this paper in place during fabrication will reduce the amount of time required for cleanup after fabrication. The sheet and plate should be stored upright, not lain on the floor.

During fabrication it is good practice to use cardboard or plastic sheets on carbon steel layout and cutting tables, forming roll aprons, and rollout benches. This will go a long way in reducing or preventing iron embedment. The use of plastic, wood, or aluminum guards on slings, hooks, and the forks of forklift trucks will further reduce the chance of iron embedding.

B. Organic Contamination

Organic contamination is the result of grease, construction markings (crayon), oil, paint, adhesive tapes, sediment, and other sticky substances being allowed to remain on the stainless. If not removed, they may cause crevice corrosion if the stainless steel is exposed to extremely corrosive atmospheres. During fabrication there is little that can be done to prevent this contamination from occurring. The only solution is to insure that all such deposits are removed during final cleanup.

The cleanup procedures to be followed will depend somewhat on the service to which the vessel is to be put. In very corrosive media, a greater degree of cleanup will be required than in relatively mild media.

Good commercial practice will always include degreasing and removal of embedded iron. A complete specification for the procurement of a vessel should include the desired cleanup procedures to be followed, even if only degreasing and removal of embedded iron are required.

C. Welding Contamination

In corrosive environments corrosion will be initiated by surface imperfections in stainless steel plate. This corrosion can occur in the presence of media to which stainless steel is normally resistant. Such imperfections can be caused by

> Weld splatter
> Welding slag from coated electrodes
> Arc strikes
> Welding stop points
> Heat tint

Weld splatter produces small particles of metal that adhere to the surface, at which point the protective film is penetrated, forming minute crevices where the film has been weakened the most. If a splatter-prevention paste is applied to either side of the joint to be welded, this problem will be eliminated. Splatter will then easily wash off with the paste during cleanup.

Whenever coated electrodes are used, there will be some slag around the welded joints. This slag is somewhat difficult to remove, but if it is not done, the small crevices formed will be points of initiation of corrosion.

Arc strikes and weld stop points are more damaging to stainless steel than embedded iron because they occur in the area where the protective film has already been weakened by the heat of welding. Weld stop points create pinpoint defects in the metal, whereas arc strikes form crevicelike imperfections in or adjacent to the heat-affected zone.

It is possible to avoid weld stop defects by employing extensions at the beginning and end of a weld (runout tabs) and by beginning just before each stop point and welding over each intermediate stop point.

An arc strike can be struck initially on a runout tab or on weld metal, provided that the filler metal will tolerate this. If the filler metal will not tolerate the striking of an arc, then the arc must be struck adjacent to it, in or near the heat-affected zone, when it is necessary to strike an arc between runout tabs.

Heat tint results in the weakening of the protective film beneath it and can be the result of the welding of intervals in a vessel or the welding of external attachments. The heat tint must be removed to prevent corrosion from taking place in the tinted area.

Welding contamination removal is best accomplished using abrasive discs and flapper wheels. Although grinding has been used, this procedure tends to overheat the surface, thereby reducing its corrosion resistance. Its use should be avoided.

7

Corrosion of Stainless Steels

The first mention of the corrosion resistance of various alloys that had been formulated in which chromium-iron alloys were prepared appeared in 1820 in a published report by J. Stodart and M. Faraday. However, the maximum chromium content was below that required for passivity. Consequently they narrowly missed discovering stainless steels.

In 1821, in France, Berthier found that iron alloyed with large amounts of chromium was more resistant to acids than unalloyed iron. However, the alloys were high in carbon, brittle, and had no value as structural materials.

During subsequent years a variety of chromium-iron alloys were developed by several investigators who took advantage of the high strength and high hardness properties imparted by the chromium. The inherent corrosion-resistance properties of the alloys was not observed, primarily because the accompanying high carbon content impaired the corrosion properties. In 1904 Guillot of France produced low carbon–chromium alloys overlapping the passive composition range. Although he studied the metallurgical structure and mechanical properties of the chromium-iron alloys and the chromium-iron-nickel alloys, he did not recognize the outstanding property of passivity.

The property of passivity, starting at a minimum of 12% chromium, was first described by Monnartz of Germany in 1908, and he published a detailed account of the chemical properties of the chromium-iron alloys in 1911.

Meanwhile H. Brearly in England was attempting to develop iron-based chromium alloys to prevent erosion and fouling in rifle barrels. During his experiments he noted their resistance to etching for metalographic examination. He observed that the 12% chromium-iron alloys did not etch with the usual nitric acid and etching reagents, and that they did not rust over long periods of exposure to the atmosphere. He called these ferritic alloys stainless steel and recognized their possible use for cutlery materials.

Simultaneously Benno Strauss and Eduard Maurer in Germany were investigating iron-chromium-nickel compositions. They observed that the austenitic alloys containing 8% nickel were resistant to acid fumes, but the alloys were impractical because they cracked during any metalworking operations. However, Strauss restored ductility when he developed an annealing heat treatment, followed by a water quench, which dissolved the chromium carbides.

Based on his experiments Monmarty postulated that the passivity in stainless steels was caused by an invisible oxide film. His theory was not universally accepted. It was not until 1930 that his theory was proven electrochemically by H. H. Uhlig of the Massachusetts Institute of Technology.

Stainless steels and similar chromium-rich alloys are characterized by their passivity. The general concept of passivity involves base metal exhibiting the corrosion behavior of a more noble metal or alloy. For example, a piece of bare steel immersed in a copper sulfate solution develops a flash plating of metallic copper by a process known as cementation. If the bare steel is first immersed in strong nitric acid, an invisible protective oxide layer is formed that prevents cementation and the steel is said to have been passivated. Passivation of ferrous alloys containing more than 10.5% chromium is conferred by the chromium addition.

As discussed in the previous chapter there are many stainless steel compositions, all which have their own set of physical, mechanical, and corrosion-resistance properties. In this chapter we discuss the various types of corrosion to which stainless steels may be susceptible, keeping in mind that all compositions are not affected to the same degree, if at all. Specific corrosion problems and resistance of specific compositions will be discussed in succeeding chapters.

Stainless steels are alloys of iron to which a minimum of 11% chromium has been added to provide a passive film to resist "rusting" when the material is exposed to weather. This film is self-forming and self-healing in environments where stainless steel is resistant. As more chromium is added to the alloy, improved corrosion resistance results. Consequently there are stainless steels with chromium contents of 15, 17, and 20% and even higher. Chromium provides resistance to oxidizing environments such as nitric acid and also provides resistance to pitting and crevice attack.

Other alloying ingredients are added to further improve the corrosion resistance and mechanical strength. Molybdenum is extremely effective in improving pitting and crevice corrosion resistance.

By the addition of copper, improved resistance to general corrosion in sulfuric acid is obtained. This will also strengthen some precipitation hardening grades. In sufficient amounts, though, copper will reduce the pitting resistance of some alloys.

The addition of nickel will provide improved resistance in reducing environments and stress corrosion cracking. Nitrogen can also be added to improve corrosion resistance to pitting and crevice attack and to improve strength.

Columbium and titanium are added to stabilize carbon. They form carbides and reduce the amount of carbon available to form chromium carbides, which can be deleterious to corrosion resistance.

It is because of all of these alloying possibilities that so many types of stainless steel exist. It should also be kept in mind that the more alloying elements used in the formulation, the greater will be the cost. Consequently it is prudent to select the specific stainless steel composition that will meet the needs of the application. For example, it is not necessary to provide additional pitting resistance if the environment of the application does not promote pitting.

I. PITTING

Pitting corrosion is a form of localized attack. It occurs when the protective film breaks down in small isolated spots such as when halide salts contact the surface. Once started, the attack may accelerate because of the differences in electric potential between the large area of passive surface and the active pit.

If appreciable attack is confined to a small area of metal, acting as an anode, the developed pits are described as deep. If the area of attack is relatively large the pits are called shallow. The ratio of deepest metal penetration to average metal penetration as determined by weight loss of the specimen is known as the *pitting factor*.

A pitting factor of 1 represents uniform corrosion.

Performance in the area of pitting and crevice corrosion is often measured using critical pitting temperature (CPT), critical crevice temperature (CCT), and pitting resistance equivalent number (PREN). As a general rule the higher the PREN, the better the resistance. Alloys having similar values may differ in actual service. The pitting resistance equivalent number is determined by the chromium, molybdenum, and nitrogen contents:

$$PREN = \%Cr + 3.3(\%Mo) + 30(\%N)$$

Table 7.1 lists the PRENs for various austenitic stainless steels.

TABLE 7.1   Pitting Resistance Equivalent
Numbers

Alloy	PREN	Alloy	PREN
654	63.09	316LN	31.08
31	54.45	316	27.90
25-6Mo	47.45	20Cb3	27.26
Al-6XN	46.96	348	25.60
20Mo-6	42.81	347	19.0
317LN	39.60	331	19.0
904L	36.51	304N	18.3
20Mo-4	36.20	304	18.0
317	33.2		

The critical pitting temperature of an alloy is the temperature of a solution at which pitting is first observed. These temperatures are usually determined in ferric chloride (10% $FeCl_3 \cdot 6H_2O$) and in an acidic mixture of chlorides and sulfates.

II. CREVICE CORROSION

Crevice corrosion is a localized type of corrosion resulting from local differences in oxygen concentration associated with deposits on the metal surface, gaskets, lap joints, or crevices under bolt or rivet heads where small amounts of liquid can collect and become stagnant.

The material responsible for the formation of a crevice need not be metallic. Wood, plastics, rubber, glass, concrete, asbestos, wax, and living organisms have all been reported to cause crevice corrosion. Once the attack begins within the crevice its progress is very rapid, and it is frequently more intense in chloride environments. For this reason the stainless steels containing molybdenum are often used to minimize the problem. However, the best solution to crevice corrosion is a design that eliminates crevices.

The critical corrosion temperature of an alloy is that temperature at which crevice corrosion is first observed when immersed in a ferric chloride solution. The critical corrosion temperatures of several alloys in 10% ferric chloride solution are as follows.

Alloy	Temperature (°F/°C)
Type 316	27/3
Alloy 825	27/3
Type 317	36/2
Alloy 904L	59/15
Alloy 220S	68/20
E-Brite	70/21
Alloy G	86/30
Alloy 625	100/38
Alloy 6NX	100/38
Alloy 276	130/55

III. STRESS CORROSION CRACKING

Stress corrosion cracking (SCC) of stainless steels is caused by the combined effects of tensile stress, corrosion, temperature, and presence of chlorides. Wet-dry or heat transfer conditions, which promote the concentration of chlorides, are particularly aggressive with respect to initiating stress corrosion cracking.

Alloy contents of stainless steels, particularly nickel, determine the sensitivity of the metal to SCC. Ferrite stainless steels, which are nickel free, and the high nickel alloys are not subject to stress corrosion cracking. An alloy with a nickel content of greater than 30% is immune to SCC. The most common grades of stainless steel (304, 304L, 316, 316L, 321, 347, 303, 302, and 301) have nickel contents in the range of 7–10% and are the most susceptible to stress corrosion cracking.

The ferritic stainless steels such as types 405 and 430 should be considered when the potential exists for stress corrosion cracking.

The corrosion resistance of ferritic stainless steels is improved by the increased addition of chromium and molybdenum, while ductility, toughness, and weldability are improved by reducing carbon and nitrogen content.

Other related corrosion phenomena are corrosion fatigue, delayed brittle fatigue, and hydrogen stress cracking. Corrosion fatigue is the result of cyclic loading in a corrosive environment. Brittle fatigue is caused by hydrogen impregnation of an alloy during processing, which leads to brittle failure when subsequently loaded. Hydrogen stress cracking results from a cathodic reaction in service.

The austenitic stainless steels resist hydrogen effects, but martensitic and precipitation hardening alloys may be susceptible to both hydrogen stress cracking and chloride stress cracking.

Sulfide ions, selenium, phosphorus, and arsenic compounds increase the likelihood of hydrogen stress cracking. Their presence should warn of a failure possibility.

Cathodic protection can also cause hydrogen stress cracking of high strength alloys in service if "overprotected." The use of cathodic protection (the coupling of hardenable stainless steels to less noble materials in corrosive environments) should be done with caution.

Only ferritic stainless steels are generally immune to both hydrogen and chloride stress cracking.

IV. INTERGRANULAR CORROSION

When austenitic stainless steels are heated or cooled through the temperature range of about 800–1650°F/427–899°C, the chromium along grain boundaries tends to combine with carbon to form chromium carbides. Called sensitization, or carbide precipitation, the effect is a depletion of chromium and the lowering of corrosion resistance in areas adjacent to the grain boundary. This is a time–temperature dependent phenomenon, as shown in Fig. 7.1.

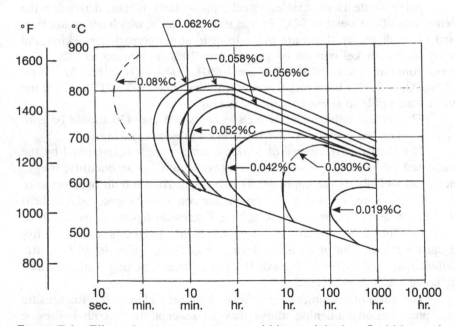

FIGURE 7.1 Effect of carbon content on carbide precipitation. Carbide precipitation forms in the areas to the right of the various carbon content curves.

Slow cooling from annealing temperature, stress relieving in the sensitization range, or welding may cause carbide precipitation. Due to the longer times at the temperature of annealing or stress relieving, it is possible that the entire piece of material will be sensitized, whereas the shorter times at welding temperatures can result in sensitization of a band, usually 1/8 to 1/4 in. wide, adjacent to but slightly removed from the weld. This region is known as the heat-affected zone (HAZ).

Intergranular corrosion depends upon the magnitude of the sensitization and the aggressiveness of the environment to which the sensitized material is exposed. Many environments do not cause intergranular corrosion in sensitized austenitic stainless steels.

For this form of attack to occur there must be a specific environment. Acids containing oxidizing agents, such as sulfuric or phosphoric acid containing ferric or cupric ions and nitric acid, as well as hot organic acids such as acetic and formic, are highly specific for this type of attack. Seawater and other high chloride waters cause severe pitting in sensitized areas, but low chloride waters (e.g., potable water) do not, except in specific situations such as might occur under the influence of microbiological corrosion.

If the carbon content is held to less than 0.030%, chromium carbide precipitation can still occur upon sensitization, but in such small amounts that no significant chromium depletion occurs. Such low carbon grades are practically immune to weld decay. However, sensitization can occur under prolonged heating in the critical temperature range, such as during service at elevated temperatures or during very prolonged thermal stress relief. Refer to Fig. 7.1. For all practicality the low carbon grades can be welded, hot formed, and even thermally stress relieved without sensitization occurring.

Sensitization can also be prevented by using stabilized stainless steels. These are stainless steels to which titanium, columbium, or columbium-titanium mixtures have been added.

Titanium and columbium additions equal to five or ten times the carbon content, respectively, permit the carbon to precipitate as titanium or columbium carbides during a sensitizing heat treatment. The carbon precipitation does not reduce the chromium content at the grain boundaries.

Three problems are presented by this approach. First, titanium-stabilized grades such as type 321 require a stabilizing anneal to tie up the carbon in the form of titanium carbides before welding.

Second, titanium does not transfer well across a welding arc and thus loses much of its effectiveness in multipass welding or cross-welding.

Third, although columbium does not have this drawback, the columbium carbides (as can the titanium carbides) can be redissolved by the heat of welding. Consequently multipass welding or cross-welding can first redissolve titanium or columbium carbides and then permit chromium carbide precipitation in the fusion zone (not the HAZ). This can cause a highly localized form of intergranular corrosion known as knife-line attack (KLA), seen particularly in alloys such as type 347, alloy 20Cb3, and alloy 825.

Intergranular attack may also occur due to mechanisms other than carbide precipitation. The ferrite phase, if present, may be selectively attacked by reducing acids such as hydrochloric or sulfuric. Its thermal conversion product, sigma phase, is selectively attacked by oxidizing acids, such as nitric.

V. HIGH TEMPERATURE CORROSION

The term high temperature is relative. In practical terms it usually means a temperature at about 35% of the absolute melting range of a given metal or alloy (or up to 60% for some nickel- and cobalt-based alloys). For the conventional austenitic grades of stainless steel, such as type 304, this would be any temperature above 1050°F (575°C).

In general, the straight chromium and austenitic varieties of stainless steel have an upper limit of about 1600°F (870°C), except the more highly alloyed grades (>20% Cr) which will tolerate slightly higher temperatures, about 2000°F (1100°C) in continuous service.

Changes can occur in the nature of the surface film of stainless steels when exposed to high temperatures. For example, at mildly elevated temperatures in an oxidizing gas a protective oxide film is formed. In an environment containing sulfur-bearing gases the film will be in the form of sulfides, which may also be protective.

In more aggressive environments, with temperatures above 1600°F (871°C) the surface film may break down with sudden increase in scaling. Depending on alloy content and environment, the film may be self-healing for a period of time followed by another breakdown.

Under extreme conditions of high temperature and corrosion, the surface film may not be protective at all. Based on this, service tests are recommended.

High temperature corrosion is considered to be electrochemical in nature, with the high temperature scale formed acting as an electrolyte. Corrosion is usually uniform in nature. The predominant effects are oxidation and carburization/decarburization. Changes in mechanical properties, specifically loss of ductility due to phase changes, also take place.

Most high temperature reactions involve oxidation because oxides are common products of reaction in the many applications where air or oxygen-rich environments are present. In clean atmospheres a thick oxide film forms which develops into a thicker scale. Oxidation phenomena are controlled by thermodynamic and kinetic factors, notably gas composition and temperature. High temperature grades of austenitic stainless steels contain at least 12% nickel. Alloys such as types 309 and 310 are widely used for good creep strength and ductility in addition to scaling resistance at high temperatures.

The nickel-rich type 310 alloy is less susceptible to sigma formation. Above 1598°F (870°C) there is little risk of embrittlement. Alloy compositions are more critical in temperature ranges of 1220 to 1598°F (650 to 870°C). Fully annealed wrought alloys are preferred.

A practical option for high temperature applications are nickel-rich alloys such as alloy 800/800H. Sigma may still be a problem for some nickel-base alloys in the range of 1450 to 1700°F (770 to 927°C).

Alloys containing molybdenum, specifically types 316 and 317 austenitic stainless steels containing 2 and 3% molybdenum, respectively, are subject to catastrophic oxidation. Under some conditions there is a selective oxidation of molybdenum with rapid loss of volatile Mo_3.

In high carbon-reducing atmospheres (e.g., carbon monoxide) at high temperatures carburization of stainless steel takes place. In oxidizing atmospheres such as steam or carbon dioxide, carbon may be selectively removed (decarburization). Usually complex gas mixtures are involved and the net result of the H_2/H_2O and CO/CO_2 is critical. Under some conditions of environment and temperature, a pitting type phenomenon called *metal dusting* takes place.

Many high temperature applications involve oxidizing conditions in which stainless steels usually perform well, within specific parameters of temperature and environment, based on their chromium content. The oxide film that causes passivation in conventional service becomes a visible scale, rich in chromic oxide Cr_2O_3. This oxide or mixed oxide provides protection against further oxidation of the substrate. However, spalling of the protective film will permit continued oxidation. High temperature reducing conditions can cause direct attack by preventing, or causing the loss of, the oxide film.

High temperature environments, like specific liquid chemical solutions, may be oxidizing or reducing. The overall nature is determined by the ratio of specific gases, vapors, or molten materials in the environment. The net effect is an algebraic sum resulting from the concentrations of specific oxidizing or reducing components. The common species encountered in gaseous media are shown in Table 7.2.

In nonfluctuating temperature service the oxidation resistance (scaling resistance) of stainless steel depends on the chromium content as shown in Fig. 7.2. Steels with less than 18% chromium, primarily ferritic grades, are limited to temperatures below 1500°F (816°C). Those containing 10–20% chromium can be used up to 1800°F (982°C), while steels having chromium

TABLE 7.2   Materials Found in Gaseous Media

Oxidizing	Reducing
Oxygen	Hydrogen
Steam	
Sulfurous oxides (SO$_2$, SO$_3$)	Hydrogen sulfide
Sulfur	Carbon disulfide
Carbon dioxide	Carbon monoxide
	Carbon
	Hydrocarbons
Chlorine	Hydrogen chloride
Oxides of nitrogen	Ammonia

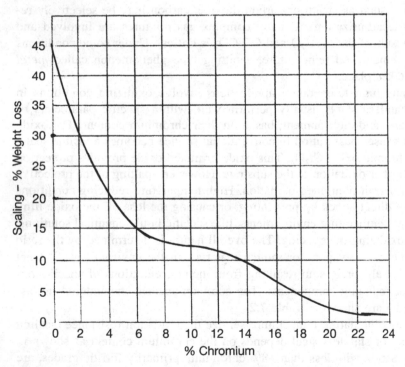

FIGURE 7.2   Effect of chromium content on scaling resistance at 1800°F (982°C).

content of at least 25% can be used up to 2000°F (1093°C). Typical of these latter steels are types 309, 310, and 416.

Based on an oxidation rate of 10 mg/cm^2 in 1000 hr, Table 7.3 provides the maximum service temperature for several stainless steels, for both non-fluctuating and intermittent service. In many processes constant temperature conditions are not maintained. Expansion and contraction differences between the base metal and the protective film (scale) during heating and cooling can cause cracking and spalling of the protective scale. This permits the oxidizing media to attack the exposed metal surface.

Higher nickel levels improve the spalling resistance of the austenitic stainless steels. This is shown in Fig. 7.3. Nickel reduces the thermal expansion differential between the alloy and oxide film, thereby reducing stresses at the alloy–oxide film interface during cooling. The cycling tem-

TABLE 7.3 Suggested Maximum Service Temperatures in Air

AISI type	Service			
	Intermittent		Continuous	
	°F	°C	°F	°C
201	1500	815	1550	845
202	1500	815	1550	845
301	1550	845	1650	900
302	1600	870	1700	925
304	1600	870	1700	925
308	1700	925	1800	980
309	1800	980	2000	1095
310	1900	1035	2100	1150
316	1600	870	1700	925
317	1600	870	1700	925
321	1600	870	1700	925
330	1900	1035	2100	1150
347	1600	870	1700	925
410	1500	815	1300	705
416	1400	760	1250	675
420	1350	735	1150	620
440	1500	815	1400	760
405	1500	815	1300	705
430	1600	870	1500	815
442	1900	1035	1800	980
446	2150	1175	1000	1095

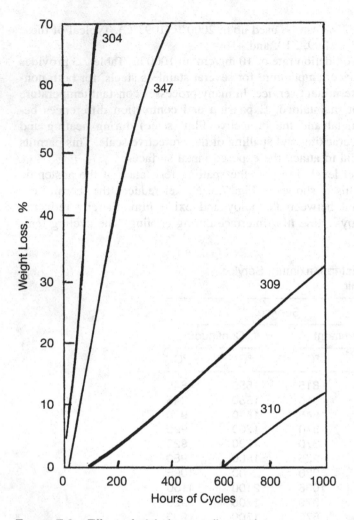

FIGURE 7.3 Effect of nickel on scaling resistance.

perature conditions in Fig. 7.3, at 1800°F (982°C), consisted of 15 min in the furnace and 5 min in air. Sheet specimens 0.031 in. (0.787 mm) thick were exposed on both sides.

As discussed previously reducing conditions can result from a high ratio of reducing to oxidizing species or from inherently reducing environments. Carburization can occur when there is an excess of carbon monoxide over carbon dioxide.

Carburization as such is not a corrosion phenomenon. However, it forms chromium carbides, thereby reducing the chromium matrix, and reduces the efficacy of the prior oxide film. Higher silicon contents will reduce the rate of carburization. Nickel in the iron-chromium-nickel alloys will improve the resistance to carburization by lowering the solubility of carbon, but not to the same degree that silicon and chromium do.

The most common corrosive condition associated with carburization is general absorption. Metal dusting, however, is a more serious form of attack, where under alternating oxidizing and reducing conditions localized high carbon areas are burned out during the oxidation period.

When hot surfaces react with active nitrogen, nitriding occurs. Since elements such as aluminum, chromium, and titanium readily form nitrides, the integrity of the oxide film is at risk. In order to form a stable protective oxide film, a nickel content in the order of 35–40% is required.

Halogens form films on stainless alloys, but their efficiency is limited because of the high volatility of metal chlorides. Chlorine in oxidizing flue gases or air will increase the corrosion compared to air alone. The attack usually entails internal corrosion and voids as well as surface attack.

Sulfur in small quantities and in various forms accelerates corrosion in many environments. The most corrosive forms are sulfur dioxide, hydrogen sulfide, and sulfur vapor, with the latter two being the most aggressive.

Sulfur attack is more severe than oxidation. Metal sulfides melt at lower temperatures than comparable oxides, and they may fuse to metal surfaces. In addition, sulfides are less likely to form tenacious, continuous protective films. Accelerated corrosion is the result of fusion and lack of adherence. The chromium content determines the resistance of stainless steel to sulfidation.

Type 316 stainless steel, when subjected to mixtures of oxygen and sulfur dioxide, in compositions ranging from 100% oxygen to 100% sulfur dioxide, at 1100 and 1600°F (593 and 871°C) did not develop a scale, only a heavy tarnish. The rate of attack was largely independent of gas composition.

Low concentrations of hydrogen sulfide can be handled satisfactorily in low chromium stainless steels. However, hydrogen under high pressure results in rapid corrosion. Under these conditions a minimum of 17% chromium is required to obtain satisfactory corrosion resistance. Type 304 stainless steel is used for this service.

Austenitic stainless steels are readily attacked by sulfur vapors. High corrosion rates are encountered at 1060°F (571°C). Liquid sulfur can be handled by austenitic stainless up to a temperature of 400°F (204°C), while stabilized grades, types 321 and 347, give satisfactory service up to 832°F (444°C).

Flue gases containing sulfur dioxide or hydrogen sulfide exhibit the same corrosivity as that of most sulfur-bearing gases. Consequently an increase in the chromium content will improve the corrosion resistance of the stainless steels, as shown in Fig. 7.4. Corrosion rates of 1 to 2 mils per year (mpy) have been reported for types 304, 321, 347, and 316 in the temperature range of 1200 to 1400°F (649 to 760°C). Service tests must be conducted for reducing flue gas environments.

Stainless steels are not suitable for the handling of molten chlorides. The molten chlorides cause intergranular attack of even the high nickel alloys by selective removal of chromium causing internal voids and martensite formation at the grain boundaries.

Stainless steels are not resistant to molten hydroxides, particularly sodium and potassium, because of chromium dissolution related to peroxide formation. However, they do perform well in molten carbonates up to 930°F (500°C). Above 1290°F (700°C) nickel-base alloys containing chromium are required.

When specific stainless steel alloys are exposed to specific molten metals there are potential problems of liquid metal embrittlement (LME) and

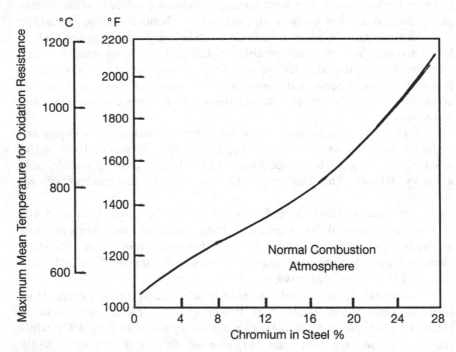

FIGURE 7.4   Effect of chromium in normal combustion atmosphere.

liquid metal cracking (LMC) development. Molten tin at 248°F (120°C) has induced LME in austenitic stainless steels. At 570°F (300°C) the fatigue limit was lowered. At about 785 to 1060°F (420 to 570°C) zinc slowly erodes unstressed 18-8 stainless steel. At 1060 to 1380°F (570 to 750°C) zinc penetrates to matrix via a Zn-Ni compound. Molten cadmium can also cause LMC of austenitic grades above 570°F (300°C).

VI. UNIFORM CORROSION

General or uniform corrosion, as found in other metals, is not to be expected in the stainless steels. The many sets of corrosion data and charts found in the literature which show various corrosion rates of stainless steel in certain environments are actually indicating that the stainless alloy, under those conditions, is fluctuating between an active and passive condition with a net result of so many mils per year loss. These may or may not be reliable figures; consequently, recommendations should be based on rates of less than 5 mpy and preferably less than 1 mpy. Under these conditions no corrosion allowance need be specified.

REFERENCES

1. PK Whitcraft. Corrosion of stainless steels. In: PA Schweitzer, Corrosion Engineering Handbook. New York: Marcel Dekker, 1996.
2. PA Schweitzer. Stainless steels. In: PA Schweitzer, ed. Corrosion and Corrosion Protection Handbook, 2nd ed. New York: Marcel Dekker, 1988.
3. CP Dillon. Corrosion Resistance of Stainless Steels. New York: Marcel Dekker, 1995.
4. PA Schweitzer. Encyclopedia of Corrosion Technology. New York: Marcel Dekker, 1998.
5. PA Schweitzer. Corrosion Resistance Tables, 4th ed. Vols. 1–3, New York: Marcel Dekker, 1995.
6. PT Lovejoy. Stainless steels. In: GT Murray, ed. Handbook of Materials Selection for Engineering Applications. New York: Marcel Dekker, 1997.
7. HH Uhlig. Corrosion and Corrosion Control. New York: John Wiley & Sons, 1963.
8. PA Schweitzer. Atmospheric Degradation and Corrosion Control. New York: Marcel Dekker, 1999.

8
Austenitic Stainless Steels

The austenitic stainless steels are the most widely used family of stainless alloys. They find application from mildly corrosive atmospheres to extremely corrosive environments. This group of alloys are nonmagnetic and are the most important for process industry applications. These stainless steels have a face-centered austenite structure from far below zero up to near melting temperatures as a result of the alloy additions of nickel and manganese. They are not hardenable by heat treatment, but can be strain hardened by cold work, which also induces a small amount of ferromagnetism.

In order to form the austenitic structure it is necessary to add about 8% nickel to the 18% chromium plateau to cause the transition from ferritic to austenitic. Compared to the ferritic structure the austenitic structure is very tough, formable, and weldable. The nickel addition also improves the corrosion resistance to mild corrodents. This includes resistance to most foods, a wide range of organic chemicals, mild inorganic chemicals, and most natural environments.

In order to further improve the corrosion resistance, molybdenum is added. This provides excellent corrosion resistance in oxidizing environments, particularly in aqueous solutions. The molybdenum aids in strengthening the passive film which forms on the surface of the stainless steel along with chromium and nickel.

Austenitic alloys also make use of the concept of stabilization. Stainless types 321 and 347 are stabilized with titanium and niobium, respectively. Another approach is also taken to avoid the effects of chromium carbide precipitation. Since the amount of chromium which will precipitate is proportional to the amount of carbon present, lowering the carbon content will prevent sensitization.

From an examination of Fig. 8.1 it can be seen that by maintaining the carbon content below about 0.035%, versus the usual 0.08% maximum, the harmful effects of chromium carbide precipitation can be avoided. This fact, along with improvements in melting technology, resulted in the development of the low carbon version of many of these alloys.

Various other elements are also added to enhance specific properties. The 200 and 300 series of stainless steels both start with the same high temperature austenite phase that exists in carbon steel, but, as mentioned previously, retain this structure to below zero. The 200 series of alloys rely mostly on manganese and nitrogen, while the 300 series utilizes nickel. Both series of stainless steel have useful levels of ductility and strength. Fabrication and welding are readily done. Grades 201 and 301, which are on the lean side of the retention elements, will transform to martensite when formed, but cool to austenite. This results in high strength parts made by stretching a low strength starting metal. Table 8.1 lists the chemical composition of the most commonly used austenitic stainless steels.

Compared to carbon steel, heat does not flow readily in stainless steel. Type 304 has a conductivity rate of 28% of that of carbon steel at 212°F (100°C) and 66% at 1200°F (649°C). This results in a temperature rise of the metal causing thermal expansion and stress and, if the component is

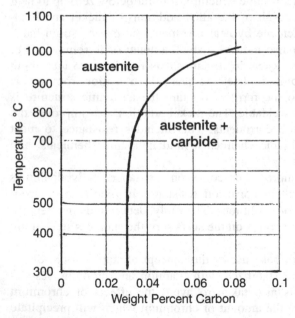

FIGURE 8.1   Solubility of carbon in austenite.

TABLE 8.1 Chemical Composition of Austenitic Stainless Steels

AISI type	C max.	Mn max.	Si max.	Cr	Ni	Others[a]
				Nominal composition (%)		
201	0.15	7.5[b]	1.00	16.00–18.00	3.50–5.50	0.25 max. N
202	0.15	10.00[c]	1.00	17.00–19.00	4.00–6.00	0.25 max. N
205	0.25	15.50[d]	0.50	16.50–18.00	1.00–1.75	0.32/0.4 max. N
301	0.15	2.00	1.00	16.00–18.00	6.00–8.00	
302	0.15	2.00	1.00	17.00–19.00	8.00–10.00	
302B	0.15	2.00	3.00[e]	17.00–19.00	8.00–10.00	
303	0.15	2.00	1.00	17.00–19.00	8.00–10.00	0.15 min. S
303(Se)	0.15	2.00	1.00	17.00–19.00	8.00–10.00	0.15 min. Se
304	0.08	2.00	1.00	18.00–20.00	8.00–12.00	
304L	0.03	2.00	1.00	18.00–20.00	8.00–12.00	
304N	0.08	2.00	1.00	18.00–20.00	8.00–10.50	0.1/0.16 N
305	0.12	2.00	1.00	17.00–19.00	10.00–13.00	
308	0.08	2.00	1.00	19.00–21.00	10.00–12.00	
309	0.20	2.00	1.00	22.00–24.00	12.00–15.00	
309S	0.08	2.00	1.00	22.00–24.00	12.00–15.00	
310	0.25	2.00	1.50	24.00–26.00	19.00–22.00	
310S	0.08	2.00	1.50	24.00–26.00	19.00–22.00	
314	0.25	2.00	3.00[f]	23.00–26.00	19.00–22.00	
316	0.08	2.00	1.00	16.00–18.00	10.00–14.00	2.00–3.00 Mo
316F	0.08	2.00	1.00	16.00–18.00	10.00–14.00	1.75–2.50 Mo
316L	0.03	2.00	1.00	16.00–18.00	10.00–14.00	2.00/3.00 Mo
316N	0.08	2.00	1.00	16.00–18.00	10.00–14.00	2.00–3.00 Mo
317	0.08	2.00	1.00	18.00–20.00	11.00–15.00	3.00–4.00 Mo
317L	0.03	2.00	1.00	18.00–20.00	11.00–15.00	3.00–4.00 Mo
321	0.08	2.00	1.00	17.00–19.00	9.00–12.00	5 × C min. Cb-Ta
330	0.08	2.00	1.5[g]	17.00–20.00	34.00–37.00	0.10 TA 0.20 Cb
347	0.08	2.00	1.00	17.00–19.00	9.00–13.00	10 × C min. Cb-Ta
348	0.08	2.00	1.00	17.00–19.00	9.00–13.00	10C min. Cb-Ta 2.0 Mo 3.0 Cu
20Cb3	0.07	0.75	1.00	20.0	29.0	8 × C Min. Cb-Ta
904L	0.02	—	—	21.0	25.5	4.7 Mo 1.5 Cu

[a]Other elements in addition to those shown are as follows: phosporus is 0.03% max. in type 205; 0.06% max. in types 201 and 202; 0.045% max. in types 301, 302, 302b, 304, 304L, 304N, 305, 308, 309, 309S, 310, 310S, 314, 316, 316N, 316L, 317, 317L, 321, 330, 347, and 348; 0.20% max. in types 303, 303(Se), and 316D. Sulfur is 0.030% max. in types 201, 202, 205, 301, 302, 302B, 304, 304L, 304N, 305, 308, 309, 309S, 310, 310X, 314, 316, 316L, 316N, 317, 317L, 321, 330, 347, and 348; 0.15% min. in type 303; and 0.10% min. in type 316D.
[b]Mn range 4.40–7.50.
[c]Mn range 7.50–10.00.
[d]Mn range 14.00–15.50.
[e]Si range 2.00–3.00.
[e]Si range 1.50–3.00.

restrained, causes unanticipated distortion. However, low thermal conductivity can be an asset in thermal insulation applications.

The thermal expansion is affected by both structure and chemistry. Austenitic stainless steels generally expand at about twice the rate of carbon steel.

The mechanical properties of stainless steels are a function of the series and product form (i.e., thin sheet, thick plate, large billets, or bars). The austenitic stainless steels have yield strengths from 30,000 or 35,000 psi up to 90,000 or 110,000 psi at their ultimate strength after total elongations of 40–60%.

Variation in the ability to work harden determines the selection of type 201 or type 301, which stretch further and reach higher flow stresses than type 304, versus type 305 which does not reach such high flow stress levels even though the elongation remains at reasonably high austenitic levels. Type 201 and 301 alloys form martensite while deforming, their high strength supporting and maintaining the continued strain to higher levels. The more highly alloyed stable austenitic alloys do not form martensite, and their properties are similar to those of type 304.

The strengths of most metals decrease with increasing temperature. Table 7.3 listed the suggested maximum operating temperatures for selected stainless steel alloys. Rupture and creep characteristics for selected austenitic stainless steels are given in Table 8.2. Stainless steel alloys are frequently used for low temperature applications in general. The austenitic stainless steels are preferred for cryogenic services. Table 8.3 lists the low temperature mechanical properties of selected stainless steels.

The 200 series austenitic stainless steels (i.e., types 201, 202, 216, etc.) are stronger than the 300 series austenitic stainless steels. Even so the 200 series alloys are not normally used by the chemical process industry because of fabrication problems. In addition they are not readily available from alloy suppliers and have very limited availability in product forms other than bar, plate, and sheet.

The austenitic stainless steels do not air harden and therefore do not require postweld heat treatment as a hardness control measure. On occasion they may be stress relieved or postweld heat treated to reduce residual stresses, thereby improving their resistance to stress corrosion cracking.

Galling may be a problem with the austenitic stainless steels. Aside from the use of lubricants and coatings, the most common way to avoid the problem is to require that the two mating surfaces have a Brinnell hardness number (BHN) difference of at least 50. A hard face weld overlay or electroless nickel plating on one of the two components will provide the hardness differential. The hardness differential in threaded connections is usually obtained by cold working one of the components. Specifying the two com-

TABLE 8.2 Rupture and Creep Characteristics for Selected Austenitic Stainless Steels

Type	Testing temp. °F	Testing temp. °C	Rupture time 100 h (ksi)	Rupture time 1000 h (ksi)	Stress Creep rate 10,000 h (ksi)	Stress 0.01% h[a] (ksi)	Extrapolated elongation at rupture in 10,000 h (%)
302	1600	871	4.70	2.80	1.75	2.50	150
	1800	982	2.45	1.55	0.96	1.30	30
	2000	1093	1.30	0.76	0.46	0.62	18
309S	1600	871	5.80	3.20	—	3.50	—
	1800	982	2.60	1.65	1.00	1.00	105
	2000	1093	1.40	0.83	0.48	0.76	42
310S	1600	871	6.60	4.00	2.50	4.00	30
	1800	982	3.20	2.10	1.35	1.75	60
	2000	1093	1.50	1.10	0.76	0.80	60
314	1600	871	4.70	3.00	1.95	2.30	110
	1800	982	2.60	1.70	1.10	1.00	120
	2000	1093	1.50	1.12	0.85	0.90	82
316	1600	871	5.00	2.70	1.40	2.60	30
	1800	982	2.65	1.25	0.60	1.20	35
	2000	1093	1.12	0.36	—	4.00	—

[a] Stress for creep rate of 0.01%/h.

ponents in different materials having appropriately different hardnesses can also solve the problem. In this case beware of galvanic action which could result in corrosion. The problem can also be alleviated by specifying one component to be a free machining grade, such as type 303 stainless.

I. TYPE 201 (S20100)

This is one of the alloys based on the substitution of manganese for nickel because of the shortage of nickel during and shortly after World War II. It was developed as a substitute for type 304 stainless steel. By adding about 4% manganese and 0.2% nitrogen, the nickel content could be lowered to about 5%. The chemical composition is shown in Table 8.1. Although the strength of this alloy is higher than that of type 304, its corrosion resistance is inferior. It does have a corrosion resistance comparable to type 301.

TABLE 8.3 Typical Mechanical Properties of Stainless Steels at Cryogenic Temperatures

Stainless steel type	Temperature (°F/°C)	Yield strength 0.2% offset (ksi)	Tensile strength (ksi)	Elongation in 2 in. (%)	Izod impact (ft-lb)
304	−40/−40	34	150	47	110
	−80/−62	34	170	39	110
	−320/−196	39	225	40	110
	−423/−252	50	243	40	110
310	−40/−40	39	95	57	110
	−80/−62	40	100	55	110
	−320/−196	74	152	54	85
	−423/−252	108	176	56	110
316	−40/−40	41	104	59	110
	−80/−62	44	116	57	
	−320/−196	75	155	59	
	−423/−252	84	210	52	
347	−40/−40	44	117	63	110
	−80/−62	45	130	57	110
	−320/−196	47	200	46	95
	−423/−252	55	228	34	
410	−40/−40	90	122	23	25
	−80/−62	94	128	22	25
	−320/−196	148	156	10	5
430	−40/−40	41	76	36	10
	−80/−62	44	81	36	8
	−320/−196	87	92	2	2

This alloy can be cold worked to high strength levels. It is nonmagnetic as annealed and becomes somewhat magnetic after cold work. The physical and mechanical properties of type 201 stainless are shown in Table 8.4.

II. TYPE 202 (S20200)

Alloy type 202 is one of the series of alloys using manganese as a replacement for nickel. As can be seen in Table 8.1 the manganese content of this alloy is greater than that of type 201, as is the chromium and nickel content. This provides improved corrosion resistance.

This alloy is capable of having its mechanical properties improved to the same degree as type 201. The mechanical and physical properties are

TABLE 8.4 Mechanical and Physical Properties of Types 201 and 202 Stainless Steel

Property	Type 201	Type 202
Modulus of elasticity $\times$ 10^6 (psi)	28.6	28.6
Tensile strength $\times$ 10^3 (psi)	95	90
Yield strength 0.2% offset $\times$ 10^3 (psi)	45	45
Elongation in 2 in. (%)	40	40
Rockwell hardness	B-90	B-90
Density (lb/in.^{3})	0.28	0.28
Specific gravity	7.7	7.7
Specific heat at 32–212°F (Btu/lb °F)	0.12	0.12
Thermal conductivity at 212°F (Btu/hr ft^2 °F)	9.4	9.4
Thermal expansion coefficient at 32–212°F $\times$ 10^{-6} (in./in. °F)		
Izod impact (ft-lb)	115	—

shown in Table 8.4. Figure 8.2 illustrates the effect of cold work on the mechanical properties of type 202 stainless.

III. TYPE 22-13-5 (S20910)

This is a nitrogen-strengthened stainless alloy having the following composition:

Carbon	0.06%
Manganese	4.00/6.00%
Phosphorus	0.040%
Sulfur	0.030%
Silicon	1.00%
Chromium	20.50/23.50%
Nickel	11.50/13.50%
Molybdenum	1.50/3.00%
Columbium	0.10/0.30%
Vanadium	0.10/0.30%
Nitrogen	0.20/0.40%
Iron	Balance

It is superior in corrosion resistance to type 316 stainless with twice the yield strength. It can be welded, machined, and cold worked using the same equipment and methods used for the conventional 300 series stainless steels. It remains nonmagnetic after severe cold work.

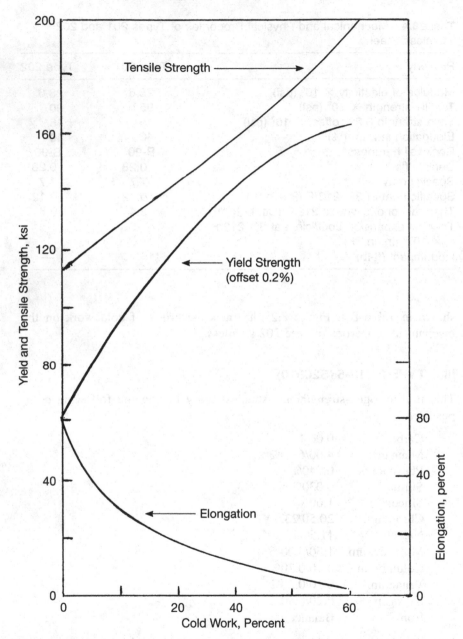

FIGURE 8.2   Effect of cold work on mechanical properties of type 202 stainless steel.

The alloy has an excellent combination of strength, ductility, toughness, corrosion resistance, and fabricability. In addition the alloy has good toughness at cryogenic temperatures and relatively high tensile and yield strengths at moderately high elevated temperatures. Mechanical and physical properties are shown in Table 8.5.

Type 22-13-5 stainless steel has very good corrosion resistance in many reducing and oxidizing acids, chlorides, and pitting environments. It has a pitting resistance equivalent number (PREN) of 45.5. In particular the alloy provides an excellent level of resistance to pitting and crevice corrosion in seawater. Resistance to intergranular attack in boiling 65% nitric acid and in ferric sulfate–sulfuric acid is excellent for both the annealed and sensitized conditions. Like other austenitic stainless steels, S20910 under certain conditions may suffer stress corrosion cracking in hot chloride environments.

This alloy also demonstrates good resistance to sulfide stress cracking at ambient temperatures.

This alloy has been used in such applications as valve shafts, taper pins, pumps, fasteners, cables, chains, screens, wire cloth, marine hardware, boat shafting, heat exchanger parts, and springs. This alloy is sometimes referred to as nitronic 50.

IV. TYPE 216L (S21603)

This is a low carbon alloy in which a portion of the nickel has been replaced by molybdenum. It has the following composition:

TABLE 8.5 Mechanical and Physical Properties of S20910 Stainless Steel

Property	
Modulus of elasticity × 10^6 (psi)	28
Tensile strength × 10^3 (psi)	210
Yield strength 0.2% offset × 10^3 (psi)	65
Elongation in 2 in. (%)	45
Rockwell hardness	B-96
Density (lb/in.^{3})	0.285
Specific gravity	7.88
Specific heat at 32–212°F (Btu/lb °F)	0.12
Thermal conductivity at 300°F (Btu/hr ft^2 °F)	108
Thermal expansion coefficient at 32–212°F × 10^{-6} (in./in. °F)	9.0
Izod impact (ft-lb)	160

Carbon	0.03%
Manganese	7.50/9.00%
Chromium	17.5/22.0%
Nickel	5.00/7.00%
Molybdenum	2.00/3.00%
Silicon	1.00%

The physical and mechanical properties are given in Table 8.6.

This alloy finds application as aircraft hydraulic lines, heat exchanger tubes, pollution control equipment, and particle accelerator tubes.

V. TYPE 301 (S30100)

This is a nitrogen-strengthened alloy which has the ability to work harden. As with the 200 series alloys it forms martensite while deforming, but retains the contained strain to higher levels. Figure 8.3 illustrates the effect of cold work on the mechanical properties. The chemical composition is shown in Table 8.1 and the mechanical and physical properties in Table 8.7. Types 301L and 301LN find application in passenger rail cars, buses, and light rail vehicles. The chemical composition of type 301L (S30103) and type 301LN (S30153) are as follows:

Alloying element	Alloy	
	301L	301LN
Carbon	0.030 max.	0.030 max.
Chromium	16.0–18.0	16.0–18.0
Manganese	2.0 max.	2.0 max.
Nitrogen	0.20 max.	0.07–0.20
Nickel	5.0–8.0	5.0–8.0
Phosphorus	0.045 max.	0.045 max.
Sulfur	0.030 max.	0.030 max.
Silicon	1.0 max.	1.0 max.

VI. TYPE 302 (S30200)

Type 302 and type 302B are nonmagnetic, extremely tough and ductile and two of the most widely used of the chromium-nickel stainless and heat-resisting steels. They are nonhardenable by heat treating.

The chemical compositions are shown in Table 8.1 and the mechanical and physical properties are given in Table 8.8.

TABLE 8.6   Mechanical and Physical Properties of Alloy
21-6-9 (S21904) Stainless Steel

Property	
Modulus of elasticity $\times$ 10^6 (psi)	28.5
Tensile strength $\times$ 10^3 (psi)	110
Yield strength 0.2% offset $\times$ 10^3 (psi)	65
Elongation in 2 in. (%)	42
Rockwell hardness	B-95
Density (lb/in.^{3})	0.283
Specific gravity	7.88
Thermal conductivity at 70°F (Btu/hr ft^2 °F)	8.0

VII.   TYPE 303 (S30300)

This is a free machining version of type 304 stainless steel for automatic machining. It is corrosion resistant to atmospheric exposures, sterilizing solutions, most organic and many inorganic chemicals, most dyes, nitric acid, and foods. The chemical composition is given in Table 8.1. Mechanical and physical properties are given in Table 8.9.

VIII.   TYPE 304 (S30400)

Type 304 stainless steels are the most widely used of any stainless steel. Although they have a wide range of corrosion resistance they are not the most corrosion resistant of the austenitic stainlesses. The chemical composition of various type 304 alloys are shown in Table 8.1.

Type 304 stainless steel is subject to intergranular corrosion as a result of carbide precipitation. Welding can cause this phenomenon, but competent welders using good welding techniques can control the problem. Depending upon the particular corrodent being handled, the effect of carbide precipitation may or may not present a problem. If the corrodent being handled will attack through intergranular corrosion, another alloy should be used.

If the carbon content of the alloy is not allowed to exceed 0.03%, carbide precipitation can be controlled. Type 304L is such an alloy. This alloy can be used for welded sections without danger of carbide precipitation.

Type 304N has nitrogen added to the alloy which improves its resistance to pitting and crevice corrosion.

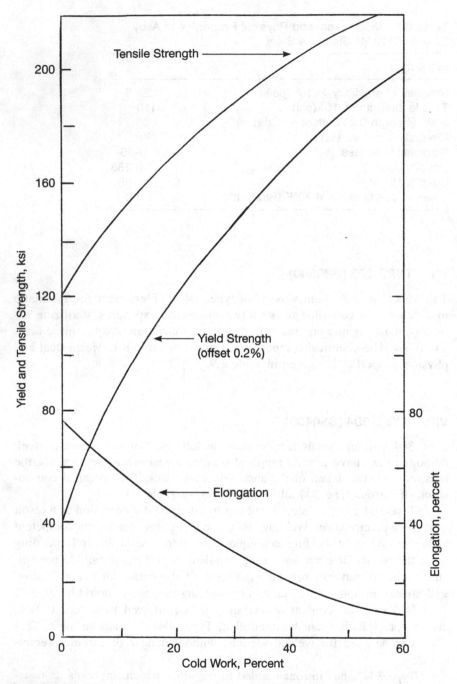

FIGURE 8.3 Effect of cold work on mechanical properties of type 301 stainless steel.

TABLE 8.7 Mechanical and Physical Properties of Type 301 Stainless Steel

Property	
Modulus of elasticity $\times 10^6$ (psi)	28
Tensile strength $\times 10^3$ (psi)	110
Yield strength 0.2% offset $\times 10^3$ (psi)	40
Elongation in 2 in. (%)	60
Rockwell hardness	B-95
Density (lb/in.^{3})	0.29
Specific gravity	8.02
Specific heat at 32–212°F (Btu/lb °F)	0.12
Thermal conductivity at 212°F (Btu/hr ft^2 °F)	93
Thermal expansion coefficient at 32–212°F $\times 10^{-6}$ (in./in. °F)	9.4

TABLE 8.8 Mechanical and Physical Properties of Types 302 and 302B Stainless Steel

Property	Type 302	Type 302B
Modulus of elasticity $\times 10^6$ (psi)	28	28
Tensile strength $\times 10^3$ (psi)	90	95
Yield strength 0.2% offset $\times 10^3$ (psi)	40	40
Elongation in 2 in. (%)	50	55
Rockwell hardness	B-85	B-85
Density (lb/in.^{3})	0.29	0.29
Specific gravity	8.02	8.02
Specific heat at 32–212°F (Btu/lb °F)	0.12	0.12
Thermal conductivity at 212°F (Btu/hr ft^2 °F)	9.3	9.3
Thermal expansion coefficient at 32–212°F $\times 10^{-6}$ (in./in. °F)	9.6	9.6

TABLE 8.9 Mechanical and Physical Properties of Types 303 and 303Se Stainless Steel

Property	Type 303	Type 303Se
Modulus of elasticity $\times 10^6$ (psi)	28	28
Tensile strength $\times 10^3$ (psi)	90	90
Yield strength 0.2% offset $\times 10^3$ (psi)	35	35
Elongation in 2 in. (%)	50	50
Rockwell hardness		
Density (lb/in.^{3})	0.29	0.29
Specific gravity	8.027	8.027
Specific heat at 32–212°F (Btu/lb °F)	9.3	9.3
Thermal conductivity at 212°F (Btu/hr ft^2 °F)		
Thermal expansion coefficient at 32–212°F $\times 10^{-6}$ (in./in. °F)		
Izod impact (ft-lb)	120	—

Types 304 and 304L stainless steel exhibit good overall corrosion resistance. They are used extensively in the handling of nitric acid. Refer to Table 8.10 for the compatibility of these alloys with selected corrodents.

The 304 series of stainless steels exhibit high temperature strength, oxidation resistance, ease of fabrication and weldability, good ductility, and good impact resistance down to at least $-216°F$ ($-183°C$). Their mechanical properties in general are excellent and are shown in Table 8.11. The allowable design stresses for types 304 and 304L stainless steel are given in Table 8.12.

These alloys can be machined, bent to angles, drawn into cups or cylindrical shapes, or stretched into parts where full use is made of the high total elongation. During bending the high levels of flow stress that are needed will result in a tendency to spring back. Overbending is used to obtain specific angles.

The effect of high flow stresses, or the basic strength levels, is the cause of other forming considerations. Lubrication should be used to ensure that the high flow stresses do not get transmitted to tooling in a way that results in galling against a die and possible fracture. Water-soluble lubricants help remove heat generated during machining operations. Solid rigid tooling and deep slow cuts will improve the machining operation. These techniques minimize the volume of metal reaching the maximum flow stresses, thus limiting heat energy.

Difficulty in drawing type 304 into any cylindrical shape can be overcome by selecting a type 305 alloy which does not work harden to the same high flow stress levels. However, if the problem is how to stretch the stainless to make a part with a corresponding reduction in thickness, then a type 201 or type 301 alloy may be a better choice.

IX. TYPE 305 (S30500)

Type 305 stainless steel is used extensively for cold heading, severe deep drawing, and spinning operations. High nickel content slows work hardening. The alloy maintains low magnetic permeability after cold working. The effect of cold working on mechanical properties is shown in Fig. 8.4. The chemical composition is shown in Table 8.1 and the mechanical and physical properties in Table 8.13. Type 305 stainless steel has the equivalent corrosion resistance of type 304 stainless steel.

X. TYPE 308 (S30800)

The chemical composition of type 308 stainless steel is given in Table 8.1. It will be noted that this alloy has an increased chromium and nickel content over that of type 304 stainless. In the annealed condition type 308 exhibits greater tensile and yield strengths than annealed type 304. The mechanical and physical properties are shown in Table 8.14. The corrosion resistance of type 308 is slightly better than that of type 304 stainless.

XI. TYPE 309 (S30900)

Types 309 and 309S are superior heat-resisting stainless alloys. They are applicable for continuous exposure to 2000°F (1093°C) and to 1800°F (982°C) for intermittent exposure. These are modifications of type 304 stainless steel. The chemical composition is shown in Table 8.1. These alloys have better creep strength than the 304 alloys. The mechanical and physical properties are given in Table 8.15.

Types 309 and 309S alloys have slightly better corrosion resistance than type 304 stainless steel. They are resistant to corrosive action of high sulfur gases, provided they are oxidizing, but poor in reducing gases like hydrogen sulfide. These alloys are excellent in resisting sulfite liquors, nitric acid, nitric–sulfuric acid mixtures, and acetic, citric, and lactic acids. Type 309S with a maximum of 0.08% carbon resists corrosion in welded parts. They may be susceptible to chloride stress corrosion cracking.

Applications include sulfur-bearing gas atmospheres, furnace parts, fire boxes, high temperature containers, and heat exchanger tubing.

XII. TYPE 310 (S31000)

This is an alloy for high temperatures. It is an improvement over types 309 and 309S. The 310 and 310S alloys have a maximum allowable temperature of 2100°F (1149°C) at continuous operation and 1900°F (1037°C) for intermittent service. Chemical compositions are shown in Table 8.1, and the mechanical and physical properties are given in Table 8.16.

These alloys have better general corrosion resistance than type 304 and type 309. They have excellent high temperature oxidation resistance and good resistance to both carburizing and reducing environments. Chloride stress corrosion cracking may cause a problem under the right conditions. Type 310S, with 0.08% maximum carbon, offers improved resistance in welded components.

XIII. TYPE 316 (S31600)

These chromium nickel grades of stainless steel have molybdenum added in the range of 2–3%. The molybdenum substantially increases resistance to pitting and crevice corrosion in systems containing chlorides and improves

TABLE 8.10 Compatibility of Types 304, 304L, and 347 Stainless Steel with Selected Corrodents[a]

Chemical	Maximum temp. °F	Maximum temp. °C	Chemical	Maximum temp. °F	Maximum temp. °C
Acetaldehyde	200	93	Benzaldehyde	210	99
Acetamide	100	38	Benzene	230	110
Acetic acid, 10%	200	93	Benzene sulfonic acid, 10%	210	99
Acetic acid, 50%	170	77	Benzoic acid	400	204
Acetic acid, 80%	170	77	Benzyl alcohol	90	32
Acetic acid, glacial	210	99	Benzyl chloride	210	99
Acetic anhydride	220	104	Borax	150	66
Acetone	190	88	Boric acid[b]	400	204
Acetyl chloride	100	38	Bromine gas, dry	X	
Acrylic acid	130	54	Bromine gas, moist	X	
Acrylonitrile	210	99	Bromine, liquid	X	
Adipic acid	210	99	Butadiene	180	82
Allyl alcohol	220	104	Butyl acetate	80	27
Allyl chloride	120	49	Butyl alcohol	200	93
Alum	X		Butyl phthalate	210	99
Aluminum acetate	210	99	Butyric acid	180	82
Aluminum chloride, aqueous	X		Calcium bisulfite[d]	300	149
Aluminum chloride, dry	150	66	Calcium carbonate	210	99
Aluminum fluoride	X		Calcium chlorate, 10%	210	99
Aluminum hydroxide	80	27	Calcium chloride[b,c]	80	27
Aluminum nitrate	80	27	Calcium hydroxide, 10%	210	99
Aluminum sulfate[b]	210	99	Calcium hydroxide, sat.	200	93
Ammonia gas	90	32	Calcium hypochlorite	X	
Ammonium carbonate	200	93	Calcium nitrate	90	32
Ammonium chloride, 10%	230	110	Calcium oxide	90	32
Ammonium chloride, 50%	X		Calcium sulfate	210	99
Ammonium chloride, sat.	X		Caprylic acid[a]	210	99
Ammonium fluoride, 10%	X		Carbon bisulfide	210	99
Ammonium fluoride, 25%	X		Carbon dioxide, dry	210	99
Ammonium hydroxide, 25%	230	110	Carbon dioxide, wet	200	93
Ammonium hydroxide, sat.	210	99	Carbon disulfide	210	99
Ammonium nitrate[c]	210	99	Carbon monoxide	570	299
Ammonium persulfate	X		Carbon tetrachloride	210	99
Ammonium phosphate, 40%	130	54	Carbonic acid	210	99
Ammonium sulfate, 10–40%	X		Cellosolve	210	99
Ammonium sulfide	210	99	Chloracetic acid, 50% water	X	
Ammonium sulfite	210	99	Chloracetic acid	X	
Amyl acetate	300	149	Chlorine gas, dry	X	
Amyl alcohol	80	27	Chlorine gas, wet	X	
Amyl chloride	150	66	Chlorine, liquid[b]	110	43
Aniline	500	260	Chlorobenzene	210	99
Antimony trichloride	X		Chloroform[c]	210	99
Aqua regia, 3:1	X		Chlorosulfonic acid	X	
Barium carbonate	80	27	Chromic acid, 10%	200	93
Barium chloride	X		Chromic acid, 50%	90	32
Barium hydroxide	230	110	Chromyl chloride	210	99
Barium sulfate	210	99	Citric acid, 15%	210	99
Barium sulfide	210	99	Citric acid, conc.	80	27

TABLE 8.10 Continued

Chemical	Maximum temp. °F	Maximum temp. °C	Chemical	Maximum temp. °F	Maximum temp. °C
Copper acetate	210	99	Nitric acid, 5%	210	99
Copper carbonate, 10%	80	27	Nitric acid, 20%	190	88
Copper chloride	X		Nitric acid, 70%	170	77
Copper cyanide	210	99	Nitric acid, anhydrous	80	27
Copper sulfate[d]	210	99	Nitrous acid, conc.	80	27
Cresol	160	71	Oleum	100	38
Cupric chloride, 5%	X		Perchloric acid, 10%	X	
Cupric chloride, 50%	X		Perchloric acid, 70%	X	
Cyclohexane	100	38	Phenol[b]	560	293
Cyclohexanol	80	27	Phosphoric acid, 50–80%[d]	120	49
Dichloroethane (ethylene dichloride)	210	99	Picric acid[b]	300	149
Ethylene glycol	210	99	Potassium bromide, 30%	210	99
Ferric chloride	X		Salicylic acid	210	99
Ferric chloride, 50% in water	X		Silver bromide, 10%	X	
Ferric nitrate, 10–50%	210	99	Sodium carbonate, 30%	210	99
Ferrous chloride	X		Sodium chloride, to 30%[b]	210	99
Fluorine gas, dry	470	243	Sodium hydroxide, 10%	210	99
Fluorine gas, moist	X		Sodium hydroxide, 50%	210	99
Hydrobromic acid, dilute	X		Sodium hydroxide, conc.	90	32
Hydrobromic acid, 20%	X		Sodium hypochorite, 20%	X	
Hydrobromic acid, 50%	X		Sodium hypochlorite, conc.	X	
Hydrochloric acid, 20%	X		Sodium sulfide, to 50%[b]	210	99
Hydrochloric acid, 38%	X		Stannic chloride	X	
Hydrocyanic acid, 10%	210	99	Stannous chloride	X	
Hydrofluoric acid, 30%	X		Sulfuric acid, 10%	X	
Hydrofluoric acid, 70%	X		Sulfuric acid, 50%	X	
Hydrofluoric acid, 100%	X		Sulfuric acid, 70%	X	
Hypochlorous acid	X		Sulfuric acid, 90%[d]	80	27
Iodine solution, 10%	X		Sulfuric acid, 98%[d]	80	27
Ketones, general	200	93	Sulfuric acid, 100%[d]	80	27
Lactic acid, 25%[b,d]	120	49	Sulfuric acid, fuming	90	32
Lactic acid, conc.[b,d]	80	27	Sulfurous acid	X	
Magnesium chloride	X		Thionyl chloride	X	
Malic acid, 50%	120	49	Toluene	210	99
Manganese chloride	X		Trichloroacetic acid	X	
Methyl chloride[b]	210	99	White liquor	100	38
Methyl ethyl ketone	200	93	Zinc chloride	X	
Methyl isobutyl ketone	200	93			
Muriatic acid	X				

[a]The chemicals listed are in the pure state or in a saturated solution unless otherwise indicated. Compatibility is shown to the maximum allowable temperature for which data are available. Incompatibility is shown by an X. When compatible, the corrosion rate is <20 mpy.
[b]Subject to pitting.
[c]Subject to stress cracking.
[d]Subject to intergranular attack (type 304).
Source: Ref. 5.

TABLE 8.11 Mechanical and Physical Properties of Types 304 and 304L Stainless Steel

Property	Type 304	Type 304L
Modulus of elasticity $\times 10^6$ (psi)	28.0	28.0
Tensile strength $\times 10^3$ (psi)	85	80
Yield strength 0.2% offset $\times 10^3$ (psi)	35	30
Elongation in 2 in. (%)	55	55
Rockwell hardness	B-80	B-80
Density (lb/in.^{3})	0.29	0.29
Specific gravity	8.02	8.02
Specific heat at 32–212°F (Btu/lb °F)	0.12	0.12
Thermal conductivity at 212°F (Btu/hr ft^2 °F)	9.4	9.4
Thermal expansion coefficient at 32–212°F $\times 10^{-6}$ (in./in. °F)	9.6	9.6
Izod impact (ft-lb)	110	110

TABLE 8.12 Allowable Design Stress for Types 304 and 304L Stainless Steel

Temperature (°F/°C)	Allowable design stress (psi)	
	304	304L
−325 to 100/−198 to 38	18750	15600
200/93	16650	15300
300/149	15000	13100
400/204	13650	11000
500/260	12500	9700
600/316	11600	9000
650/343	11200	8750
700/371	10800	8500
750/399	10400	8300
800/427	10000	8100
850/454	9700	
900/482	9400	
950/150	9100	
1000/538	8800	
1050/560	8500	
1100/593	7500	
1150/621	5750	
1200/649	4500	
1250/677	3250	
1300/704	2450	
1350/732	1800	
1400/760	1400	
1450/788	1000	
1500/816	750	

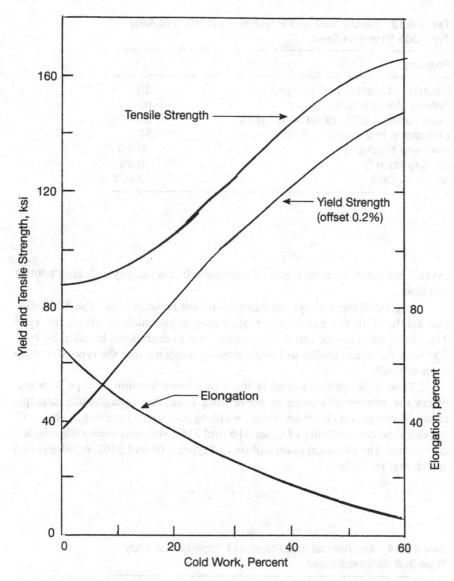

FIGURE 8.4 Effect of cold work on mechanical properties of type 305 stainless steel.

TABLE 8.13   Mechanical and Physical Properties of Alloy
Type 305 Stainless Steel

Property	
Modulus of elasticity $\times 10^6$ (psi)	28
Tensile strength $\times 10^3$ (psi)	85
Yield strength 0.2% offset $\times 10^3$ (psi)	35
Elongation in 2 in. (%)	50
Rockwell hardness	B-80
Density (lb/in.^{3})	0.29
Specific gravity	8.027

overall resistance to most types of corrosion in chemically reducing neutral solutions.

In general these alloys are more corrosion resistant than type 304 stainless steels. With the exception of oxidizing acids, such as nitric, the type 316 alloys will provide satisfactory resistance to corrodents handled by type 304 with the added ability to handle some corrodents that the type 304 alloy cannot handle.

Type 316L stainless steel is the low carbon version of type 316 and offers the additional feature of preventing excessive intergranular precipitation of chromium chlorides during welding and stress relieving. Table 8.17 provides the compatibility of types 316 and 316L stainless steel with selected corrodents. The chemical compositions of types 316 and 316L stainless steel are shown in Table 8.1.

TABLE 8.14   Mechanical and Physical Properties of Alloy
Type 308 Stainless Steel

Property	
Modulus of elasticity $\times 10^6$ (psi)	28
Tensile strength $\times 10^3$ (psi)	115
Yield strength 0.2% offset $\times 10^3$ (psi)	80
Elongation in 2 in. (%)	40
Rockwell hardness	B-80

TABLE 8.15 Mechanical and Physical Properties of Types 309 and 309S Stainless Steel

Property	Type 309	Type 309S
Modulus of elasticity $\times 10^6$ (psi)	29	29
Tensile strength $\times 10^3$ (psi)	90	90
Yield strength 0.2% offset $\times 10^3$ (psi)	45	45
Elongation in 2 in. (%)	45	45
Rockwell hardness	B-85	B-85
Density (lb/in.^{3})	0.29	0.29
Specific gravity	8.02	8.02
Specific heat at 32–212°F (Btu/lb °F)	0.12	0.12
Thermal conductivity at 212°F (Btu/hr ft^2 °F)	8	8
Thermal expansion coefficient at 32–212°F $\times 10^{-6}$ (in./in. °F)	8.3	8.3

In the annealed condition these alloys are nonhardenable and nonmagnetic, but are slightly magnetic when cold worked. They have higher tensile and creep strength at elevated temperature than type 304 stainless. The mechanical and physical properties are shown in Table 8.18 and the maximum allowable design stresses are given in Table 8.19.

TABLE 8.16 Mechanical and Physical Properties of Types 310 and Type 310S Stainless Steel

Property	Type 310	Type 310S
Modulus of elasticity $\times 10^6$ (psi)	29	29
Tensile strength $\times 10^3$ (psi)	95	95
Yield strength 0.2% offset $\times 10^3$ (psi)	45	45
Elongation in 2 in. (%)	45	45
Rockwell hardness	B-85	B-85
Density (lb/in.^{3})	0.28	0.28
Specific gravity	7.7	7.7
Thermal conductivity (Btu/ft hr °F)		
at 70°F		8.0
at 1500°F		10.8

TABLE 8.17   Compatibility of Types 316, and 316L Stainless Steel with
Selected Corrodents[a]

Chemical	Maximum temp. °F	Maximum temp. °C	Chemical	Maximum temp. °F	Maximum temp. °C
Acetaldehyde	210	99	Barium hydroxide	400	204
Acetamide	340	171	Barium sulfate	210	99
Acetic acid, 10%	420	216	Barium sulfide	210	99
Acetic acid, 50%	400	204	Benzaldehyde	400	204
Acetic acid, 80%	230	110	Benzene	400	204
Acetic acid, glacial	400	204	Benzene sulfonic acid, 10%	210	99
Acetic anhydride	380	193	Benzoic acid	400	204
Acetone	400	204	Benzyl alcohol	400	204
Acetyl chloride	400	204	Benzyl chloride	210	99
Acrylic acid	120	49	Borax	400	204
Acrylonitrile	210	99	Boric acid	400	204
Adipic acid	210	99	Bromine gas, dry	X	
Allyl alcohol	400	204	Bromine gas, moist	X	
Allyl chloride	100	38	Bromine, liquid	X	
Alum	200	93	Butadiene	400	204
Aluminum acetate	200	93	Butyl acetate	380	193
Aluminum chloride, aqueous	X		Butyl alcohol	400	204
Aluminum chloride, dry	150	66	n-Butylamine	400	204
Aluminum fluoride	90	32	Butyl phthalate	210	99
Aluminum hydroxide	400	204	Butyric acid	400	204
Aluminum nitrate	200	93	Calcium bisulfide	60	16
Aluminum sulfate[b]	210	99	Calcium bisulfite	350	177
Ammonia gas	90	32	Calcium carbonate	205	96
Ammonium bifluoride, 10%	90	32	Calcium chloride[b]	210	99
Ammonium carbonate	400	204	Calcium hydroxide, 10%	210	99
Ammonium chloride, 10%	230	110	Calcium hypochlorite	80	27
Ammonium chloride, 50%	X		Calcium nitrate	350	177
Ammonium chloride, sat.	X		Calcium oxide	80	27
Ammonium fluoride, 10%	90	32	Calcium sulfate	210	99
Ammonium fluoride, 25%	X		Caprylic acid[a]	400	204
Ammonium hydroxide, 25%	230	110	Carbon bisulfide	400	204
Ammonium hydroxide, sat.	210	99	Carbon dioxide, dry	570	299
Ammonium nitrate[b]	300	149	Carbon dioxide, wet	200	93
Ammonium persulfate	360	182	Carbon disulfide	400	204
Ammonium phosphate, 40%	130	54	Carbon monoxide	570	299
Ammonium sulfate, 10–40%	400	204	Carbon tetrachloride[b,c]	400	204
Ammonium sulfide	390	171	Carbonic acid	350	177
Ammonium sulfite	210	99	Cellosolve	400	204
Amyl acetate	300	149	Chloracetic acid, 50% water	X	
Amyl alcohol	400	204	Chloracetic acid	X	
Amyl chloride	150	66	Chlorine gas, dry	400	204
Aniline	500	260	Chlorine gas, wet	X	
Antimony trichloride	X		Chlorine, liquid dry	120	49
Aqua regia, 3:1	X		Chlorobenzene, ELC only	260	127
Barium carbonate	80	27	Chloroform[b]	210	99
Barium chloride[c]	210	99	Chlorosulfonic acid	X	

TABLE 8.17 Continued

Chemical	Maximum temp.		Chemical	Maximum temp.	
	°F	°C		°F	°C
Chromic acid, 10%[d]	400	204	Methyl ethyl ketone	330	166
Chromic acid, 50%[d]	150	49	Methyl isobutyl ketone	350	177
Chromyl chloride	210	99	Muriatic acid	X	
Citric acid, 15%[c]	200	93	Nitric acid, 5%[e]	210	99
Citric acid, conc.[c]	380	193	Nitric acid, 20%[e]	270	132
Copper acetate	210	99	Nitric acid, 70%[e]	400	204
Copper carbonate, 10%	80	27	Nitric acid, anhydrous[e]	110	43
Copper chloride	X		Nitrous acid, conc.	80	27
Copper cyanide	210	99	Oleum	80	27
Copper sulfate	400	204	Perchloric acid, 10%	X	
Cresol	100	38	Perchloric acid, 70%	X	
Cupric chloride, 5%	X		Phenol	570	299
Cupric chloride, 50%	X		Phosphoric acid, 50–80%[e]	400	204
Cyclohexane	400	204	Picric acid	400	204
Cyclohexanol	80	27	Potassium bromide, 30%[c]	350	177
Dichloroethane (ethylene dichloride)	400	204	Salicylic acid	350	177
Ethylene glycol	340	171	Silver bromide, 10%	X	
Ferric chloride	X		Sodium carbonate	350	177
Ferric chloride, 50% in water	X		Sodium chloride, to 30%[b]	350	177
Ferric nitrate, 10–50%	350	177	Sodium hydroxide, 10%	350	177
Ferrous chloride	X		Sodium hydroxide, 50%[a]	350	177
Fluorine gas, dry	420	216	Sodium hydroxide, conc.	350	177
Fluorine gas, moist	X		Sodium hypochlorite, 20%	X	
Hydrobromic acid, dilute	X		Sodium hypochlorite, conc.	X	
Hydrobromic acid, 20%	X		Sodium sulfide, to 50%	190	88
Hydrobromic acid, 50%	X		Stannic chloride	X	
Hydrochloric acid, 20%	X		Stannous chloride, 10%	210	99
Hydrochloric acid, 38%	X		Sulfuric acid, 10%	X	
Hydrocyanic acid, 10%	210	99	Sulfuric acid, 50%	X	
Hydrofluoric acid, 30%	X		Sulfuric acid, 70%	X	
Hydrofluoric acid, 70%	X		Sulfuric acid, 90%[e]	80	27
Hydrofluoric acid, 100%	80	27	Sulfuric acid, 98%[e]	210	99
Hypochlorous acid	X		Sulfuric acid, 100%[e]	210	99
Iodine solution, 10%	X		Sulfuric acid, fuming	210	99
Ketones, general	250	121	Sulfurous acid[e]	150	66
Lactic acid, 25%	210	99	Thionyl chloride	X	
Lactic acid, conc.[c,e]	300	149	Toluene	350	177
Magnesium chloride, 50%[b,c]	210	99	Trichloroacetic acid	X	
Malic acid	250	121	White liquor	100	38
Manganese chloride, 30%	210	99	Zinc chloride	200	93
Methyl chloride, dry	350	177			

[a]The chemicals listed are in the pure state or in a saturated solution unless otherwise indicated. Compatibility is shown to the maximum allowable temperature for which data are available. Incompatibility is shown by an X. When compatible, the corrosion rate is <20 mpy.
[b]Subject to stress cracking.
[c]Subject to pitting.
[d]Subject to crevice attack.
[e]Subject to intergranular corrosion.
Source: Ref. 5.

TABLE 8.18 Mechanical and Physical Properties of Types 316 and 316L
Stainless Steel

Property	Type 316	Type 316L
Modulus of elasticity $\times$ 10^6 (psi)	28	28
Tensile strength $\times$ 10^3 (psi)	75	70
Yield strength 0.2% offset $\times$ 10^3 (psi)	30	25
Elongation in 2 in. (%)	50	50
Rockwell hardness	B-80	B-80
Density (lb/in.^{3})	0.286	0.286
Specific gravity	7.95	7.95
Specific heat at 32–212°F (Btu/lb °F)	0.12	0.12
Thermal conductivity (Btu/hr ft^2 °F)		
at 70°F	9.3	9.3
at 1500°F	12.4	12.4
Thermal expansion coefficient at 32–212°F $\times$ 10^{-6} (in./in. °F)	8.9	8.9
Izod impact (ft-lb)	110	110

Type 316H stainless steel has a higher carbon content for better high
temperature creep properties and to meet requirements of ASME Section
VIII, Table UHA-21, Footnote 8. This alloy is used in temperatures over
1832°F (1000°C). It has a chemical composition as follows:

Chromium	16.0/18.0%
Nickel	10.0/14.0%
Molybdenum	2.0/3.0%
Carbon	0.04/0.10%
Iron	Balance

The corrosion resistance of type 316H stainless steel is the same as
type 316 stainless except after long exposure to elevated temperatures where
intergranular corrosion may be more severe. It may also be susceptible to
chloride stress corrosion cracking. Mechanical and physical properties are
listed in Table 8.20.

Type 316N is a high nitrogen type 316 stainless steel. The chemical
composition is shown in Table 8.1. It has a higher strength than type 316
and greater ASME Section VIII allowables. Corrosion resistance is the same
as type 316, and it may be susceptible to chloride stress corrosion cracking.
Mechanical and physical properties are given in Table 8.20.

It is used in tubing and applications where minimum wall pressure
requirements are critical.

TABLE 8.19 Maximum Allowable Design Stress for
Types 316 and 316L Stainless Steel

Temperature (°F/°C)	Allowable design stress (psi)	
	316	316L
−325 to 100/−198 to 38	18750	15600
200/93	18750	15600
300/149	17900	14500
400/204	17500	12000
500/260	17200	11000
600/316	17100	10150
650/343	17050	9800
700/371	17000	9450
750/399	16900	9100
800/427	16750	8800
850/450	16500	
900/482	16000	
950/150	15100	
1000/538	14000	
1050/560	12200	
1100/593	10400	
1150/621	8500	
1200/649	6800	
1250/677	5300	
1300/704	4000	
1350/732	3000	
1400/760	2350	
1450/788	1850	
1500/816	1500	

Type 316LN stainless steel is a low carbon, high nitrogen type 316 stainless. It has the following composition:

Chromium	16.0/18.0%
Nickel	10.0/15.0%
Molybdenum	2.0/3.0%
Carbon	0.035%
Nitrogen	0.10/0.16%
Iron	Balance

TABLE 8.20 Mechanical and Physical Properties of Types 316H, 316N, and 316LN Stainless Steel

Property	Type 316H	Type 316N	Type 316LN
Modulus of elasticity $\times 10^6$ (psi)	28	28	28
Tensile strength $\times 10^3$ (psi)	75	80	75
Yield strength 0.2% offset $\times 10^3$ (psi)	30	35	30
Elongation in 2 in. (%)	35	35	35
Rockwell hardness	B-90	B-90	B-90
Density (lb/in.^{3})	0.286	0.286	0.286
Specific gravity	7.88	7.88	7.88
Specific heat at 32–212°F (Btu/lb °F)			
Thermal conductivity at 212°F (Btu/hr ft^2 °F)	9.3	9.3	9.3

Type 316LN stainless has the same high temperature strength and ASME allowables as type 316, but the weldability of type 316L. The corrosion resistance is the same as type 316 stainless and may be susceptible to chloride stress corrosion cracking. Mechanical and physical properties are given in Table 8.20.

XIV. TYPE 317 (S31700)

Type 317 stainless steel contains greater amounts of molybdenum, chromium, and nickel than type 316. The chemical composition is shown in Table 8.1. As a result of the increased alloying elements these alloys offer higher resistance to pitting and crevice corrosion than type 316 in various process environments encountered in the process industry. However, they may still be subject to chloride stress corrosion cracking. The alloy is non-hardenable and nonmagnetic in the annealed condition, but becomes slightly magnetic when cold worked. Type 317 stainless steel finds application in the chemical, petroleum, and pulp and paper industries for heat exchangers, evaporators, and condenser tubes.

Type 317L is a low carbon version of the basic alloy, which offers the additional advantage of preventing intergranular precipitation of chromium carbide during welding and stress relieving. The chemical composition is shown in Table 8.1.

Type 317L has improved pitting resistance over that of type 316L, but still may be subject to chloride stress corrosion cracking. The compatibility of type 317 and type 317L stainless steel with selected corrodents is shown in Table 8.21.

The mechanical and physical properties of types 317 and 317L stainless steel are given in Table 8.22, and the maximum allowable design stress of type 317 stainless is given in Table 8.23.

TABLE 8.21 Compatibility of Types 317 and 317L Stainless Steel with Selected Corrodents[a]

Chemical	Maximum temp. °F/°C	Chemical	Maximum temp. °F/°C
Acetaldehyde	150/66	Copper sulfate	70/21
Acetic acid, 10%	232/111	Ferric chloride	70/21
Acetic acid, 50%	232/111	Hydrochloric acid, 20%	X
Acetic acid, 80%	240/116	Hydrochloric acid, 38%	X
Acetic acid, glacial	240/116	Hydrofluoric acid, 30%	X
Acetic anhydride	70/21	Hydrofluoric acid, 70%	X
Acetone	70/21	Hydrofluoric acid, 100%	X
Aluminum chloride, aqueous	X	Iodine solution, 10%	70/21
Aluminum chloride, dry	X	Lactic acid, 25%	70/21
Aluminum sulfate, 50–55%	225/107	Lactic acid, conc.	330/166
Ammonium nitrate, 66%	70/21	Magnesium chloride, 30%	70/21
Ammonium phosphate	80/27	Nitric acid, 5%	70/21
Ammonium sulfate, 10–40%	100/38	Nitric acid, 20%	210/99
Benzene	100/38	Nitric acid, 70%	210/99
Boric acid	210/99	Phenol	70/21
Bromine gas, dry	X	Phosphoric acid, 50–80%	140/60
Bromine gas, moist	X	Sodium carbonate	210/99
Bromine liquid	X	Sodium chloride, 30%	X
Butyl alcohol, 5%	195/91	Sodium hydroxide, 10%	210/99
Calcium chloride	210/99	Sodium hydroxide, 50%	70/21
Calcium hypochlorite	70/21	Sodium hydrochlorite, 20%	70/21
Carbon tetrachloride	70/21	Sodium hypochorite, conc.	70/21
Carbonic acid	70/21	Sodium sulfide, to 50%	210/99
Chloracetic acid, 78%	122/50	Sulfuric acid, 10%	120/49
Chlorine, liquid	X	Sulfuric acid, 50%	X
Chlorobenzene	265/129	Sulfuric acid, 70%	X
Chromic acid, 10%	X	Sulfuric acid, 90%	X
Chromic acid, 50%	X	Sulfuric acid, 98%	X
Citric acid, 15%	210/99	Sulfuric acid, 100%	X
Citric acid, conc.	210/99	Sulfurous acid	X

[a]The chemicals listed are in the pure state or in a saturated solution unless otherwise indicated. Compatibility is shown to the maximum allowable temperature for which data are available. Incompatibility is shown by an X. When compatible, the corrosion rate is <20 mpy.
Source: Ref. 5.

TABLE 8.22 Mechanical and Physical Properties of Types 317 and 317L
Stainless Steel

Property	Type 317	Type 317L
Modulus of elasticity $\times 10^6$ (psi)	28.0	28.0
Tensile strength $\times 10^3$ (psi)	75	75
Yield strength 0.2% offset $\times 10^3$ (psi)	30	30
Elongation in 2 in. (%)	35	35
Rockwell hardness	B-85	B-85
Density (lb/in.^{3})	0.286	0.286
Specific gravity	7.88	7.88
Specific heat at 32–212°F (Btu/lb °F)	0.12	0.12
Thermal conductivity (Btu/hr ft^2 °F)		
at 70°F	9.3	9.3
at 1500°F	12.4	12.4
Thermal expansion coefficient at 32–212°F $\times 10^{-6}$ (in./in. °F)	9.2	9.2
Izod impact (ft-lb)	110	110

TABLE 8.23 Maximum Allowable Design Stress for
Type 317 Stainless Steel

Temperature (°F/°C)	Allowable stress (psi)
−325 to 100/−198 to 38	18750
200/93	18750
300/149	17900
400/204	17500
500/260	17200
600/316	17100
650/343	17050
700/371	17000
750/399	16900
800/247	16750
850/450	16500
900/482	16000
950/516	15100
1000/538	14000
1050/560	12200
1100/593	10400
1150/621	8500
1200/649	6800
1250/677	5600
1300/704	4000
1350/732	3000
1400/760	2350
1450/788	1850
1500/816	1500

Type 317L stainless steel is used for welding, brazing, and other short-term exposures to high temperatures.

Type 317LM stainless steel is a low carbon, high molybdenum form of type 317. It has better corrosion resistance than types 317L, 316L, or 304L, and the best chloride resistance of the 300 series stainless steels. It may be susceptible to chloride stress corrosion cracking. The chemical composition of type 317LM is as follows:

Chromium	18.0/20.0%
Nickel	13.0/17.0%
Molybdenum	4.0/5.0%
Nitrogen	0.1% max.
Carbon	0.03% max.
Iron	Balance

This alloy is used for flue gas applications and other heat exchanger tubing subject to higher acid chlorides.

Type 317LMN is a low carbon, high molybdenum, high nitrogen type 317 stainless steel. It has a composition of

Chromium	17.0/20.0%
Nickel	13.0/17.0%
Molybdenum	4.0/5.0%
Nitrogen	0.1/0.2%
Carbon	0.03% max.
Iron	Balance

The corrosion resistance of this alloy is the same as for type 317LM with the advantage of preventing chromium carbide precipitation during welding or stress relieving, and it has the strength of type 317 stainless. It is used where weldability and strength are important. The mechanical and physical properties of type 317LM and type 317LMN are shown in Table 8.24.

XV. TYPE 321 (S32100)

By alloying austenitic alloys with a small amount of an element having a higher affinity for carbon than does chromium, carbon is restrained from diffusing to the grain boundaries, and any carbon that reaches the boundary reacts with the element instead of with the chromium. These are known as stabilized grades. Type 321 is such an alloy which is stabilized by the addition of titanium. Its chemical composition is shown in Table 8.1.

TABLE 8.24 Mechanical and Physical Properties of Types 317LM and 317LMN Stainless Steel

Property	Type 317LM	Type 317LMN
Modulus of elasticity $\times 10^6$ (psi)	28	28
Tensile strength $\times 10^3$ (psi)	75	75
Yield strength 0.2% offset $\times 10^3$ (psi)	30	30
Elongation in 2 in. (%)	35	35
Rockwell hardness	B-90	B-90
Density (lb/in.^{3})	0.286	0.286
Specific gravity	7.88	7.88
Thermal conductivity (Btu/hr ft^2 °F)		
at 70°F	9.3	9.3
at 1500°F	12.4	12.4

The mechanical and physical properties are given in Table 8.25, and the maximum allowable design stress is shown in Table 8.26. This alloy has excellent weldability in the field.

Type 321 stainless steel can be used wherever type 316 is suitable, with improved corrosion resistance, particularly in the presence of nitric acid. This alloy is particularly useful in high temperature service in the carbide precipitation range and for parts heated intermittently between 800–1650°F (427–899°C). Even with the overall improved corrosion resistance it still may be susceptible to chloride stress corrosion cracking. Table 8.27 provides the compatibility of type 321 with selected corrodents.

Applications include use in exhaust manifolds, expansion joints, high temperature chemical process heat exchanger tubes and recuperator tubes.

Type 321H is a high carbon type 321 stainless steel with better high temperature creep properties and meets the requirements of ASME Section VIII, Table UHA-21, Footnote 8. It has good weldability in the field.

The corrosion resistance of type 321H is the same as the corrosion resistance of type 321 and may be susceptible to chloride stress corrosion cracking. It has the following chemical composition:

Chromium 17.0/20.0%
Nickel   9.0/13.0%
Carbon   0.04/0.10%
Titanium   4 × Carbon min., 0.60% max.
Iron   Balance

Type 321H stainless steel is used in applications where temperatures exceed 1000°F (538°C).

TABLE 8.25 Mechanical and Physical Properties of Type 321 Stainless Steel

Modulus of elasticity $\times 10^6$ (psi)	29
Tensile strength $\times 10^3$ (psi)	75
Yield strength 0.2% offset $\times 10^3$ (psi)	30
Elongation in 2 in. (%)	35
Rockwell hardness	B-85
Density (lb/in.^{3})	0.286
Specific gravity	7.92
Specific heat at 32–212°F (Btu/lb °F)	0.12
Thermal conductivity (Btu/hr ft^2 °F)	
at 70°F	9.3
at 1500°F	12.8
Thermal expansion coefficient at 32–212°F $\times 10^{-6}$ (in./in. °F)	9.3
Izod impact (ft-lb)	110

TABLE 8.26 Maximum Allowable Design Stress for Type 321 Stainless Steel

Temperature (°F/°C)	Allowable design stress (psi)
−325 to 100/−198 to 38	18750
200/93	18750
300/149	17000
400/204	15800
500/260	15200
600/316	14900
700/371	14800
750/399	14700
800/247	14550
850/450	14300
900/482	14100
950/510	13850
1000/538	13500

TABLE 8.27 Compatibility of Type 321 Stainless Steel with Selected Corrodents[a]

Chemical	Maximum temp. °F/°C	Chemical	Maximum temp. °F/°C
Acetic acid, 10%	X	Copper sulfate	70/21
Acetic acid, 50%	X	Ferric chloride	X
Acetic acid, 80%	X	Hydrochloric acid, 20%	X
Acetic acid, glacial	X	Hydrochloric acid, 38%	X
Acetic anhydride	70/21	Hydrofluoric acid, 30%	X
Alum	X	Hydrofluoric acid, 70%	X
Aluminum chloride, aqueous	X	Hydrofluoric acid, 100%	X
Aluminum chloride, dry	X	Iodine solution, 10%	X
Aluminum sulfate	70/21	Lactic acid, 25%	70/21
Ammonium phosphate	70/21	Lactic acid, conc.	70/21
Ammonium sulfate, 10–40%	70/21	Magnesium chloride	X
Benzene	100/38	Nitric acid, 5%	70/21
Boric acid	210/99	Nitric acid, 20%	210/99
Bromine gas, dry	X	Nitric acid, 70%	210/99
Bromine gas, moist	X	Phenol	X
Bromine, liquid	X	Phosphoric acid, 50–80%	70/21
Calcium chloride	X	Sodium carbonate	70/21
Calcium hypochlorite	X	Sodium chloride, 30%	X
Carbon tetrachloride	X	Sodium hydroxide, 10%	70/21
Carbonic acid	70/21	Sodium hydroxide, 50%	70/21
Chloracetic acid, 78%	X	Sodium hydrochlorite, 20%	X
Chlorine, liquid	X	Sodium hypochorite, conc.	X
Chromic acid, 10%	X	Sodium sulfide, to 50%	70/21
Chromic acid, 50%	X	Sulfuric acid, 98%	X
Citric acid, 15%	70/21	Sulfuric acid, 100%	X
Citric acid, conc.	70/21	Sulfurous acid	X

[a]The chemicals listed are in the pure state or in a saturated solution unless otherwise indicated. Compatibility is shown to the maximum allowable temperature for which data are available. Incompatibility is shown by an X. When compatible, the corrosion rate is <20 mpy.

XVI. TYPE 329 (S32900)

Type 329 stainless steel is listed under the austenitic stainless steels while in actuality it is really the basic material of duplex stainless steels. We will consider it as an austenitic stainless steel. It has the following chemical composition:

Chromium	26.5%
Nickel	4.5%
Molybdenum	1.5%
Carbon	0.05%
Iron	Balance

Type 329 stainless possesses higher strength values than those typically found in the austenitic stainlesses, as can be seen in Table 8.28.

The general corrosion resistance of type 329 stainless is slightly above that of 316 stainless in most media. In addition, since the nickel content is low, it has good resistance to chloride stress corrosion cracking.

XVII. TYPE 347 (S34700)

Type 347 stainless steel is a columbium- (niobium-) stabilized alloy. Its chemical composition will be found in Table 8.1. Being stabilized it will resist carbide precipitation during welding and intermittent heating to 800–1650°F (427–899°C) and has good high temperature scale resistance. Basically this alloy is equivalent to type 304 stainless steel with the added protection against carbide precipitation. Type 304L also offers this protection but is limited to a maximum operating temperature of 800°F (427°C), while type 347 can be operated to 1000°F (538°C).

The mechanical and physical properties will be found in Table 8.29 and the maximum allowable design stresses in Table 8.30.

TABLE 8.28 Mechanical and Physical Properties of Types 329 and 330 Stainless Steel

Property	Type 329	Type 330
Modulus of elasticity $\times$ 10^6 (psi)		28.5
Tensile strength $\times$ 10^3 (psi)	105	80
Yield strength 0.2% offset $\times$ 10^3 (psi)	80	38
Elongation in 2 in. (%)	25	40
Hardness	Brinell 230	Rockwell B-80
Density (lb/in.^{3})	0.280	0.289
Specific gravity	7.7	8.01
Specific heat at 32–212°F (Btu/lb °F)		8.0
Thermal expansion coefficient at 32–212°F $\times$ 10^{-6} (in./in. °F)		
Izod impact (ft-lb)	90	

TABLE 8.29 Mechanical and Physical Properties of Type 347
Stainless Steel

Modulus of elasticity $\times 10^6$ (psi)	29.0
Tensile strength $\times 10^3$ (psi)	75
Yield strength 0.2% offset $\times 10^3$ (psi)	30
Elongation in 2 in. (%)	35
Rockwell hardness	B-85
Density (lb/in.^{3})	0.285
Specific gravity	7.92
Thermal conductivity (Btu/hr ft^2 °F)	
at 70–212°F	9.3
at 1500°F	12.8
Thermal expansion coefficient at 32–212°F $\times 10^{-6}$ (in./in. °F)	9.3
Izod impact (ft-lb)	110

In general the corrosion resistance of type 347 is equivalent to that of
type 304 stainless steel and may be susceptible to chloride stress corrosion
cracking.

Applications include exhaust manifolds, expansion joints, high temperature heat exchanger tubes, and recuperation tubes.

TABLE 8.30 Maximum Allowable Design Stress for
Type 347 Stainless Steel

Temperature (°F/°C)	Allowable design stress (psi)
−325 to 100/−198 to 38	18750
200/93	18750
300/149	17000
400/204	15800
500/260	15200
600/316	14900
650/343	14850
700/371	14800
750/399	14700
800/247	14550
850/450	14300
900/482	14100
950/516	13850
1000/538	13500

TABLE 8.31 Mechanical and Physical Properties of
Type 347H Stainless Steel

Property	
Modulus of elasticity $\times$ 10^6 (psi)	29
Tensile strength $\times$ 10^3 (psi)	75
Yield strength 0.2% offset $\times$ 10^3 (psi)	30
Elongation in 2 in. (%)	35
Rockwell hardness	B-90
Density (lb/in.^{3})	0.285
Specific gravity	7.88
Thermal conductivity (Btu/hr ft^2 °F)	
at 70°F	9.3
at 1500°F	12.8

Types 347H stainless steel is a high carbon type 347 for better high temperature creep properties and to meet requirements of ASME Section VIII, Table UHA 21, Footnote 8. The chemical composition is as follows:

Chromium	17/20%
Nickel	9/13%
Carbon	0.04/0.01%
Columbium + tantalum	8 $\times$ carbon min., 1.0% max.
Iron	Balance

Type 347H has the same corrosion resistance as type 347 and may be susceptible to chloride stress corrosion cracking. It has excellent weldability

TABLE 8.32 Mechanical and Physical Properties of
Type 348 Stainless Steel

Property	
Modulus of elasticity $\times$ 10^6 (psi)	29
Tensile strength $\times$ 10^3 (psi)	95
Yield strength 0.2% offset $\times$ 10^3 (psi)	40
Elongation in 2 in. (%)	45
Rockwell hardness	B-85
Density (lb/in.^{3})	0.285
Specific gravity	7.88
Thermal conductivity at 212°F (Btu/hr ft^2 °F)	9.3

in the field and is used in applications where temperatures exceed 1000°F (538°C).

Mechanical and physical properties will be found in Table 8.31.

XVIII. TYPE 348 (S34800)

Type 348 stainless is the same as type 347 except that the tantalum content is restricted to a maximum of 0.10%. The chemical composition is as follows:

Chromium	17.0/20.0%
Nickel	9.0/13.0%
Carbon	0.08% max.
Columbium + tantalum	10 × carbon min.
	1.0% max. (0.01% max. tantalum)
Iron	Balance

Table 8.32 lists the mechanical and physical properties of type 348 stainless.

In general the corrosion resistance is the same as that of type 347 stainless and may be subject to chloride stress corrosion cracking.

This material is used in nuclear application where tantalum is undesirable because of high neutron cross-section.

Type 348H stainless steel is a high carbon version of type 348 designed to provide better high temperature creep properties and to meet the requirements of ASME Section VIII, Table UHA-21, Footnote 8.

It finds application in nuclear environments at temperatures over 1000°F (538°C) where tantalum is undesirable because of high neutron cross-section.

REFERENCES

1. CP Dillon. Corrosion Control in the Chemical Process Industry. St. Louis: Materials Technology Institute of the Chemical Process Industries, 1986.
2. PA Schweitzer. Stainless steel. In: PA Schweitzer, ed. Corrosion and Corrosion Protection Handbook, 2nd ed. New York: Marcel Dekker, 1988.
3. CP Dillon. Corrosion Resistance of Stainless Steels. New York: Marcel Dekker, 1995.
4. PA Schweitzer. Encyclopedia of Corrosion Technology. New York: Marcel Dekker, 1998.
5. PA Schweitzer. Corrosion Resistance Tables, 4th ed., Vols. 1–3, New York: Marcel Dekker, 1995.
6. PT Lovejoy. Stainless steels. In: GT Murray, ed. Handbook of Materials Selection for Engineering Applications. New York: Marcel Dekker, 1997.

7. PA Schweitzer. Corrosion Resistant Piping Systems. New York: Marcel Dekker, 1994.
8. PK Whitcraft. Corrosion of stainless steels. In: PA Schweitzer, ed. Corrosion Engineering Handbook. New York: Marcel Dekker, 1996.
9. GT Murray. Introduction to Engineering Materials. New York: Marcel Dekker, 1993.
10. DA Hansen, RB Puyear. Materials Selection for Hydrocarbon and Chemical Plants. New York: Marcel Dekker, 1996.

9

Superaustenitic Stainless Steels

The classification of superaustenitic stainless steels came about during the 1970s and 1980s. Carpenter Steel's introduction of alloy 20 in 1951 as a cast material was the foundation for this class of materials. In 1965 Carpenter introduced the wrought product 20Cb3. This alloy became popular as an intermediate step between 316 stainless steel and the more highly alloyed nickel base materials. It was a cost effective way to combat chloride stress corrosion cracking.

Because of the high nickel content of 20Cb3 it received a nickel base alloy UNS designation as UNS N08020. However, since the major constituent is iron, it is truly a stainless steel.

The superaustenitic term is derived from the fact that the composition plots high above the austenite–ferrite boundary on the Schaeffler diagram. Unlike the 300 series stainless alloys there is no chance of developing ferrite in this material.

In a similar time frame another superaustenitic alloy, RA330, was developed. It also was assigned a nickel base number UNS N08330. Many of the superaustenitic alloys have been assigned nickel base identification numbers, but they are truly stainless steels.

The initial alloys that were developed exhibited good general corrosion resistance to strong acids, but their pitting resistance was only slightly better than that of type 316L. In order to improve the pitting resistance and crevice corrosion resistance the molybdenum content was increased. One of the first introduced was 904L (UNS N08904), which increased the molybdenum content to 4% and reduced the nickel content to 25%. The reduction in nickel content was a cost saving factor, with minimal loss of general corrosion resistance and maintained sufficient resistance to chloride stress corrosion cracking.

Improvements continued with the alloying addition of nitrogen to offset the tendency for the formation of the sigma phase and increasing of the molybdenum content to 6%. This concept was introduced with two alloys, 254SM0 (S31254) and Al-6XN (N08367).

Performance in the area of pitting and crevice corrosion is often measured using critical pitting temperature (CPT), critical crevice temperature (CCT), and pitting resistance equivalent numbers (PREN). As a general rule the higher the PREN, the better resistance to pitting. The pitting resistance number is determined by the chromium, molybdenum, and nitrogen contents.

$$PREN = \%Cr + 3.3 \times \%Mo + 30 \times \%N.$$

The PREN for various austenitic stainless steels can be found in Table 9.1.

Another method used to compare the resistance of alloys to localized attack are their respective critical pitting and critical crevice temperatures. These are the temperatures at which pitting and crevice attack are initiated. Critical temperatures for several alloys are given in Table 9.2.

The various superaustenitic alloys are discussed in detail in the following sections.

I. ALLOY 20Cb3 (N08020)

This alloy was originally developed to provide improved corrosion resistance to sulfuric acid. However, it has found wide application throughout the chemical process industry. The alloy's composition is as follows:

Chemical	Weight percent
Chromium	20.0
Nickel	33.5
Silicon	1.00 max.
Manganese	0.75 max.
Carbon	0.07 max.
Columbium/tantalum	8 × %carbon min.
Iron	Balance

The alloy is stabilized with columbium and tantalum and has a high nickel content, approximately 33%.

Alloy 20Cb3 is weldable, machinable, and cold formable and has min-

TABLE 9.1 Pitting Resistance of
Selected Alloys

Alloy	UNS	Pitting resistance equivalent
654	S32654	63.09
31	N08031	54.45
825	N08825	51.9
686		51.0
625	N06625	50.7
25-6Mo	N08926	47.45
A16XN	N08367	46.96
926	N08926	46.45
254SMo	S31254	45.8
20Mo6	N08026	42.81
317LN	S31753	39.6
904L	N08904	36.51
20Mo4	N08024	36.2
317	S31700	33.12
316LN	S31653	31.08
315	S35315	29.5
316	S31600	27.9
20Cb3	N08020	27.26
348	S34800	25.6
800	N08800	21.0
810	N08810	21.0
347	S34700	19.0
331	N08331	19.0
330	N08330	18.5
304	S30400	18.0

imum carbide precipitation due to welding. The mechanical and physical properties are given in Table 9.3. Refer to Table 9.4 for elevated tensile properties and Table 9.5 for maximum allowable design stress values.

This alloy is particularly useful in the handling of sulfuric acid. It is resistant to stress corrosion cracking in sulfuric acid at a variety of temperatures and concentrations. The resistance of 20Cb3 to chloride stress corrosion cracking is also increased over type 304 and type 316 stainless steels. The alloy also exhibits excellent resistance to sulfide stress cracking and consequently finds many applications in the oil industry.

TABLE 9.2 Critical Pitting and Crevice Temperatures of Selected Alloys

	Critical temperature (°F/°C)	
Alloy	Pitting attack	Crevice attack
Inconel[a] alloy 686	>185/>85	>185/>85
UNS N06059	>185/>85	>185/>85
Inconel[a] alloy 622	≥185/≥85	≥185/≥85
UNS N06022	>185/>85	>136/>58
Alloy C-276	>185/>85	113/44
Alloy 625	>185/>85	95/35
Alloy 25-6Mo	158/70	86/30
Alloy 825	86/30	41/5
UNS S31600	68/20	<32/<0
317LM		36/2.5

[a]Inconel is the trademark of Inco Alloys International.

TABLE 9.3 Mechanical and Physical Properties of 20Cb-3 Stainless Steel

Modulus of elasticity $\times 10^6$ (psi)	28
Tensile strength $\times 10^3$ (psi)	91
Yield strength 0.2% offset $\times 10^3$ (psi)	45
Elongation in 2 in. (%)	45
Rockwell hardness	B-86
Density (lb/in.^{3})	0.292
Specific gravity	8.08
Specific heat (Btu/lb °F)	0.12
Thermal conductivity (Btu/h ft^2 °F)	
at 122°F	7.05
at 212°F	7.57
at 392°F	8.50
at 572°F	9.53
at 752°F	10.50
Thermal expansion coefficient $\times 10^{-6}$ (in./in. °F)	
at 77–212°F	8.16
at 77–842°F	8.84
at 77–1652°F	9.53
Charpy V-notch impact strength (ft-lb)	200

TABLE 9.4 Elevated Temperature Tensile Properties of 20Cb-3 Stainless Steel

Temperature (°F/°C)	0.2% yield strength (ksi)	Ultimate tensile strength (ksi)
Room	45	91
200/93	40	86
400/204	35	83
600/316	33	80
800/427	30	79
1000/538	28	77
1400/760	26	45
1600/871	19	29

In high concentrations of chlorides, alloy 20Cb3 is vulnerable to pitting and crevice attack. For improved resistance to these types of corrosion the 2% molybdenum must be increased to 4 or 6% as has been done in alloy 20Mo-4 and 20Mo-6. Table 9.6 contains the compatibility of alloy 20Cb3 with selected corrodents.

This alloy finds application in the manufacture of synthetic fibers, heavy chemicals, organic chemicals, pharmaceuticals, and food processing equipment.

TABLE 9.5 Maximum Allowable Design Stress Values for 20Cb3

Temperature (°F/°C)	Stress value (psi)
100/38	14450
200/93	14450
300/149	14280
400/204	13515
500/260	13175
600/316	12835
650/343	12665
700/371	12495
750/399	12325
800/427	12155

TABLE 9.6   Compatibility of Type 20Cb3 Stainless Steel with Selected Corrodents[a]

Chemical	Maximum temp.		Chemical	Maximum temp.	
	°F	°C		°F	°C
Acetaldehyde	200	93	Barium sulfate	210	99
Acetamide	60	16	Barium sulfide	210	99
Acetic acid, 10%	220	104	Benzaldehyde	210	99
Acetic acid, 50%	300	149	Benzene	230	110
Acetic acid, 80%	300	149	Benzene sulfonic acid, 10%	210	99
Acetic acid, glacial	300	149	Benzoic acid	400	204
Acetic anhydride	180	82	Benzyl alcohol	210	99
Acetone	220	104	Benzyl chloride	230	110
Acetyl chloride	210	99	Borax	100	38
Acrylonitrile	210	99	Boric acid	130	54
Adipic acid	210	99	Bromine gas, dry	80	27
Allyl alcohol	300	149	Bromine gas, moist	X	
Allyl chloride	200	93	Butadiene	180	82
Alum	200	93	Butyl acetate	300	149
Aluminum acetate	60	16	Butyl alcohol	90	32
Aluminum chloride, aqueous	120	43	Butyl phthalate	210	99
Aluminum chloride, dry	120	43	Butyric acid	300	149
Aluminum fluoride	X		Calcium bisulfide	300	149
Aluminum hydroxide	80	27	Calcium carbonate	210	99
Aluminum nitrate	80	27	Calcium chlorate	90	32
Aluminum sulfate	210	99	Calcium chloride	210	99
Ammonia gas	90	32	Calcium hydroxide, 10%	210	99
Ammonium bifluoride	90	32	Calcium hydroxide, sat.	210	99
Ammonium carbonate	310	154	Calcium hypochlorite	90	32
Ammonium chloride, 10%	230	110	Calcium oxide	80	27
Ammonium chloride, 50%	170	77	Calcium sulfate	210	99
Ammonium chloride, sat.[b]	210	99	Caprylic acid	400	204
Ammonium fluoride, 10%	90	32	Carbon bisulfide	210	99
Ammonium fluoride, 25%	90	32	Carbon dioxide, dry	570	299
Ammonium hydroxide, 25%	90	32	Carbon dioxide, wet	400	204
Ammonium hydroxide, sat.	210	99	Carbon disulfide	210	99
Ammonium nitrate[b]	210	99	Carbon monoxide	570	299
Ammonium persulfate	210	99	Carbon tetrachloride	210	99
Ammonium phosphate	210	99	Carbonic acid	570	299
Ammonium sulfate, 10–40%	210	99	Cellosolve	210	99
Ammonium sulfide	210	99	Chloracetic acid	80	27
Ammonium sulfite	210	99	Chlorine gas, dry	400	204
Amyl acetate	310	154	Chlorine gas, wet	X	
Amyl alcohol	160	71	Chlorobenzene, dry	100	38
Amyl chloride	130	54	Chloroform	210	99
Aniline	500	260	Chlorosulfonic acid	130	54
Antimony trichloride	200	93	Chromic acid, 10%	130	54
Aqua regia, 3:1	X		Chromic acid, 50%	140	60
Barium carbonate	90	32	Chromyl chloride	210	99
Barium chloride, 40%	210	99	Citric acid, 15%	210	99
Barium hydroxide, 50%	230	110	Citric acid, conc.	210	99

TABLE 9.6 Continued

Chemical	Maximum temp. °F	Maximum temp. °C	Chemical	Maximum temp. °F	Maximum temp. °C
Copper acetate	100	38	Muriatic acid	X	
Copper carbonate	90	32	Nitric acid, 5%	210	99
Copper chloride	X		Nitric acid, 20%	210	99
Copper cyanide	210	99	Nitric acid, 70%	210	99
Copper sulfate	210	99	Nitric acid, anhydrous	80	27
Cupric chloride, 5%	60	16	Nitrous acid, conc.	90	32
Cupric chloride, 50%	X		Oleum	110	43
Cyclohexane	200	93	Perchloric acid, 10%	100	38
Cyclohexanol	80	27	Perchloric acid, 70%	110	43
Dichloroethane (ethylene dichloride)	210	99	Phenol	570	299
Ethylene glycol	210	99	Phosphoric acid, 50–80%	210	99
Ferric chloride	X		Picric acid	300	149
Ferric chloride, 50% in water	X		Potassium bromide, 30%	210	99
Ferric nitrate, 10–50%	210	99	Salicylic acid	210	99
Ferrous chloride	X		Silver bromide, 10%	90	32
Fluorine gas, dry	570	299	Sodium carbonate	570	299
Fluorine gas, moist	X		Sodium chloride, to 30%[b]	210	99
Hydrobromic acid, dilute	X		Sodium hydroxide, 10%	300	149
Hydrobromic acid, 20%	X		Sodium hydroxide, 50%[c]	300	149
Hydrobromic acid, 50%	X		Sodium hydroxide, conc.	200	93
Hydrochloric acid, 20%	X		Sodium hypochlorite, 30%	90	32
Hydrochloric acid, 38%	X		Sodium sulfide, to 50%	200	93
Hydrocyanic acid, 10%	210	99	Stannic chloride	X	
Hydrofluoric acid, 30%	190	88	Stannous chloride, 10%	90	32
Hydrofluoric acid, 70%	X		Sulfuric acid, 10%	200	93
Hydrofluoric acid, 100%	80	27	Sulfuric acid, 50%	110	43
Iodine solution, 10%	X		Sulfuric acid, 70%	120	49
Ketones, general	100	38	Sulfuric acid, 90%	100	38
Lactic acid, 25%[b]	210	99	Sulfuric acid, 98%	300	149
Lactic acid, conc., air free	300	149	Sulfuric acid, 100%	300	149
Magnesium chloride	200	93	Sulfuric acid, fuming	210	99
Malic acid, 50%	160	71	Sulfurous acid	360	182
Manganese chloride, 40%	210	99	Toluene	210	99
Methyl chloride	210	99	White liquor	100	38
Methyl ethyl ketone	200	93	Zinc chloride	210	99
Methyl isobutyl ketone	210	99			

[a]The chemicals listed are in the pure state or in a saturated solution unless otherwise indicated. Compatibility is shown to the maximum allowable temperature for which data are available. Incompatibility is shown by an X. When compatible the corrosion rate is <20 mpy.
[b]Material subject to intergranular corrosion.
[c]Material subject to stress cracking.
Source: Ref. 1.

II.   20 Mo-4 (N08024)

This alloy is similar to alloy 20Cb3 but with 4% molybdenum content in place of the 2% providing improved pitting and crevice corrosion resistance over alloy 20Cb3. The chemical composition is as follows:

Chemical	Weight percent
Nickel	35/40
Chromium	22.5/25.0
Molybdenum	3.5/5.0
Copper	0.5/1.5
Columbium	0.15/0.35
Carbon	0.03 max.
Iron	Balance

The mechanical and physical properties are given in Table 9.7.

Alloy 20 Mo-4 has outstanding corrosion resistance to chloride pitting and crevice corrosion with good resistance to sulfuric acid and various other acidic environments.

Applications include heat exchangers, chemical process equipment, and wet process phosphoric acid environments.

III.   ALLOY 20Mo-6 (N08026)

Of the three grades of alloy 20 this offers the highest level of pitting and crevice corrosion resistance. Mechanical and physical properties will be found in Table 9.8 and the maximum allowable stress values in Table 9.9.

TABLE 9.7   Mechanical and Physical Properties of Alloy 20Mo4 Stainless Steel

Modulus of elasticity $\times 10^6$ (psi)	27.0
Tensile strength $\times 10^3$ (psi)	89
Yield strength 0.2% offset $\times 10^3$ (psi)	38
Elongation in 2 in. (%)	41
Rockwell hardness	B-80
Density (lb/in.^{3})	0.293
Specific gravity	8.02
Thermal conductivity (Btu/ft^2 hr °F)	
at 70°F	6.9
at 1500°F	10.6

TABLE 9.8 Mechanical and Physical Properties of 20-Mo6
Stainless Steel

Modulus of elasticity $\times$ 10^6 (psi)	27.0
Tensile strength $\times$ 10^3 (psi)	88
Yield strength 0.2% offset $\times$ 10^3 (psi)	45
Elongation in 2 in. (%)	50
Rockwell hardness	B-85
Density (lb/in.^{3})	0.294
Specific gravity	8.133
Specific heat (Btu/hr °F, 32–212°F)	0.11
Thermal conductivity (Btu/ft^2 hr °F)	
at 122°F	6.99
at 212°F	7.51
at 392°F	8.55
at 572°F	9.53
at 752°F	10.52
Thermal expansion coefficient $\times$ 10^{-6} (in./in. °F)	
at 77–212°F	8.22
at 77–392°F	8.29
at 77–752°F	8.73
at 77–1112°F	9.29
at 77–1652°F	9.86

TABLE 9.9 Maximum Allowable
Design Stress Values for 20Mo-6
Stainless Steel

Temperature (°F/°C)	Allowable stress (psi)
100/38	17000
200/93	17000
300/149	16100
400/204	14900
500/260	13800
600/316	13000
700/371	12400
800/427	11800

Alloy 20Mo-6 is resistant to corrosion in hot chloride environments and is also resistant to oxidizing media. This alloy is designed for applications where better pitting and crevice corrosion resistance than 20Cb3 offers is required.

The alloy is melted with low carbon to provide a high level of resistance to intergranular corrosion. It also possesses excellent resistance to chloride stress corrosion cracking. When in contact with sulfuric acid, excellent resistance is shown at 176°F (80°C) with the exception of concentrations in the range of approximately 75–97 wt%. In boiling sulfuric acid, 20Mo-6 stainless has good resistance to general corrosion only in relatively dilute solutions. At approximately 10% concentration of boiling sulfuric acid, the corrosion rate becomes excessive.

The alloy is highly resistant to phosphoric acid, both wet process plant acid and reagent grade concentrated phosphoric acid.

20Mo-6 has the following chemical composition:

Chemical	Weight percent
Chromium	22.00/26.00
Nickel	33.00/37.20
Molybdenum	5.00/6.70
Silicon	0.03/0.50
Manganese	1.00
Phosphorus	0.03
Carbon	0.03
Iron	Balance

IV. ALLOY 904L (N08904)

This is a fully austenitic low carbon chromium stainless steel with additives of molybdenum and copper. The chemical composition will be found in Table 8.1. Its high nickel and chromium contents make alloy 904L resistant to corrosion in a wide variety of both oxidizing and reducing environments. Molybdenum and copper are included in the alloy for increased resistance to pitting and crevice corrosion and to general corrosion in reducing acids. Other advantages of the alloy's composition are sufficient nickel for resistance to chloride ion stress corrosion cracking and low carbon content for resistance to intergranular corrosion.

TABLE 9.10 Mechanical and Physical Properties of 904L
Stainless Steel

Modulus of elasticity $\times 10^6$ (psi)	28.5
Tensile strength $\times 10^3$ (psi)	75
Yield strength 0.2% offset $\times 10^3$ (psi)	32
Elongation in 2 in. (%)	35
Rockwell hardness	B-90
Density (lb/in.^{3})	0.29
Specific gravity	7.9
Specific heat (Btu/lb °F)	0.105
Thermal conductivity (Btu-in./ft^2 hr °F)	
at 68°F	90
at 200°F	94
at 400°F	97
at 600°F	105
at 800°F	113
at 1000°F	121
Thermal expansion coefficient $\times 10^{-6}$ (in./in. °F)	
at 200°F	8.4
at 600°F	8.8
at 800°F	9.0
at 1000°F	9.2
at 1500°F	10.0
at 1800°F	10.3
Charpy V-notch impact (ft-lb)	125

The alloy's outstanding attributes are resistance to nonoxidizing acids along with resistance to pitting, crevice corrosion, and stress corrosion cracking in such media as stack gas condensate and brackish water.

Alloy 904L is especially suited for handling sulfuric acid. Hot solutions at moderate concentrations represent the most corrosive conditions. It also has excellent resistance to phosphoric acid.

At high temperatures 904L may be subject to stress corrosion cracking. The mechanical and physical properties are shown in Table 9.10.

Alloy 904L finds applications in piping systems, pollution control equipment, heat exchangers, and bleaching systems.

V. ALLOY 800 (N08800)

The composition of alloy 800 is as follows:

Chemical	Weight percent
Nickel	30.0/35.0
Chromium	19.0/23.0
Aluminum	0.15/0.6
Titanium	0.15/0.6
Carbon	0.10 max.
Iron	Balance

This alloy is used primarily for its oxidation resistance and strength at elevated temperatures. It is particularly useful for high temperature applications because the alloy does not form the embrittling sigma phase after long exposures at 1200–1600°F (649–871°C). High creep and rupture strengths are other factors that contribute to its performance in many other applications. It resists sulfidation, internal oxidation, scaling, and carburization. Table 9.11 provides the mechanical and physical properties, while maximum allowable design stresses will be found in Table 9.12.

At moderate temperatures the general corrosion resistance of alloy 800 is similar to that of the other austenitic nickel-iron-chromium alloys. However, as the temperature increases, alloy 800 continues to exhibit good corrosion resistance, while other austenitic alloys are unsatisfactory for the service.

Alloy 800 has excellent resistance to nitric acid at concentrations up to about 70%. It resists a variety of oxidizing salts, but not halide salts. It also has good resistance to organic acids, such as formic, acetic, and propionic. Alloy 800 is particularly suited for the handling of hot corrosive gases such as hydrogen sulfide.

In aqueous service, alloy 800 has general resistance that falls between type 304 and type 316 stainless steels. Thus, the alloy is not widely used for aqueous service. The stress corrosion cracking resistance of alloy 800, while not immune, is better than that of the 300 series of stainless steels and may be substituted on that basis. Table 9.13 provides the compatibility of alloy 800 with selected corrodents.

Applications include heat exchangers, process piping, steam generators, and heating element cladding.

Alloy 800H is a controlled version of alloy 800. The carbon content is maintained between 0.05 and 0.1% to provide the alloy with better elevated temperature creep and stress rupture properties. It is solution annealed to assure the improved creep and stress-to-rupture properties.

TABLE 9.11 Mechanical and Physical Properties of Alloy 800

Modulus of elasticity $\times 10^6$ (psi)	28.5
Tensile strength $\times 10^3$ (psi)	75
Yield strength 0.2% offset $\times 10^3$ (psi)	26.6
Elongation in 2 in. (%)	30
Brinell hardness	152
Density (lb/in.^{3})	0.287
Specific gravity	7.94
Specific heat (Btu/lb °F)	0.11
Thermal conductivity (Btu-in./hr/ft^2/°F/in.)	
at 0°F	75
at 70°F	80
at 100°F	83
at 200°F	89
at 400°F	103
at 600°F	115
at 1200°F	152
at 1600°F	181
Coefficient of thermal expansion $\times 10^{-6}$ (in./in.°F)	
at 70–200°F	7.9
at 70–400°F	8.8
at 70–600°F	9.0
at 70–800°F	9.2
at 70–1000°F	9.4
at 70–1200°F	9.6
at 70–1400°F	9.9
at 70–1600°F	10.2

Applications include superheater and reheater tubing, headers, and furnace tubing as well as applications in the refining and heat treatment industries.

Alloy 800AT is similar to alloy 800 but has higher levels of aluminum and titanium. It is used for thermal processing applications, chemical and petrochemical piping, pigtails, and outlet manifolds.

VI. ALLOY 825 (N08825)

Alloy 825 is very similar to alloy 800, but the composition has been modified to improve its aqueous corrosion resistance.

Chemical	Weight percent
Nickel	38/46
Chromium	19.5/23.5
Molybdenum	2.5/3.5
Copper	1.5/3.0
Titanium	0.6/1.2
Aluminum	0.2 max.
Iron	Balance

The higher nickel content of alloy 825 compared to alloy 800 makes it resistant to chloride stress corrosion cracking. Addition of molybdenum and copper gives resistance to pitting and to corrosion in reducing acid

TABLE 9.12   Allowable Design Stress for Alloy 800

Operating temperature (°F/°C)	Allowable design stress (psi)
100/38	15600
200/93	13400
300/149	12100
400/204	11100
500/260	10400
600/316	10000
650/343	9800
700/371	9600
750/399	9500
800/427	9300
850/454	9200
900/482	9100
950/510	9000
1000/538	8900
1050/566	8800
1100/593	8800
1150/621	8700
1200/649	7100
1250/677	5400
1300/704	4150
1350/732	3250
1400/760	2500
1450/788	1900
1500/816	1500

environments such as sulfuric or phosphoric acid solutions. Alloy 825 is resistant to pure sulfuric acid solution up to 40% by weight at boiling temperatures and at all concentrations at a maximum temperature of 150°F (60°C). In dilute solutions the presence of oxidizing salts such as cupric or ferric actually reduces the corrosion rates. It has limited use in hydrochloric or hydrofluoric acids.

The chromium content of alloy 825 gives it resistance to various oxidizing environments such as nitrates, nitric acid solutions, and oxidizing salts. The alloy is not fully resistant to stress corrosion cracking when tested in magnesium chloride, but it has good resistance in neutral chloride environments.

If localized corrosion is a problem with the 300 series stainless steels, alloy 825 may be substituted. Alloy 825 also provides excellent resistance to corrosion by seawater. The compatibility of alloy 825 with selected corrodents is shown in Table 9.13.

Table 9.14 contains the mechanical and physical properties of alloy 825.

Applications include the nuclear industry, chemical processing, and pollution control systems.

VII. TYPE 330 (N08330)

This is a nickel-chromium-iron alloy with the addition of silicon. Refer to Table 8.1 for its chemical composition. Type 330 stainless has good strength at elevated temperatures, good thermal stability, and excellent resistance to carburizing and oxidizing atmospheres. It is weldable and machinable. This alloy has been used in low stress applications to temperatures as high as 2250°F (1230°C) and has moderate resistance to creep to 1600°F (870°C). Mechanical and physical properties will be found in Table 8.28.

Type 330 stainless resists the absorption of carbon and nitrogen, making it an excellent choice for furnace components. Overall it exhibits good corrosion resistance.

VIII. AL-6XN (N08367)

Alloy Al-6XN is the registered trademark of Allegheny Ludlum Corporation and has the UNS designation of N08367. The typical and specified chemical compositions of this alloy are given in Table 9.15.

Alloy Al-6XN was originally designed to resist seawater. However, it has proven to also be resistant to a wide range of corrosive environments.

The high strength and corrosion resistance of this alloy make it a better choice than the more expensive nickel base alloys in applications where

TABLE 9.13 Compatibility of Alloy 800 and Alloy 825 with Selected Corrodents[a]

Chemical	Maximum temp. °F	Maximum temp. °C	Chemical	Maximum temp. °F	Maximum temp. °C
Acetic acid, 10%[b]	200	93	Chloroform	90	32
Acetic acid, 50%[b]	220	104	Chlorosulfonic acid	X	
Acetic acid, 80%[b]	210	99	Chromic acid, 10%[b]	210	99
Acetic acid, glacial[b]	220	104	Chromic acid, 50%	X	
Acetic anhydride	230	110	Citric acid, 15%	210	99
Acetone	210	99	Citric acid, conc.[b]	210	99
Acetyl chloride	210	99	Copper acetate	90	32
Aluminum acetate	60	16	Copper carbonate	90	32
Aluminum chloride, aqueous	60	16	Copper chloride, 5%[b]	80	27
Aluminum fluoride, 5%	80	27	Copper cyanide	210	99
Aluminum hydroxide	80	27	Copper sulfate	210	99
Aluminum sulfate	210	99	Cupric chloride, 5%	X	
Ammonium carbonate	190	88	Ferric chloride	X	
Ammonium chloride, 10%[b]	230	110	Ferric chloride, 50% in water	X	
Ammonium chloride, sat.	200	93	Ferric nitrate, 10–50%	90	32
Ammonium hydroxide, sat.	110	43	Ferrous chloride[b,c]	90	32
Ammonium nitrate	90	32	Fluorine gas, dry	X	
Ammonium persulfate	90	32	Fluorine gas, moist	X	
Ammonium sulfate, 10–40%	210	99	Hydrobromic acid, 20%	X	
Ammonium sulfite	210	99	Hydrobromic acid, 50%	X	
Amyl acetate[b]	200	93	Hydrochloric acid, 20%[b]	90	32
Amyl chloride	90	32	Hydrochloric acid, 38%	X	
Aniline	90	32	Hydrocyanic acid, 10%	60	16
Antimony trichloride	90	32	Hydrofluoric acid, 30%	X	
Barium carbonate	90	32	Hydrofluoric acid, 70%	X	
Barium sulfate	90	32	Hydrofluoric acid, 100%	X	
Benzene	190	88	Magnesium chloride, 1–5%	170	77
Benzoic acid, 5%	90	32	Malic acid	170	77
Borax	190	88	Magnanese chloride, 10–50%	210	99
Boric acid, 5%	210	99	Muriatic acid[b]	90	32
Bromine gas, dry[b]	90	32	Nitric acid, 5%	90	32
Butyl acetate[b]	90	32	Nitric acid, 20%	60	16
Butyric acid, 5%	90	32	Nitric acid, anhydrous	210	99
Calcium carbonate	90	32	Phenol	90	32
Calcium chlorate	80	27	Picric acid	90	32
Calcium chloride[b,c]	60	16	Potassium bromide, 5%	90	32
Calcium hydroxide, 10%	200	93	Salicylic acid	90	32
Calcium hypochlorite	X		Silver bromide, 10%[b]	90	32
Calcium sulfate	90	32	Sodium carbonate	90	32
Carbon monoxide	570	299	Sodium chloride[c]	200	93
Carbon tetrachloride	90	32	Sodium hydroxide, 10%	90	32
Carbonic acid	90	32	Sodium hydroxide, conc.	90	32
Chloracetic acid	X		Sodium sulfide, to 50%	90	32
Chlorine gas, dry[b]	90	32	Stannic chloride	X	
Chlorine gas, wet	X		Stannous chloride, 5%	90	32
Chlorobenzene	90	32	Sulfuric acid, 10%[b]	230	110

TABLE 9.13 Continued

Chemical	Maximum temp.		Chemical	Maximum temp.	
	°F	°C		°F	°C
Sulfuric acid, 50%[b]	210	99	Sulfuric acid, 100%[b]	230	110
Sulfuric acid, 70%[b]	150	66	Sulfuric acid, fuming	X	
Sulfuric acid, 90%[b]	180	82	Sulfurous acid[a]	370	188
Sulfuric acid, 98%[b]	220	104	Zinc chloride, 5%	140	60

[a]The chemicals listed are in the pure state or in a saturated solution unless otherwise indicated. Compatibility is shown to the maximum allowable temperature for which data are available. Incompatibility is shown by an X. When compatible the corrosion rate is <20 mpy.
[b]Applicable to alloy 825 only.
[c]Material subject to pitting.
Source: Ref. 1.

excellent formability, weldability, strength, and corrosion resistance are essential.

It is also a cost-effective alternative to less expensive alloys, such as type 316, that do not have the strength or corrosion resistance required to minimize life cycle costs in certain applications.

TABLE 9.14 Mechanical and Physical Properties of Alloy 825

Modulus of elasticity × 10^6 (psi)	28.3
Tensile strength × 10^3 (psi)	85
Yield strength 0.2% offset × 10^3 (psi)	35
Elongation in 2 in. (%)	30
Brinell hardness	150
Density (lb/in.^{3})	0.294
Specific gravity	8.14
Coefficient of thermal expansion × 10^{-6} (in./in. °F)	
at 80–200°F	7.8
at 80–400°F	8.3
at 80–600°F	8.5
at 80–800°F	8.7
at 80–1000°F	8.8
at 80–1200°F	9.1
at 80–1400°F	9.5
at 80–1600°F	9.7

TABLE 9.15 Typical and Specified Chemical Composition
of Alloy Al6XN

Chemical element	Composition (wt%)	
	Typical Al6XN alloy	UNS N08367 specification
Carbon	0.02	0.03 max.
Manganese	0.40	2.00 max.
Phosphorus	0.020	0.040 max.
Sulfur	0.001	0.030 max.
Silicon	0.40	1.00 max.
Chromium	20.5	20.00/22.00
Nickel	24.0	23.50/25.00
Molybdenum	6.2	6.00/7.00
Nitrogen	0.22	0.18/0.25
Copper	0.2	0.75 max.
Iron	Balance	Balance

The high nickel and molybdenum contents provide improved resistance to chloride stress corrosion cracking. Copper has been kept to a residual level for improved performance in seawater. The high alloy composition resists crevice corrosion and pitting in oxidizing chloride solutions.

The low carbon content of the alloy defines it as an L grade, providing resistance to intergranular corrosion in the as-welded condition.

Wrought alloy Al-6XN is approved by the ASME for use to 800°F (427°C) in unfired pressure vessels under the ASME Boiler and Pressure Vessel Code, Section VIII, Division 1. The mechanical and physical properties are shown in Table 9.16, while Tables 9.17 and 9.18 list the ASME maximum allowable stress values for several Al-6XN alloy product forms. Note that the stress values shown in Table 9.18 are higher than those in Table 9.17. Using alloy Al-6XN at these stresses is permissible according to the ASME code, but may result in dimensional change. The higher values are not recommended for applications such as flanges or gasketed joints where slight distortions may cause leakage or malfunction.

Alloy Al-6XN is stronger with comparable ductility as compared with the more conventional austenitic stainless steels. Table 9.19 shows typical transverse tensile properties at room temperature for Al-6XN sheet 0.026 to 0.139 in. thick in comparison to other austenitic stainless flat-rolled products.

Excellent toughness is also demonstrated at subzero temperatures, and strength is maintained at elevated temperatures.

TABLE 9.16   Mechanical and Physical Properties of Al-6XN[a]
Stainless Steels

Modulus of elasticity $\times 10^6$ (psi)	27
Tensile strength $\times 10^3$ (psi)	107.5
Yield strength 0.2% offset $\times 10^3$ (psi)	54.7
Elongation in 2 in. (%)	50
Density (lb/in.^{3})	0.291
Specific gravity	8.06
Rockwell hardness	B-90
Thermal conductivity at 68–212°F (Btu/hr °F)	7.9
Coefficient of thermal expansion $\times 10^{-6}$ (in./in. °F)	
at 68–212°F	8.5
at 68–932°F	8.9
at 68–1472°F	10.0
Izod impact (ft-lb)	140

[a]Registered trademark of Allegheny Ludlum Corporation.

The corrosion resistant properties of alloy Al-6XN show exceptional resistance to pitting, crevice attack, and stress cracking in high chlorides and general resistance in various acid, alkaline, and salt solutions found in chemical processing and other industrial environments. Excellent resistance is shown to oxidizing chlorides, reducing solutions, and seawater corrosion. Sulfuric, nitric, phosphoric, acetic, and formic acids can be handled at various concentrations and a variety of temperatures. The material is also approved for contact with foods. Refer to Table 9.20 for the compatibility of alloy Al-6XN with selected corrodents.

Alloy Al-6XN finds applications as

Chemical process vessels and pipelines
Condensers, heat exchangers
Power plant flue gas scrubbers
Distillation columns
Service water piping in nuclear power plants
Food processing equipment

IX.   254SMO (S31254)

This is a superaustenitic stainless steel in the 6 moly alloy family which is designed for maximum resistance to pitting and crevice corrosion. Its chemical makeup is as follows:

TABLE 9.17 Maximum Allowable Design Stress Values in Tension (ksi)—Dimensionally Stable Under Cited Conditions

For metal temperature not exceeding		Seamless pipe and tube <3/16 in. thick sheet and strip 40YS/104TS	3/16–3/4 in. thick plate 45YS/100TS	Forgings, bars, rods, wire >3/4 in. 45YS/95TS	Welded tube[a] 45YS/100TS	Welded pipe[a] 45YS/95TS
°F	°C					
100	38	26.0	25.0	23.8	21.3	20.2
200	93	26.0	25.0	23.8	21.3	20.2
300	149	24.3	23.7	22.5	20.1	19.1
400	204	22.7	22.2	21.4	18.9	18.2
500	260	20.9	20.4	20.4	17.3	17.3
600	316	19.9	19.5	19.5	16.6	16.6
700	371	19.3	18.6	18.6	15.8	15.8
750	399	18.7	18.3	18.3	15.6	15.6
800	427	18.4	18.0	18.0	15.3	15.3

YS = minimum yield strength: 0.2% offset; TS = minimum tensile strength.
[a] These values were obtained by multiplying those for plate, sheet, and strip by a factor of 0.85.

TABLE 9.18 Maximum Allowable Design Stress Values in Tension (ksi)—Dimensionally Unstable Under Cited Conditions

For metal temperature not exceeding		Seamless pipe and tube <3/16 in. thick sheet and strip 40YS/104TS	3/16–3/4 in. thick plate 45YS/100TS	Forgings, bars, rods, wire >3/4 in. 45YS/95TS	Welded tube[a] 45YS/100TS	Welded pipe[a] 45YS/95TS
°F	°C					
100	38	26.2	25.0	23.8	21.3	20.2
200	93	26.0	25.0	23.8	21.3	20.2
300	149	24.5	23.7	22.5	20.1	19.1
400	204	23.5	22.6	21.4	19.2	18.2
500	260	22.8	22.0	20.9	18.7	17.7
600	316	22.3	21.7	20.4	18.4	17.3
650	343	22.1	21.3	20.2	18.1	17.1
700	371	21.9	21.1	20.0	17.9	17.0
750	399	21.8	21.0	19.9	17.8	16.9
800	427	21.7	20.9	19.8	17.7	16.8

YS = minimum yield strength: 0.2% offset; TS = minimum tensile strength.
[a]These values were obtained by multiplying those for plate, sheet, and strip by a factor of 0.85.

Chemical	Weight percent
Carbon	0.02
Chromium	19.5–20.5
Nickel	17.5–18.5
Molybdenum	6.0–6.5
Nitrogen	0.18–0.22
Copper	0.50–1.00
Iron	Balance

The alloy has a PREN of 45.8. A value above 33 is considered necessary for pitting and crevice resistance to ambient seawater. With its high levels of chromium, molybdenum, and nitrogen, S31254 is especially suited for high chloride environments such as brackish water, seawater, pulp mill bleach plants, and other high chloride process streams.

The mechanical and physical properties of alloy 254SMO will be found in Table 9.21. Its strength is nearly equivalent to that of the duplex stainless steels.

X. 25-6Mo (N08926)

This alloy 25-6Mo is produced by Inco International. It is also known as 1925 hMo and has been assigned UNS N08926. Typical and specified compositions of this alloy are shown in Table 9.22.

These alloys have higher mechanical properties than those of the austenitic stainless steels such as type 316L. They also have higher design values than lower strength materials, enabling the use of thinner sections. Refer to Table 9.23 for the mechanical and physical properties of alloy 25-6Mo.

One of the outstanding attributes of alloy 26-6Mo is its resistance to environments containing chlorides or other halides. It is especially suited for

TABLE 9.19 Comparison of Transverse Tensile Properties of Selected Alloys

Property at room temparature	Type 316	Alloy 904L	Alloy Al-6XN
Yield strength, 0.2% offset (ksi)	45	42	53
Ultimate tensile strength (ksi)	88	86	108
Elongation in 2 in. (%)	57	43	47
Rockwell hardness	B-81	B-79	B-88

TABLE 9.20 Compatibility of Al-6XN Stainless Steel with Selected Corrodents[a]

Chemical	Maximum temp. (°F/°C)
Acetic acid, 20%	210/99
Acetic acid, 80%	217/103
Formic acid, 45%	220/104
Formic acid, 50%	220/104
Nitric acid, 10%	194/90
Nitric acid, 65%	241/116
Oxalic acid, 10%	210/99
Phosphoric acid, 20%	210/99
Phosphoric acid, 85%	158/76
Sulfamic acid, 10%	210/99
Sulfuric acid, 10%	X
Sulfuric acid, 60%	122/50
Sulfuric acid, 95%	86/30
Sodium bisulfate, 10%	210/99
Sodium hydroxide, 50%	210/99

[a]Compatibility is shown to the maximum allowable temperature for which data are available. Incompatibility is shown by an X. When compatible, the corrosion rate is <20 mpy.

applications in high chloride environments such as brackish water, seawater, caustic chlorides, and pulp mill bleach systems.

The alloy offers excellent resistance to pitting and crevice corrosion, having a PREN of 47.45. The critical pitting temperature for alloy 25-6Mo is 140°F (60°C) or higher while the critical crevice temperature is 90°F (32.5°C).

TABLE 9.21 Mechanical and Physical Properties of Alloy 254SMO (S31254)

Yield strength $\times 10^3$ (psi)	44
Tensile strength, 0.2% offset $\times 10^3$ (psi)	94
Elongation in 2 in. (%)	35
Density (lb/in.^{3})	0.289
Specific gravity	7.93
Impact strength at room temperature (ft-lb)	88
Brinell hardness	210

TABLE 9.22 Typical and Specified Composition of
Alloy 25-6Mo

Chemical	Alloy 25-6Mo (wt%)	UNS N08926 (wt%)
Carbon	0.02 max.	0.02 max.
Chromium	19.0–21.0	20.0–21.0
Nickel	24.0–26.0	24.5–25.5
Molybdenum	6.0–7.0	6.0–6.8
Nitrogen	0.15–0.25	0.18–0.20
Copper	0.5–1.5	0.8–1.0
Manganese	2.0 max.	2.0 max.
Phosphorus	0.030 max.	0.030 max.
Sulfur	0.010 max.	0.010 max.
Silicon	0.050 max.	0.050 max.
Iron	Balance	Balance

TABLE 9.23 Mechanical and Physical Properties of Alloy 25-6Mo
(N08926)

Modulus of elasticity $\times 10^6$ (psi)	28.1
Tensile strength $\times 10^3$ (psi)	100
Yield strength 0.2% offset $\times 10^3$ (psi)	48
Elongation in 2 in. (%)	42
Density (lb/in.^{3})	0.296
Specific gravity	8.16
Thermal conductivity (Btu/ft^2 hr °F)	116
Coefficient of thermal expansion $\times 10^{-6}$ (in./in.°F) at 68–212°F	8.4

TABLE 9.24 Mechanical and Physical Properties of Alloy
654SMO (S32654)

Tensile strength $\times 10^3$ (psi)	108
Yield strength 0.2% offset $\times 10^3$ (psi)	62
Elongation in 2 in. (%)	40
Density (lb/in.^{3})	0.289
Specific gravity	7.93
Impact strength at room temperature (ft-lb)	130

In brackish and wastewater systems, microbially influenced corrosion can occur, especially in systems where equipment has been idle for extended periods. A 6% molybdenum alloy offers protection from manganese-bearing, sulfur-bearing, and generally reducing types of bacteria. Because of its resistance to microbially influenced corrosion, alloy 25-6Mo is being used in the wastewater piping systems of power plants.

In saturated sodium chloride environments and pH values of 6–8 alloy 25-6Mo exhibits a corrosion rate of less than 1 mpy. Even under more aggressive oxidizing conditions involving sodium chlorate, alloy 25-6Mo maintains a corrosion rate of less than 1 mpy and shows no pitting even at temperatures up to boiling.

XI. ALLOY 31 (N08031)

The chemical composition of this alloy is

Chemical	Weight percent
Carbon	0.02 max.
Nickel	31
Chromium	27
Molybdenum	6.5
Copper	1.8
Nitrogen	0.20
Iron	Balance

With the 6.5% molybdenum content alloy 31 exhibits excellent resistance to pitting and crevice corrosion in neutral and acid solutions. The high chromium content of 27% imparts superior resistance to corrosive attack by oxidizing media. It has a PREN of 54.45.

XII. 654SMO (S32654)

Alloy 654SMO has about double the strength of type 316L stainless steel. Refer to Table 9.24 for the mechanical and physical properties.

This alloy contains 7+% of molybdenum, which provides it with a corrosion resistance associated with nickel-based alloys. The composition will be found in Table 9.25.

Alloy 654 has better resistance to localized corrosion than other superaustenitic alloys. Indications are that alloy 654 is as corrosion resistant as alloy C-276, based on tests in filtered seawater, bleach plants, and other

TABLE 9.25 Chemical Composition
of Alloy 254SMO (S32654)

Chemical	Weight percent
Carbon	0.02 max.
Chromium	24.0
Nickel	22.0
Molybdenum	7.3
Nitrogen	0.5
Copper	0.5
Manganese	3.0
Iron	Balance

TABLE 9.26 Chemical Composition
of Alloy 686 (N06686)

Chemical	Weight percent
Chromium	19.0–23.0
Molybdenum	15.0–17.0
Tungsten	3.0–4.0
Titanium	0.02–0.25
Iron	5.0 max.
Carbon	0.01 max.
Manganese	0.75 max.
Sulfur	0.02 max.
Silicon	0.08 max.
Phosphorus	0.04 max.
Nickel	Balance

TABLE 9.27 Mechanical and Physical Properties of Alloy 686
(N06686) at 70°F/20°C

Modulus of elasticity $\times 10^6$ (psi)	30
Tensile strength $\times 10^3$ (psi)	104
Yield strength 0.2% $\times 10^3$ (psi)	52.8
Elongation in 2 in. (%)	71
Density (lb/in.^{3})	0.315
Specific heat (Btu/lb °F)	0.089
Coefficient of thermal expansion $\times 10^{-6}$ (in./in.°F)	6.67
Impact strength (ft-lb)	299

TABLE 9.28 Mechanical Properties of Alloy 686 (N06686) at Elevated Temperatures

Property	Temperature (°F/°C)					
	200/93	400/204	600/315	800/427	1000/538	1200/649
Modulus of elasticity × 10^6 (psi)	29.7	28.5	28.0	26.9	26.0	24.6
Tensile strength × 10^3 (psi)	100.2	92.1	87.3	82.6	79.1	
Yield strength 0.2% offset × 10^3 (psi)	46.8	42.1	41.7	32.6	37.9	
Elongation in 2 in. (%)	69	67	60	69	61	
Specific heat (Btu/lb °F)	0.092	0.098	0.104	0.111	0.116	0.122
Coefficient of thermal expansion × 10^{-6} (in./in.°F)	6.67	6.81	7.00	7.17	7.15	0.122

aggressive chloride environments. It is intended to compete with titanium in the handling of high chloride environments.

XIII. INCONEL ALLOY 686 (NO6686)

Inconel alloy 686 is an austenitic, nickel-chromium-molybdenum-tungsten alloy. The chemical composition of this alloy will be found in Table 9.26.

This highly alloyed material has good mechanical strength. It is most often used in the annealed condition. Since alloy 686 is a solid-solution alloy it cannot be strengthened by heat treatment, but strain hardening by cold work will greatly increase the strength of the alloy. Typical room temperature properties in the annealed temper are given in Table 9.27. Tensile properties at elevated temperature will be found in Table 9.28. Exposure to high temperatures for long periods of time can have an embrittling effect on the alloy.

The alloy's composition provides resistance to general corrosion, stress corrosion cracking, pitting, and crevice corrosion in a broad range of aggressive environments. The high nickel and molybdenum contents provide good corrosion resistance in reducing environments, while the high chromium level imparts resistance to oxidizing media. The molybdenum and tungsten also aid resistance to localized corrosion such as pitting, while the low carbon content and other composition controls helps minimize grain boundary precipitation to maintain resistance to corrosion in heat-affected zones of welded joints.

The ability of alloy 686 to resist pitting can be seen from its pitting resistance equivalent, which is 51.

Alloy 686 has excellent resistance to mixed acids as well as reducing and oxidizing acids and to mixed acids containing high concentrations of halides. Good resistance has been shown to mixed acid media having pH levels of 1 or less and chloride levels in excess of 100,000 ppm.

REFERENCES

1. PA Schweitzer. Corrosion Resistance Tables, 4th ed., Vols. 1–3, New York: Marcel Dekker, 1995.
2. PK Whitecraft. Corrosion of stainless steels. In: PA Schweitzer, ed. Corrosion Engineering Handbook. New York: Marcel Dekker, 1996.
3. CP Dillon. Corrosion Resistance of Stainless Steels, New York: Marcel Dekker, 1995.
4. PA Schweitzer. Encyclopedia of Corrosion Technology, New York: Marcel Dekker, 1998.

10

Ferritic Stainless Steels

The ferritic stainless steels are the simplest of the stainless steel family of alloys since they are principally iron-chromium alloys. They are magnetic, have body-centered cubic atomic structures, and possess mechanical properties similar to those of carbon steel, though are less ductile.

This class of alloys usually contains 15–18% chromium, although they can go as low as 11% in special cases, under the influence of other alloying elements, or as high as about 30%. Continued additions of chromium will improve corrosion resistance in more severe environments. Chromium additions are particularly beneficial in terms of resistance in oxidizing environments, at both moderate and elevated temperatures. Addition of chromium is the most cost-effective means of increasing corrosion resistance of steel.

These materials are historically known as 400 series stainless as they are identified with numbers beginning with 400 when the American Institute for Iron and Steel (AISI) had the authority to designate alloy compositions. Under the new UNS system the old three-digit numbers were retained, such as the old 405, a basic 12% chromium, balance iron material, which is now designated S40500.

The strength of ferritic stainless steels can be increased by cold working but may not be by heat treatment. These alloys have the ability to be worked hot or cold. Their impact resistance and weldability are rather poor. Welding may result in brittleness and/or reduced corrosion resistance, unless proper pre- and postweld heat treatments are employed. This effect and the relatively low toughness at low temperatures have limited their acceptance for structural applications. Generally, toughness in the annealed condition decreases as the chromium content increases. Molybdenum tends to decrease ductility. As a result, ferritic stainless steels are used for nonstructural applications such as kitchen sinks and automotive, appliance, and luggage trim, which require good resistance to corrosion and bright, highly polished finishes.

In comparison to low carbon steels such as SAE 1010, the standard numbered AISI ferritic stainless steels (such as type 430) exhibit somewhat higher yield and tensile strengths and low elongations; thus they are not as formable as the low carbon steels. The proprietary ferritic stainless steels, on the other hand, with lower carbon levels, have improved ductility and formability comparable with those of low carbon steels. Because of the higher strength levels, the ferritic stainless steels require slightly more power to form.

Cleanliness, in terms of low sulfur and silicon contents, is important to good formability of the ferritic types because inclusions can act as initiation sites for cracks during forming.

These alloys are useful in high temperature situations, with type 446 exhibiting useful oxidation (scaling) resistance through about 2100°F (1148°C). Ferritic alloys containing more than about 18% chromium are susceptible to an embrittlement phenomenon. This is due to the formation of a secondary phase and is termed 885°F/475°C embrittlement, after the temperature which causes the rapid formation. These materials are not brittle in this temperature range but lose ductility when cooled to room temperature. Consequently these alloys are limited to a maximum operating temperature of 650°F (343°C).

Corrosion resistance is rated good, although ferritic alloys do not resist reducing acids such as hydrochloric. Mildly corrosive conditions and oxidizing media are handled satisfactorily. Type 430 finds wide application in nitric acid plants. Increasing the chromium content to 24 and 30% improves the resistance to oxidizing conditions at elevated temperatures. These alloys are useful for all types of furnace parts not subject to high stress. Because the oxidation resistance is independent of the carbon content, soft forgeable alloys, low in carbon, can be rolled into plates, shapes, and sheets.

Ferritic stainless steels offer useful resistance to mild atmospheric corrosion and most fresh waters. They will corrode with exposure to seawater atmospheres.

I. TYPE 405 (S40500)

This is a nonhardenable 12% chromium stainless steel. The chemical composition is given in Table 10.1. Type 405 stainless is designed for use in the as-welded condition, however heat treatment improves corrosion resistance. The low chromium favors less sensitivity to 855°F/475°C embrittlement and

TABLE 10.1   Chemical Composition of Ferritic Stainless Steels

AISI type	Nominal composition (%)				
	C max.	Mn max.	Si max.	Cr	Other[a]
405	0.08	1.00	1.00	11.50–14.50	0.10–0.30 Al
430	0.12	1.00	1.00	14.00–18.00	
430F	0.12	1.25	1.00	14.00–18.00	0.15 S min.
430(Se)	0.12	1.25	1.00	14.00–18.00	0.15 Se min.
444	0.025	1.00	1.00 max.	17.5–19.5	1.75–2.50 Mo
446	0.20	1.50	1.00	23.00–17.00	0.25 max. N
XM-27[b]	0.002	0.10	0.20	26.00	

[a] Elements in addition to those shown are as follows: phosphorus—0.06% max. in type 430F and 430(Se), 0.015% in XM-27; sulfur—0.03% max. in types 405, 430, 444, and 446; 0.15% min. type 430F, 0.01% in XM-27; nickel—1.00% max. in type 444, 0.15% in XM-27; titanium + niobium—0.80% max. in type 444; copper—0.02% in XM-27; nitrogen—0.010% in XM-27.
[b] E-Brite 26-1 trademark of Allegheny Ludlum Industries Inc.

sigma phase formation. Mechanical and physical properties are shown in Table 10.2. Rupture and creep characteristics of type 405 will be found in Table 10.3.

Type 405 stainless steel is resistant to nitric acid, organic acids, and alkalies. It will be attacked by sulfuric, hydrochloric, hydrofluoric, and phosphoric acids as well as seawater. It is resistant to chloride stress corrosion cracking.

TABLE 10.2   Mechanical and Physical Properties of Alloy 405 (S40500) Stainless Steel

Modulus of elasticity $\times 10^6$ (psi)	29.0
Tensile strength $\times 10^3$ (psi)	60
Yield strength 0.2% offset $\times 10^3$ (psi)	30
Elongation (%)	20
Rockwell hardness	B-85
Density (lb/in.³)	0.279
Specific gravity	7.6
Thermal conductivity (Btu/ft hr °F)	
at 70°F (20°C)	15.6
at 1500°F (815°C)	17.9

TABLE 10.3 Rupture and Creep Characteristics of Type 405 (S40500)
Stainless Steel

Temperature (°F/°C)	Stress for rupture (ksi) in		Stress for creep (ksi) rate/hr of	
	1000 hr	10,000 hr	0.0001%	0.00001%
900/482	25.0	22.0	43.0	14.0
1000/538	16.0	12.0	8.0	4.5
1100/593	6.8	4.7	2.0	0.5
1200/649	3.8	2.5		
1300/704	2.2	1.4		
1400/760	1.2	0.7		
1500/861	0.8	0.4		

Applications include heat exchanger tubes in the refining industry and other areas where exposure may result in the 885°F (475°C) or sigma temperature range. It has an allowable maximum continuous operating temperature of 1300°F (705°C) with an intermittent allowable temperature of 1500°F (815°C).

II. TYPE 409 (S40900)

Type 409 stainless is an 11% chromium alloy stabilized with titanium. Its chemical composition will be found in Table 10.4. The material can be

TABLE 10.4 Chemical Composition of Alloy 409
(S40900)

Chemical	Weight percent
Carbon	0.08
Manganese	1
Silicon	1
Chromium	10.5–11.75
Nickel	0.5
Phosphorus	0.045
Sulfur	0.045
Titanium	6 × %C min. to 0.75% max.
Iron	Balance

welded in the field; however, heat treatment improves corrosion resistance. Table 10.5 lists the mechanical and physical properties of the alloy. It has a maximum allowable continuous operating temperature of 1300°F (705°C) with an allowable intermittent operating temperature of 1500°F (815°C). It cannot be hardened by heat treatment.

The primary application for alloy 409 is in the automotive industry as mufflers, catalytic converters, and tail pipes. It has proven an attractive replacement for carbon steel because it combines economy and good resistance to oxidation and corrosion.

III. TYPE 430 (S43000)

This is the most widely used of the ferritic stainless steels. It combines good heat resistance and mechanical properties. The chemical composition will be found in Table 10.6. In continuous service type 430 may be operated to a maximum temperature of 1500°F (815°C) and 1600°F (870°C) in intermittent service. However, they are subject to 885°F/475°C embrittlement and loss of ductility at subzero temperatures. Rupture and creep characteristics will be found in Table 10.7.

Type 430 stainless is resistant to chloride stress corrosion cracking and elevated sulfide attack. Applications are found in nitric acid services, water and food processing, automobile trim, heat exchangers in petroleum and chemical processing industries, reboilers for desulfurized naphtha, heat exchangers in sour water strippers, and hydrogen plant effluent coolers. The compatibility of type 430 stainless steel with selected corrodents is provided in Table 10.8.

TABLE 10.5 Mechanical and Physical Properties of Alloy 409 (S40900) Stainless Steel

Modulus of elasticity $\times 10^6$ (psi)	29.0
Tensile strength $\times 10^3$ (psi)	55
Yield strength 0.2% offset $\times 10^3$ (psi)	30
Elongation (%)	20
Rockwell hardness	B-85
Density (lb/in.^{3})	0.28
Specific gravity	7.7
Thermal conductivity (Btu/ft hr °F)	
at 70°F (20°C)	14.4
at 1500°F (815°C)	16.6

TABLE 10.6 Physical and Mechanical Properties of Ferritic Stainless Steels

Property	Type of alloy		
	430	444	XM-27
Modulus of elasticity $\times$ 10^6 (psi)	29	29	
Tensile strength $\times$ 10^3 (psi)	60	60	70
Yield strength 0.2% offset $\times$ 10^3 (psi)	35	40	56
Elongation in 2 in. (%)	20	20	30
Hardness	Rockwell B-165	Brinell 217	Rockwell B-83
Density (lb/in.^{3})	0.278	0.28	0.28
Specific gravity	7.75	7.75	7.66
Specific heat (32–212°F) (Btu/lb °F)	0.11	0.102	0.102
Thermal conductivity (Btu/lb °F)			
at 70°F (20°C)	15.1	15.5	
at 1500°F (815°C)	15.2		
Coefficient of thermal expansion $\times$ 10^{-6} (in./in. °F)			
at 32–212°F	6.0	6.1	5.9

Stainless steel type 430F is a modification of type 430. The carbon content is reduced to 0.065%, manganese to 0.80%, and silicon to 0.3/0.7%, while 0.5% molybdenum and 0.60% nickel have been added. This is an alloy used extensively in solenoid armatures and top plugs. It has also been used in solenoid cores and housings operating in corrosive environments.

TABLE 10.7 Rupture and Creep Characteristics of Type 430 (S43000) Stainless Steel

Temperature (°F/°C)	Stress for rupture (ksi) in		Stress for creep (ksi) rate/hr of	
	1000 hr	10,000 hr	0.0001%	0.00001%
800/482	—	—	23.0	17.5
900/482	30.0	24.0	15.4	12.0
1000/538	17.5	13.5	8.6	6.7
1100/593	9.1	6.5	4.3	3.4
1200/649	5.0	3.4	1.2	1.5
1300/704	2.8	2.2	1.4	0.9
1400/760	1.7	0.7	0.9	0.6
1500/861	0.9	0.5	0.6	0.3

TABLE 10.8 Compatibility of Ferritic Stainless Steels with Selected Corrodents

	Type of alloy		
Chemical	430 (°F/°C)	444 (°F/°C)	XM-27 (°F/°C)
Acetic acid, 10%	70/21	200/93	200/93
Acetic acid, 50%	X	200/93	200/93
Acetic acid, 80%	70/21	200/93	130/54
Acetic acid, glacial	70/21		140/60
Acetic anhydride, 90%	150/66		300/149
Aluminum chloride, aqueous	X		110/43
Aluminum hydroxide	70/21		
Aluminum sulfate	X		
Ammonia gas	212/100		
Ammonium carbonate	70/21		
Ammonium chloride, 10%			200/93
Ammonium hydroxide, 25%	70/21		
Ammonium hydroxide, sat.	70/21		
Ammonium nitrate	212/100		
Ammonium persulfate, 5%	70/21		
Ammonium phosphate	70/21		
Ammonium sulfate, 10–40%	X		
Amyl acetate	70/21		
Amyl chloride	X		
Aniline	70/21		
Antimony trichloride	X		
Aqua regia, 3:1			X
Barium carbonate	70/21		
Barium chloride	70/21[a]		
Barium sulfate	70/21		
Barium sulfide	70/21		
Benzaldehyde			210/99
Benzene	70/21		
Benzoic acid	70/21		
Borax, 5%	200/93		
Boric acid	200/93[a]		
Bromine gas, dry	X		
Bromine gas, moist	X		
Bromine, liquid	X		
Butyric acid	200/93		
Calcium carbonate	200/93		
Calcium chloride	X		
Calcium hypochlorite	X		

TABLE 10.8   Continued

Chemical	Type of alloy		
	430 (°F/°C)	444 (°F/°C)	XM-27 (°F/°C)
Calcium sulfate	70/21		
Carbon bisulfide	70/21		
Carbon dioxide, dry	70/21		
Carbon monoxide	1600/871		
Carbon tetrachloride, dry	212/100		
Carbonic acid	X		
Chloracetic acid, 50% water	X		
Chloracetic acid	X		
Chlorine gas, dry	X		
Chlorine gas, wet	X		
Chloroform, dry	70/21		
Chromic acid, 10%	70/21		120/49
Chromic acid, 50%	X		X
Citric acid, 15%	70/21	200/93	200/93
Citric acid, concentrated	X		
Copper acetate	70/21		
Copper carbonate	70/21		
Copper chloride	X		X
Copper cyanide	212/100		
Copper sulfate	212/100		
Cupric chloride, 5%	X		
Cupric chloride, 50%	X		
Ethylene glycol	70/21		
Ferric chloride	X		80/27
Ferric chloride, 10% in water			75/25
Ferric nitrate, 10–50%	70/21		
Ferrous chloride	X		
Fluorine gas, dry	X		
Fluorine gas, moist	X		
Hydrobromic acid, dilute	X		
Hydrobromic acid, 20%	X		
Hydrobromic acid, 50%	X		
Hydrochloric acid, 20%	X		
Hydrochloric acid, 38%	X		
Hydrocyanic acid, 10%	X		
Hydrofluoric acid, 30%	X		X
Hydrofluoric acid, 70%	X		X
Hydrofluoric acid, 100%	X		X
Iodine solution, 10%	X		

TABLE 10.8 Continued

Chemical	Type of alloy		
	430 (°F/°C)	444 (°F/°C)	XM-27 (°F/°C)
Lactic acid, 20%	X	200/93	200/93
Lactic acid, conc.	X		
Magnesium chloride			200/93
Malic acid	200/93		
Muriatic acid	X		
Nitric acid, 5%	70/21	200/93	320/160
Nitric acid, 20%	200/93	200/93	320/160
Nitric acid, 70%	70/21	X	210/99
Nitric acid, anhydrous	X	X	
Nitrous acid, 5%	70/21		
Phenol	200/93		
Phosphoric acid, 50–80%	X	200/93	200/93
Picric acid	X		
Silver bromide, 10%	X		
Sodium chloride	70/21 [a]		
Sodium hydroxide, 10%	70/21	212/100	200/93
Sodium hydroxide, 50%		X	180/82
Sodium hydroxide, conc.		X	
Sodium hypochlorite, 30%			90/32
Sodium sulfide, to 50%	X		
Stannic chloride	X		
Stannous chloride, 10%			90/32
Sulfuric acid, 10%	X	X	X
Sulfuric acid, 50%	X	X	X
Sulfuric acid, 70%		X	X
Sulfuric acid, 90%		X	X
Sulfuric acid, 98%		X	280/138
Sulfuric acid, 100%	70/21	X	
Sulfuric acid, fuming		X	
Sulfurous acid, 5%	X		360/182
Toluene			210/99
Trichloroacetic acid	X		
Zinc chloride, 20%	70/21 [a]		200/93

The chemicals listed are in the pure state or in a saturated solution unless otherwise indicated. Compatibility is shown to the maximum allowable temperature for which data are available. Incompatibility is shown by an X. A blank space indicates that data are unavailable. When compatible, the corrosion rate is <20 MPY.
[a] Pitting may occur.
Source: Ref. 4.

As supplied it offers good permeability and low residual magnetism. It is available in three hardenesses:

1. Rockwell B-75: lowest degree of residual magnetism
2. Rockwell B-82/91: slightly higher residual magnetism
3. Rockwell B-91 minimum: optimum machinability condition, but parts must be annealed to provide desired soft magnetic properties

Type 430F stainless is another modification of the basic type 430 which should be considered when making machined articles from a 17% chromium steel. The composition has been altered by increasing the manganese content to 1.25% and the phosphorus content to 0.06%, with the sulfur content at 0.15% minimum and addition of 0.60% of molybdenum. This material will not harden by heat treatment. It has been used in automatic screw machines for parts requiring good corrosion resistance such as aircraft parts and gears.

Type 430FR alloy has the same chemical composition as type 430F except for increasing the silicon content to 1.00/1.50 wt%. The alloy has been used for solenoid valve magnetic core components which must combat corrosion from atmospheric fresh water and corrosive environments. This grade has a higher electrical resistivity than 430F solenoid quality, which reduces eddy current losses of the material.

IV. TYPE 439L (S43035)

The composition of type 439L stainless will be found in Table 10.9. This alloy is nonhardenable through heat treatment and has excellent ductility and weldability. It resists intergranular attack and formation of martensite in

TABLE 10.9 Chemical Composition of Alloy 439L (S43035)

Chemical	Weight percent
Carbon	0.07 max.
Manganese	1.00 max.
Silicon	1.00 max.
Chromium	17.0–19.0
Nitrogen	0.50
Titanium	12 × %C min.
Aluminum	0.15 max.

TABLE 10.10 Mechanical and Physical Properties of Alloy 439L (S43035) Stainless Steel

Modulus of elasticity $\times 10^6$ (psi)	29.0
Tensile strength $\times 10^3$ (psi)	60
Yield strength 0.2% offset $\times 10^3$ (psi)	30
Elongation (%)	20
Rockwell hardness	B-90
Density (lb/in.^{3})	0.280
Specific gravity	7.7
Thermal conductivity (Btu/ft hr °F)	
at 70°F (29°C)	15.1
at 1500°F (815°C)	15.2

the as-welded, heat-affected zone, but is subject to 885°F/475°C embrittlement. Refer to Table 10.10 for the mechanical and physical properties.

Alloy 439L is resistant to chloride stress corrosion, organic acids, alkalies, and nitric acid. It will be attacked by sulfuric, hydrochloric, hydrofluoric, and phosphoric acids as well as seawater.

Applications include heat exchangers, condensers, feedwater heaters, tube oil coolers, and moisture separator reheaters.

V. TYPE 444 (S44400)

Table 10.1 provides the chemical composition of this alloy, while the mechanical and physical properties will be found in Table 10.6. The allowable stress values are shown in Table 10.11.

This is a low carbon alloy with molybdenum added to improve chloride pitting resistance. It is virtually immune to chloride stress corrosion cracking. The alloy is subject to 885°F/475°C embrittlement and loss of ductility at subzero temperatures.

TABLE 10.11 Allowable Stress Values of Ferritic Stainless Steels

SS type	\multicolumn{7}{c}{Allowable stress value (ksi) at temperature not exceeding (°F/°C)}
	100/38	200/93	300/149	400/204	500/260	600/316	650/343
444	12.8	12.2	11.8	11.3	10.9	10.5	10.4
XM-27	12.75	12.75	12.5	12.5	12.5	12.5	12.5

The chloride pitting resistance of this alloy is similar to that of type 316 stainless steel and superior to that of types 430 or 439L. Like all ferritic stainless steels type 444 relies on a passive film to resist corrosion, but exhibits rather high corrosion rates when activated. This characteristic explains the abrupt transition in corrosion rates that occur at particular acid concentrations. For example, it is resistant to very dilute solutions of sulfuric acid at boiling temperature, but corrodes rapidly at higher concentrations. The corrosion rates of type 444 in strongly concentrated sodium hydroxide solutions are also higher than those for austenitic stainless steels. The compatibility of type 444 alloy with selected corrodents will be found in Table 10.8. The corrosion resistance of type 444 is generally considered equal to that of type 304.

This alloy is used for heat exchangers in chemical, petroleum, and food processing industries as well as piping.

VI. TYPE 446 (S44600)

Type 446 is a heat resisting grade of ferritic stainless steel. It has a maximum temperature rating of 2000°F (1095°C) for continuous service and a maximum temperature rating of 2150°F (1175°C) for intermittent service. Table 10.1 lists the chemical composition and Table 10.6 the mechanical and physical properties.

This nonhardenable chromium steel exhibits good resistance to reducing sulfurous gases and fuel-ash corrosion. The rupture and creep characteristics of type 446 stainless will be found in Table 10.12.

TABLE 10.12 Rupture and Creep Characteristics of Type 446 (S44600) Stainless Steel

Temperature (°F/°C)	Stress for rupture (ksi) in		Stress for creep (ksi) rate/hr of	
	1000 hr	10,000 hr	0.0001%	0.00001%
800/427	—	—	31.0	27.0
900/482	—	—	16.4	13.0
1000/538	17.9	13.5	6.1	4.5
1100/593	5.6	3.0	2.8	1.8
1200/649	4.0	2.2	1.4	0.8
1300/704	2.7	1.6	0.7	0.3
1400/760	1.8	1.1	0.3	0.1
1500/861	1.2	0.8	0.1	0.05

Alloy S44600 has good general corrosion resistance in mild atmospheric environments, fresh water, mild chemicals, and mild oxidizing conditions.

Applications have included furnace parts, kiln linings, and annealing boxes.

REFERENCES

1. CP Dillon. Corrosion Control in the Chemical Process Industry. St. Louis, Mo: Materials Technology Institute of the Chemical Process Industries, 1986.
2. CP Dillon. Corrosion Resistance of Stainless Steels, Marcel Dekker, New York, 1995.
3. PA Schweitzer. Encyclopedia of Corrosion Technology, New York: Marcel Dekker, 1998.
4. PA Schweitzer. Corrosion Resistance Tables, 4th ed., Vols. 1–3. New York: Marcel Dekker, 1995.
5. PA Schweitzer. Corrosion Resistant Piping Systems. New York: Marcel Dekker, 1994.
6. PK Whitcraft. Corrosion of stainless steels. In: Corrosion Engineering Handbook, PA Schweitzer, ed. New York: Marcel Dekker, 1996.
7. GT Murray. Introduction to Engineering Materials. New York: Marcel Dekker, 1993.

11

Superferritic Stainless Steels

Ferritic stainless alloys are noted for their ability to resist chloride stress corrosion cracking, which is one of their most useful features in terms of corrosion resistance. Consequently development efforts during the 1970s were undertaken to produce ferritic stainlesses that would also possess a high level on general and localized pitting resistance as well.

The first significant alloy developed commercially to meet these requirements contained 26% chromium and 1% molybdenum. In order to obtain the desired corrosion resistance and acceptable fabrication characteristics, the material had to have very low interstitial element contents. To achieve these levels the material was electron beam rerefined under a vacuum. It was known as E-Brite alloy. Carbon plus nitrogen contents were maintained at levels below 0.02%.

The E-Brite alloy (S44627) was termed a superferritic because of its high level of corrosion resistance for a ferritic material and partly because it is located so far into the ferritic zone on the Schaeffler diagram. For a period of years the usage of this alloy grew. Finally its benefits for the construction of pressure vessels were overshadowed by the difficult nature of fabrication and a concern over its toughness. Due to the very low level of interstitial elements the alloy had a tendency to absorb these elements during welding processes. Increases in oxygen plus nitrogen to levels much over 100 ppm resulted in poor toughness. Even without these effects, the alloy could exhibit a ductile-to-brittle transition temperature (DBIT) around room temperature. Other superferritic alloys were also developed.

The chemical composition of selected superferritic alloys are shown in Table 11.1, while Table 11.2 lists mechanical and physical properties. These alloys exhibit excellent localized corrosion resistance. Although the

TABLE 11.1   Chemical Composition of Selected Superferritic Stainless
Steels

Alloy	C	Cr	Ni	Mo	N	Other
S44627	0.002	26.0	—	1.0	0.010	
S44660	0.02	26.0	2.5	3.0	0.025	Ti + Cb 0.5
S44800	0.005	29.0	2.2	4.0	0.01	

Values are in wt%.

superferritic materials alloyed with some nickel have improved mechanical
toughness and are less sensitive to contamination from interstitial elements,
their availability is still limited to thicknesses below approximately 0.200
in. This is related to the formation of embrittling phases during cooling from
annealing temperatures. Section thicknesses over these levels cannot be
cooled quickly enough to avoid a loss of toughness.

Toughness is determined primarily by means of high impact Charpy
V-notch tests which measure the absorption of energy of a given thickness
of material at a given temperature. The values at different temperatures are
then delineated in a nil ductility transition temperature diagram (NDTT)
(Fig. 11.1). Other types of tests have also been devised to study this
phenomenon.

Nil ductility is the phenomenon in which an otherwise ductile metal
or alloy becomes brittle as the temperature decreases, as shown in Fig. 11.1.

TABLE 11.2   Mechanical and Physical Properties of Selected
Superferritic Alloys

	Alloy		
Property	S44627	S44660	S44800
Modulus of elasticity $\times 10^6$ (psi)		31.5	
Tensile strength $\times 10^3$ (psi)	75	95	90
Yield strength 0.2% offset $\times 10^3$ (psi)	60	80	75
Elongation (%)	25	30	25
Density (lb/in.^{3})	0.280	0.280	0.277
Specific gravity	7.7	7.7	7.63
Thermal conductivity (Btu/ft hr °F) at 70°F/20°C		9.2	
Rockwell hardness	B-84	B-95	

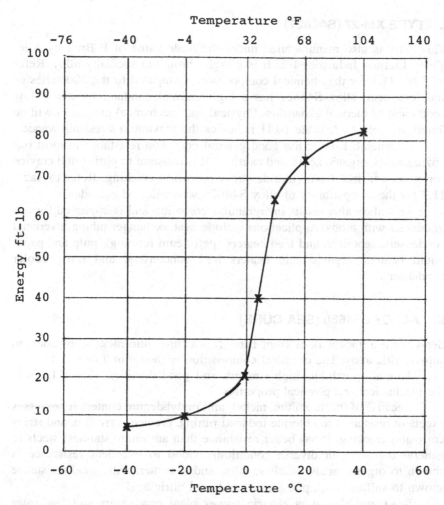

FIGURE 11.1 Nil ductility transition temperature diagram.

Steels are particularly susceptible to this phenomenon, but so are such corrosion resistant alloys as the low-interstitial ferritic stainless steels. The exact nil ductility transition temperature will vary with the thickness of the material, but is also influenced by compositional variables and heat treatment. Special impact tests have been devised to determine the NDTT for a specific set of conditions, including flaw size (which can be very critical). Note that this problem is distinct from the behavior of such brittle materials as cast iron or 14.5% silicon iron.

I. TYPE XM-27 (S44627)

This alloy is also manufactured under the trade name of E-Brite by Allegheny Ludlum Industries Inc. It is a high chromium specialty alloy. Refer
to Table 11.1 for the chemical composition. Compared to the 300 series of
stainless steel, alloy S44627 has a high thermal conductivity and a low
coefficient of thermal expansion. Physical and mechanical properties will be
found in Table 11.2. Table 10.11 indicates the maximum stress allowance.

In general, E-Brite has good general corrosion resistance in most oxidizing acids, organic acids, and caustics. It is resistant to pitting and crevice
corrosion and free from chloride stress corrosion cracking. Refer to Table
11.3 for the compatibility of alloy S44627 with selected corrodents.

This alloy also resists intergranular corrosion and is approved for use
in contact with foods. Applications include heat exchanger tubing, overhead
condensers, reboilers, and feed heaters (petroleum refining), pulp and paper
liquid heaters, organic acid heaters and condensers, and nitric cooler
condensers.

II. ALLOY S44660 (SEA-CURE)

Sea-Cure is a trademark of Trent Tube. It is a chromium-nickel-molybdenum
superferritic alloy. The chemical composition is shown in Table 11.1.

This alloy exhibits high strength and good ductility. Table 11.2 lists
the mechanical and physical properties.

Because of its chromium, nickel, and molybdenum content it possesses
excellent resistance to chloride-induced pitting, crevice corrosion, and stress
corrosion cracking. It has better resistance than austenitic stainless steels to
general corrosion in diverse conditions. Good to excellent resistance is
shown to organic acids, alkalies, salts, and seawater, with good resistance
shown to sulfuric acid, phosphoric acid, and nitric acid.

Sea-Cure is used in electric power plant condensers and feedwater
heaters and heat exchangers in chemical, petrochemical, and refining
applications.

III. ALLOY S44735 (29-4C)

The chemical composition of alloy 29-4C is shown in Table 11.4. This alloy
has improved general corrosion resistance and improved resistance to chloride pitting and stress corrosion cracking in some environments. The absence
of nickel reduces the cost.

Applications are found in the utility industry, chemical processing
equipment, household condensing furnaces, and vent pipes.

TABLE 11.3 Compatibility of E-Brite Alloy S44627 with Selected Corrodents

Chemical	°F/°C	Chemical	°F/°C
Acetic acid, 10%	200/93	Methylene chloride	X
Acetic acid, 20%	200/93	Nitric acid, 5%*	310/154
Acetic acid, 50%	200/93	Nitric acid, 10%*	310/154
Acetic acid, 80%	130/54	Nitric acid, 20%*	320/160
Acetic acid, glacial	140/60	Nitric acid, 30%*	320/160
Acetic anhydride*	300/149	Nitric acid, 40%*	200/93
Ammonium chloride, 10%*	200/93	Nitric acid, 50%*	200/93
Aqua regia, 3:1	X	Nitric acid, 70%*	200/93
Beer	160/71	Oxalic acid, 10%	X
Beet sugar liquors	120/49	Phosphoric acid, 25–50%*	210/99
Benzaldehyde*	210/99	Sodium chlorite	90/32
Bromine water, 1%	80/27	Sodium hydroxide, 10%	200/93
Calcium hydroxide, 50%*	210/99	Sodium hydroxide, 15%	200/93
Chromic acid, 10%	130/54	Sodium hydroxide, 30%	200/93
Chromic acid, 30%	90/32	Sodium hydroxide, 50%	180/82
Chromic acid, 40%	80/27	Sodium hypochlorite, 30%*	90/32
Chromic acid, 50%	X	Stearic acid	210/99
Citric acid, 10%	200/93	Sulfamic acid	100/38
Citric acid, 25%	210/99	Sulfur dioxide, wet	550/293
Copper chloride, 5%	100/38	Sulfuric acid, 10%	X
Ethylene chloride*	210/99	Sulfuric acid, 30–90%	X
Ferric chloride	80/27	Sulfuric acid, 95%	150/66
Fluosilicic acid	X	Sulfuric acid, 98%	280/138
Formic acid, 80%	210/99	Sulfurous acid, 5%*	210/99
Hydrochloric acid	X	Tartaric acid, 50%	210/99
Lactic acid, 80%	200/93	Toluene	210/99

The chemicals listed are in the pure state or in a saturated solution unless otherwise indicated. Compatibility is shown to the maximum allowable temperature for which data are available. Incompatibility is shown by an X. When compatible, corrosion rate is <2 mpy except for those marked with an *, whose corrosion rate is <20 mpy.
Source: Ref. 1.

IV. ALLOY S44800 (29-4-2)

The chemical composition of alloy 29-4-2 is shown in Table 11.1, and the mechanical properties will be found in Table 11.2. Applications are found in chemical processing equipment and the utility industry for use in corrosive environments.

TABLE 11.4   Chemical Composition of Alloy S44735
(29-4C)

Chemical	Weight percent
Carbon	0.03 max.
Manganese	0.30 max.
Silicon	1.0 max.
Chromium	28.0–30.0
Nickel	1.0
Phosphorus	0.03
Molybdenum	3.60–4.20
Titanium + niobium	6 (%C + %N): 0.045 N min.
Iron	Balance

This alloy has improved resistance to chloride pitting and stress corrosion cracking and improved general corrosion resistance in some environments.

V.   ALLOY S44700 (29-4)

This is a chromium-nickel-molybdenum alloy, with its composition shown in Table 11.5. It has excellent resistance to chloride pitting and stress corrosion cracking environments. It finds use in the chemical processing and utility industries.

TABLE 11.5   Chemical Composition of
Alloy S44700 (29-4)

Chemical	Weight percent
Carbon	0.010 max.
Manganese	0.30 max.
Chromium	28.0–30.0
Nickel	0.15
Molybdenum	3.50–4.20
Silicon	0.02 max.
Copper	0.15
Nitrogen	0.02
Iron	Balance

REFERENCES

1. PA Schweitzer. Corrosion Resistance Tables, 4th ed., Vols. 1–3, New York: Marcel Dekker, 1995.
2. PK Whitcraft. Corrosion of stainless steels. In: PA Schweitzer, ed. Corrosion Engineering Handbook. New York: Marcel Dekker, 1996.
3. CP Dillon. Corrosion Control in the Chemical Process Industries, 2nd ed. St. Louis, Mo, Materials Technology Institute of the Chemical Process Industries, 1994.

12
Precipitation Hardening Stainless Steels

This family of stainless alloys utilizes a thermal treatment to intentionally precipitate phases which cause a strengthening of the alloy. The principle of precipitation hardening is that a supercooled solid solution (solution annealed) changes its metallurgical structure on aging. The principal advantage is that products can be fabricated in the annealed condition and then strengthened by a relatively low temperature (900–1500°F/462–620°C) treatment, minimizing the problems associated with high temperature treatments. Strength levels of up to 260 ksi (tensile) can be achieved—exceeding even those of the martensitic stainless steels—while corrosion resistance is usually superior—approaching that of type 304 stainless steel. Ductility is similar to corresponding martensitic grades at the same strength level.

The precipitating phase is generated through an alloy addition of one or more of the following: niobium, titanium, copper, molybdenum, or aluminum. The metallurgy is such that the material can be solution treated, i.e., all alloying elements are in solid solution and the material is in its softest or annealed state. In this condition the material can be machined, formed, and welded to desired configuration. After fabrication the unit is exposed to an elevated temperature cycle (aging) which precipitates the desired phases to cause an increase in mechanical properties.

Precipitation hardening stainless steels have high strength and relatively good ductility and corrosion resistance at high temperatures. These steels can attain very high strength levels. They reach these high strengths by precipitation of intermetallic compounds via the same mechanism as that found in aluminum alloys. These compounds are usually formed from iron or nickel with titanium, aluminum, molybdenum, and copper. Typical compounds are Ni_3Al, Ni_3Ti, and Ni_3Mo. Chromium contents are in the range

209

of 13 to 17%. These steels have been around for several decades but are now being recognized as a real alternative to the other stainless steels. They have the good characteristics of the austenitic steels plus strength approaching that of the martensitic steels. One of the early problems centered around forging difficulties, but these problems have been overcome to some extent.

Precipitation hardenable (PH) stainless steels are themselves divided into three alloy types: martensitic, austenitic, and semicaustenitic. An illustration of the relationship between these alloys is shown in Fig. 12.1. The martensitic and austenitic PH stainless steels are directly hardened by thermal treatment. The semiaustenitic stainless steels are supplied as an unstable austenitic, which is the workable condition, and must be transformed to martensite before aging.

On average the general corrosion resistance is below that of type 304 stainless. However, the corrosion resistance of type PH 15-7 Mo alloy approaches that of type 316 stainless. The martensitic and semiaustenitic

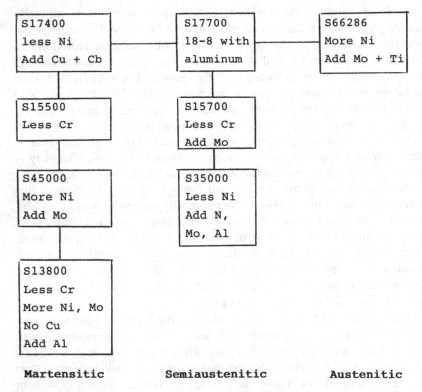

FIGURE 12.1 Precipitation hardening stainless steels.

grades are resistant to chloride stress cracking. These materials are susceptible to hydrogen embrittlement.

The PH steels find a myriad of uses in small forged parts and even in larger support members in aircraft designs. They have been considered for landing gears. Many golf club heads are made from these steels by investment casting techniques, and the manufacturers proudly advertise these clubs as being made from 17-4 stainless steel. Applications also include fuel tanks, landing gear covers, pump parts, shafting bolts, saws, knives, and flexible bellows type expansion joints.

I. PH 13-8 Mo (S13800)

PH 13-8Mo is a registered trademark of Armco Inc. It is a martensitic precipitation/age hardening stainless steel capable of high strength and hardness along with good levels of resistance to both general corrosion and stress corrosion cracking. The chemical composition is shown in Table 12.1.

Generally this alloy should be considered where high strength, toughness, corrosion resistance, and resistance to stress corrosion cracking are required in a stainless steel showing minimal directionality in properties. Mechanical and physical properties will be found in Table 12.2.

II. ALLOY 15-5PH (S15500)

Alloy 15-5PH, a martensitic precipitation hardening stainless steel, is a trademark of Armco Inc. It provides a combination of high strength, good

TABLE 12.1 Chemical
Composition of Alloy PH-13-8Mo
(S13800)

Chemical	Weight percent
Carbon	0.05
Manganese	0.10
Phosphorus	0.010
Sulfur	0.008
Silicon	0.10
Chromium	12.5–13.25
Nickel	7.5–8.50
Molybdenum	2.00–2.50
Aluminum	0.90–1.35
Nitrogen	0.010
Iron	Balance

TABLE 12.2 Mechanical and Physical Properties of Alloy
PH-13-8Mo (S13800)

Tensile strength × 10³ (psi)	160
Yield strength 0.2% offset × 10³ (psi)	120
Elongation in 2 in. (%)	17
Rockwell hardness	C-33
Density (lb/in.³)	0.28
Specific gravity	7.7
Specific heat (J kg K)	
at 212°F (100°C)	8.1
at 932°F (500°C)	12.7
Mean coefficient of thermal expansion × 10⁻⁶ (in./in. °F)	
at 32–212°F	5.9
at 32–600°F	6.2
at 32–1000°F	6.6

corrosion resistance, good mechanical properties at temperatures up to 600°F
(316°C) and good toughness in both the longitudinal and transverse directions in both the base metal and welds. Short-time, low-temperature heat
treatments minimize distortion and scaling. The chemical composition is
shown in Table 12.3.

As supplied from the mill in condition A, 15-5PH stainless steel can
be heat treated at a variety of temperatures to develop a wide range of
properties. For condition A the metal is solution treated to 1900 ± 25°F

TABLE 12.3 Chemical Composition of Alloy
15-5PH (S15500)

Chemical	Weight percent
Carbon	0.07 max.
Manganese	1.00 max.
Phosphorus	0.04 max.
Sulfur	0.03 max.
Silicon	1.00 max.
Chromium	14.0–15.50
Nickel	3.50–5.50
Copper	2.50–4.50
Columbium + tantalum	0.15–0.45
Iron	Balance

(1038 ± 14°C) and air cooled below 90°F (37°C). Eight standard heat treatments have been developed for the material. Table 12.4 outlines the times and temperatures required.

Alloy 15-5PH in condition A exhibits useful mechanical properties. Tests at Kure Beach, NC, show excellent stress corrosion resistance after 14 years of exposure. Condition A material has been used successfully in numerous applications. The hardness and tensile properties fall within the range of those for conditions H1100 and H1150.

However, in critical applications, alloy 15-5PH should be used in the precipitation-hardened condition rather than in condition A. Heat treating to the hardened condition, especially at the higher end of the temperature range, stress relieves the structure and may provide more reliable resistance to stress corrosion cracking than condition A. Refer to Table 12.5 for the mechanical and physical properties of alloy 15-5PH in various conditions.

The general level of corrosion resistance of alloy 15-5PH exceeds that of types 410 and 431, and is approximately equal to that of alloy 17-4PH. Very little rusting is experienced when exposed to 5% salt fog at 95°F (35°C) for a period of 500 hr. When exposed to seacoast atmospheres rust gradually develops. This is similar to other precipitation hardening stainless steels. The general level of corrosion resistance of alloy 15-5PH stainless steel is best in the fully hardened condition, and decreases slightly as the aging temperature is increased.

III. ALLOY 17-4PH (S17400)

Alloy 17-4PH is a trademark of Armco Inc. It is a martensitic-hardening stainless steel that has a combination of high strength, good corrosion resistance, good mechanical properties at temperatures up to 600°F (316°C), good toughness in both base metal and welds, and short-time, low-temperature heat treatments that minimize warpage and scaling. The chemical composition will be found in Table 12.6.

As supplied from the mill in condition A, 17-4PH stainless steel can be heat treated at a variety of temperatures to develop a wide range of properties. Condition A material has been solution treated at 1900 ± 25°F (1038 ± 14°C) and air cooled below 90°F (32°C).

Alloy 17-4PH stainless steel exhibits useful mechanical properties in condition A. Excellent stress corrosion resistance has been exhibited by this alloy after 14 years exposure at Kure Beach, NC. Condition A material has been used successfully in numerous applications. The hardness and tensile properties fall within the range of those for conditions H1100 and H1150.

However, in critical applications alloy 17-4PH stainless steel should be used in the precipitation-hardened condition, rather than in condition A.

TABLE 12.4   Heat Treatments for Alloy 15-5PH (S15500)

Condition	Heat to ±15°F/8.4°C (°F/°C)	Time at temperature (hr)	Type of cooling
H900	900/482	1	Air
H925	925/496	4	Air
H1025	1025/551	4	Air
H1075	1075/580	4	Air
H1100	1100/593	4	Air
H1150	1150/621	4	Air
H1150 + 1150	1150/621	4 followed by 4	Air
H1150M	1400/760	2 followed by 4	Air
	1150/621		Air

Heat treating to the hardened condition, especially at the higher end of the temperature range, stress relieves the structure and may provide more reliable resistance to stress corrosion cracking than in condition A. The heat treatments for alloy 17-4PH are shown in Table 12.7.

Alloy 17-4PH stainless steel has excellent mechanical properties. This material is recommended for applications requiring high strength and hardness as well as corrosion resistance. Refer to Table 12.8 for the mechanical and physical properties.

After being exposed to elevated temperatures (750°F (399°C) for an extended period of time and tested at room temperature after exposure, a slight increase in strength and a slight loss of toughness can be detected. However, the properties of condition H1150 can be restored by heat treating at 1150°F (621°C) for 4 hr after original exposure. By taking advantage of this reaging treatment, the service life of parts exposed at elevated temperatures can be extended indefinitely.

Alloy 17-4PH stainless steel has excellent corrosion resistance. It withstands attack better than any of the standard hardenable stainless steels and is comparable to type 304 in most media. It is equivalent to type 304 when exposed in rural or mild industrial atmospheres. However, when exposed in a seacoast atmosphere it will gradually develop overall light rusting and pitting in all heat treated conditions.

This alloy is suitable for use in pump and motor shafting provided it is operated continuously. As with other stainless steels, crevice attack will occur when exposed to stagnant seawater for any length of time.

Table 12.9 shows the compatibility of alloy 17-4PH with selected corrodents. A more comprehensive listing will be found in Ref. 1.

TABLE 12.5 Mechanical and Physical Properties of Alloy 15-5PH Stainless Steel

Property		Condition					
	A	H900	H925	H1025	H1075	H1150M	H1150
Modulus of elasticity in torsion × 10^6 (psi)		11.2		11.0	10.0		10.0
Tensile stress × 10^3 (psi)	185	209	181	174	162	136	150
Yield stress 0.2% offset × 10^3 (psi)	160	201	175	171	160	111	140
Elongation in 2 in. (%)	8.4	10.1	12.2	12.2	12.8	18.8	14.6
Rockwell hardness	C-35	C-46	C-41	C-40	C-38	C-31	C-36
Impact resistance (in.–lbs/in.^{2})	3265	2857		3974		5616	4626
Density (lb/in.^{3})	0.28	0.282			0.283		0.284
Coefficient of thermal expansion × 10^{-6} (in./in./°F)							
at −100–70°F		5.8	6.3				6.1
at 70–200°F	6.0	6.0	6.5				6.6
at 70–400°F	6.0	6.0	6.6				6.9
at 70–600°F	6.2	6.3	6.8				7.1
at 70–800°F	6.3	6.5					7.2
at 70–900°F							7.3

TABLE 12.6   Chemical Composition of Alloy
17-4PH (S17400)

Chemical	Weight percent
Carbon	0.07 max.
Manganese	1.00 max.
Phosphorus	0.04 max.
Sulfur	0.03 max.
Silicon	1.00 max.
Chromium	15.00–17.50
Nickel	3.00–5.00
Copper	3.00–5.00
Columbium + tantalum	0.15–0.45
Iron	Balance

IV. ALLOY 17-7PH (S17400)

This is a semiaustenitic stainless steel. In the annealed or solution annealed condition it is austenitic (nonmagnetic), and in the aged or cold worked condition it is martensitic (magnetic). The chemical composition is shown in Table 12.10.

The alloy exhibits high strength in all conditions. Refer to Table 12.11 for the mechanical and physical properties. Service over 1050°F (565°C)

TABLE 12.7   Heat Treatments for Alloy 17-4PH

Condition	Heat to ±15°F/8.4°C (°F/°C)	Time at temperature (hr)	Type of cooling
H900	900/482	1	Air
H925	925/496	4	Air
H1025	1026/551	4	Air
H1075	1075/580	4	Air
H1100	1100/593	4	Air
H1150	1150/621	4	Air
H1150 + 1150	1150/621	4 followed by 4	Air
	1150/621		Air
H1150M	1400/760	2 followed by 4	Air
	1150/621		Air

TABLE 12.8 Mechanical and Physical Properties of Alloy 17-4PH (S17400)

Property	Condition						
	A	H900	H925	H1025	H1075	H1150	H1150M
Tensile stress × 10³ (psi)	160	210	200	185	175	160	150
Yield strength 0.2% offset × 10³ (psi)	145	200	195	170	165	150	130
Elongation in 2 in. (%)	5.7	7.0	8.0	8.0	8.0	11.0	12.0
Rockwell hardness	C-35	C-45	C-43	C-38	C-37	C-35	C-33
Density (lb/in.³)	0.280	0.282			0.283	0.284	
Coefficient of thermal expansion × 10⁻⁶ (in./in./°F)							
at 70–200°F	6.0	6.0			6.3	6.6	
at 70–600°F	6.2	6.3			6.6	7.1	
at 70–800°F	6.3	6.5			6.8	7.2	
Specific heat (Btu/lb °F) at 32–212°F	0.11	0.11					

TABLE 12.9   Compatibility of 17-4PH Stainless Steel with Selected Corrodents

Chemical	°F/°C	Chemical	°F/°C
Acetic acid, 20%	200/93	Calcium chloride	110/43
Acetic acid, glacial	X	Calcium hypochlorite	X
Acetyl chloride	110/43	Calcium sulfate	150/54
Acetylene	110/43	Carbon dioxide, dry	210/99
Allyl alcohol	90/32	Carbon dioxide, wet	210/99
Aluminum fluoride	X	Carbon monoxide	230/110
Aluminum hydroxide	80/27	Carbon tetrachloride	150/66
Aluminum nitrate	110/43	Chloric acid, 20%	X
Aluminum potassium sulfate	X	Chlorine liquid	X
Aluminum sulfate	X	Chlorosulfonic acid	X
Ammonia, anhydrous	270/132	Chromic acid, 10%	X
Ammonium bifluoride	X	Chromic acid, 30%	X
Ammonium carbonate	110/43	Chromic acid, 40%	X
Ammonium chloride	X	Chromic acid, 50%	X
Ammonium hydroxide, 10%	210/99	Ethyl alcohol	170/77
Ammonium nitrate	130/54	Ethyl chloride, dry	210/99
Ammonium persulfate	130/54	Ferric nitrate	150/66
Amyl acetate	90/32	Ferrous chloride	X
Amyl alcohol	90/32	Fluorine gas, dry	230/110
Amyl chloride	90/32	Formic acid, 10%	180/82
Aniline	170/71	Heptane	130/54
Aniline hydrochloride	X	Hydrobromic acid	X
Antimony trichloride	X	Hydrochloric acid	X
Argon	210/99	Hydrocyanic acid	X
Arsenic acid	130/54	Hydrogen sulfide, wet	X
Barium hydroxide	110/43	Iodine	X
Barium sulfate	130/54	Magnesium chloride	X
Beer	110/43	Magnesium hydroxide	140/66
Beet sugar liquors	110/43	Magnesium nitrate	130/54
Benzene	130/54	Magnesium sulfate	130/54
Benzene sulfonic acid	X	Methylene chloride	130/54
Benzoic acid	150/66	Phenol	130/54
Benzyl alcohol	110/43	Phosphoric acid, 5%	200/93
Boric acid	110/43	Phosphoric acid, 10%	200/93
Bromine gas, dry	X	Phosphoric acid, 25–50%	200/93
Bromine gas, moist	X	Phosphoric acid, 70%	X
Bromine liquid	X	Phthalic acid	270/132
Butyl cellosolve	140/66		

The chemicals listed are in the pure state or in a saturated solution unless otherwise indicated. Compatibility is shown to the maximum allowable temperature for which data are available. Incompatibility is shown by an X. When compatible, the corrosion rate is less than 20 mpy.
Source: Ref. 1.

TABLE 12.10 Chemical Composition
of Alloy 17-7PH (S17700)

Chemical	Weight percent
Carbon	0.09 max.
Aluminum	0.75–1.5
Chromium	16.0–18.0
Nickel	6.5–7.75
Iron	Balance

will cause overaging. Overaging may occur at lower temperatures depending on tempering temperature selected.

In the aged condition the alloy is resistant to chloride cracking. Its corrosion resistance in general is on a par with that of type 304 stainless steel.

V. ALLOY 350 (S35000)

This is a chromium-nickel-molybdenum stainless alloy hardenable by martensitic transformation and precipitation hardening. The chemical composition is shown in Table 12.12.

Alloy 350 normally contains 5–10% delta ferrite, which aids weldability. When heat treated it has high strength. However, to achieve optimum properties a complex heat treatment is required including two subzero (−100°F/−73°C) exposures. Unless cooled to subzero temperatures prior to aging the alloy may be subject to intergranular attack. Mechanical and physical properties are shown in Table 12.13.

In general the corrosion resistance of alloy 350 is similar to that of type 304 stainless steel.

This alloy is used where high strength and corrosion resistance at room temperatures are essential.

VI. ALLOY 355 (S35500)

Alloy 355 is a chromium-nickel-molybdenum stainless alloy hardenable by martensitic transformation and precipitation hardening. The chemical composition is shown in Table 12.14.

Depending on the heat treatment the alloy may be austenitic with formability similar to other austenitic stainless steels. Other heat treatments yield a martensitic structure with high strength. Table 12.15 lists the mechanical and physical properties.

TABLE 12.11 Mechanical and Physical Properties of 17-7PH Stainless Steel

Modulus of elasticity $\times 10^6$ (psi)	
annealed	30.5
aged	32.5
Tensile strength $\times 10^3$ (psi)	
annealed	133
aged	210
Yield strength 0.2% offset $\times 10^3$ (psi)	
annealed	42
aged	190
Elongation in 2 in. (%)	
annealed	19
Rockwell hardness	
annealed	B-85
aged	C-48
Density (lb/in.^{3})	0.282
Thermal conductivity (Btu/ft hr °F)	
at 70°F (20°C)	9.75
at 1500°F (815°C)	12.2

TABLE 12.12 Chemical Composition of Alloy 350 (S35000)

Chemical	Weight percent
Carbon	0.07–0.11
Manganese	0.50–1.25
Phosphorus	0.04
Sulfur	0.03
Silicon	0.50
Chromium	16.00–17.00
Nickel	4.00–5.00
Molybdenum	2.50–3.25
Nitrogen	0.07–0.13
Iron	Balance

TABLE 12.13 Mechanical and Physical Properties of Alloy 350 (S35000)

Modulus of elasticity $\times 10^6$ (psi)	
aged	29.4
Tensile strength $\times 10^3$ (psi)	
annealed	160
aged	200
Yield strength 0.2% offset $\times 10^3$ (psi)	
annealed	60
aged	85
Elongation (%)	
annealed	30
aged	12
Rockwell hardness	
annealed	B-95
aged	C-30
Density (lb/in.^{3})	0.286
Thermal conductivity (Btu/ft-hr °F)	
at 70°F (20°C)	8.4
at 1500°F (815°C)	12.2

TABLE 12.14 Chemical Composition of Alloy 355 (S35500)

Chemical	Weight percent
Carbon	0.10–0.15
Manganese	0.50–1.25
Phosphorus	0.04
Sulfur	0.03
Silicon	0.05
Chromium	15.00–16.00
Nickel	4.00–5.00
Molybdenum	2.50–3.25
Nitrogen	0.07–0.13
Iron	Balance

TABLE 12.15 Mechanical and Physical Properties
of Alloy 355 (S35500)

Modulus of elasticity × 10⁶ (psi)	
aged	29.4
Tensile strength × 10³ (psi)	
annealed	182
aged	220
Yield strength 0.2% offset × 10³ (psi)	
annealed	167
aged	185
Elongation (%)	
annealed	16
aged	12
Rockwell hardness	
annealed	C-40
aged	C-48
Density (lb/in.³)	0.286
Thermal conductivity (Btu/ft-hr °F)	
at 70°F (20°C)	8.75
at 1500°F (815°C)	12.0

The alloy exhibits better corrosion resistance than other quench hardenable martensitic stainless steels. Service over 1000°F (538°C) will cause
overaging. Overaging may occur at lower temperatures depending on the
tempering temperature selected. Overaged material is susceptible to intergranular corrosion. A subzero treatment during heat treatment removes this
susceptibility.

Alloy 355 finds application where high strength is required at intermediate temperatures.

VII. CUSTOM 450 (S45000)

Custom 450 is a trademark of Carpenter Technology Corp. It is a martensitic
age-hardenable stainless steel with very good corrosion resistance and moderate strength. Table 12.16 contains its chemical composition.

The alloy has high strength, good ductility and toughness, and is easily
fabricated. Refer to Table 12.17 for the mechanical and physical properties.
Unlike alloy 17-4, Custom 450 can be used in the solution annealed
condition.

TABLE 12.16 Chemical Composition of
Custom 450 (S45000)

Chemical	Weight percent
Carbon	0.05
Manganese	2.00
Phosphorus	0.03
Sulfur	0.03
Silicon	1.00
Chromium	14.00–16.00
Nickel	5.00–7.00
Molybdenum	0.50–1.00
Copper	1.25–1.75
Columbium	8 × %C min.
Iron	Balance

The corrosion resistance of Custom 450 stainless is similar to that of type 304 stainless steel. Custom 450 alloy is used in applications where type 304 is not strong enough or type 410 is insufficiently corrosion resistant.

VIII. CUSTOM 455 (S45500)

Custom 455 is a registered trademark of Carpenter Technology Corp. It is a martensitic, age-hardenable stainless steel which is relatively soft and formable in the annealed condition. A single-step aging treatment develops exceptionally high yield strength with good ductility and toughness. It has the highest strength and highest hardness capability (approximately Rockwell C-50). The chemical composition is shown on Table 12.18.

Custom 455 exhibits high strength with corrosion resistance better than type 410 and approaching type 430. Table 12.19 provides the mechanical and physical properties of Custom 455 stainless steel. Service over 1050°F (565°C) will cause overaging. Overaging may occur at lower temperatures depending on the tempering temperature selected.

The alloy may be susceptible to hydrogen embrittlement under some conditions. Custom 455 stainless steel should be considered when ease of fabrication, high strength, and corrosion resistance are required. Custom 455 alloy is suitable to be used in contact with nitric acid and alkalies. It also resists chloride stress corrosion cracking. Materials such as sulfuric acid, phosphoric acid, hydrochloric acid, hydrofluoric acid, and seawater will attack Custom 455.

TABLE 12.17 Mechanical and Physical Properties of Custom 450 Stainless Steel

Modulus of elasticity $\times 10^3$ (psi)	
annealed	28
aged	29
Tensile strength $\times 10^3$ (psi)	
annealed	142
aged	196
Yield strength 0.2% offset $\times 10^3$ (psi)	
annealed	118
aged	188
Elongation in 2 in. (%)	
annealed	13
aged	14
Rockwell hardness	
annealed	C-28
aged	C-42.5
Density (lb/in.^{3})	0.28

TABLE 12.18 Chemical Composition of Custom 455 (S45500)

Chemical	Weight percent
Carbon	0.05
Manganese	0.50
Phosphorus	0.040
Sulfur	0.030
Silicon	0.50
Chromium	11.00–12.50
Nickel	7.50–9.50
Titanium	0.80–1.40
Columbium + tantalum	0.10–0.50
Copper	1.50–2.50
Molybdenum	0.50
Iron	Balance

TABLE 12.19 Mechanical and Physical Properties of Custom 455 Stainless Steel

Modulus of elasticity × 10⁶ (psi)	
aged	29
Tensile strength × 10³ (psi)	
annealed	140
aged	230
Yield strength 0.2% offset × 10³ (psi)	
annealed	115
aged	220
Elongation in 2 in. (%)	
annealed	12
aged	10
Rockwell hardness	
annealed	C-31
aged	C-48
Density (lb/in.³)	0.28
Thermal conductivity (Btu/ft hr °F)	
at 70°C (20°C)	10.4
at 1500°F (815°C)	14.3

IX. ALLOY 718 (N07718)

Alloy 718 is a precipitation-hardened, nickel-base alloy designed to display exceptionally high yield, tensile, and creep rupture properties up to 1300°F (704°C). It can also be used as low as −423°F (−253°C). Table 12.20 shows the chemical composition.

The alloy is readily fabricated and has excellent resistance to postweld cracking. Physical and mechanical properties will be found in Table 12.21.

Excellent oxidation resistance is displayed up to 1800°F (952°C). Alloy 718 is resistant to sulfuric acid, organic acids, and alkalies. It is also resistant to chloride stress corrosion cracking. Hydrochloric, hydrofluoric, phosphoric, and nitric acids will attack the alloy as well as seawater.

This alloy has been used for jet engines and high speed airframe parts such as wheels, buckets, and spacers and high temperature bolts and fasteners.

X. ALLOY A286 (S66286)

Alloy A286 is an austenitic precipitation hardenable stainless steel. Its chemical composition will be found in Table 12.22. The mechanical properties

TABLE 12.20 Chemical Composition of Alloy
718 (N07718)

Chemical	Weight percent
Carbon	0.10
Manganese	0.35
Silicon	0.35
Phosphorus	0.015
Sulfur	0.015
Chromium	17.00–21.00
Nickel + cobalt	50.00–55.00
Molybdenum	2.80–3.30
Columbium + tantalum	4.75–5.50
Titanium	0.65–1.15
Aluminum	0.35–0.85
Boron	0.001–0.006
Copper	0.015
Iron	Balance

TABLE 12.21 Mechanical and Physical Properties of
Alloy 718 (N07718)

Modulus of elasticity $\times 10^6$ (psi)	
aged	29.0
Tensile strength $\times 10^3$ (psi)	
annealed	140
aged	180
Yield strength 0.2% offset $\times 10^3$ (psi)	
annealed	115
aged	220
Elongation (%)	
annealed	12
aged	10
Rockwell hardness	
annealed	C-31
aged	C-48
Density (lb/in.^{3})	0.296
Thermal conductivity (Btu/ft-hr °F)	
at 70°F (20°C)	6.6
at 1500°F (815°C)	13.9

TABLE 12.22 Chemical Composition of
Alloy A286 (S66286)

Chemical	Weight percent
Carbon	0.08
Manganese	2.00
Silicon	1.00
Chromium	13.50–16.00
Nickel	24.00–27.00
Molybdenum	1.00–2.30
Titanium	1.90–2.30
Vanadium	0.10–0.50
Aluminum	0.35
Boron	0.003–0.010
Iron	Balance

of alloy 286 are retained at temperatures up to 1300°F/704°C, having high strength, a notched rupture strength superior to any other alloy with comparable high temperature properties, and a high ductility in notched specimens. The physical and mechanical properties are given in Table 12.23. The alloy is nonmagnetic.

Alloy A286 has excellent resistance to sulfuric and phosphoric acids and good resistance to nitric acid and organic acids. It is also satisfactory for use with salts, seawater, and alkalies.

TABLE 12.23 Mechanical and Physical Properties of
Alloy A286 (S66286)

Modulus of elasticity $\times 10^6$ (psi)	28.8
Tensile strength $\times 10^3$ (psi)	130
Yield strength 0.2% offset $\times 10^3$ (psi)	85
Elongation (%)	15
Rockwell hardness	C-25
Density (lb/in.^{3})	0.286
Thermal conductivity (Btu/ft-hr °F)	
at 70°F (20°C)	8.7
at 1500°F (815°C)	13.8

The alloy has been used for gas turbine components and applications requiring high strength and corrosion resistance.

XI. ALLOY X-750 (N07750)

This is a precipitation hardening alloy highly resistant to chemical corrosion and oxidation. The chemical composition is shown in Table 12.24. Alloy NO7750 exhibits excellent properties down to cryogenic temperatures and good corrosion and oxidation resistance up to 1300°F (704°C). When exposed to temperatures above 1300°F (704°C) overaging results with a loss of strength. It also has excellent relaxation resistance. Table 12.25 shows the mechanical and physical properties.

Alloy X-750 is resistant to sulfuric, hydrochloric, phosphoric, and organic acids; alkalies; salts; and seawater. It is also resistant to chloride stress corrosion cracking. Hydrofluoric and nitric acids will attack the alloy.

The alloy finds applications where strength and corrosion resistance are important, for example, as high temperature structural members for jet engine parts, heat-treating fixtures, and forming tools.

XII. PYROMET ALLOY 31

Pyromet Alloy 31 is a trademark of Carpenter Technology. It is a precipitation-hardenable superalloy which exhibits corrosion resistance and strength to 1500°F (816°C). It is resistant to sour brines and hot sulfidation attack.

TABLE 12.24 Chemical Composition of Alloy X-750 (N07750)

Chemical	Weight percent
Carbon	0.08
Nickel + columbium	70.00
Chromium	14.00–17.00
Manganese	0.30
Sulfur	0.010
Silicon	0.50
Copper	0.05
Columbium + tantalum	0.70–1.20
Titanium	2.25–2.70
Aluminum	0.40–1.00
Iron	5.0–9.0

Applications include hardware in coal gasification units. It has a chemical composition as follows:

Chemical	Weight percent
Carbon	0.04
Manganese	0.20
Silicon	0.20
Phosphorus	0.015
Sulfur	0.015
Chromium	27.7
Nickel	55.5
Molybdenum	2.0
Titanium	2.5
Aluminum	1.5
Columbium	1.1
Boron	0.005
Iron	Balance

XIII. ALLOY CTX-1

Pyromet Alloy CTX-1 is a trademark of Carpenter Technology. The alloy is a high-strength, precipitation-hardening superalloy having a low coefficient of expansion with high strength at temperatures to 1200°F (649°C). Applications include gas turbine engine components and hot-work dies.

If exposed to atmospheric conditions above 1000°F (538°C) a protective coating must be applied to the alloy. The chemical composition is as follows:

TABLE 12.25 Mechanical and Physical Properties of Alloy X-750 (N07750)

Property	Aged	Annealed
Modulus of elasticity × 10⁶ (psi)	31.0	
Tensile strength × 10³ (psi)	165	130
Yield strength 0.2% offset × 10³ (psi)	105	60
Elongation (%)	20	40
Rockwell hardness	C-32	
Density (lb/in.³)	0.299	0.299
Thermal conductivity (Btu/ft-lb hr °F)		
at 70°F (20°C)	6.9	6.9
at 1500°F (815°C)	13.2	13.2

Chemical	Weight percent
Carbon	0.05
Manganese	0.50
Silicon	0.50
Phosphorus	0.015
Sulfur	0.015
Chromium	0.50
Molybdenum	0.20
Copper	0.50
Nickel	38.00–40.00
Columbium and tantalum	2.50–3.50
Titanium	1.25–1.75
Aluminum	0.70–1.20
Boron	0.0075
Cobalt	14.00–16.00
Iron	Balance

XIV. PYROMET ALLOY CTX-3

This is a low-expansion, high-strength, precipitation-hardenable superalloy. It has significant improvement in notched stress rupture strength over Pyromet CTX-1. As with Alloy CTX-1 a protective coating must be applied if the alloy is to be exposed at atmospheric conditions above 1000°F (538°C). Applications include gas turbine components. It has the following chemical composition:

Chemical	Weight percent
Carbon	0.05
Manganese	0.50
Silicon	0.50
Phosphorus	0.015
Sulfur	0.015
Chromium	0.50
Nickel	37.00–39.00
Copper	0.50
Cobalt	13.00–15.00
Columbium and tantalum	4.50–5.50
Titanium	1.25–1.75
Aluminum	0.25
Boron	0.012
Iron	Balance

XV. PYROMET ALLOY CTX-909

Alloy CTX-909 is a high-strength, precipitation-hardenable superalloy which offers significant improvements over Alloys CTX-1 and CTX-3 due to its combination of tensile properties and stress rupture strength to 1200°F (649°C) in the recrystallized condition combined with the use of common age-hardening treatments.

The alloy exhibits a low and relatively constant coefficient of thermal expansion over a broad temperature range, a high hot hardness, and good thermal fatigue resistance. As with the other CTX alloys a protective coating is required if the alloy is exposed to atmospheric conditions above 1000°F (538°C).

The chemical composition is as follows:

Chemical	Weight percent
Carbon	0.06
Manganese	0.50
Silicon	0.40 nom.
Phosphorus	0.015
Sulfur	0.015
Chromium	0.50
Nickel	38.00 nom.
Cobalt	14.00 nom.
Titanium	1.60 nom.
Columbium + tantalum	4.90 nom.
Aluminum	0.15
Copper	0.50
Boron	0.012
Iron	Balance

XVI. PYROMET ALLOY V-57

This an iron-base, austenitic, precipitation-hardening alloy for parts requiring high strength and good corrosion resistance at operating temperatures to 1400°F (760°C). It is produced by Carpenter Technology. Chemically it has the following composition:

Chemical	Weight percent
Carbon	0.08
Manganese	0.35
Silicon	0.50
Phosphorus	0.015
Sulfur	0.015
Chromium	13.50–16.00
Nickel	22.50–28.50
Molybdenum	1.00–1.50
Titanium	2.70–3.20
Vanadium	0.50
Aluminum	0.10–0.35
Boron	0.005–0.012
Iron	Balance

XVII. THERMOSPAN ALLOY

Thermospan alloy is a trademark of Carpenter Technology. It is a precipitation-hardenable superalloy having an excellent combination of tensile properties and stress rupture strength in the recrystallized condition with the use of common solution and age-hardening treatments. The alloy also exhibits a low coefficient of thermal expansion over a broad temperature range, high tensile and rupture strengths, and good thermal fatigue.

As a result of the chromium addition, significant improvements in environmental resistance over that of the CTX alloys is realized.

The alloy should be considered for all applications in which other current low-expansion superalloys are presently being used, such as compressor and exhaust casings, seals, and other gas turbine engine components.

The alloy has the following composition:

Chemical	Weight percent
Carbon	0.05
Manganese	0.50
Silicon	0.30
Phosphorus	0.015
Sulfur	0.015
Chromium	5.50
Nickel	25.0
Cobalt	29.0
Titanium	0.80
Columbium	4.80
Aluminum	0.50
Copper	0.50
Boron	0.01
Iron	Balance

REFERENCES

1. PA Schweitzer. Corrosion Resistance Tables, 4th ed., Vols. 1–3. New York: Marcel Dekker, 1995.
2. PD Whitcraft. Corrosion of stainless steels. In: PA Schweitzer, ed. Corrosion Engineering Handbook. New York: Marcel Dekker, 1996.
3. GT Murray. Introduction to Engineering Materials. New York: Marcel Dekker, 1993.

13

Martensitic Stainless Steels

The martensitic grades are so named because when heated above the critical temperature, 1600°F (870°C), and cooled rapidly, a metallurgical structure known as martensite is obtained. In the hardened condition the steel has very high strength and hardness, but to obtain optimal corrosion resistance, ductility, and impact strength, the steel is given a stress relieving or tempering treatment, usually in the range of 300–700°F (149–371°C). These alloys are hardenable because of the phase transformation from body-centered cubic to body-centered tetragonal. As with the low-alloy steels, this transformation is thermally controlled.

The Fe-Cr phase diagram suggests that the maximum chromium content would be about 12.7%. But the carbon content expands the γ region to the extent that larger chromium contents are possible. Common alloys are 410, containing 12% chromium and low carbon and alloy 440 of 17% chromium with a high carbon content. The martensitic stainless steels are the strongest of all stainless steels, having strength to 275 ksi. But at such high strength levels they lack ductility.

A hardening temperature range depends upon the composition, but in general the higher the quenching temperature, the harder the article. Oil quenching is preferable, but with thin and intricate shapes, hardening by cooling in air should be undertaken.

Tempering at 800°F (425°C) does not reduce the hardness of the part, and in this condition these alloys show an exceptional resistance to fruit and vegetable acids, lye, ammonia, and other corrodents to which cutlery may be subjected.

The martensitic stainless steels fall into two main groups that are associated with two ranges of mechanical properties: low carbon compositions with a maximum hardness of about Rockwell C-45 and the higher carbon compositions, which can be hardened to Rockwell C-60. (The maximum hardness of both groups in the annealed condition is about Rockwell C-24.) A carbon content of approximately 0.15% forms the dividing line between the two groups.

With a low carbon content the chromium content must also be low or the materials will not harden. At the higher carbon levels the chromium content can be raised to about 18%. However, because of potential problems of carbide precipitation high chromium martensitic stainless steels are not usually tempered to the same degree as the low carbon types.

The low carbon class contains types 410, 416, and 430. Properties, performance, heat treatments, and fabrication of these three stainless steels are similar, except for type 416, which is a free-machining grade which has better machinability.

Types 440A, 440B, and 440C are high carbon alloys.

There are three types that do not fit into either category: types 420, 414, and 431. The minimum carbon content for type 420 is 0.15%, but it is usually produced to a carbon specification of 0.3–0.4%. Although type 420 will not harden to such high values as the 440 types, it can be tempered without substantial loss in corrosion resistance. Consequently a combination of hardness and ductility can be achieved which is suitable for cutlery.

Types 414 and 431 contain 1.25–2.50% nickel, which is enough to make them austenitic at ambient temperature. By adding nickel to the composition two purposes are achieved. The addition of nickel permits a higher chromium content which improves corrosion resistance and enhances notch toughness.

If toughness is important in the application, martensitic stainless steels should not be heat treated or used in the range of 800–1050°F (427–566°C) since they are subject to temper brittleness. Tempering is usually performed above this temperature.

Toughness of the martensitic grades of stainless steel tends to decrease as the hardness increases. Because of this, high-strength (high carbon) type 440A has lower toughness than type 410. Nickel, however, increases toughness, and type 414 has a higher level of toughness than type 410 at the same strength level.

Martensitic grades have a ductile–brittle transition temperature at which notch ductility drops very suddenly. The transition temperature is near room temperature, and at low temperature, about −300°F (−184°C), they become very brittle. This effect depends on composition, heat treatment, and other variables.

If notch ductility is critical at room temperature or below, and the steel is to be used in the hardened condition, careful evaluation is required. If the material is to be used much below room temperature, the chances are that quenched and tempered type 410 will not be satisfactory. While its notch ductility is better in the annealed condition down to $-100°F$ ($-73°C$), another type of stainless steel would probably be a better choice.

The fatigue properties of the martensitic stainless steels depend on heat treatment and design. A notch, for example, in a structure, or the effect of a corrosive environment can do more to reduce fatigue limit than alloy composition or heat treatment.

Abrasion, or wear, resistance is another important property. In most cases, the harder the material, the greater the abrasion resistance. In applications where corrosion occurs, however, such as in coal handling operations, this may not be the case since the oxide film is continuously removed, resulting in a high corrosion rate.

Other mechanical properties of martensitic stainless steels, such as compressive yield shear strength, are generally similar to those of carbon and alloy steels at the same strength levels.

The moduli of the martensitic stainless steels (29×10^6 psi) are slightly less than the modulus of carbon steel (30×10^6 psi).

Since the densities of the martensitic stainless steels are slightly lower than those of the carbon and alloy steels they have an excellent vibration damping capacity.

Moderate corrosion resistance, relatively high strength, and good fatigue properties after suitable heat treatment are usually the reasons for selecting the martensitic stainless steels.

High carbon martensitic stainless steels are not usually recommended for welded applications, although type 410 can be welded with relative ease. Hardening heat treatments should follow forming operations because of the poor forming qualities of the hardened steels.

Type 410 is used for fasteners, machinery parts, press plates, and similar items. If greater hardness of higher toughness is required type 414 may be used, and for better machinability types 416 or 416Se are used. Springs, flatware, knife blades, and hand tools are often made from type 420, while 431 is frequently used for aircraft parts requiring high yield strength and resistance to shock. Types 440A and 440B are used for cutlery while type 440C finds application in valve parts requiring good wear resistance.

I. TYPE 410 (S41000)

Type 410 stainless steel is heat treatable and is the most widely used of the martensitic stainless steels. Its chemical composition is shown in Table 13.1. This alloy, when heat treated, has high strength properties with good ductility. Type 410 stainless has a maximum operating temperature of 1300°F

TABLE 13.1 Chemical
Composition of Type 410
Stainless Steel

Chemical	Weight percent
Carbon	0.15
Manganese	1.00
Phosphorus	0.040
Sulfur	0.030
Silicon	1.00
Chromium	11.50–13.50
Iron	Balance

(705°C) for continuous service, but for intermittent service may be operated at a maximum of 1500°F (815°C). Table 13.2 shows the mechanical and physical properties of type 410 stainless.

With time and temperature, changes in metallurgical structure can be expected for almost any steel or alloy. In martensitic stainless steels softening occurs when exposed to temperatures approaching or exceeding the original tempering temperature. Type 410 stainless, which is a 12% chromium alloy, has been known to display brittle tendencies after extended periods in the same temperature range. This phenomenon is called 885°F embrittlement, which has been discussed previously. The rupture and creep characteristics of type 410 stainless are shown in Table 13.3.

Alloys for low temperature service must have suitable engineering properties such as yield and tensile strength and ductility. Many metals may have satisfactory "room temperature" characteristics but do not perform adequately at low temperatures. Low temperature brittle fracture can occur without any warning such as stretching, sagging, bulging, or other indication of plastic failure. Alloys that are ordinarily ductile may suddenly fail at very low levels of stress. Table 13.4 shows the mechanical properties of type 410 stainless at cryogenic temperatures. Note that the yield strength and tensile strength increase as the temperature decreases, but the toughness (Izod impact) drops suddenly.

Type 410 stainless steel is used where corrosion is not severe such as air, fresh water, some chemicals, and food acids. Table 13.5 provides the compatibility of type 401 stainless steel with selected corrodents.

Applications include valve and pump parts, fasteners, cutlery, turbine parts, bushings, and heat exchangers.

Type 410 double tempered is a quenched and double tempered variation conforming to NACE and API specifications for parts used in hydrogen

TABLE 13.2   Mechanical and Physical Properties of Type 410
Stainless Steel

Modulus of elasticity $\times 10^6$ (psi)	29
Tensile strength $\times 10^3$ (psi)	
annealed	70
heat treated	150
Yield strength 0.2% offset $\times 10^3$ (psi)	
annealed	45
heat treated	115
Elongation in 2 in. (%)	
annealed	25
heat treated	15
Toughness (ft-lb)	
annealed	33
heat treated	49
Density (lb/in.^{3})	0.28
Specific gravity	7.75
Specific heat (32–212°F) (Btu/lb °F)	0.11
Coefficient of thermal expansion $\times 10^{-6}$ (in./in. °F)	
at 32–212°F	5.5
Thermal conductivity (Btu/ft^2/hr/°F/in.)	173
Brinell hardness	
annealed	150
heat treated	410

TABLE 13.3   Rupture and Creep Characteristics of Type 410
(S41000) Stainless Steel

Temperature (°F/°C)	Stress for rupture (ksi) in		Stress for creep (ksi) rate/hr of	
	1,000 hr	10,000 hr	0.0001%	0.00001%
800/427	54.0	42.5	43.0	19.5
900/482	34.0	26.0	29.0	13.8
1000/538	19.0	13.0	9.2	7.2
1100/593	10.8	6.9	4.2	3.4
1200/649	4.9	3.5	2.0	1.2
1300/704	2.5	1.5	1.0	0.6
1400/760	1.2	0.6	0.8	0.4

TABLE 13.4   Mechanical Properties of Type 410 Stainless Steel at
Cryogenic Temperatures

Test temp. (°F/°C)	Yield strength 0.2% offset (ksi)	Tensile strength (ksi)	Elongation in 2 in. (%)	Izod impact (ft-lb)
−40/−40	90	122	23	25
−80/−62	94	128	22	25
−320/−196	148	158	10	5

sulfide service. Type 410S has a lower carbon content (0.8%) and a nitrogen
content of 0.60%.

II.   TYPE 414 (S41400)

Type 414 stainless is a nickel-bearing chromium stainless steel. The chemical composition is shown in Table 13.6. By adding nickel the hardenability
is increased but not enough to make it austenitic at ambient temperatures.
By adding nickel the chromium content can be increased, which leads to
improved corrosion resistance. The nickel addition also increases notch
toughness. Type 414 can be heat treated to somewhat higher tensile and
impact strengths than type 410. The mechanical and physical properties are
given in Table 13.7.

Type 414 stainless steel is resistant to mild atmospheric corrosion,
fresh water, and mild chemical exposures. Applications include high strength
nuts and bolts.

III.   TYPE 416 (S41600)

Type 416 stainless steel is a low carbon class martensitic alloy, a free-machining variation of type 410 stainless steel. The chemical composition is
shown in Table 13.8. It has a maximum continuous service operating temperature of 1250°F (675°C) and an intermittent maximum operating temperature of 1400°F (760°C). Table 13.9 lists the mechanical and physical
properties of type 416 stainless steel.

Type 416Se has selenium added to the composition and the sulfur
quantity reduced to improve the machinability of any of the stainless steels.
Refer to Table 13.10 for the chemical composition of type 416Se and Table
13.11 for the mechanical and physical properties.

These alloys exhibit useful corrosion resistance to natural food acids,
basic salts, water, and most natural atmospheres.

TABLE 13.5 Compatibility of Type 410 Stainless Steel with Selected Corrodents[a]

Chemical	Maximum temp.		Chemical	Maximum temp.	
	°F	°C		°F	°C
Acetaldehyde	60	16	Borax	150	66
Acetamide	60	16	Boric acid	130	54
Acetic acid, 10%	70	21	Bromine gas, dry	X	
Acetic acid, 50%	70	21	Bromine gas, moist	X	
Acetic acid, 80%	70	21	Bromine, liquid	X	
Acetic acid, glacial	X		Butadiene	60	16
Acetic anhydride	X		Butyl acetate	90	32
Acetone	210	99	Butyl alcohol	60	16
Acrylonitrile	110	43	Butyric acid	150	66
Allyl alcohol	90	27	Calcium bisulfite	X	
Alum	X		Calcium carbonate	210	99
Aluminium chloride, aqueous	X		Calcium chloride[b]	150	66
Aluminum chloride, dry	150	66	Calcium hydroxide, 10%	210	99
Aluminum fluoride	X		Calcium hypochlorite	X	
Aluminum hydroxide	60	16	Calcium sulfate	210	99
Aluminum nitrate	210	99	Carbon bisulfide	60	16
Aluminum oxychloride	X		Carbon dioxide, dry	570	299
Aluminum sulfate	X		Carbon dioxide, wet	570	299
Ammonium bifluoride	X		Carbon disulfide	60	16
Ammonium carbonate	210	99	Carbon monoxide	570	299
Ammonium chloride, 10%[b]	230	110	Carbon tetrachloride[b]	210	99
Ammonium chloride, 50%	X		Carbonic acid	60	16
Ammonium chloride, sat.	X		Chloracetic acid	X	
Ammonium hydroxide, sat.	70	21	Chlorine gas, dry	X	
Ammonium nitrate	210	99	Chloride gas, wet	X	
Ammonium persulfate, 5%	60	16	Chloride, liquid	X	
Ammonium phosphate, 5%	90	32	Clorobenzene, dry	60	16
Ammonium sulfate, 10–40%	60	16	Chloroform	150	66
Ammonium sulfite	X		Chlorosulfonic acid	X	
Amyl acetate[b]	60	16	Chromic acid, 10%	X	
Amyl alcohol	110	43	Chromic acid, 50%	X	
Amyl chloride	X		Citric acid, 15%	210	99
Aniline	210	99	Citric acid, 50%	140	60
Antimony trichloride	X		Copper acetate	90	32
Barium carbonate, 10%	210	99	Copper carbonate	80	27
Barium chloride[b]	60	16	Copper chloride	X	
Barium hydroxide	230	110	Copper cyanide	210	99
Barium sulfate	210	99	Copper sulfate	210	99
Barium sulfide	70	21	Cupric chloride, 5%	X	
Benzaldehyde			Cupric chloride, 50%	X	
Benzene	230	110	Cyclohexane	80	27
Benzoic acid	210	99	Cyclohexanol	90	32
Benzyl alcohol	130	54	Ethylene glycol	210	99

TABLE 13.5 Continued

Chemical	Maximum temp. °F	Maximum temp. °C	Chemical	Maximum temp. °F	Maximum temp. °C
Ferric chloride	X		Perchloric acid, 10%	X	
Ferric chloride, 50% in water	X		Perchloric acid, 70%	X	
Ferric nitrate, 10–50%	60	16	Phenol[b]	210	99
Ferrous chloride	X		Phosphoric acid, 50–80%	X	
Fluorine gas, dry	570	299	Picric acid	60	16
Fluorine gas, moist	X		Potassium bromide, 30%	210	99
Hydrobromic acid, dilute	X		Salicylic acid	210	99
Hydrobromic acid, 20%	X		Silver bromide, 10%	X	
Hydrobromic acid, 50%	X		Sodium carbonate, 10–30%	210	99
Hydrochloric acid, 20%	X		Sodium chloride[b]	210	99
Hydrochloric acid, 38%	X		Sodium hydroxide, 10%	210	99
Hydrocyanic acid, 10%	210	99	Sodium hydroxide, 50%	60	16
Hydrofluoric acid, 30%	X		Sodium hypochlorite, 20%	X	
Hydrofluoric acid, 70%	X		Sodium hypochlorite, conc.	X	
Hydrofluoric acid, 100%	X		Sodium sulfide, to 50%	X	
Ketones, general	60	16	Stannic chloride	X	
Lactic acid, 25%	60	16	Stannous chloride	X	
Lactic acid, conc.	60	16	Sulfuric acid, 10%	X	
Magnesium chloride, 50%	210	99	Sulfuric acid, 50%	X	
Malic acid	210	99	Sulfuric acid, 70%	X	
Methyl chloride, dry	210	99	Sulfuric acid, 90%	X	
Methyl ethyl ketone	60	16	Sulfuric acid, 98%	X	
Muriatic acid	X		Sulfuric acid, 100%	X	
Nitric acid, 5%	90	32	Sulfurous acid	X	
Nitric acid, 20%	160	71	Toluene	210	99
Nitric acid, 70%	60	16	Trichloroacetic acid	X	
Nitric acid, anhydrous	X		Zinc chloride	X	
Nitrous acid, conc.	60	16			

[a]The chemicals listed are in the pure state or in a saturated solution unless otherwise indicated. Compatibility is shown to the maximum allowable temperature for which data are available. Incompatibility is shown by an X. When compatible, the corrosion rate is <20 mpy.
[b]Material is subject to pitting.
Source: Ref. 1.

IV. TYPE 420 (S42000)

Type 420 stainless steel is a hardenable 12% chrome stainless steel with higher strength, hardness, and wear resistance than type 410. Table 13.12 shows the chemical composition and Table 13.13 the mechanical and physical properties. This alloy has been used for cutlery, surgical instruments, magnets, molds, shafts, valves, and other products.

TABLE 13.6 Chemical
Composition of Type 414
Stainless Steel

Chemical	Weight percent
Carbon	0.15
Manganese	1.00
Phosphorus	0.040
Sulfur	0.030
Silicon	1.00
Chromium	11.50–13.50
Nickel	1.25–2.50
Iron	Balance

TABLE 13.7 Mechanical and Physical Properties of Type 414
Stainless Steel

Modulus of elasticity $\times 10^6$ (psi)	29
Tensile strength $\times 10^3$ (psi)	
annealed	70
heat treated	200
Yield strength 0.2% offset $\times 10^3$ (psi)	
annealed	45
heat treated	150
Elongation in 2 in. (%)	
annealed	25
heat treated	17
Density (lb/in.^{3})	0.28
Specific gravity	7.75
Specific heat (32–212°F) (Btu/lb °F)	0.11
Thermal expansion coefficient $\times 10^{-6}$ (in./in. °F)	
at 32–212°F	6.1
Thermal conductivity (Btu/ft^2/hr/°F/in.)	173
Rockwell hardness	
annealed	C-22
heat treated	C-44

TABLE 13.8   Chemical
Composition of Type 416
Stainless Steel

Chemical	Weight percent
Carbon	0.15
Manganese	1.25
Phosphorus	0.060
Silicon	1.00
Chromium	12.00–14.00
Molybdenum	0.60[a]
Iron	Balance

[a]May be added at manufacturer's
option.

TABLE 13.9   Mechanical and Physical Properties
of Type 416 Stainless Steel

Modulus of elasticity $\times 10^6$ (psi)	29
Tensile strength $\times 10^3$ (psi)	
annealed	75
heat treated	150
Yield strength 0.2% offset $\times 10^3$ (psi)	
annealed	40
heat treated	115
Elongation in 2 in. (%)	
annealed	30
heat treated	15
Toughness (ft-lb)	
annealed	33
heat treated	49
Density (lb/in.^{3})	0.276
Specific gravity	7.74
Specific heat (32–212°F) (Btu/lb °F)	0.11
Rockwell hardness	
annealed	B-82
heat treated	C-43

TABLE 13.10 Chemical
Composition of Type 416Se
Stainless Steel

Chemical	Weight percent
Carbon	0.15
Manganese	1.25
Phosphorus	0.060
Sulfur	0.060
Silicon	1.00
Chromium	12.00–14.00
Selenium	0.15 min.
Iron	Balance

Type 420F (S42020) stainless is a free-machining version of type 420. It is hardenable and also exhibits higher strength, hardness, and wear resistance than type 410. The chemical composition will be found in Table 13.14 and the mechanical and physical properties in Table 13.15.

V. TYPE 422 (S42200)

This alloy is designed for service temperatures to 1200°F (649°C). It is a high carbon martensitic alloy whose composition is shown in Table 13.16. It exhibits good resistance to scaling and oxidation in continuous service at 1200°F (649°C), with high strength and toughness. Mechanical and physical properties are shown in Table 13.17.

Type 422 stainless is used in steam turbines for blades and bolts.

VI. TYPE 431 (S43100)

The addition of nickel to type 431 provides improved corrosion resistance and toughness (impact strength). Table 13.18 shows the chemical composition, while Table 13.19 shows the mechanical and physical properties.

This alloy finds application as fasteners and fittings, for structural components exposed to marine atmospheres, and for highly stressed aircraft components.

VII. TYPE 440A (S44002)

Type 440A stainless is a high carbon chromium steel providing stainless properties with excellent hardness. Because of the high carbon content type

TABLE 13.11 Mechanical and Physical Properties
of Type 416Se Stainless Steel

Modulus of elasticity $\times$ 10^6 (psi)	29
Tensile strength $\times$ 10^3 (psi)	
annealed	75
heat treated	150
Yield strength 0.2% offset $\times$ 10^3 (psi)	
annealed	40
heat treated	115
Elongation in 2 in. (%)	
annealed	30
heat treated	15
Density (lb/in.^{3})	0.28
Specific gravity	7.75
Specific heat (32–212°F) (Btu/lb °F)	0.11
Rockwell hardness	
annealed	B-82
heat treated	C-43

440A exhibits lower toughness than type 410. The chemical composition is shown in Table 13.20. Type 440 has a lower carbon content than type 440B or 440C and consequently results in a lower hardness but greater toughness. The mechanical and physical properties are contained in Table 13.21. When heat treated, a Rockwell hardness of C-56 can be obtained.

VIII. TYPE 440B (S44003)

When heat treated, this high carbon chromium steel attains a hardness of Rockwell C-58, intermediate between types 440A and 440C with a comparable intermediate toughness. Table 13.22 shows the chemical composition and Table 13.23 the mechanical and physical properties.

Type 440B has been used for cutlery, hardened balls, and similar parts.

IX. TYPE 440C (S44004)

Type 440C stainless steel is a high carbon chromium steel that can attain the highest hardness (Rockwell C-60) of the 440 series stainless steels. In the hardened and stress relieved condition, type 440C has maximum hardness together with high strength and corrosion resistance. It also has good abrasion resistance. The chemical composition is shown in Table 13.24 and the mechanical and physical properties in Table 13.25.

TABLE 13.12 Chemical Composition of Type 420 Stainless Steel

Chemical	Weight percent
Carbon	0.15 min.
Manganese	1.50
Phosphorus	0.040
Sulfur	0.030
Silicon	1.50
Chromium	12.00–14.00
Iron	Balance

TABLE 13.13 Mechanical and Physical Properties of Type 420 Stainless Steel

Modulus of elasticity $\times 10^6$ (psi)	29
Tensile strength $\times 10^3$ (psi)	
annealed	95
heat treated	250
Yield strength 0.2% offset $\times 10^3$ (psi)	
annealed	50
heat treated	200
Elongation in 2 in. (%)	
annealed	25
heat treated	8
Toughness, heat treated (ft-lb)	15
Density (lb/in.^{3})	0.28
Specific gravity	7.75
Specific heat (32–212°F) (Btu/lb °F)	0.11
Coefficient of thermal expansion $\times 10^{-6}$ (in./in. °F)	
at 32–212°F	5.7
Thermal conductivity (Btu/ft^2/hr/°F/in.)	173
Rockwell hardness	
annealed	B-92
heat treated	C-54

TABLE 13.14 Chemical Composition of Type 420F (S42020) Stainless Steel

Chemical	Weight percent
Carbon	0.15 min.
Manganese	1.25
Phosphorus	0.060
Sulfur	0.15 min.
Silicon	1.00
Chromium	12.00–14.00
Molybdenum	0.60
Iron	Balance

TABLE 13.15 Mechanical and Physical Properties of Type 420F (42020) Stainless Steel

Modulus of elasticity $\times 10^6$ (psi)	29
Tensile strength $\times 10^3$ (psi)	
annealed	95
heat treated	250
Yield strength 0.2% offset $\times 10^3$ (psi)	
annealed	55
heat treated	200
Elongation in 2 in. (%)	
annealed	22
heat treated	8
Toughness, heat treated (ft-lb)	15
Density (lb/in.^{3})	0.28
Specific gravity	7.75
Specific heat (32–212°F) (Btu/lb °F)	0.11
Coefficient of thermal expansion $\times 10^{-6}$ (in./in. °F)	
at 32–212°F	5.7
Rockwell hardness	
annealed	B-92
heat treated	C-54

TABLE 13.16 Chemical Composition of Type 422 Stainless Steel

Chemical	Weight percent
Carbon	0.2–0.25
Manganese	1.00
Phosphorus	0.025
Sulfur	0.025
Silicon	0.75
Chromium	11.00–13.00
Nickel	0.5–1.00
Molybdenum	0.75–1.25
Vanadium	0.15–0.30
Tungsten	0.75–1.25
Iron	Balance

TABLE 13.17 Mechanical and Physical Properties of Type 422 Stainless Steel

Tensile strength, heat treated, $\times 10^3$ (psi)	145
Yield strength 0.2% offset, heat treated, $\times 10^3$ (psi)	125
Elongation in 2 in., heat treated (%)	16
Specific heat (32–212°F) (Btu/lb °F)	0.11
Brinell hardness, heat treated	320

TABLE 13.18 Chemical Composition of Type 431 Stainless Steel

Chemical	Weight percent
Carbon	0.20
Manganese	1.00
Phosphorus	0.040
Sulfur	0.030
Silicon	1.00
Chromium	15.00–17.00
Nickel	1.25–2.50
Iron	Balance

TABLE 13.19   Mechanical and Physical Properties of Type 431
Stainless Steel

Modulus of elasticity $\times 10^6$ (psi)	29
Tensile strength $\times 10^3$ (psi)	
annealed	125
heat treated	196
Yield strength 0.2% offset $\times 10^3$ (psi)	
annealed	95
heat treated	150
Elongation in 2 in. (%)	
annealed	25
heat treated	20
Toughness, heat treated (ft-lb)	25
Density (lb/in.^{3})	0.28
Specific gravity	7.75
Specific heat (32–212°F) (Btu/lb °F)	0.11
Coefficient of thermal expansion $\times 10^{-6}$ (in./in. °F)	
at 32–212°F	6.5
Thermal conductivity (Btu/ft^2/hr/°F/in.)	140
Rockwell hardness	
annealed	C-24
heat treated	C-41

TABLE 13.20   Chemical
Composition of Type 440A
Stainless Steel

Chemical	Weight percent
Carbon	0.60–0.75
Manganese	1.00
Phosphorus	0.040
Sulfur	0.030
Silicon	1.00
Chromium	16.00–18.00
Molybdenum	0.75
Iron	Balance

TABLE 13.21 Mechanical and Physical Properties of Type 440A Stainless Steel

Modulus of elasticity $\times 10^6$ (psi)	29
Tensile strength $\times 10^3$ (psi)	
annealed	105
heat treated	260
Yield strength 0.2% offset $\times 10^3$ (psi)	
annealed	60
heat treated	240
Elongation in 2 in. (%)	
annealed	20
heat treated	5
Toughness, heat treated (ft-lb)	8
Specific heat (32–212°F) (Btu/lb °F)	0.11
Rockwell hardness	
annealed	B-95
heat treated	C-56

This stainless steel is used principally in bearing assemblies, including bearing balls and races.

X. ALLOY 440-XH

This alloy is produced by Carpenter Technology, having a nominal composition as follows:

Chemical	Weight percent
Carbon	1.60
Manganese	0.50
Silicon	0.40
Chromium	16.00
Nickel	0.35
Molybdenum	0.80
Vanadium	0.45
Iron	Balance

This is a high carbon, high chromium, corrosion resistant alloy which can be described as either a high hardness type 440C stainless steel or a corrosion resistant D2 tool steel. It possesses corrosion resistance equivalent to

TABLE 13.22 Chemical
Composition of Type 440B
Stainless Steel

Chemical	Weight percent
Carbon	0.75–0.95
Manganese	1.00
Phosphorus	0.040
Sulfur	0.030
Silicon	1.00
Chromium	16.00–18.00
Molybdenum	0.75
Iron	Balance

type 440C stainless but can attain a maximum hardness of Rockwell C-64, approaching that of tool steel.

XI. TYPE 440F OR 440F-Se

This high carbon chromium steel is designed to provide stainless properties with maximum hardness, approximately Rockwell C-60 after heat treatment. However, the addition of sulfur to type 440F, or the addition of selenium to

TABLE 13.23 Mechanical and Physical Properties
of Type 440B Stainless Steel

Modulus of elasticity $\times 10^6$ (psi)	29
Tensile strength $\times 10^3$ (psi)	
annealed	107
heat treated	280
Yield strength 0.2% offset $\times 10^3$ (psi)	
annealed	62
heat treated	270
Elongation in 2 in. (%)	
annealed	18
heat treated	3
Specific heat (32–212°F) (Btu/lb °F)	0.11
Rockwell hardness	
annealed	B-96
heat treated	C-55

TABLE 13.24 Chemical Composition of Type 440C Stainless Steel

Chemical	Weight percent
Carbon	0.95–1.2
Manganese	1.00
Phosphorus	0.040
Sulfur	0.030
Silicon	1.00
Chromium	16.00–18.00
Molybdenum	0.75
Iron	Balance

type 440F-Se, makes these two grades free machining. Either of these two types should be considered for machined parts which require higher hardness values than possible with other free-machining grades.

XII. 13Cr-4N (F6NM)

F6NM is a high nickel, low carbon martensitic stainless with higher toughness and corrosion resistance than type 410 and superior weldability. It has

TABLE 13.25 Mechanical and Physical Properties of Type 440C Stainless Steel

Modulus of elasticity $\times$ 10^6 (psi)	29
Tensile strength $\times$ 10^3 (psi)	
annealed	110
heat treated	285
Yield strength 0.2% offset $\times$ 10^3 (psi)	
annealed	65
heat treated	275
Elongation in 2 in. (%)	
annealed	14
heat treated	2
Toughness, heat treated (ft-lb)	5
Specific heat (32–212°F) (Btu/lb °F)	0.11
Rockwell hardness	
annealed	B-97
heat treated	C-60

been used in oil field applications as a replacement for type 410. F6NM has
a chemical composition as follows:

Chemical	Weight percent
Carbon	0.05
Manganese	0.50–1.00
Phosphorus	0.030
Sulfur	0.030
Silicon	0.30–0.60
Chromium	12.00–14.00
Nickel	3.50–4.50
Molybdenum	0.40–0.70
Iron	Balance

REFERENCES

1. PA Schweitzer. Corrosion Resistance Tables, 4th ed., Vols. 1–3. New York: Marcel Dekker, 1995.
2. PK Whitcraft. Corrosion of stainless steels. In: PA Schweitzer, ed. Corrosion Engineering Handbook, New York: Marcel Dekker, 1996.
3. PA Schweitzer. Stainless steel. In: PA Schweitzer, ed. Corrosion and Corrosion Protection Handbook, 2nd ed. New York: Marcel Dekker, 1988.
4. GT Murray. Introduction to Engineering Materials. New York: Marcel Dekker, 1993.

14

Duplex Stainless Steels

The duplex stainless steels are those alloys whose microstructures are a mixture of austenite and ferrite. These alloys were developed to improve the corrosion resistance of the austenitic stainlesses, particularly in the areas of chloride stress corrosion cracking and in maintaining corrosion resistance after welding. The original duplex stainlesses developed did not meet all of the criteria desired. Consequently, additional research was undertaken.

Duplex stainless steels have been available since the 1930s. The first-generation duplex stainless steels, such as type 329 (S32900), have a good general corrosion resistance because of their high chromium and molybdenum contents. When welded, however, these grades lose the optimal balance of austenite and ferrite, and consequently corrosion resistance and toughness are reduced. While these properties can be restored by a postweld heat treatment, most of the applications of the first-generation duplexes use fully annealed material without further welding. Since these materials do not meet all of the criteria of duplex stainless steels they have been included in the chapter on austenitic stainless steels.

In the 1970s, this problem was made manageable through the use of nitrogen as an alloy addition. The introduction of argon-oxygen decarburization (AOD) technology permitted the precise and economical control of nitrogen in stainless steel. Although nitrogen was first used because it was an inexpensive austenite former, replacing some nickel, it was quickly found that it had other benefits. These include improved tensile properties and pitting and crevice corrosion resistance.

The original duplex stainless steels did not have nitrogen added specifically as an alloying ingredient. By adding 0.15–0.25% nitrogen, the chromium partitioning between the two phases is reduced, resulting in the pitting

255

and crevice corrosion resistance of the austenite being improved. This nitrogen addition also improves the weldability of the stainless steel without losing any of its corrosion resistance.

Nitrogen also causes austenite to form from ferrite at a higher temperature, allowing for restoration of an acceptable balance of austenite and ferrite after a rapid thermal cycle in the heat-affected zone (HAZ) after welding. This nitrogen enables the use of duplex grades in the as-welded condition and has created the second generation of duplex stainless steels.

The duplex grades characteristically contain molybdenum and have a structure approximately 50% ferrite and 50% austenite because of the excess of ferrite-forming elements such as chromium and molybdenum. The duplex structure, combined with molybdenum, gives them improved resistance to chloride-induced corrosion (pitting, crevice corrosion, and stress corrosion cracking), in aqueous environments particularly.

However, the presence of ferrite is not an unmixed blessing. Ferrite may be attacked selectively in reducing acids, sometimes aggravated by a galvanic influence of the austenite phase, while the sigma phase produced by thermal transformation (as by heat of welding) is susceptible to attack by strong oxidizing acids. The duplex structure is subject to 885°F (475°C) embrittlement and has poor NDIT properties. Except for temper embrittlement these problems can be minimized through corrosion testing and impact testing.

Because the stainless steels are a mixture of austenite and ferrite, it is only logical that their physical properties would lie between the comparable properties of these microstructures. The duplexes have better toughness than ferritic grades and higher yield strengths than the austenitics.

Since the duplexes contain a large amount of ferrite, they are magnetic. However, unlike the ferritics, they have a high degree of toughness along with their high strength.

Because the duplexes have a higher yield strength than the austenitics, they can provide certain economic advantages. Money can be saved using thinner-walled sections for piping and vessels without sacrificing operating pressures. Conversely piping and equipment manufactured from these stainless steels using conventional wall thicknesses can be operated at higher pressures.

Although more formable than the ferritic alloys, they are not as ductile as the austenitic family of alloys. Welding requires more care than with the austenitic alloys due to a greater tendency to compositional segregation and sensitivity to weld heat imput.

Due to the high chromium contents, duplex alloys are sensitive to 885°F (475°C) embrittlement. This generally limits their usage to 600°F (313°C) maximum for pressure vessels. Due to the presence of nickel, chromium, and molybdenum they are also susceptible to the formation of σ phase. This is a brittle phase which forms islands in the matrix and will affect mechanical properties and corrosion resistance due to alloy depletion. The σ phase forms in the temperature range of 1100°F (593°C) to 1600°F (882°C) and most rapidly at about 1450°F (788°C). The deleterious effects of σ phase formation are not obvious at the elevated temperature but can become a factor at room temperature. The formation of σ phase in these alloys is sufficiently rapid to have an effect on properties due to slow cooling (air) after anneal. A measurable effect as a result of exposure in this temperature range due to welding has been demonstrated.

The high chromium and molybdenum contents of the duplex stainless steels are particularly important in providing resistance in oxidizing environments and are also responsible for the exceptionally good pitting and crevice corrosion resistance, especially in chloride environments. In general these stainless steels have greater pitting resistance than type 316, and several have an even greater resistance than alloy 904L. The critical crevice corrosion temperature of selected duplex stainless steels in 10% $FeCl_3 \cdot 6H_2O$ having a pH of 1 are shown below:

UNS number	Temperature (°F/°C)
S32900	41/5
S31200	41/5
S31260	50/10
S32950	60/15
S31803	63.5/17.5
S32250	72.5/22.5

The resistance to crevice corrosion of the duplexes is superior to the resistance of the 300 series austenitics. They also provide an appreciably greater resistance to stress corrosion cracking. Like 20Cb3, the duplexes are resistant to chloride stress corrosion cracking in chloride-containing process streams and cooling water. However, under very severe conditions, such as boiling magnesium chloride, the duplexes will crack, as will alloy 20Cb3.

To achieve the desired microstructure, the nickel content of the duplexes is below that of the austenitics. Because the nickel content is a factor for providing corrosion resistance in reducing environments, the duplexes show less resistance in these environments than do the austenitics. However, the high chromium and molybdenum contents partially offset this loss, and consequently they can be used in some reducing environments, particularly dilute and cooler solutions. Although their corrosion resistance is good, the boundary between acceptable and poor performance is sharper than with austenitic materials. As a result, they should not be used under conditions that operate close to the limits of their acceptability.

Duplex stainless steels are known best for the following performance characteristics:

1. Lower life-cycle cost
2. High resistance to stress corrosion cracking
3. Excellent resistance to pitting and crevice corrosion
4. High resistance to erosion and general corrosion in many environments
5. Very high mechanical strength
6. Low thermal expansion
7. Good weldability

Included among the duplex stainless steels are the following:

Alloy	UNS	Alloy	UNS
2206	S31803	329	S32900
3RE60	S31500	7Mo Plus	S32950
255	S32550	Z100	S32760
44LN	S31200	DP3W	S32740
DP-3	S31260	45D	J93345
2304	S32304	CD4MCu	J93370
2507	S32750	U-50	S32404

Of these the four most commonly used are alloy 2205 (S31803), 7M0 Plus (S32950), Z100 (S32760), and 255 (S32550). Each of these alloys will be discussed in detail.

I.   2205 (S31803)

Alloy 2205 exhibits an excellent combination of both strength and corrosion resistance. The chemical composition is shown in Table 14.1.

The approximate 50/50 ferrite-austenite structure provides excellent chloride pitting and stress corrosion cracking resistance, with roughly twice the yield strength of the standard austenitic grades.

The high chromium and molybdenum contents, coupled with the nitrogen addition, provide general corrosion pitting and crevice corrosion resistance, superior to those of type 316L and 317L.

TABLE 14.1 Chemical Composition of Alloy 2205 Stainless Steel

Chemical	Weight percent
Carbon	0.03 max.
Manganese	2.00 max.
Phosphorus	0.03 max.
Sulfur	0.02 max.
Silicon	1.00 max.
Chromium	21.00–23.00
Nickel	4.50–6.50
Molybdenum	2.50–3.50
Nitrogen	0.14–0.20
Iron	Balance

When compared to type 316 stainless steel, alloy 2205 demonstrates superior erosion-corrosion resistance. It is not subject to intergranular corrosion in the welded condition. The mechanical and physical properties of alloy 2205 will be found in Table 14.2.

Alloy 2205 resists oxidizing mineral acids and most organic acids in addition to reducing acids, chloride environments, and hydrogen sulfide.

To achieve the desired microstructure, the nickel content of the duplexes is below that of the austenitics. Because the nickel content is a factor for providing corrosion resistance in reducing environments, the duplexes show less resistance in these environments than do the austenitics. However,

TABLE 14.2 Mechanical and Physical Properties of Alloy 2205

Modulus of elasticity $\times 10^3$ (psi)	29.0
Tensile strength $\times 10^3$ (psi)	90
Yield strength 0.2% offset $\times 10^3$ (psi)	65
Elongation in 2 in. (%)	25
Rockwell hardness	C-31
Density lb/in^3	0.283
Thermal conductivity (Btu/ft hr °F) at 70°F (20°C)	10.0
Mean coefficient of thermal expansion $\times 10^{-6}$ (in./in./°F)	
at 68–212°F (20–100°C)	7.6
at 68–390°F (20–199°C)	7.9
at 68–570°F (20–298°C)	8.2

the high chromium and molybdenum contents partially offset this loss and consequently they can be used in some reducing environments, particularly dilute and cooler solutions. Although their corrosion resistance is good, the boundary between acceptable and poor performance is sharper than with austenitic materials. As a result, they should not be used under conditions that operate close to the limits of their acceptability.

Duplex stainless steels are known best for the following performance characteristics:

1. Lower life-cycle cost
2. High resistance to stress corrosion cracking
3. Excellent resistance to pitting and crevice corrosion
4. High resistance to erosion and general corrosion in many environments

The following corrosion rates have been reported for alloy 2205:

Solution	Corrosion rate (mpy)
1% Hydrochloric acid, boiling	0.1
10% Sulfuric acid, 150°F/66°C	1.2
10% Sulfuric acid, boiling	206
30% Phosphoric acid, boiling	1.6
85% Phosphoric acid, 150°F/66°C	0.4
65% Nitric acid, boiling	21
10% Acetic acid, boiling	0.1
20% Acetic acid, boiling	0.1
20% Formic acid, boiling	1.3
45% Formic acid, boiling	4.9
3% Sodium chloride, boiling	0.1

Alloy 2205 will be attacked by hydrochloric and hydrofluoric acids. Applications are found primarily in oil and gas field piping applications, condensers, reboilers, and heat exchangers.

II.   7-Mo PLUS (S32950)

7-Mo Plus stainless steel is a trademark of Carpenter Technology. It is a two-phase (duplex) alloy with approximately 45% austenite distributed within a ferrite matrix. Alloy S32950 displays good resistance to chloride stress corrosion cracking, pitting corrosion, and general corrosion in many severe environments. The yield strength of annealed 7-Mo Plus stainless is

TABLE 14.3 Chemical Composition of Type 7-Mo Plus Stainless Steel

Chemical	Weight percent
Carbon	0.03 max.
Manganese	2.00 max.
Phosphorus	0.035 max.
Sulfur	0.010 max.
Silicon	0.60 max.
Chromium	26.00–29.00
Nickel	3.50–5.20
Molybdenum	1.00–2.50
Nitrogen	0.15–0.35
Iron	Balance

greater than twice that of typical austenitic stainless steels. The chemical composition is shown in Table 14.3.

This alloy is subject to 885°F/475°C embrittlement when exposed for extended period of times between about 700–1000°F (371–538°C). Table 14.4 shows the impact strength of alloy 7-Mo Plus at room temperature after having been held at elevated temperatures.

TABLE 14.4 Typical V-Notch Impact Strength of 7-Mo Plus Stainless Steel at Room Temperature After Holding at Elevated Temperatures

Holding temp. (°F/°C)	10 hr (ft-lb)	100 hr (ft-lb)	500 hr (ft-lb)	1000 hr (ft-lb)
Strip 0.250 in. thick, hot rolled, annealed				
500/260	105	98	103	103
600/316	98	100	90	74
700/371	85	23	5	3
Bar 2.50 in. round, annealed				
600/316	—	34	31	28
700/371	—	17	11	10
900/482	9	4	—	—

TABLE 14.5 Mechanical and Physical Properties of Type 7-Mo Plus Stainless Steel

Modulus of elasticity $\times 10^6$ (psi)	29.0
Tensile strength $\times 10^3$ (psi)	90
Yield strength 0.2% offset $\times 10^3$ (psi)	70
Elongation in 2 in. (%)	20
Rockwell hardness	C-30.5
Density (lb/in.^{3})	0.280
Specific gravity	7.74
Specific heat, (75–212°F) (Btu/lb °F)	0.114
Thermal conductivity (Btu/h °F)	
at 70°F (20°C)	8.8
at 1500°F (815°C)	12.5
Coefficient of thermal expansion $\times 10^{-6}$ (in./in. °F)	
at 75–400°F	6.39
at 75–600°F	6.94
at 75–800°F	7.49
at 75–1000°F	7.38
Charpy V-notch impact at 75°F (20°C) (ft-lb)	101

    7-Mo Plus is also subject to precipitation of sigma phase when exposed between 1250–1550°F (677–843°C) for extended periods. Sigma phase increases strength and hardness but decreases ductility and corrosion resistance.

    Refer to Table 14.5 for the mechanical and physical properties of type 7-Mo Plus stainless steel. The tensile properties of type 7-Mo Plus stainless

TABLE 14.6 Typical Elevated Temperature Tensile Properties of 7-Mo Plus Bar (2.50 in. round, annealed)

Test temp. (°F/°C)	Yield strength 0.2% offset (ksi)	Tensile strength (ksi)	Elongation (%)
73/23	82	110	38
200/93	74	99	35
300/149	69	94	36
400/204	61	92	36
500/260	58	93	37
600/316	57	96	34
700/371	56	96	33

TABLE 14.7 Typical Impact Strength of Type 7-Mo Plus
Stainless Steel at Cryogenic Temperatures

Test temp. (°F/°C)	Impact strength (ft-lb)	Fracture mode
20/−7	61	Mixed 67% ductile
0/−18	36	Mixed 50% ductile
−75/−59	19	Mixed 33% ductile
−150/−101	5	Brittle

steel when operating at an elevated temperature are shown in Table 14.6, and the effect on impact strength when operating at cryogenic temperature is shown in Table 14.7.

The general corrosion resistance of 7-Mo Plus stainless is superior to that of stainless steels such as type 304 and type 316 in many environments. Because of its high chromium content, it has good corrosion resistance in strongly oxidizing media such as nitric acid. Molybdenum extends the corrosion resistance into the less oxidizing environments. Chromium and molybdenum impart a high level of resistance to pitting and crevice corrosion. It has a PREN of 40.

TABLE 14.8 Chemical
Composition of Zeron 100
(S32760) Stainless Steel

Chemical	Weight percent
Carbon	0.03 max.
Manganese	1.00 max.
Phosphorus	0.03 max.
Sulfur	0.01 max.
Silicon	1.00 max.
Chromium	24.0–26.0
Nickel	6.0–8.0
Molybdenum	3.0–4.0
Copper	0.5–1.0'
Nitrogen	0.2–0.3
Tungsten	0.5–1.0
Iron	Balance

Alloy S32950 exhibits excellent resistance to nitric acid, phosphoric acid, organic acids, alkalies, seawater, and chloride stress corrosion cracking. It is not suitable for service in hydrochloric or hydrofluoric acids or some salts.

III. ZERON 100 (S32760)

Zeron 100 is a trademark of Weir Materials Limited of Manchester, England. Table 14.8 details the chemical composition of Zeron 100, which is tightly controlled by Weir Materials, while the chemical composition of S32760 is a broad compositional range.

Zeron 100 is a highly alloyed duplex stainless steel for use in aggressive environments. In general its properties include high resistance to pitting and crevice corrosion, resistance to stress corrosion cracking in both chloride

TABLE 14.9 Mechanical and Physical Properties of Zeron 100 (S32760) Stainless Steel

Modulus of elasticity $\times 10^6$ (psi)	27.6
Tensile strength $\times 10^3$ (psi)	104
Yield strength 0.2% offset $\times 10^3$ (psi)	80
Elongation in 2 in. (%)	25
Rockwell hardness	C-28
Specific heat (Btu/lb °F)	
at 68°F	0.115
at 100°F	0.116
at 200°F	0.119
at 300°F	0.122
at 400°F	0.125
at 500°F	0.128
Thermal conductivity (Btu/hr ft^2 °F)	
at 68°F	7.5
at 100°F	7.7
at 200°F	8.3
at 300°F	8.9
at 400°F	9.5
at 500°F	10.1
Density (lb/ft^3)	489
Coefficient of thermal expansion $\times 10^{-6}$ (in./in. °F)	
at 70–200°F	7.0
at 70–400°F	7.4
at 70–600°F	7.7

TABLE 14.10 Mechanical Properties of Zeron 100 at Elevated Temperatures

Product form	Temperature (°F/°C)	Yield strength 0.2% offset (ksi)	Tensile strength (ksi)
Plates up to 1 in.	68/20	80	109
	122/50	73	105
	212/100	68	102
	302/150	65	99
	392/200	62	97
	482/250	58	94
	572/300	56	92
Plates 1.25 to 1.75 in.	68/20	80	109
	122/50	68	102
	212/100	62	97
	302/150	58	90
	392/200	55	88
	482/250	54	87
	572/300	52	86

TABLE 14.11 Chemical Composition of Ferralium 255 (S32550) Stainless Steel

Chemical	Weight percent
Carbon	0.04
Manganese	1.50
Phosphorus	0.04
Sulfur	0.03
Silicon	1.00
Chromium	24.0–27.0
Nickel	4.5–6.5
Molybdenum	2.9–3.9
Copper	1.5–2.5
Nitrogen	0.1–0.25
Iron	Balance

TABLE 14.12 Mechanical and Physical Properties of
Ferralium 255 (S32550) Stainless Steel

Modulus of elasticity $\times 10^6$ (psi)	31
Tensile strength $\times 10^3$ (psi)	110
Yield strength 0.2% offset $\times 10^3$ (psi)	80
Elongation (%)	15
Toughness (ft-lb)	140
Coefficient of thermal expansion $\times 10^{-6}$ (in./in. °F) at 68–212°F (20–100°C)	6.1
Thermal conductivity (Btu/ft hr °F/in.) at 68°F (20°C)	7.8

and sour environments, resistance to erosion-corrosion and corrosion fatigue, and excellent mechanical properties.

The mechanical and physical properties of Zeron 100 are shown in Table 14.9. Zeron 100 exhibits good impact strength (toughness), having a minimum of 51 ft-lb at −50°F/−46°C for standard products. There is not true ductile brittle transition, just a gradual decrease in impact energy as the temperature is lowered. The impact energy varies according to product type and production route. The impact strength of welded Zeron 100 is slightly less than that of the parent metal. Zeron 100 also exhibits good mechanical properties at elevated temperatures; however, it is not recommended for uses which involve extended exposure to temperatures greater than 572°F (300°C) as there is a substantial reduction in toughness. See Table 14.10 for properties at elevated temperatures.

Zeron 100 is highly resistant to corrosion in a wide range of organic and inorganic acids. Its excellent resistance to many nonoxidizing acids is the result of the copper content.

A high resistance to pitting and crevice corrosion is also exhibited by Zeron 100. It has a PREN of 48.2. Intergranular corrosion is not a problem since the alloy is produced to a low carbon specification and water quenched from solution annealing, which prevents the formation of any harmful precipitates and eliminates the risk of intergranular corrosion.

Resistance is also exhibited to stress corrosion cracking in chloride environments and process environments containing hydrogen sulfide and carbon dioxide.

IV. FERRALIUM 255 (S32550)

The chemical composition of Ferralium 255 is shown in Table 14.11. This is a duplex alloy with austenite distributed within a ferrite matrix. Mechanical and physical properties are given in Table 14.12. This alloy has a maximum service temperature of 500°F (260°C).

Ferralium 255 exhibits good general corrosion resistance to a variety of media, with a high level of resistance to chloride pitting and stress corrosion cracking. The following corrosion rates for Ferralium 255 have been reported:

Solution	Corrosion rate (mpy)
1% Hydrochloric acid, boiling	0.1
10% Sulfuric acid, 150°F (66°C)	0.2
10% Sulfuric acid, boiling	40
30% Phosphoric acid, boiling	0.2
85% Phosphoric acid, 150°F (66°C)	0.1
65% Nitric acid, boiling	5
10% Acetic acid, boiling	0.2
20% Formic acid, boiling	0.4
3% Sodium chloride, boiling	0.4

15

Nickel and High Nickel Alloys

The nickel-based alloys show a wider range of application than any other class of alloys. These alloys are used as corrosion resistant alloys, heating elements, controlled expansion alloys, creep-resistant alloys in turbines and jet engines, and high temperature corrosion resistant alloys.

The austenitic stainless steels were developed and utilized early in the 1900s, whereas the development of the nickel-based alloys did not begin until about 1930. Initially some of the alloys were produced only as castings, and later the wrought versions developed. Since that time there has been a steady progression of different or improved alloys emerging from the laboratories of nickel base alloy producers. Many of these find their major usage in the high temperature world of gas turbines, furnaces, and the like, but several are used primarily by the chemical industry for aqueous corrosion service.

Historically, the use of these alloys was typically reserved for those applications where it was adjudged that nothing else would work. At one time the primary factor in the selection of construction materials was initial cost. Very little thought was given to the possible maintenance and downtime associated with the equipment. Today the increasing costs of maintenance and downtime have placed greater emphasis on the reliable performance of the process equipment. The annual amortized cost of the equipment over the expected life is now important in the material selection.

In the environmental series nickel is nobler than iron but more active than copper. Reducing environments, such as dilute sulfuric acid, find nickel more corrosion resistant than iron but not as resistant as copper or nickel-copper alloys. The nickel molybdenum alloys are more corrosion resistant to reducing environment than nickel or nickel-copper alloys.

While nickel can form a passive film in some environments, it is not a particularly stable film; therefore nickel cannot generally be used in oxidizing media, such as nitric acid. When alloyed with chromium a much improved stable passive film results, producing a greater corrosion resistance to a variety of oxidizing environments. However, these alloys are subject to attack in media containing chloride or other halides, especially if oxidizing agents are present; corrosion will be in the form of pitting. The corrosion resistance can be improved by adding molybdenum and tungsten.

One of the most important attributes of nickel with respect to the formation of corrosion resistant alloys is its metallurgical compatibility with a number of other metals, such as copper, chromium, molybdenum, and iron. A survey of the binary-phase diagrams for nickel and these other elements shows considerable solid solubility, and thus one can make alloys with a wide variety of composition. Nickel alloys are, in general, all austenitic alloys; however, they can be subject to precipitation of intermetallic and carbide phases when aged. In some alloys designed for high temperature service, intermetallic and carbide precipitation reactions are encouraged to increase properties. However, for corrosion applications the precipitation of second phases usually promotes corrosion attack. The problem is rarely encountered because the alloys are supplied in the annealed condition, and the service temperatures rarely approach the level required for sensitization.

In iron-chromium-nickel austenitic stainless steels, minimization of carbide precipitation can be achieved by lowering the carbon content to a maximum of about 0.03%. As the nickel content is increased from the nominal 8% in these alloys to that of the majority element (i.e., more than 5%), the nature of the carbide changes from predominantly $M_{23}C_6$ to M_6C, and the carbon solubility decreases by a factor of 10. It was therefore very difficult in the past to produce an L grade material because of the state of the art of melting. Many alloys were produced with carbon stabilizers to tie up the carbon, but with varying degrees of success. Changes in melting techniques were developed to overcome the problem. The transfer of alloys from air induction or vacuum induction melting to air arc plus argon-oxygen decarburization has provided a means for producing nickel alloys comparable to the L grades of stainless steels.

While general corrosion is important, one of the major reasons that nickel-based alloys are specified for many applications is their excellent resistance to localized corrosion, such as pitting, crevice corrosion, and stress corrosion cracking. In many environments, austenitic stainless steels do not exhibit general attack but suffer from significant localized attack, after causing excessive downtime and/or expensive repair and replacement.

In general the localized corrosion resistance of alloys is improved by the addition of molybdenum. However, molybdenum content alone does not solve the problem. For example, alloy B-2 has the highest molybdenum content (26.5%) and is not recommended for most localized corrosion service. Chromium, which is present in alloy B-2, in residual quantities, also plays an important role because the environments are normally oxidizing in nature.

The nickel-based alloys are sometimes referred to as superalloys. They have been defined as those possessing good high temperature strength and oxidation resistance and are alloys of nickel, cobalt, and iron which contain larger amounts of chromium (25 to 30%) for oxidation resistance. Classifications include iron-nickel, nickel, and cobalt-based alloys. For many years cobalt-based superalloys held the edge, but because of the precarious availability of cobalt from South Africa, the nickel-based superalloys have replaced many of the cobalt-based alloys.

The physical metallurgy of these alloys is the result of the precipitation of a very fine distribution of small particles, primarily Ni_3Al and Ni_3Ti, which have the generic name gamma prime in a gamma matrix. The nickel-iron alloys also have a γ phase in the form of a Ni_3Nb compound. In actuality the superalloys are really dispersion-hardened alloys since they achieve their strength by a fine dispersion of these compounds. Even though these compound particles are often obtained via an aging or precipitation heat treatment, they do not develop the coherency strains that the true precipitation hardening alloys do. These particles resist dislocation motion and thereby strengthen the base metal. In addition these particles resist growth at elevated temperatures. For this reason they have found application in the turbines and hot components of jet aircraft. Dispersion hardening alloys do not overage as readily as precipitation hardening alloys.

Not all nickel-based alloys are used for high temperature applications, the Monels and some solid–solution Inconels are the most notable exceptions.

The nickel and high nickel alloys will be discussed individually.

I. NICKEL 200 AND NICKEL 201

This family is represented by nickel alloys 200 (N02200) and 201 (NO2201). The chemical composition is shown in Table 15.1. Commercially pure nickel is a white magnetic metal very similar to copper in its other physical and mechanical properties. Refer to Tables 15.2 and 15.3.

The Curie point—the temperature at which it loses its magnetism— varies with the type and quantity of alloy additions, rising with increased iron and cobalt additions and falling as copper, silicon, and most other elements are added. Nickel is also an important alloying element in other families of corrosion resistant materials.

TABLE 15.1 Chemical Composition of Nickel 200 and Nickel 201

Chemical	Weight percent, max.	
	Nickel 200	Nickel 201
Carbon	0.1	0.02
Copper	0.25	0.25
Iron	0.4	0.4
Nickel	99.2	99.0
Silicon	0.15	0.15
Titanium	0.1	0.1

TABLE 15.2 Mechanical and Physical Properties of Nickel 200 and Nickel 201

Property	Nickel 200	Nickel 201
Modulus of elasticity $\times\ 10^6$ (psi)	28	30
Tensile strength $\times\ 10^3$ (psi)	27	58.5
Yield strength 0.2% offset $\times\ 10^3$ (psi)	21.5	15
Elongation in 2 in. (%)	47	50
Brinell hardness	105	87
Density (lb/in.^{3})	0.321	0.321
Specific gravity	8.89	8.89
Specific heat (Btu/lb °F)	0.109	0.109
Thermal conductivity (Btu/hr ft^2/°F/in.)		
at 0–70°F	500	569
at 70–200°F	465	512
at 70–400°F	425	460
at 70–600°F	390	408
at 70–800°F	390	392
at 70–1000°F	405	410
at 70–1200°F	420	428
Coefficient of thermal expansion $\times\ 10^{-6}$ (in./in./°F)		
at 0–70°F	6.3	
at 70–200°F	7.4	7.3
at 70–400°F	7.7	
at 70–600°F	8.0	
at 70–800°F	8.3	
at 70–1000°F	8.5	
at 70–1200°F	8.7	

TABLE 15.3 Allowable Design Stress for Nickel 200 and Nickel 201

Operating temperature (°F/°C)	Allowable stress (psi)			
	4 in. diameter pipe		5- to 12-in. diameter pipe	
	Nickel 200	Nickel 201	Nickel 200	Nickel 201
100/38	10000	8000	8000	6700
200/93	10000	7700	8000	6400
300/149	10000	7500	8000	6300
400/204	10000	7500	8000	6200
500/260	10000	7500	8000	6200
600/316	10000	7500	8000	6200
700/371		7400		6200
800/427		7200		5900
900/482		4500		4500
1000/538		3000		3000
1100/593		2000		2000
1200/649		1200		1200

Alloy 201 is the low carbon version of alloy 200. Alloy 200 is subject to the formation of a grain boundary graphitic phase, which reduces ductility tremendously. Consequently nickel alloy 200 is limited to a maximum operating temperature of 600°F (315°C). For applications above this temperature alloy 201 should be used.

The corrosion resistance of alloy 200 and alloy 201 are the same. They exhibit outstanding resistance to hot alkalies, particularly caustic soda. Excellent resistance is shown at all concentrations at temperatures up to and including the molten state. Below 50% the corrosion rates are negligible, usually being less than 0.2 mpy even in boiling solutions. As concentrations and temperatures increase, corrosion rates increase very slowly. Impurities in the caustic, such as chlorates and hypochlorites, will determine the corrosion rate.

Nickel is not subject to stress corrosion cracking in any of the chloride salts, and it exhibits excellent general resistance to nonoxidizing halides. Oxidizing acid chlorides, such as ferric, cupric, and mercuric, are very corrosive and should be avoided.

Nickel 201 finds application in the handling of hot, dry chlorine and hydrogen chloride gas on a continuous basis up to 1000°F (540°C). The resistance is attributed to the formation of a nickel chloride film. Dry fluorine and bromine can be handled in the same manner. The resistance will decrease when moisture is present.

Nickel exhibits excellent resistance to most organic acids, particularly fatty acids such as stearic and oleic, if aeration is not high.

Nickel is not attacked by anhydrous ammonia or ammonium hydroxide in concentrations of 1% or less. Stronger concentrations cause rapid attack.

Nickel also finds application in the handling of food and synthetic fibers because of its ability to maintain product purity. The presence of nickel ions is not detrimental to the flavor of food products, and it is not toxic. Unlike iron and copper, nickel will not discolor organic chemicals such as phenol and viscous rayon.

Refer to Table 15.4 for the compatibility of nickel 200 and nickel 201 with selected corrodents.

In addition to alloy 200 there are a number of alloy modifications developed for increased strength, hardness, resistance to galling, and improved corrosion resistance. Other alloys in this family are not specifically used for their corrosion resistance. Alloy 270 is a high-purity, low-inclusion version of alloy 200. Alloy 301 (also referred to by the tradename of Duranickel) is a precipitation-hardenable alloy containing aluminum and titanium. Alloy 300 (also called by the tradename Permanickel) is a moderately precipitation-hardenable alloy containing titanium and magnesium that also possesses higher thermal and electrical conductivity.

II. MONEL ALLOY 400 (NO4400)

The first nickel alloy, invented in 1905, was approximately two-thirds nickel and one-third copper. The present equivalent of that alloy, Monel 400, remains one of the widely used nickel alloys. Refer to Table 15.5 for the chemical composition.

Nickel-copper alloys offer somewhat higher strength than unalloyed nickel, with no sacrifice of ductility. The thermal conductivity of alloy 400, although lower than that of nickel, is significantly higher than that of nickel alloys containing substantial amounts of chromium or iron. The alloying of 30–33% copper with nickel, producing alloy 400, provides an alloy with many of the characteristics of chemically pure nickel but improves others.

Nickel-copper alloy 400 is a solid solution binary alloy, combining high strength (comparable to structural steel) and toughness over a wide range, with excellent resistance to many corrosive environments. It is strengthened by cold working. The alloy can be used at temperatures up to 800°F (427°C) and as high as 1000°F (538°C) in sulfur-free oxidizing atmospheres. It also has excellent mechanical properties at subzero temperatures since it does not undergo a ductile-to-brittle transition even at such

TABLE 15.4 Compatibility of Nickel 200 and Nickel 201 with Selected Corrodents[a]

Chemical	Maximum temp.		Chemical	Maximum temp.	
	°F	°C		°F	°C
Acetaldehyde	200	93	Barium sulfide	110	43
Acetic acid, 10%	90	32	Benzaldehyde	210	99
Acetic acid, 50%	90	32	Benzene	210	99
Acetic acid, 80%	120	49	Benzene sulfonic acid, 10%	190	88
Acetic acid, glacial	X		Benzoic acid	400	204
Acetic anhydride	170	77	Benzyl alcohol	210	99
Acetone	190	88	Benzyl chloride	210	99
Acetyl chloride	100	38	Borax	200	93
Acrylic acid			Boric acid	210	99
Acrylonitrile	210	99	Bromine gas, dry	60	16
Adipic acid	210	99	Bromine gas, moist	X	
Allyl alcohol	220	104	Bromine, liquid		
Allyl chloride	190	88	Butadiene	80	27
Alum	170	77	Butyl acetate	80	27
Aluminum acetate			Butyl alcohol	200	93
Aluminum chloride, aqueous	300	149	n-Butylamine		
Aluminum chloride, dry	60	16	Butyl phthalate	210	99
Aluminum fluoride	90	32	Butyric acid	X	
Aluminum hydroxide	80	27	Calcium bisulfide		
Aluminum nitrate			Calcium bisulfite	X	
Aluminum oxychloride			Calcium carbonate		
Aluminum sulfate	210	99	Calcium chlorate	140	60
Ammonia gas	90	32	Calcium chloride	80	27
Ammonium bifluoride			Calcium hydroxide, 10%	210	99
Ammonium carbonate	190	88	Calcium hydroxide, sat.	200	93
Ammonium chloride, 10%	230	110	Calcium hypochlorite	X	
Ammonium chloride, 50%	170	77	Calcim nitrate		
Ammonium chloride, sat.	570	299	Calcium oxide	90	32
Ammonium fluoride, 10%	210	99	Calcium sulfate	210	99
Ammonium fluoride, 25%	200	93	Caprylic acid[b]	210	99
Ammonium hydroxide, 25%	X		Carbon bisulfide	X	
Ammonium hydroxide, sat.	320	160	Carbon dioxide, dry	210	99
Ammonium nitrate	90	32	Carbon dioxide, wet	200	93
Ammonium persulfate	X		Carbon disulfide	X	
Ammonium phosphate, 30%	210	99	Carbon monoxide	570	290
Ammonium sulfate, 10–40%	210	99	Carbon tetrachloride	210	99
Ammonium sulfide			Carbonic acid	80	27
Ammonium sulfite	X		Cellosolve	210	99
Amyl acetate	300	149	Chloracetic acid, 50% water		
Amyl alcohol			Chloracetic acid	210	99
Amyl chloride	90	32	Chlorine gas, dry	200	93
Aniline	210	99	Chlorine gas, wet	X	
Antimony trichloride	210	99	Chlorine, liquid		
Aqua regia, 3:1	X		Chlorobenzene	120	49
Barium carbonate	210	99	Chloroform	210	99
Barium chloride	80	27	Chlorosulfonic acid	80	27
Barium hydroxide	90	32	Chromic acid, 10%	100	38
Barium sulfate	210	99	Chromic acid, 50%	X	

TABLE 15.4   Continued

Chemical	Maximum temp. °F	Maximum temp. °C	Chemical	Maximum temp. °F	Maximum temp. °C
Chromyl chloride	210	99	Methyl ethyl ketone		
Citric acid, 15%	210	99	Methyl isobutyl ketone	200	93
Citric acid, conc.	80	27	Muriatic acid	X	
Copper acetate	100	38	Nitric acid, 5%	X	
Copper carbonate	X		Nitric acid, 20%	X	
Copper chloride	X		Nitric acid, 70%	X	
Copper cyanide	X		Nitric acid, anhydrous	X	
Copper sulfate	X		Nitrous acid, conc.	X	
Cresol	100	38	Oleum		
Cupric chloride, 5%	X		Perchloric acid, 10%	X	
Cupric chloride, 50%	X		Perchloric acid, 70%		
Cyclohexane	80	27	Phenol, sulfur free	570	299
Cyclohexanol	80	27	Phosphoric acid, 50–80%	X	
Dichloroacetic acid			Picric acid	80	27
Dichloroethane (ethylene dichloride)	X		Potassium bromide, 30%		
Ethylene glycol	210	99	Salicyclic acid	80	27
Ferric chloride	X		Silver bromide, 10%		
Ferric chloride, 50% in water	X		Sodium carbonate, to 30%	210	99
Ferric nitrate, 10–50%	X		Sodium chloride, to 30%	210	99
Ferrous chloride	X		Sodium hydroxide, 10%^{c}	210	99
Ferrous nitrate			Sodium hydroxide, 50%^{c}	300	149
Fluorine gas, dry	570	290	Sodium hydroxide, conc.	200	93
Fluorine gas, moist	60	16	Sodium hypochlorite, 20%	X	
Hydrobromic acid, dilute	X		Sodium hypochlorite, conc.	X	
Hydrobromic acid, 20%	X		Sodium sulfide, to 50%	X	
Hydrobromic acid, 50%	X		Stannic chloride	X	
Hydrochloric acid, 20%	80	27	Stannous chloride, dry	570	299
Hydrochloric acid, 38%	X		Sulfuric acid, 10%	X	
Hydrocyanic acid, 10%			Sulfuric acid, 50%	X	
Hydrofluoric acid, 30%^{c}	170	77	Sulfuric acid, 70%	X	
Hydrofluoric acid, 70%^{c}	100	38	Sulfuric acid, 90%	X	
Hydrofluoric acid, 100%^{c}	120	49	Sulfuric acid, 98%	X	
Hypochlorous acid	X		Sulfuric acid, 100%	X	
Iodine solution, 10%			Sulfuric acid, fuming	X	
Ketones, general	100	38	Sulfurous acid	X	
Lactic acid, 25%	X		Thionyl chloride	210	99
Lactic acid, conc.	X		Toluene	210	99
Magnesium chloride	300	149	Trichloroacetic acid	80	27
Malic acid	210	99	White liquor		
Manganese chloride, 37%	90	32	Zinc chloride, to 80%	200	93
Methyl chloride	210	99			

^{a}The chemicals listed are in the pure state or in a saturated solution unless otherwise indicated. Compatibility is shown to the maximum allowable temperature for which data are available. Incompatibility is shown by an X. When compatible, corrosion rate is <20 mpy.
^{b}Material subject to pitting.
^{c}Material subject to stress cracking.
Source: Ref. 6.

TABLE 15.5 Chemical Composition of Monel Alloys

	Weight percent		
Chemical	400 (N04400)	405 (N04405)	K-500 (N05500)
Carbon	0.2 max.	0.3 max.	0.1 max.
Manganese	2.0 max.	2.0 max.	0.8 max.
Silicon	0.5 max.	0.5 max.	0.2 max.
Sulfur	0.015 max.	0.020–0.060	—
Nickel	63.0–70.0	63.0–70.0	63.0 min.
Iron	2.50 max.	2.50 max.	1.0
Copper	Balance	Balance	27.0–33.0
Columbium	—	—	2.3–3.15
Titanium	—	—	0.35–0.85

temperatures. The alloy is readily fabricated and is virtually immune to chloride ion stress corrosion cracking in typical environments. Generally its corrosion resistance is very good in reducing environments, but poor in oxidizing conditions. Physical and mechanical properties of alloy 400 will be found in Table 15.6 and allowable design stresses in Table 15.7.

Alloy 400 is readily hot or cold worked. The hot working range is 1700°F (927°C) to 2100°F (1149°C). Optimum working temperature is 2000°F (1093°C). Finished fabrications can be produced to a rather wide range of mechanical properties by proper control of the amount of hot and/or cold work and by the selection of proper thermal treatments. Refer to Table 15.8 for the effect of cold work on typical room temperature tensile properties.

The general corrosion resistance of alloy 400 in the nonoxidizing acids, such as sulfuric, hydrochloric, and phosphoric, is improved over that of pure nickel. The influence of oxidizers is the same as for nickel. The alloy is not resistant to oxidizing media such as nitric acid, ferric chloride, chromic acid, wet chlorine, sulfur dioxide, or ammonia.

Alloy 400 does have excellent resistance to hydrofluoric acid solutions at all concentrations and temperatures, as shown in Fig. 15.1. Again aeration or the presence of oxidizing salts increases the corrosion rate. This alloy is widely used in HF alkylation, is comparatively insensitive to velocity effects, and is widely used for critical parts such as bubble caps or valves that are in contact with flowing acid Monel 400 is subject to stress corrosion cracking in moist, aerated hydrofluoric or hydrofluorosilicic acid vapor. However, cracking is unlikely if the metal is completely immersed in the acid.

TABLE 15.6   Mechanical and Physical Properties of
Monel 400

Modulus of elasticity $\times$ 10^6 (psi)	26
Tensile strength $\times$ 10^3 (psi)	70
Yield strength 0.2% offset $\times$ 10^3 (psi)	25–28
Elongation in 2 in. (%)	48
Brinell hardness	130
Density (lb/in.^{3})	0.318
Specific gravity	8.84
Specific heat, at 32–212°F (Btu/lb °F)	0.102
Thermal conductivity (Btu/hr/ft^2/in./°F)	
at 70°F	151
at 200°F	167
at 500°F	204
Coefficient of thermal expansion	
$\times$ 10^{-6} (in./in. °F)	
at 70–200°F	7.7
at 70–400°F	8.6
at 70–500°F	8.7
at 70–1000°F	9.1

TABLE 15.7   Allowable Design Stress for Monel 400

	Allowable design stress $\times$ 10^3 (psi)	
	≤4-in. diameter	5- to 12-in. diameter
100/38	17.5	16.6
200/93	16.5	14.6
300/149	15.5	13.6
400/204	14.8	13.2
500/260	14.7	13.1
600/316	14.7	13.1
700/371	14.7	13.1
800/427	14.5	13.1
900/482	8.0	8.0

TABLE 15.8   Effect of Cold Work on Typical Room Temperature
Tensile Properties

Percent cold worked	Yield strength 0.2% offset (ksi)	Tensile strength (ksi)	Elongation (%)	Reduction of area (%)
Starting material 0.250 in. round unannealed				
0	40	82	54	80
5	80	89	39	79
10	85	95	35	80
15	95	103	25	78
20	101	110	20	79
30	116	121	21	73
40	126	131	18	72
Starting material 0.250 in. round annealed, 1733°F (945°C) 30 min. W.Q. reducing atmosphere				
0	25	72	57	82
5	69	81	42	79
10	74	83	37	82
15	85	90	30	83
20	92	99	23	86
30	105	111	21	82
40	112	117	19	79

Water handling, including seawater and brackish waters, is a major area of application. It gives excellent service under high velocity conditions, as in propellers, propeller shafts, pump shafts, impellers, and condenser tubes. The addition of iron to the composition improves the resistance to cavitation and erosion in condenser tube applications. Alloy 400 can pit in stagnant seawater, as does nickel 200; however, the rates are considerably lower. The absence of chloride stress corrosion cracking is also a factor in the selection of the alloy for this service.

Alloy 400 undergoes negligible corrosion in all types of natural atmospheres. Indoor exposures produce a very light tarnish that is easily removed by occasional wiping. Outdoor surfaces that are exposed to rain produce a thin gray-green patina. In sulfurous atmospheres a smooth, brown adherent film forms.

Monel 400 exhibits stress corrosion cracking in high temperatures, concentrated caustic, and in mercury. Refer to Table 15.9 for the compati-

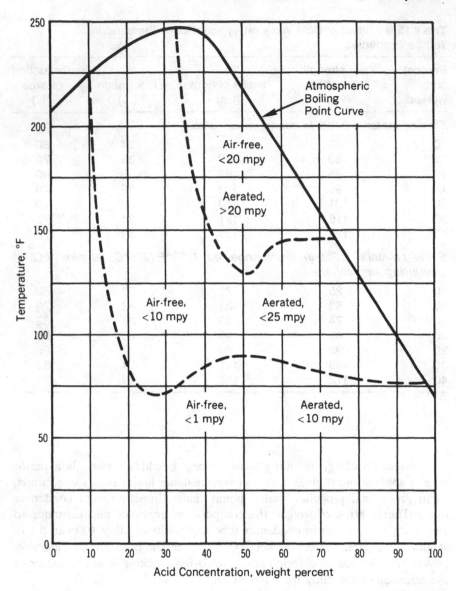

FIGURE 15.1 Isocorrosion diagram for alloy 400 in hydrofluoric acid. (From Ref. 3.)

TABLE 15.9 Compatibility of Monel 400 with Selected Corrodents[a]

Chemical	Maximum temp. °F	Maximum temp. °C	Chemical	Maximum temp. °F	Maximum temp. °C
Acetaldehyde	170	77	Benzene sulfonic acid, 10%	210	99
Acetamide	340	171	Benzoic acid	210	99
Acetic acid, 10%	80	27	Benzyl alcohol	400	204
Acetic acid, 50%	200	93	Benzyl chloride	210	99
Acetic acid, 80%	200	93	Borax	90	32
Acetic acid, glacial	290	143	Boric acid	210	99
Acetic anhydride	190	88	Bromine gas, dry	120	49
Acetone	190	88	Bromine gas, moist	X	
Acetyl chloride	400	204	Butadiene	180	82
Acrylonitrile	210	99	Butyl acetate	380	193
Adipic acid	210	99	Butyl alcohol	200	93
Allyl alcohol	400	204	Butyl phthalate	210	99
Allyl chloride	200	93	Butyric acid	210	99
Alum	100	38	Calcium bisulfide	60	16
Aluminum acetate	80	27	Calcium bisulfite	X	
Aluminum chloride, aqueous	X		Calcium carbonate	200	93
Aluminum chloride, dry	150	66	Calcium chlorate	140	60
Aluminum fluoride	90	32	Calcium chloride	350	177
Aluminum hydroxide	80	27	Calcium hydroxide, 10%	210	99
Aluminum sulfate	210	99	Calcium hydroxide, sat.	200	93
Ammonia gas	X		Calcium hypochlorite	X	
Ammonium bifluoride	400	204	Calcium oxide	90	32
Ammonium carbonate	190	88	Calcium sulfate	80	27
Ammonium chloride, 10%	230	110	Caprylic acid[b]	210	99
Ammonium chloride, 50%	170	77	Carbon bisulfide	X	
Ammonium chloride, sat.	570	299	Carbon dioxide, dry	570	299
Ammonium fluoride, 10%	400	204	Carbon dioxide, wet[b]	400	204
Ammonium fluoride, 25%	400	204	Carbon disulfide	X	
Ammonium hydroxide, 25%	X		Carbon monoxide	570	299
Ammonium hydroxide, sat.	X		Carbon tetrachloride	400	204
Ammonium nitrate	X		Carbonic acid	X	
Ammonium persulfate	X		Cellosolve	210	99
Ammonium phosphate, 30%	210	99	Chloracetic acid, 50% water	180	82
Ammonium sulfate, 10–40%	400	204	Chloracetic acid	X	
Ammonium sulfite	90	32	Chlorine gas, dry	570	299
Amyl acetate	300	149	Chlorine gas, wet	X	
Amyl alcohol	180	82	Chlorine, liquid	150	66
Amyl chloride	400	204	Chlorobenzene, dry	400	204
Aniline	210	99	Chloroform	210	99
Antimony trichloride	350	177	Chlorosulfonic acid	80	27
Aqua regia, 3:1	X		Chromic acid, 10%	130	54
Barium carbonate	210	99	Chromic acid, 50%	X	
Barium chloride	210	99	Chromyl chloride	210	99
Barium hydroxide	80	27	Citric acid, 15%	210	99
Barium sulfate	210	99	Citric acid, conc.	80	27
Barium sulfide	X		Copper acetate	X	
Benzaldehyde	210	99	Copper carbonate	X	
Benzene	210	99	Copper chloride	X	

TABLE 15.9 Continued

Chemical	Maximum temp. °F	°C	Chemical	Maximum temp. °F	°C
Copper cyanide	X		Nitric acid, 5%	X	
Copper sulfate	X		Nitric acid, 20%	X	
Cresol	100	38	Nitric acid, 70%	X	
Cupric chloride, 5%	X		Nitric acid, anhydrous	X	
Cupric chloride, 50%	X		Nitrous acid, conc.	X	
Cyclohexane	180	82	Oleum	X	
Cyclohexanol	80	27	Perchloric acid, 10%	X	
Dichloroethane (ethylene dichloride)	200	93	Perchloric acid, 70%	X	
Ethylene glycol	210	99	Phenol	570	299
Ferric chloride	X		Phosphoric acid, 50–80%	X	
Ferric chloride, 50% in water	X		Picric acid	X	
Ferric nitrate, 10–50%	X		Potassium bromide, 30%, air free	210	99
Ferrous chloride	X		Salicylic acid	210	99
Ferrous nitrate			Silver bromide, 10%	80	27
Fluorine gas, dry	570	299	Sodium carbonate	210	99
Fluorine gas, moist	X		Sodium chloride, to 30%	210	99
Hydrobromic acid, dilute	X		Sodium hydroxide, 10%[c]	350	177
Hydrobromic acid, 20%	X		Sodium hydroxide, 50%[c]	300	149
Hydrobromic acid, 50%	X		Sodium hydroxide, conc.	350	177
Hydrochloric acid, 20%	80	27	Sodium hypochlorite, 20%	X	
Hydrochloric acid, 38%	X		Sodium hypochlorite, conc.	X	
Hydrocyanic acid, 10%	80	27	Sodium sulfide, to 50%	210	99
Hydrofluoric acid, 30%[c]	400	204	Stannic chloride	X	
Hydrofluoric acid, 70%[c]	400	204	Stannous chloride, dry	570	299
Hydrofluoric acid, 100%[c]	210	99	Sulfuric acid, 10%	X	
Hypochlorous acid	X		Sulfuric acid, 50%	80	27
Iodine solution, 10%	X		Sulfuric acid, 70%	80	27
Ketones, general	100	38	Sulfuric acid, 90%	X	
Lactic acid, 25%	X		Sulfuric acid, 98%	X	
Lactic acid, conc.	X		Sulfuric acid, 100%	X	
Magnesium chloride, 50%	350	177	Sulfuric acid, fuming	X	
Malic acid	210	99	Sulfurous acid	X	
Manganese chloride, 40%	100	38	Thionyl chloride	300	149
Methyl chloride	210	99	Toluene	210	99
Methyl ethyl ketone	200	93	Trichloroacetic acid	170	77
Methyl isobutyl ketone	200	93	White liquor	X	
Muriatic acid	X		Zinc chloride, to 80%	200	93

[a] The chemicals listed are in the pure state or in a saturated solution unless otherwise indicated. Compatibility is shown to the maximum allowable temperature for which data are available. Incompability is shown by an X. When compatible, the corrosion rate is <20 mpy.
[b] Not for use with carbonated beverages.
[c] Material is subject to stress cracking.
Source: Ref. 6.

bility of Monel 400 with selected corrodents. A more detailed compilation will be found in Ref. 6.

Monel alloy 405 is a higher sulfur grade in which the sulfur content is increased over that of alloy 400 in order to improve machinability. Refer to Table 15.5 for the chemical composition. The corrosion resistance of this alloy is essentially the same as alloy 400. Mechanical properties are shown in Table 15.10.

Monel alloy K-500 (NO5500) is an age-hardenable alloy which combines the excellent corrosion resistance characteristics of alloy 400 with the added advantage of increased strength and hardness. Chemical composition will be found in Table 15.5. Age hardening increases its strength and hardness; however, still higher properties can be achieved when the alloy is cold worked prior to the aging treatment. Alloy K-500 has good mechanical properties over a wide range of temperatures. Strength is maintained up to about 1200°F (649°C), and the alloy is strong, tough, and ductile at temperatures as low as −423°F (−235°C). It also has low permeability and is nonmagnetic to −210°F (−134°C). Refer to Table 15.11 for mechanical and physical properties.

Typical applications include pump shafts, impellers, electronic components, doctor blades and scrapers, oil well drill collars and instruments, springs, and valve trim.

III. ALLOY B-2

Alloy B was originally developed to resist hydrochloric acid up to the atmospheric boiling point. However, because of susceptibility to intergranular attack in the heat-affected zone after welding in some environments, a low carbon variant, alloy B-2, was developed, and is replacing alloy B in most applications. The chemical composition is shown in Table 15.12.

This alloy is uniquely different from other corrosion resistant alloys because it does not contain chromium. Molybdenum is the primary alloying

TABLE 15.10 Mechanical and Physical Properties of Monel Alloy 405

Tensile strength × 10^3 (psi)	70
Yield strength 0.2% offset × 10^3 (psi)	25
Elongation in 2 in. (%)	28
Brinell hardness	110–140
Density (lb/in.³)	0.318
Specific gravity	8.48

Table 15.11 Mechanical and Physical Properties of
Monel Alloy K-500

Tensile strength × 10³ (psi)	100
Yield strength 0.2% offset × 10³ (psi)	50
Elongation in 2 in. (%)	35
Brinell hardness	161
Density (lb/in.³)	0.318
Specific gravity	8.48
Specific heat (J/kgK)	418
Thermal conductivity (W/mK)	17.4
Coefficient of thermal expansion (m/mK) at 20–93°C	13.7

element and provides significant corrosion resistance to reducing environments.

Alloy B-2 has improved resistance to knifeline and heat-affected zone attack. It also resists formation of grain-boundary precipitates in weld heat-affected zone. Refer to Table 15.13 for the physical and mechanical properties of alloy B-2 and Table 15.14 for the maximum allowable design stress.

Alloy B-2 has excellent elevated-temperature (1650°F/900°C) mechanical properties because of the high molybdenum content and has been used for mechanical components in reducing environments and vacuum furnaces. Because of the formation of the intermetallic phases Ni_3Mo and Ni_4Mo after long aging, the use of alloy B-2 in the temperature range 1110–1560°F (600–850°C) is not recommended, regardless of environment.

Alloy B-2 is recommended for service in handling all concentrations of hydrochloric acid in the temperature range of 158–212°F (70–100°C) and for handling wet hydrogen chloride gas, as shown in Fig. 15.2.

Alloy B-2 has excellent resistance to pure sulfuric acid at all concentrations and temperatures below 60% acid and good resistance to 212°F

Table 15.12 Chemical
Composition of Alloy B-2

Chemical	Weight percent
Molybdenum	26.0–30.0
Chromium	1.0 max.
Iron	2.0 max.
Nickel	Balance

TABLE 15.13 Mechanical and Physical Properties of Alloy B-2

Modulus of elasticity $\times 10^6$ (psi)	31.4
Tensile strength $\times 10^3$ (psi)	110
Yield strength 0.2% offset $\times 10^3$ (psi)	60
Elongation in 2 in. (%)	60
Brinell hardness	210
Density (lb/in.^{3})	0.333
Specific gravity	9.22
Specific heat, at 212°F (Btu/lb °F)	0.093
Thermal conductivity (Btu/ft^2/in. hr °F)	
at 32°F	77
at 212°F	85
at 392°F	93
at 572°F	102
at 752°F	111
at 932°F	120
at 1112°F	130
Coefficient of thermal expansion $\times 10^{-6}$ (in./in. °F)	
at 68–200°F	5.7
at 68–600°F	6.2
at 68–1000°F	6.5

TABLE 15.14 Allowable Design Stress for Alloy B-2

Operating temperature (°F/°C)	Allowable design stress (psi)
100/38	25000
200/93	25000
300/149	24750
400/204	22750
500/260	21450
600/316	20750
650/343	20100

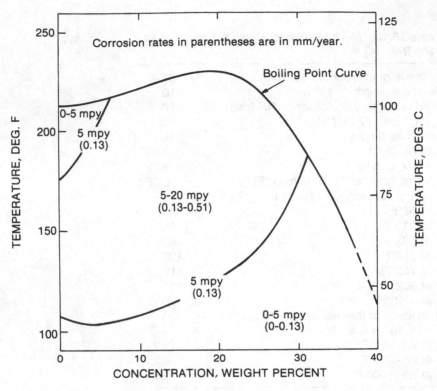

FIGURE 15.2 Isocorrosion diagram for alloy B-2 in hydrochloric acid. (From Ref. 3.)

(100°C) above 60% acid, as shown in Fig. 15.3. The alloy is resistant to a number of phosphoric acids and numerous organic acids such as acetic, formic, and cresylic. It is also resistant to many chloride-bearing salts (non-oxidizing), such as aluminum chloride, magnesium chloride, and antimony chloride.

Since alloy B-2 is nickel rich (approximately 70%), it is resistant to chloride-induced stress corrosion cracking. Because of its high molybdenum content it is highly resistant to pitting attack in most acid chloride environments.

Alloy B-2 is not recommended for elevated temperature service except in very specific circumstances. Since there is no chromium in the alloy, it scales heavily at temperatures above 1400°F (760°C). A nonprotective layer of molybdenum trioxide forms and results in a heavy green oxidation scale.

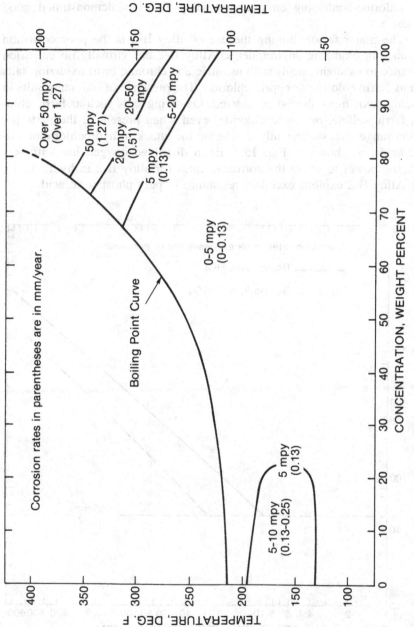

FIGURE 15.3 Isocorrosion diagram for alloy B-2 in sulfuric acid. (From Ref. 3.)

In a chlorine-containing environment alloy B-2 has demonstrated good resistance.

The major factor limiting the use of alloy B-2 is the poor corrosion resistance in oxidizing environments. Alloy B-2 has virtually no corrosion resistance to oxidizing acids such as nitric and chromic or to oxidizing salts such as ferric chloride or cupric chloride. The presence of oxidizing salts in reducing acids must also be considered. Oxidizing salts such as ferric chloride, ferric sulfate, or cupric chloride, even when present in the parts per million range, can significantly accelerate the attack in hydrochloric or sulfuric acids, as shown in Fig. 15.4. Even dissolved oxygen has sufficient oxidizing power to affect the corrosion rates for alloy B-2 in hydrochloric acid. Alloy B-2 exhibits excellent resistance to pure phosphoric acid.

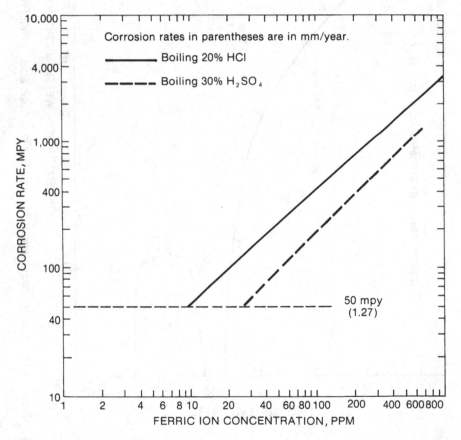

FIGURE 15.4 Effect of ferric ions on corrosion rate of alloy B-2. (From Ref. 3.)

Stress corrosion cracking has been observed in alloy B-2 in 20% magnesium chloride solution at temperatures exceeding 500°F (260°C). Other environments in which stress corrosion cracking of this alloy has been observed include high purity water at 350°F (170°C), molten lithium at 315°F (157°C), oxygenated deionized water at 400°F (204°C), 1% hydrogen iodide at 62–450°F (17–232°C), and 10% hydrochloric acid at 400°F (204°C). In some environments, such as concentrated ammonia at 77–140°F (25–60°C) cracking has been observed if the alloy was aged at 1382°F (750°C) for 24 hr before the test. Precipitation of an ordered intermetallic phase Ni_4Mo has been attributed to be the cause of the increased embrittlement.

Table 15.15 shows the compatibility of alloy B-2 with selected corrodents. Reference 6 contains a more extensive listing.

IV. ALLOY 600 (N06600)

Alloy 600, also known as Inconel, is a nickel base alloy with about 16% chromium and 7% iron that is used primarily to resist corrosive atmospheres at elevated temperatures. The chemical composition will be found in Table 15.16.

Alloy 600 has excellent mechanical properties and a combination of high strength and good workability. It performs well in temperatures from cryogenic to 1200°F (649°C) and is readily fabricated and welded. Refer to Table 15.17 for the mechanical and physical properties of alloy 600 and to Table 15.18 for allowable design stresses.

Although resistant to oxidation, the presence of sulfur in the environment can significantly increase the rate of attack. The mode of attack is generally intergranular, and therefore the attack proceeds more rapidly and the maximum use temperature is restricted to about 600°F (315°C).

Inconel has excellent resistance to dry halogens at elevated temperatures and has been used successfully for chlorination equipment at temperatures up to 1000°F (538°C). Where arrangements can be made for cooling the metal surface, the alloy can be used at even higher gas temperatures.

Resistance to stress corrosion cracking is imparted to alloy 600 by virtue of its nickel base. The alloy therefore finds considerable use in handling water environments where stainless steels fail by cracking. Because of its resistance to corrosion in high purity water, it has a number of uses in nuclear reactors, including steam generator tubing and primary water piping. The lack of molybdenum in the alloy precludes its use in applications where pitting is the primary mode of failure.

In certain high temperature caustic applications where sulfur is present, alloy 600 is substituted for alloy 201 because of its improved resistance. Inconel is, however, subject to stress corrosion cracking in high-temperature,

TABLE 15.15 Compatibility of Alloy B-2 and Alloy C-276 with
Selected Corrodents[a]

Chemical	Maximum temp. (°F/°C)	
	Alloy B-2	Alloy C-276
Acetaldehyde	80/27	140/60
Acetamide		60/16
Acetic acid, 10%	300/149	300/149
Acetic acid, 50%	300/149	300/149
Acetic acid, 80%	300/149	300/149
Acetic acid, glacial	560/293	560/293
Acetic anhydride	280/138	280/138
Acetone	200/93	200/93
Acetyl chloride	80/27	
Acrylic acid	210/99	
Acrylonitrile	210/99	210/99
Adipic acid		210/99
Allyl alcohol		570/299
Allyl chloride	200/93	
Alum	150/66	150/66
Aluminum acetate	60/16	60/16
Aluminum chloride, aqueous	300/149	210/99
Aluminum chloride, dry	210/99	210/99
Aluminum fluoride	80/27	80/27
Aluminum sulfate	210/99	210/99
Ammonia gas	200/93	200/93
Ammonium bifluoride		380/193
Ammonium carbonate	300/149	300/149
Ammonium chloride, 10%	210/99	210/99
Ammonium chloride, 50%	210/99	210/99
Ammonium chloride, sat.	570/299	570/299
Ammonium fluoride, 10%	210/99	210/99
Ammonium fluoride, 25%		210/99
Ammonium hydroxide, 25%	210/99	570/299
Ammonium hydroxide, sat.	210/99	570/299
Ammonium persulfate	X	
Ammonium sulfate, 10–40%	80/27	200/93
Ammonium sulfite		100/38
Amyl acetate	340/171	340/171
Amyl alcohol		180/82
Amyl chloride	210/99	90/32
Aniline	570/299	570/299
Antimony trichloride	210/99	210/99
Aqua regia, 3:1	X	X
Barium carbonate	570/299	570/299
Barium chloride	570/299	210/99
Barium hydroxide	270/132	270/132

TABLE 15.15 Continued

Chemical	Maximum temp. (°F/°C)	
	Alloy B-2	Alloy C-276
Barium sulfate	80/27	
Benzaldehyde	210/99	210/99
Benzene	210/99	210/99
Benzene sulfonic acid, 10%	210/99	210/99
Benzoic acid	210/99	
Benzyl alcohol	210/99	210/99
Benzyl chloride	210/99	
Borax	120/49	120/49
Boric acid	570/299	570/299
Bromine gas, dry	60/16	60/16
Bromine gas, moist		60/16
Bromine liquid		180/82
Butadiene	300/149	300/149
Butyl acetate	200/93	200/93
Butyl alcohol	210/99	200/93
n-Butylamine	210/99	210/99
Butyric acid	280/138	280/138
Calcium bisulfite		80/27
Calcium carbonate	210/99	210/99
Calcium chlorate		210/99
Calcium chloride	350/177	350/177
Calcium hydroxide, 10%	210/99	170/77
Calcium hydroxide, sat.	210/99	
Calcium hypochlorite	X	
Calcium nitrate	210/99	210/99
Calcium oxide		90/32
Calcium sulfate, 10%	320/160	320/160
Caprylic acid	300/149	300/149
Carbon bisulfide	180/82	210/99
Carbon dioxide, dry	570/299	570/299
Carbon dioxide, wet	570/299	200/93
Carbon disulfide	180/82	300/149
Carbon monoxide	570/299	570/299
Carbon tetrachloride	300/149	300/149
Carbonic acid	80/27	80/27
Cellosolve	210/99	210/99
Chloracetic acid, 50% water		210/99
Chloracetic acid	370/188	300/149
Chlorine gas, dry	200/93	570/299
Chlorine gas, wet	X	220/104
Chlorine, liquid		110/43
Chlorobenzene	350/177	350/177
Chloroform	210/99	210/99

TABLE 15.15   Continued

Chemical	Maximum temp. (°F/°C)	
	Alloy B-2	Alloy C-276
Chlorosulfonic acid	230/110	230/110
Chromic acid, 10%	130/54	210/99
Chromic acid, 50%	X	210/99
Chromyl chloride	210/99	210/99
Citric acid, 15%	210/99	210/99
Citric acid, conc.	210/99	210/99
Copper acetate	100/38	100/38
Copper carbonate	90/32	90/32
Copper chloride	200/93	200/93
Copper cyanide	150/66	150/66
Copper sulfate	210/99	210/99
Cresol	210/99	210/99
Cupric chloride, 5%	60/16	210/99
Cupric chloride, 50%	210/99	210/99
Cyclohexane	210/99	210/99
Cyclohexanol	80/27	80/27
Dichloroethane	230/110	230/110
Ethylene glycol	570/299	570/299
Ferric chloride	90/32	90/32
Ferric chloride, 50% in water	X	
Ferric nitrate, 10–50%	X	
Ferrous chloride	280/138	280/138
Fluorine gas, dry	80/27	150/66
Fluorine gas, moist		570/299
Hydrobromic acid, dilute	210/99	
Hydrobromic acid, 20%	210/99	90/32
Hydrobromic acid, 50%	260/127	90/32
Hydrochloric acid, 20%	140/60	150/66
Hydrochloric acid, 38%	140/60	90/32
Hydrofluoric acid, 30%	140/60	210/99
Hydrofluoric acid, 70%	110/43	200/93
Hydrofluoric acid, 100%	80/27	210/99
Hypochlorous acid	90/32	80/27
Iodine solution, 10%		180/82
Ketones, general	180/82	100/38
Lactic acid, 25%	250/121	210/99
Lactic acid, conc.	250/121	210/99
Magnesium chloride	300/149	300/149
Malic acid	210/99	210/99
Manganese chloride, 40%	210/99	210/99
Methyl chloride	210/99	90/32
Methyl ethyl ketone	210/99	210/99
Methyl isobutyl ketone	200/93	200/93

TABLE 15.15 Continued

Chemical	Maximum temp. (°F/°C)	
	Alloy B-2	Alloy C-276
Muriatic acid	90/32	90/32
Nitric acid, 5%	X	210/99
Nitric acid, 20%	X	160/71
Nitric acid, 70%	X	200/93
Nitric acid, anhydrous	X	80/27
Nitrous acid, conc.	X	X
Oleum, to 25%	110/43	140/60
Perchloric acid, 70%		220/104
Phenol	570/299	570/299
Phosphoric acid, 50–80%	210/99	210/99
Picric acid	220/104	300/149
Potassium bromide, 30%	90/32	90/32
Salicylic acid	80/27	250/121
Silver bromide, 10%	90/32	90/32
Sodium carbonate	570/299	210/99
Sodium chloride, to 30%	210/99	210/99
Sodium hydroxide, 10%[b]	240/116	230/110
Sodium hydroxide, 50%	250/121	210/99
Sodium hydroxide, conc.	200/93	120/49
Sodium hypochlorite, 20%	X	X
Sodium hypochlorite, conc.	X	X
Sodium sulfide, to 50%	210/99	210/99
Stannic chloride, to 50%	210/99	210/99
Stannous chloride[c]	570/299	210/99
Sulfuric acid, 10%	210/99	200/93
Sulfuric acid, 50%	230/110	230/110
Sulfuric acid, 70%	290/143	290/143
Sulfuric acid, 90%	190/88	190/88
Sulfuric acid, 98%	280/138	210/99
Sulfuric acid, 100%	290/143	190/88
Sulfuric acid, fuming	210/99	90/32
Sulfurous acid	210/99	370/188
Toluene	210/99	210/99
Trichloroacetic acid	210/99	210/99
White liquor	100/38	100/38
Zinc chloride	60/16	250/121

[a]The chemicals listed are in the pure state or in a saturated solution unless otherwise indicated. Compatibility is shown to the maximum allowable temperature for which data are available. Incompatibility is shown by an X. A blank space indicates that data are unavailable. When compatible, corrosion rate is <20 mpy.
[b]Alloy B-2 is subject to stress cracking.
[c]Alloy B-2 is subject to pitting.
Source: Ref. 6.

TABLE 15.16 Chemical
Composition of Alloy 600
(N06600)

Chemical	Weight percent
Nickel	72.0 min.
Chromium	14.0–17.0
Iron	6.0–10.0
Carbon	0.15 max.
Copper	0.50 max.
Manganese	1.0 max.
Sulfur	0.015 max.
Silicon	0.5 max.

TABLE 15.17 Mechanical and Physical Properties of
Alloy 600

Modulus of elasticity $\times 10^6$ (psi)	30–31
Tensile strength $\times 10^3$ (psi)	80
Yield strength 0.2% offset $\times 10^3$ (psi)	30–35
Elongation in 2 in. (%)	40
Rockwell hardness	B-120–170
Density (lb/in.^{3})	0.306
Specific gravity	8.42
Specific heat (Btu/lb °F)	0.106
Thermal conductivity (Btu/h/ft^2/°F/in.)	
at 70°F	103
at 200 °F	109
at 400 °F	121
at 600°F	133
at 800°F	145
at 1000°F	158
at 1200°F	172
Coefficient of thermal expansion	
$\times 10^{-6}$ (in./in./°F)	
at 70–200°F	7.4
at 70–400°F	7.7
at 70–600°F	7.9
at 70–800°F	8.1
at 70–1000°F	8.4
at 70–1200°F	8.6

TABLE 15.18 Allowable Design Stress for
Alloy 600

Operating temp. (°F/°C)	Allowable design stress (psi)	
	≤4-in. pipe	≥5-in. pipe
100/38	20000	20000
200/93	19300	19100
300/149	18800	18200
400/204	18500	17450
500/260	18500	16850
600/316	18500	16100
700/371	18500	15600
800/427	18500	15300
900/482	16000	14900
1000/538	7000	7000
1100/593	3000	3000
1200/649	2000	2000

high-concentration alkalies. For that reason the alloy should be stress relieved prior to use and the operating stresses kept to a minimum. Alloy 600 is almost entirely resistant to attack by solutions of ammonia over the complete range of temperatures and concentrations.

The alloy exhibits greater resistance to sulfuric acid under oxidizing conditions than either nickel 200 or alloy 400. The addition of oxidizing salts to sulfuric acid tends to passivate alloy 600 which makes it suitable for use with acid mine waters or brass pickling solutions, where alloy 400 cannot be used. Table 15.19 provides the compatibility of alloy 600 with selected corrodents. Reference 6 provides a more comprehensive listing.

V. ALLOY 625 (N06625)

Alloy 625, also known as Inconel alloy 625, is used both for its high strength and aqueous corrosion resistance. The strength of alloy 625 is primarily a solid solution effect from molybdenum and columbium. Alloy 625 has excellent weldability. The chemical composition is shown in Table 15.20.

Because of its combination of chromium, molybdenum, carbon, and columbium + tantalum, the alloy retains its strength and oxidation resistance to elevated temperatures. Elevated tensile properties are given in Table 15.21.

TABLE 15.19 Compatibility of Alloy 600 and Alloy 625 with Selected Corrodents[a]

Chemical	Maximum temp. °F	Maximum temp. °C	Chemical	Maximum temp. °F	Maximum temp. °C
Acetaldehyde	140	60	Bromine gas, dry	60	16
Acetic acid, 10%	80	27	Bromine gas, moist	X	
Acetic acid, 50%	X		Butadiene	80	27
Acetic acid, 80%	X		Butyl acetate	80	27
Acetic acid, glacial	220	104	Butyl alcohol	80	27
Acetic anhydride	200	93	n-Butylamine		
Acetone	190	88	Butyl phthalate	210	99
Acetyl chloride	80	27	Butyric acid	X	
Acrylonitrile	210	99	Calcium bisulfite	X	
Adipic acid	210	99	Calcium carbonate	90	32
Allyl alcohol	200	93	Calcium chlorate	80	27
Allyl chloride	150	66	Calcium chloride	80	27
Alum	200	93	Calcium hydroxide, 10%	210	99
Aluminum acetate	80	27	Calcium hydroxide, sat.	90	32
Aluminum chloride, aqueous	X		Calcium hypochlorite	X	
Aluminum chloride, dry	X		Calcium sulfate[c]	210	99
Aluminum fluoride	80	27	Caprylic acid	230	110
Aluminum hydroxide	80	27	Carbon bisulfide	80	27
Aluminum sulfate	X		Carbon dioxide, dry	210	99
Ammonium carbonate	190	88	Carbon dioxide, wet	200	93
Ammonium chloride, 10%[b]	230	110	Carbon disulfide	80	27
Ammonium chloride, 50%	170	77	Carbon monoxide	570	299
Ammonium chloride, sat.	200	93	Carbon tetrachloride	210	99
Ammonium fluoride, 10%	90	32	Carbonic acid	210	99
Ammonium fluoride, 25%	90	32	Cellosolve	210	99
Ammonium hydroxide, 25%	80	27	Chloracetic acid	X	
Ammonium hydroxide, sat.	90	32	Chlorine gas, dry	90	32
Ammonium nitrate	X		Chlorine gas, wet	X	
Ammonium persulfate	80	27	Chlorobenzene	210	99
Ammonium phosphate, 10%	210	99	Chloroform	210	99
Ammonium sulfate, 10–40%[c]	210	99	Chromic acid, 10%	130	54
Ammonium sulfide			Chromic acid, 50%	90	32
Ammonium sulfite	90	32	Chromyl chloride	210	99
Amyl acetate	300	149	Citric acid, 15%	210	99
Amyl chloride	X		Citric acid, conc.	210	99
Aniline	210	99	Copper acetate	100	38
Antimony trichloride	90	32	Copper carbonate	80	27
Aqua regia, 3:1	X		Copper chloride	X	
Barium carbonate	80	27	Copper cyanide	80	27
Barium chloride	570	299	Copper sulfate	80	27
Barium hydroxide	90	32	Cresol	100	38
Barium sulfate	210	99	Cupric chloride, 5%	X	
Benzaldehyde	210	99	Cupric chloride, 50%	X	
Benzene	210	99	Cyclohexanol	80	27
Benzoic acid, 10%	90	32	Dichloroethane (ethylene dichloride)	200	93
Benzyl alcohol	210	99	Ethylene glycol	210	99
Benzyl chloride	210	99	Ferric chloride	X	
Borax	90	32	Ferric chloride, 50% in water	X	
Boric acid	80	27	Ferric nitrate, 10–50%	X	

TABLE 15.19 Continued

Chemical	Maximum temp.		Chemical	Maximum temp.	
	°F	°C		°F	°C
Ferrous chloride		X	Phenol	570	299
Fluorine gas, dry	570	299	Phosphoric acid, 50–80%	190	88
Fluorine gas, moist	60	16	Picric acid	X	
Hydrobromic acid, dilute	90	32	Potassium bromide, 30%	210	99
Hydrobromic acid, 20%	80	27	Salicylic acid	80	27
Hydrobromic acid, 50%		X	Sodium carbonate, to 30%	210	99
Hydrochloric acid, 20%	80	27	Sodium chloride, to 30%	210	99
Hydrochloric acid, 38%		X	Sodium hydroxide, 10%	300	149
Hydrofluoric acid, 30%		X	Sodium hydroxide, 50%ᵇ	300	149
Hydrofluoric acid, 70%		X	Sodium hydroxide, conc.	80	27
Hydrofluoric acid, 100%	120	49	Sodium hypochlorite, 20%	X	
Lactic acid, 25%	210	99	Sodium hypochlorite, conc.	X	
Lactic acid, conc.	90	32	Sodium sulfide, to 50%	210	99
Magnesium chloride, 50%	130	54	Stannic chloride	X	
Malic acid	210	99	Stannous chloride, dry	570	299
Manganese chloride, 37%		X	Sulfuric acid, 10%	X	
Methyl chloride	210	99	Sulfuric acid, 50%	X	
Methyl ethyl ketone	210	99	Sulfuric acid, 70%	X	
Methyl isobutyl ketone	200	93	Sulfuric acid, 90%	X	
Muriatic acid		X	Sulfuric acid, 98%	X	
Nitric acid, 5%	90	32	Sulfuric acid, 100%	X	
Nitric acid, 20%	80	27	Sulfuric acid, fuming	X	
Nitric acid, 70%		X	Sulfurous acid	90	32
Nitric acid, anhydrous		X	Toluene	210	99
Nitrous acid, conc.		X	Trichloroacetic acid	80	27
Oleum		X	Zinc chloride, dry	80	27

ᵃThe chemicals listed are in the pure state or in a saturated solution unless otherwise indicated. Compatibility is shown to the maximum allowable temperature for which data are available. Incompatibility is shown by an X. When compatible, corrosion rate is <20 mpy.
ᵇMaterial is subject to stress cracking.
ᶜMaterial subject to pitting.
Source: Ref. 6.

This alloy finds application where strength and corrosion resistance are required. It exhibits exceptional fatigue strength and superior strength and toughness at temperatures ranging from cryogenic to 2000°F (1093°C). The columbium and tantalum stabilization makes the alloy suitable for corrosion service in the as-welded condition. It also has excellent resistance to chloride stress corrosion cracking. Refer to Table 15.22 for the mechanical and physical properties of alloy 625 and Table 15.23 for the maximum allowable design stress.

TABLE 15.20 Chemical Composition of
Alloy 625 (N06625)

Chemical	Weight percent
Chromium	20.0–23.0
Molybdenum	8.0–10.0
Cobalt	1.00 max.
Columbium + tantalum	3.15–4.15
Aluminum	0.40 max.
Titanium	0.40 max.
Carbon	0.10 max.
Iron	5.00 max.
Manganese	0.50 max.
Silicon	0.50 max.
Phosphorus	0.015 max.
Sulfur	0.015 max.
Nickel	Balance

Resistance to aqueous solutions is good in organic acids, sulfuric and hydrochloric acid at temperatures below 150°F (65°C), as well as a variety of other applications. Satisfactory resistance has also been exhibited to hydrofluoric acid. Although nickel base alloys are not normally used in nitric acid service, alloy 625 is resistant to mixtures of nitric-hydrofluoric acids, in which stainless steel loses its resistance.

TABLE 15.21 Elevated Temperature Tensile Properties of Alloy 625 Bar with 1800°F (982°C) Anneal

Temperature (°F/°C)	Tensile strength $\times 10^3$ (psi)	Yield strength 0.2% offset $\times$ 10^3 (psi)	Elongation in 2 in. (%)
70/21	144	84	44
400/204	134	66	45
600/315	132	63	42.5
800/427	131	61	45
1000/538	130	60.5	48
1200/649	119	60	34
1400/760	78	58.5	59
1600/871	40	39	117

TABLE 15.22 Mechanical and Physical Properties of Alloy 625

Modulus of elasticity $\times 10^6$ (psi)	30.1
Tensile strength $\times 10^3$ (psi)	100–120
Yield strength 0.2% offset $\times 10^3$ (psi)	60
Elongation in 2 in. (%)	30
Brinell hardness	192
Density (lb/in.^{3})	0.305
Specific gravity	8.44
Specific heat (Btu/lb °F)	0.098
Thermal conductivity (Btu-in./ft^2 h °F)	
at −250°F	50
at −100°F	58
at 0°F	64
at 70°F	68
at 100°F	70
at 200°F	75
at 400°F	87
at 600°F	98
at 1000°F	121
at 1400°F	144
Coefficient of thermal expansion $\times 10^{-6}$ in./in. °F	
at 70–200°F	7.1
at 70–400°F	7.3
at 70–600°F	7.4
at 70–800°F	7.6
at 70–1000°F	7.8
at 70–1200°F	8.2
at 70–1400°F	8.5
at 70–1600°F	8.8

TABLE 15.23 Maximum Allowable Design Stress for Alloy 625, Welded Construction Through 4-in. Thickness

Temperature (°F/°C)	Stress (ksi)	Temperature (°F/°C)	Stress (ksi)
100/38	27.5	800/427	24.6
200/93	27.5	900/482	24.05
300/149	27.5	1000/538	23.7
400/204	26.8	1050/566	23.6
500/260	26.1	1100/593	23.4
600/315	25.45	1150/621	21.0
700/371	25.0	1200/649	13.2

Field operating experience has shown that alloy 625 exhibits excellent resistance to phosphoric acid solutions, including commercial grades that contain fluorides, sulfates, and chlorides that are used in the production of superphosphoric acid (72% P_2O_5).

Refer to Table 15.19 for the compatibility of alloy 625 with selected corrodents. Reference 1 contains a more extensive listing.

Elevated-temperature applications include ducting systems, thrust reverser assemblies, and afterburners. Use of this alloy has been considered in the high-temperature, gas-cooled reactor; however, after long aging in the temperature range of 1100–1400°F (590–760°C), the room temperature ductility is significantly reduced.

Alloy 625 has also been used in preheaters for sulfur dioxide scrubbing systems in coal-fired power plants and bottoms of electrostatic precipitators that are flushed with seawater.

VI. CUSTOM AGE 625 PLUS (N07716)

Custom Age 625 Plus is a trademark of Carpenter Technology. It is a precipitation-hardenable nickel base alloy which in many environments displays corrosion resistance similar to that of alloy 625 and superior to that of alloy 718. The chemical composition is given in Table 15.24.

This alloy provides high levels of strength while maintaining corrosion resistance even in applications where large-section size or intricate shape

TABLE 15.24 Chemical
Composition of Custom Age 625
Plus Alloy (N07716)

Chemical	Weight percent
Carbon	0.03 max.
Manganese	0.020 max.
Phosphorus	0.015 max.
Sulfur	0.010 max.
Silicon	0.20 max.
Chromium	19.00–22.00
Nickel	59.00–63.00
Molybdenum	7.00–9.5
Columbium	2.75–4.00
Titanium	1.00–1.60
Aluminum	0.35 max.
Iron	Balance

precludes warm or cold working. It offers exceptional resistance to pitting, crevice, and general corrosion as well as stress corrosion cracking in the age-hardened (high-strength) condition. It has good strength up to about 1000°F (538°C). Physical and mechanical properties will be found in Table 15.25. A yield strength (0.2% offset) of over 120,000 psi can be obtained by aging without prior warm or cold working. The precipitation hardening ability is particularly important in applications where large-section sizes of intricate shapes prevent warm or cold working.

Applications include

Deep sour gas wells
Refineries
Chemical process industry environments
High-temperature, high-purity nuclear water

VII. ALLOY C-276 (N10276)

Hastelloy alloy C-276 is a low carbon (0.01% maximum) and silicon (0.08% maximum) version of Hastelloy C. The chemical composition is given in

TABLE 15.25 Mechanical and Physical Properties of Custom Age 625 Plus (N07716)

Modulus of Elasticity $\times$ 10^6 (psi)	30
Tensile strength $\times$ 10^3 (psi)	180–190
Yield strength 0.2% offset $\times$ 10^3 (psi)	126–139
Elongation (%)	31–36
Rockwell hardness	C36–C39
Toughness (ft-lb)	69–75
Density (lb/in.^{3})	0.304
Specific gravity	8.40
Coefficient of thermal expansion $\times$ 10^{-6} (in./in. °F)	
at 77–212°F	7.1[a]/7.0[b]
at 77–392°F	7.3[a]/7.2[b]
at 77–572°F	7.5[a]/7.4[b]
at 77–752°F	7.6[a]/7.5[b]
at 77–932°F	7.8[a]/7.8[b]
at 77–1112°F	8.0[a]/8.0[b]
at 77–1292°F	8.3[a]/8.3[b]

[a]Solution annealed 1900°F (1038°C); 3 hr, air cooled.
[b]Solution annealed plus aged 1350°F (732°C); 8 hr, furnace cooled to 1150°F (621°C); 8 hr, air cooled.

Table 15.26. Alloy C-276 was developed to overcome the corrosion problems associated with the welding of alloy C. When used in the as-welded condition, alloy C was often susceptible to serious intergranular corrosion attack in many oxidizing and chlroide-containing environments. The low carbon and silicon content of alloy C-276 prevents precipitation of continuous grain-boundary precipitates in the weld heat-affected zone. Thus alloy C-276 can be used in most applications in the as-welded condition without suffering severe intergranular attack. Table 15.27 shows the mechanical and physical properties of alloy C-276, and Table 15.28 shows the allowable design stress values.

Alloy C-276 is extremely versatile because it possesses good resistance to both oxidizing and reducing media, including conditions with ion contamination. When dealing with acid chloride salts, the pitting and crevice corrosion resistance of the alloy make it an excellent choice.

Alloy C-276 has exceptional corrosion resistance to many process materials including highly oxidizing, neutral and acid chlorides, solvents, chlorine, formic and acetic acids, and acetic anhydride. It also resists highly corrosive agents such as wet chlorine gas, hypochlorite, and chlorine solutions.

Exceptional corrosion resistance is exhibited in the presence of phosphoric acid at all temperatures below the boiling point of phosphoric acid, when concentrations are less than 65% by weight. Corrosion rates of less than 5 mpy were recorded. At concentrations above 65% by weight and up to 85%, alloy C-276 displays similar corrosion rates, except at temperatures between 240°F (116°C) and the boiling point, where corrosion rates may be erratic and may reach 25 mpy.

Isocorrosion diagrams for alloy C-276 have been developed for a number of inorganic acids, for example, for sulfuric, see Fig. 15.5. Rather than having one or two acid systems in which the corrosion resistance is exceptional, as with alloy B-2, alloy C-276 is a good compromise material for a number of systems. For example, in sulfuric acid coolers handling 98% acid from the absorption tower, alloy C-276 is not the optimum alloy for the process-side corrosion, but it is excellent for the water-side corrosion and allows the use of brackish water or seawater. Concentrated sulfuric acid is used to dry chlorine gas. The dissolved chlorine will accelerate the corrosion of alloy B-2, but alloy C-276 has performed quite satisfactorily in a number of chlorine-drying installations.

Alloy C-276 has been indicated as a satisfactory material for scrubber construction where problems of localized attack have occurred with other alloys because of pH, temperature, or chloride content. Refer to Table 15.15 for the compatibility of alloy C-276 with selected corrodents and Ref. 6 for a more comprehensive listing.

TABLE 15.26 Chemical Composition of Alloy C-276 (N10276)

Chemical	Weight percent
Carbon	0.01 max.
Manganese	0.5
Silicon	0.08 max.
Chromium	15.5
Nickel	57
Molybdenum	16
Tungsten	3.5
Iron	5.5

TABLE 15.27 Mechanical and Physical Properties of Alloy C-276

Modulus of elasticity $\times 10^6$ (psi)	29.8
Tensile strength $\times 10^3$ (psi)	100
Yield strength 0.2% offset $\times 10^3$ (psi)	41
Elongation in 2 in. (%)	40
Brinell hardness	190
Density (lb/in.^{3})	0.321
Specific gravity	8.89
Specific heat (Btu/lb °F)	0.102
Thermal conductivity (Btu/ft^2/hr °F/in.)	
at -270°F	50
at 0°F	65
at 100°F	71
at 200°F	77
at 400°F	90
at 600°F	104
at 800°F	117
at 1000°F	132
at 1200°F	145
Coefficient of thermal expansion $\times 10^{-6}$ (in./in. °F)	
at 75–200°F	6.2
at 75–400°F	6.7
at 75–600°F	7.1
at 75–800°F	7.3
at 75–1000°F	7.4
at 75–1200°F	7.8
at 75–1400°F	8.3
at 75–1600°F	8.8

TABLE 15.28 Allowable Design Stress for
Alloy C-276 Seamless Pipe

Operating temp. (°F/°C)	Allowable design stress (psi)
100/38	25,000
200/93	24,500
300/149	24,000
400/204	22,500
500/260	20,750
600/316	19,400
700/371	18,750
800/427	16,750
900/482	16,750
1000/538	15,400

VIII. ALLOY G (N06007) AND ALLOY G-3 (N06985)

Alloy G is a high nickel austenitic stainless steel having the following chemical composition:

Chemical	Weight percent
Chromium	22
Nickel	45
Iron	20
Molybdenum	6.5
Copper	2
Carbon	0.05 max.
Columbium	2.0

It is intended to use alloy G in the as-welded condition, even under the circumstance of multipass welding. The columbian addition provides better resistance in highly oxidizing environments than does titanium additions. Because of the nickel base the alloy is resistant to chloride-induced stress corrosion cracking. The 2% copper addition improves the corrosion resistance of the alloy in reducing acids, such as sulfuric and phosphoric. Alloy G will also resist combinations of sulfuric acid and halides.

Alloy G resists pitting, crevice corrosion, and intergranular corrosion. Applications include heat exchangers, pollution control equipment, and various applications in the manufacture of phosphoric and sulfuric acids.

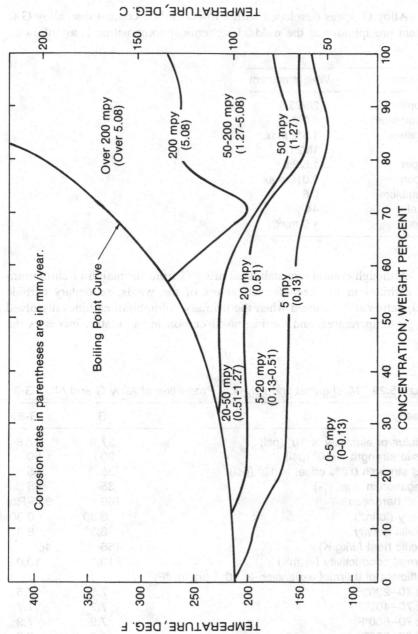

FIGURE 15.5 Isocorrosion diagram for Hastelloy C-276 in sulfuric acid. (From Ref. 3.)

Alloy G-3 was developed with a lower carbon content than alloy G to prevent precipitation at the welds. Its chemical composition is as follows:

Chemical	Weight percent
Chromium	22/23.5
Molybdenum	6.0/8.0
Tungsten	1.5% max.
Iron	18/21
Copper	1.5/2.5
Carbon	0.015 max.
Columbium	0.8
Nickel	44
Silicon	1.0 max.

Although columbium stabilized alloy G from formation of chromium-rich carbides in the heat-affected zones of the welds, secondary carbide precipitation still occurred when the primary columbium carbides dissolved at high temperatures, and the increased carbon in the matrix increases the

TABLE 15.29 Mechanical and Physical Properties of Alloy G and Alloy G-3

Property	G	G-3
Modulus of elasticity $\times 10^6$ (psi)	27.8	27.8
Tensile strength $\times 10^3$ (psi)	90	90
Yield strength 0.2% offset $\times 10^3$ (psi)	35	35
Elongation in 2 in. (%)	35	45
Brinell hardness	169	885(Rb)
Density (lb/in.^{3})	0.30	0.30
Specific gravity	8.31	8.31
Specific heat (J/kg K)	456	464
Thermal conductivity (W/mK)	10.1	10.0
Coefficient of thermal expansion $\times 10^{-6}$ (in./in. °F)		
at 70–200°F	7.5	7.5
at 70–400°F	7.7	7.7
at 70–600°F	7.9	7.9
at 70–800°F	8.3	8.3
at 70–1000°F	8.7	8.7
at 70–1200°F	9.1	9.1

tendency of the alloy to precipitate intermetallic phases. Alloy G-3 has lower carbon (0.015% maximum versus 0.05% maximum for alloy G) and lower columbium (0.3% maximum versus 2% for alloy G). The alloy also possesses slightly higher molybdenum (7 versus 5% for alloy G).

The corrosion resistance of alloy G-3 is about the same as that of alloy G, however the thermal stability is much better. The mechanical and physical properties of alloy G and alloy G-3 are shown in Table 15.29. Refer to Table 15.30 for the compatibility of alloy G and alloy G-3 with selected corrodents.

IX. ALLOY G-30 (N06030)

This alloy has a higher chromium content than alloy G, which gives it a higher resistance to oxidizing environments than other alloys in this series. It has the following composition:

Chemical	Weight percent
Chromium	28.0/31.5
Molybdenum	4.0/6.0
Tungsten	1.5/4.0
Iron	13.0/17.0
Copper	1.0/2.4
Columbium	0.30/1.50
Nickel + cobalt	Balance

Alloy G-30 possesses excellent corrosion resistance in the as-welded condition. In acid mixtures such as nitric plus hydrofluoric and sulfuric plus nitric acids, alloy G-30 shows the highest resistance of this class of alloys. The mechanical and physical properties of alloy G-30 will be found in Table 15.31.

Applications include pipe and tubing in phosphoric acid manufacture, sulfuric acid manufacture, and fertilizer and pesticide manufacture. The alloy is also used in the evaporators of commercial wet process phosphoric manufacturing systems. This process contains complex mixtures of phosphoric, sulfuric, and hydrofluoric acids and various oxides. Under these conditions the corrosion rate for alloy G-30 was 6 mpy as compared to 16 mpy for alloy G-3 and 625.

TABLE 15.30 Compatibility of Alloy G and Alloy G-3
with Selected Corrodents

Chemical	Temperature (°F/°C)
Ammonium chloride, 28%	180/82
Calcium carbonate	120/49
Calcium chloride, 3–20%	220/104
Chlorine gas, wet	80/27
Chlorobenzene, 3–60%	100/38
Fluorosilicic acid, 3–12%	180/82
Hydrofluoric acid	X
Hydrofluorosilicic acid, 10–50%	160/71
Kraft liquor	80/27
Lime slurry	140/60
Lithium chloride, 30%	260/127
Magnesium hydroxide	210/99
Magnesium sulfate	210/99
Mercury	250/121
Nitric acid, 10%	250/121
Nitric acid, 20%	250/121
Nitric acid, 40%	250/121
Nitric acid, 50%	180/82
Nitric acid, 70%	180/82
Nitrous oxide	560/293
Oleum	240/116
Phosphoric acid, 50–80%	210/99
Potassium chloride, 10%	230/110
Sodium chlorate	80/27
Sodium chloride	210/99
Sodium hydroxide, conc.	X
Sodium hypochlorite, conc.	90/32
Sodium sulfide, 3–20%	120/49
Sodium dioxide, wet	130/54
Sulfuric acid, 10%	250/121
Sulfuric acid, 30%	210/99
Sulfuric acid, 70%	X
Sulfuric acid, 98%	270/131

The chemicals listed are in the pure state or in a saturated
solution unless otherwise indicated. Compatibility is shown to
the maximum allowable temperature for which data are available. Incompatibility is shown by an X. When compatible, the
corrosion rate is less that 20 mpy.

TABLE 15.31 Mechanical and Physical Properties of Alloy G-30

Modulus of elasticity $\times 10^6$ (psi)	29.3
Tensile strength $\times 10^3$ (psi)	100
Yield strength 0.2% offset $\times 10^3$ (psi)	47
Elongation (%)	55
Rockwell hardness	B-85
Density (lb/in.^{3})	0.297
Specific gravity	8.2
Thermal conductivity (Btu/ft^2 hr °F)	
at 70°F (20°C)	5.9
at 1500°F (815°C)	12.4

X. ALLOY C-22 (NO6022)

Hastelloy alloy C-22 is a versatile nickel-chromium-molybdenum alloy with better overall corrosion resistance than other nickel-chromium-molybdenum alloys including C-276, C-4, and alloy 625. The chemical composition is shown in Table 15.32.

Alloy C-22 resists the formation of grain-boundary precipitates in the weld heat-affected zone. Consequently it is suitable for most chemical process applications in the as-welded condition. Refer to Table 15.33 for the mechanical and physical properties of alloy C-22.

TABLE 15.32 Chemical Composition of Alloy C-22 (N06022)

Chemical	Weight percent
Carbon	0.015 max.
Manganese	0.50 max.
Phosphorus	0.025 max.
Sulfur	0.010 max.
Chromium	20.0–22.5
Molybdenum	12.5–14.5
Cobalt	2.5 max.
Tungsten	2.5–3.5
Iron	2.0–6.0
Silicon	0.08 max.
Vanadium	0.35 max.
Nickel	Balance

TABLE 15.33   Mechanical and Physical Properties
of Alloy C-22 (N06022)

Modulus of elasticity $\times$ 10^6 (psi)	29.9
Tensile strength $\times$ 10^3 (psi)	115
Yield strength 0.2% offset $\times$ 10^3 (psi)	60
Elongation (%)	55
Rockwell hardness	B-87
Density (lb/in.^{3})	0.314
Specific gravity	8.69
Thermal conductivity (Btu/ft^2 hr °F)	
at 70°F (20°C)	5.8
at 1500°F (816°C)	12.3

Although alloy C-276 is a versatile alloy its main limitations are in oxidizing environments containing low amounts of halides and in environments containing nitric acid. In addition the thermal stability of the alloy was not sufficient to enable it to be used as a casting.

Alloy C-22 was invented to improve the resistance to oxidizing environments, such as nitric acid, and also to improve the thermal stability sufficiently to enable it to be used as casting. The higher chromium level in this alloy not only makes it superior in oxidizing environments containing nitric acid, but also improves the pitting resistance over that of alloy C-276.

Alloy C-22 has outstanding resistance to pitting, crevice corrosion, and stress corrosion cracking. It has excellent resistance to oxidizing aqueous media including acids with oxidizing agents, wet chlorine, and mixtures containing nitric or oxidizing acids with chloride ions. The alloy also has outstanding resistance to both reducing and oxidizing media and because of its versatility can be used where "upset" conditions are likely to occur or in multipurpose plants.

Alloy C-22 has exceptional resistance to a wide variety of chemical process environments, including strong oxidizers such as ferric and cupric chlorides, hot contaminated media (organic and inorganic), chlorine, formic and acetic acids, acetic anhydride, seawater, and brine solutions. The compatibility of alloy C-22 with selected corrodents will be found in Table 15.34.

The areas of application of alloy C-22 are many of the same as those for alloy C-276. It is being used in pulp and paper bleaching systems, pollution control systems, and various areas in the chemical process industry.

TABLE 15.34 Compatibility of Alloy C-22 with Selected Corrodents

Corrodent	Weight percent	Temperature (°F/°C)	Average corrosion rate (mpy)
Acetic acid	99	Boiling	Nil
Ferric chloride	10	Boiling	1.0
Formic acid	88	Boiling	0.9
Hydrochloric acid	1	Boiling	2.5
Hydrochloric acid	1.5	Boiling	11
Hydrochloric acid	2	194/90	Nil
Hydrochloric acid	2	Boiling	61
Hydrochloric acid	2.5	194/90	0.3
Hydrochloric acid	2.5	Boiling	84
Hydrochloric acid	10	Boiling	400
Hydrofluoric acid	2	158/70	9.4
Hydrofluoric acid	5	158/70	19
Phosphoric acid, reagent grade	55	Boiling	12
Phosphoric acid, reagent grade	85	Boiling	94
Nitric acid	10	Boiling	0.8
Nitric acid	65	Boiling	5.3
Nitric acid + 1% HCl	5	Boiling	0.5
Nitric acid + 2.5% HCl	5	Boiling	1.6
Sulfuric acid	10	Boiling	11
Sulfuric acid	20	150/66	0.2
Sulfuric acid	20	174/79	1.2
Sulfuric acid	20	Boiling	33
Sulfuric acid	30	150/66	0.6
Sulfuric acid	30	174/79	3.3
Sulfuric acid	30	Boiling	64
Sulfuric acid	40	100/38	0.1
Sulfuric acid	40	150/66	0.5
Sulfuric acid	40	174/79	6.4
Sulfuric acid	50	100/38	0.2
Sulfuric acid	50	150/66	1.0
Sulfuric acid	50	174/79	16
Sulfuric acid	60	100/38	0.1
Sulfuric acid	70	100/38	Nil
Sulfuric acid	80	100/38	Nil

TABLE 15.35 Chemical Composition
of Alloy C-2000

Chemical	Weight percent
Carbon	0.01 max.
Manganese	0.050 max.
Phosphorus	0.025 max.
Sulfur	0.010 max.
Silicon	0.080 max.
Chromium	22.00–24.00
Molybdenum	15.00–17.00
Copper	1.30–1.90
Cobalt	2.00 max.
Iron	3.00 max.
Aluminum	0.50 max.
Nickel	Balance

XI. HASTELLOY ALLOY C-2000

Hastelloy Alloy C-2000 is a trademark of Haynes International. It is one of
the nickel-chromium-molybdenum alloys. The chemical composition is
shown in Table 15.35.

Alloy C-2000 exhibits outstanding resistance to oxidizing media with
superior resistance to reducing environments. The mechanical and physical
properties are shown in Table 15.36.

In the family of nickel-chromium-molybdenum alloys, a high chromium content is required for resistance to oxidizing media, such as ferric

TABLE 15.36 Mechanical and Physical
Properties of Alloy C-2000

Tensile strength $\times$ 10³ (psi)	109
Yield strength 0.2% offset $\times$ 10³ (psi)	52.4
Elongation (%)	66.2
Thermal conductivity (in./ft² hr °F)	
at 70°F	63
at 200°F	74
at 600°F	99
at 1000°F	133
at 1400°F	180
at 1800°F	191

ions, cupric ions, or dissolved oxygen. Reducing environments, however, such as dilute hydrochloric or sulfuric acids, require a high content of molybdenum plus tungsten. Metallurgical stability limitations dictate that you cannot optimize both.

Alloy C-2000 solves this alloy dilemma. A high chromium content is combined with both molybdenum and copper contents, sufficient to provide outstanding resistance to reducing environments, with no sacrifice of metallurgical stability.

Alloy C-2000 also exhibits pitting resistance and crevice corrosion resistance superior to that of alloy C-276. Its critical pitting temperature is 230°F (110°C) and its critical crevice temperature is 203°F (95°C). Some typical uniform corrosion rates are as follows:

Chemical	Weight percent	Temperature (°F/°C)	Corrosion rate (mpy)
Hydrofluoric acid	20	174/79	19
Phosphoric acid	50	Boiling	1
Acetic acid	75	Boiling	33
	99	Boiling	0.1
Formic acid	88	Boiling	0.4
Chromic acid	10	Boiling	44

XII. ALLOY X (NO6002)

Alloy X is a nonmagnetic heat- and corrosion-resistant nickel-based alloy. The chemical composition will be found in Table 15.37.

TABLE 15.37 Chemical Composition of Alloy X (N06002)

Chemical	Weight percent
Chromium	20.5–23.00
Molybdenum	8.0–10.0
Iron	17.0–20.0
Tungsten	0.2–1.0
Carbon	0.05–0.15
Cobalt	0.5–2.5
Nickel	Balance

Alloy X possesses a combination of high strength and excellent oxidation resistance. Its oxidation resistance is due to the formation of a complex chromium oxide spinel that provides good resistance up to temperatures of 2150°F (1177°C). The high temperature strength and resistance to warpage and distortion provide outstanding performance as distributor plates and catalyst grid supports. Table 15.38 provides the mechanical and physical properties of alloy X.

Alloy X has excellent resistance to nitric acid, organic acids, alkalies, salts, seawater, chloride cracking, and good to excellent resistance to phosphoric acid and sulfuric acid, with good resistance in hydrochloric and hydrofluoric acids.

The catalyst regenerator for high density polyethylene is constructed of alloy X, since high temperatures and pressures are required to revitalize the catalysts. Unfortunately, this continued temperature cycling eventually reduces the room temperature ductility in alloy X so that repair welding becomes difficult without solution annealing.

Alloy X also finds application in gas turbine components, high temperature heat exchangers, afterburner components, and furnace hardware.

XIII. ALLOY C-4 (NO6455)

Alloy C-4 was developed for improved stability relative to precipitation of both carbides and intermetallic phases. The chemical composition is shown in Table 15.39. By controlling these secondary phases excellent high temperature stability is achieved to the point that the corrosion resistance and mechanical properties in the thermally aged condition are similar to the

TABLE 15.38   Mechanical and Physical
Properties of Alloy X (N06002)

Modulus of elasticity $\times 10^6$ (psi)	29.0
Tensile strength $\times 10^3$ (psi)	100
Yield strength 0.2% offset $\times 10^3$ (psi)	40
Elongation (%)	35
Rockwell hardness	B-92.5
Density (lb/in.^{3})	0.297
Specific gravity	8.221
Thermal conductivity (in./ft^2 hr °F)	
at 70°F (20°C)	5.2
at 1500°F (815°C)	11.2

TABLE 15.39 Chemical Composition
of Alloy C-4 (N06455)

Chemical	Weight percent
Chromium	14.0–16.0
Molybdenum	14.0–17.0
Titanium	0.07 max.
Iron	3.0 max.
Nickel	Balance

annealed condition properties. The mechanical and physical properties are shown in Table 15.40.

Examples can be taken from various chemical processing applications in which oxidizing and reducing conditions can cause serious intergranular corrosion of a sensitized (precipitated) microstructure. This sensitization can be the result of welding, improper anneal, stress relief, thermomechanical processing, or operation of process equipment in the sensitizing range.

Alloy C-4 alleviates this problem since it can be subjected to temperatures in the normal sensitizing range of 1022–1994°F (550–1090°C) for extended periods without experiencing the severe corrosion attack that is found with the common austenitic alloys.

With the exception of iron and tungsten the composition of alloys C-4 and C-276 are approximately the same. Consequently, the corrosion resistance of the two alloys is approximately the same. In a strongly reducing medium, such as hydrochloric acid, alloy C-4 has a slightly higher rate of

TABLE 15.40 Mechanical and Physical Properties
of Alloy C-4 (N06455)

Modulus of elasticity $\times$ 10^6 (psi)	30.8
Tensile strength $\times$ 10^3 (psi)	115
Yield strength $\times$ 10^3 (psi)	55
Elongation (%)	56
Rockwell hardness	B-90
Density (lb/in.^{3})	0.312
Specific gravity	8.64
Thermal conductivity (Btu/ft^2 hr °F)	
at 70°F (20°C)	5.8
at 1500°F (815°C)	11.8

TABLE 15.41 Chemical Composition
of Hastelloy Alloy S

Chemical	Weight percent
Chromium	15.5
Molybdenum	14.5
Iron	1.0
Carbon	0.01
Silicon	0.4
Manganese	0.5
Lanthanum	0.02
Nickel + cobalt	Balance

corrosion than alloy C-276, but in an oxidizing medium the rates are reversed.

Alloy C-4 offers excellent corrosion resistance to nitric acid, hydrochloric acid, organic acids, alkalies, salts, seawater, and chloride stress corrosion cracking. Good to excellent resistance is exhibited in sulfuric, hydrofluoric, and phosphoric acids.

XIV. HASTELLOY ALLOY S

In 1973 Hastelloy alloy S was developed for gas turbine applications requiring oxidation resistance, good alloy stability, and a low thermal expansion coefficient. The chemical composition will be found in Table 15.41. Its composition is similar to that of alloys C-4 and C-276, and it has similar corrosion resistance. However, the carbon content may prevent its use in some aqueous media in the as-welded condition. Refer to Table 15.42 for

TABLE 15.42 Mechanical and Physical Properties
of Hastelloy Alloy S

Tensile stress $\times 10^3$ (psi)	116
Yield stress 0.2% offset $\times 10^3$ (psi)	49
Elongation (%)	65
Brinell hardness	180
Density (g/cm^3)	8.76
Thermal expansion $\times 10^{-6}$ (m/m K)	
at 23–93°C	11.5
Thermal conductivity (W/mK)	10.8

the mechanical and physical properties. After 10,000 hr of aging in the temperature ranges typically encountered in this application, the alloy S welds exhibited 80% of their original ductility.

XV. HAYNES ALLOY 556 (R30556)

Haynes alloy 556 exhibits useful resistance to a wide variety of high temperature corrosive atmospheres as well as molten salts. The presence of approximately 18% cobalt results in more resistance to sulfidation than many nickel-based alloys. Table 15.43 shows the chemical composition of alloy 556.

The alloy has good oxidation resistance and fabrication properties and excellent high temperature strength. Mechanical and physical properties will be found in Table 15.44.

In pure oxidation the alloy shows good resistance, but it is superceded in performance by other alloys such as alloy X and 214. In chlorine-bearing oxidizing environments, the alloy shows better resistance than alloys 800H and X, but not as good as alloy 214.

Alloy 556 exhibits excellent resistance to hydrochloric, phosphoric, and organic acids, alkalies, salts, seawater, and chlorine stress corrosion cracking. It also offers good resistance to sulfuric acid, hydrofluoric acid, and nitric acid.

TABLE 15.43 Chemical Composition of Haynes Alloy 556 (R30556)

Chemical	Weight percent
Nickel	19.0–22.5
Chromium	21.0–23.0
Molybdenum	2.5–4.0
Tungsten	2.0–3.5
Carbon	0.05–0.15
Silicon	0.2–0.8
Cobalt	16.0–21.0
Manganese	0.5–2.00
Aluminum	0.1–0.5
Tantalum	0.3–1.25
Zirconium	0.001–0.1
Lanthanum	0.005
Nitrogen	0.1–0.3
Iron	Balance

TABLE 15.44 Mechanical and Physical Properties
of Haynes Alloy 556 (R30556)

Modulus of elasticity $\times 10^6$ (psi)	29.7
Tensile strength $\times 10^3$ (psi)	116.4
Yield strength 0.2% offset $\times 10^3$ (psi)	54.6
Elongation (%)	51.4
Rockwell hardness	B-95
Density (g/mm³)	0.295
Specific gravity	8.23
Specific heat (J/kgK)	475
Thermal expansion $\times 10^{-6}$ (m/mK)	
at 20–93°C	14.7
Thermal conductivity (Btu/ft² hr °F)	
at 70°F (20°C)	6.4
at 1500°F (815°C)	12.3

Typical applications include internals of municipal waste incinerators
and refractory anchors in refinery tail gas burning units.

XVI. ALLOY 214

The chemical composition of alloy 214 will be found in Table 15.45. This
is a nickel-based alloy with excellent resistance to 2200°F (1204°C). The
excellent oxidation resistance is the result of a tenacious aluminum oxide
film which protects the metal during prolonged exposure. Mechanical and
physical properties will be found in Table 15.46.

The alloy possesses the highest oxidation resistance to both static and
dynamic environments among the nickel-based alloys. The alumina film also
lends superior resistance to carburizing environments and complex environ-

TABLE 15.45 Chemical
Composition of Alloy 214

Chemical	Weight percent
Chromium	16.0
Iron	3.0
Aluminum	4.5
Yttrium	Trace
Nickel	Balance

TABLE 15.46 Mechanical and Physical Properties of Alloy 214

Modulus of elasticity $\times$ 10^6 (psi)	31.6
Tensile strength $\times$ 10^3 (psi)	135
Yield strength 0.2% offset $\times$ 10^3 (psi)	83
Elongation (%)	42
Rockwell hardness	C-20
Density (lb/in.^{3})	0.291
Specific gravity	8.05
Thermal expansion $\times$ 10^{-6} (m/mK) at 20–93°C	13.3

ments containing chlorine and oxygen. However, as typical of many high temperature alloys, alloy 214 does not possess good resistance to aqueous chloride solutions, so dew point conditions must be avoided.

Alloy 214 exhibits excellent resistance to nitric and organic acids and alkalies. It is not recommended for use with sulfuric, hydrochloric, hydrofluoric, or phosphoric acids or salts or seawater.

Applications include radiant tubes, high temperature heat exchangers, honeycomb seals in turbine engines, and mesh belts for supporting chinaware being heated in a kiln.

XVII. ALLOY 230 (NO6230)

Alloy 230 has excellent high temperature strength and outstanding resistance to oxidizing environments up to 2100°F (1150°C). Refer to Table 15.47 for the chemical composition.

This alloy exhibits resistance to nitriding, excellent long-term thermal stability, and low thermal expansion. It is also resistant to grain coarsening at high temperatures. Good resistance to carburization is also displayed. Because of its nickel matrix the alloy does not possess adequate resistance to sulfadizing environments. Mechanical and physical properties are shown in Table 15.48.

Excellent resistance is shown to phosphoric acid, organic acids, alkalies, salts, seawater, and chloride stress corrosion cracking, while good to excellent resistance is shown to sulfuric and nitric acids. It is not suitable for use with hydrochloric acid.

Because of its nitridizing resistance and high creep strength it has found application as a catalyst grid support in the manufacture of nitric acid.

TABLE 15.47 Chemical Composition
of Alloy 230 (N06230)

Chemical	Weight percent
Chromium	22.0
Tungsten	14.0
Molybdenum	2.0
Iron	3.0 max.
Cobalt	5.0 max.
Aluminum	0.3
Carbon	0.10
Lanthanum	0.02
Boron	0.005
Nickel	Balance

XVIII.   ALLOY RA 333 (NO6333)

Alloy RA 333 is a registered trademark of Rolled Alloys, Inc. It is a high
chromium nickel base alloy with extreme temperature corrosion resistance
and strength. The chemical composition is shown in Table 15.49.

Alloy RA 333 is one of the few materials that can withstand corrosive
conditions ranging from aqueous to white heat. The alloy has been used for
dampers and refractory anchors in 13% SO_2/SO_3 at 1800°F (982°C) and for
refinery flare tips. Other features include high temperature SO_x, hot salt
resistance, practical immunity to chloride ion and to polythionic acid stress
corrosion cracking, good resistance to sulfuric acid, and excellent oxidation
and carburization resistance at elevated temperatures. Refer to Table 15.50
for the mechanical and physical properties.

TABLE 15.48 Mechanical and Physical Properties
of Alloy 230 (N06230)

Modulus of elasticity $\times$ 10^{-6} (psi)	30.6
Tensile strength $\times$ 10^3 (psi)	128
Yield strength 0.2% offset $\times$ 10^3 (psi)	62
Elongation (%)	45
Rockwell hardness	B-93
Density (lb/in.^{3})	0.319
Specific gravity	8.83
Thermal conductivity (Btu/ft^2 hr °F)	
at 70°F (20°C)	5.2

TABLE 15.49 Chemical
Composition of Alloy RA333
(N06333)

Chemical	Weight percent
Nickel	44.0–47.0
Chromium	24.0–27.0
Molybdenum	2.50–4.00
Cobalt	2.50–4.00
Tungsten	2.50–4.00
Carbon	0.08 max.
Silicon	0.75–1.50
Manganese	2.00 max.
Phosphorus	0.03 max.
Sulfur	0.03 max.
Iron	Balance

XIX. ALLOY 102 (NO6102)

This is a nonmagnetic nickel-chromium base alloy strengthened with refractory metals. The chemical composition is given in Table 15.51. It possesses excellent corrosion properties, strength, ductility, and toughness and
has outstanding structural stability. Refer to Table 15.52 for the mechanical
and physical properties.

Alloy 102 exhibits excellent resistance to phosphoric, nitric, and organic acids, alkalies, salts, seawater, and chloride stress corrosion cracking.
It has good resistance to sulfuric acid and acceptable resistance to hydrochloric and hydrofluoric acids.

XX. ALLOY H-9M

This alloy can be considered either as a modification of alloy G-3 or of
alloy 625. The purpose of the modification is to improve the localized corrosion resistance of both alloys G-3 and 625. Alloy H-9M has higher molybdenum content than alloy G-3 and higher tungsten content than alloy
625. In order to increase the localized corrosion resistance, copper has been
eliminated from alloy H-9M. The critical pitting temperatures of these alloys
in an oxidizing, acidic chloride mixture are shown in Table 15.53. The critical pitting temperature indicates the temperature above which pitting is
observed in the solution, and the higher the temperature, the better the alloy
in pitting resistance.

TABLE 15.50 Mechanical and Physical Properties of Alloy RA333 (N06333)

Temperature (°F/°C)	Tensile strength × 10³ (psi)	Yield strength 0.2% offset × 10³ (psi)	Elongation (%)	Coefficient of thermal expansion × 10⁻⁶ (in./in.°F)	Thermal conductivity (Btu/ft² hr °F)	Modulus of elasticity × 10⁶ (psi)
−300/−184	154	82.4	47	—	—	—
70/20	107	47	48	—	6.4	29.2
400/204	96	37	49	—	8.1	28.1
600/316	92.5	32.5	54	—	9.1	27.0
1000/538	85.4	30.8	53	8.6	11.3	24.6
1200/649	73.6	30.7	43	9.0	12.4	23.4

TABLE 15.51 Chemical
Composition of Alloy 102
(N06102)

Chemical	Weight percent
Chromium	14.0–16.0
Columbium	2.75–3.25
Molybdenum	2.75–3.75
Tungsten	2.75–3.75
Iron	5.0–9.0
Aluminum	0.3–0.6
Titanium	0.4–0.7
Boron	0.003–0.008
Carbon	0.08 max.
Nickel	Balance

TABLE 15.52 Mechanical and Physical Properties
of Alloy 102 (N06102)

Modulus of elasticity $\times 10^6$ (psi)	29.7
Tensile strength $\times 10^3$ (psi)	120
Yield strength 0.2% offset $\times 10^3$ (psi)	60
Elongation (%)	35
Rockwell hardness	B-98
Density (lb/in.^{3})	0.309
Specific gravity	8.55
Thermal conductivity (Btu/ft^2 hr °F)	
at 70°F (20°C)	6.5
at 1500°F (815°C)	11.3

TABLE 15.53 Critical Pitting Temperature of
Alloy H-9M in Comparison to Other Alloys[a]

Alloy	Critical pitting temp. (°F/°C)
H-9M	103/95
625	194/90
G-3	167/75

[a]Solution 4% NaCl + 0.1% $Fe_2(SO_4)_3$ + 0.01 M
HCl.

REFERENCES

1. GT Murray. Introduction to Engineering Materials. New York: Marcel Dekker, 1993.
2. CP Dillon. Corrosion Control in the Chemical Process Industry. St. Louis, MO: Materials Technology Institute of the Chemical Process Industries, 1994.
3. GF Hodge. Nickel and high-nickel alloys. In: PA Schweitzer, ed. Corrosion and Corrosion Protection Handbook, 1st ed. New York: Marcel Dekker, 1983.
4. N Sridhas, GF Hodge. Nickel and high-nickel alloys. In: PA Schweitzer, ed. Corrosion and Corrosion Protection Handbook, 2nd ed. New York: Marcel Dekker, 1996.
5. PA Schweitzer. Corrosion Resistant Piping Systems. New York: Marcel Dekker, 1994.
6. PA Schweitzer. Corrosion Resistance Tables, 4th ed., Vols. 1–3. New York: Marcel Dekker, 1995.
7. PA Schweitzer. Encyclopedia of Corrosion Technology, New York: Marcel Dekker, 1998.

16

Comparative Corrosion Resistance of Stainless Steel and High Nickel Alloys

The corrosion tables on the following pages are arranged alphabetically according to corrodent. The chemicals listed are in the pure state or in a saturated solution unless otherwise indicated. Compatibility is shown to the maximum allowable temperature for which data are available. Symbols used to designate specific corrosion rates are as follows:

E indicates that the corrosion rate rate is <2 mpy.

G indicates that the corrosion rate is between 2 and 20 mpy.

S indicates that the corrosion rate is between 20 and 50 mpy.

U indicates that the corrosion rate is >50 mpy and therefore not recommended for this service.

Further information regarding the corrosion of specific materials by certain corrodents is provided by the following symbols. In the tables, the symbols follow the applicable material:

Symbol	Meaning
1	Material is subject to pitting.
2	Material is subject to stress cracking.
3	Material is subject to crevice attack.
4	Applicable to alloy 825 only.
5	Material is subject to intergranular corrosion.
6	Material not to be used with carbonated beverages.
7	Corrodent must be acid free.
8	Corrodent must be acid free and the material passivated.
9	Corrodent must be alkaline
10	Material is subject to stress cracking when corrodent is wet.
11	Corrodent must be sulfur free.
ELC	Material must be low carbon grade.

Corrosion rate is shown as a function of temperature. The use of the temperature scale is explained by the following example.

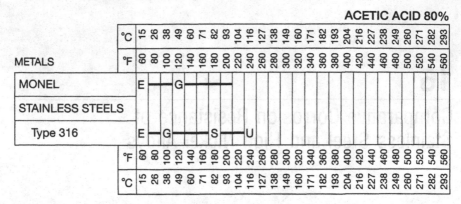

ACETIC ACID 80%

From the above it is seen that Monel has a corrosion rate varying with temperature as follows:

<2 mpy between 60 and 120°F (E——)
<20 mpy between 120 and 210°F (G——)
No data beyond 210°F

Type 316 stainless steel has a corrosion rate varying with temperature as follows:

<2 mpy between 60 and 100°F (E——)
<20 mpy between 100 and 180°F (G——)
<50 mpy between 180 and 240°F (S——)
>50 mpy above 240°F (U)

In reading the temperature scale note that the vertical lines refer to temperatures midway between the temperatures cited. (Refer to example given above.)

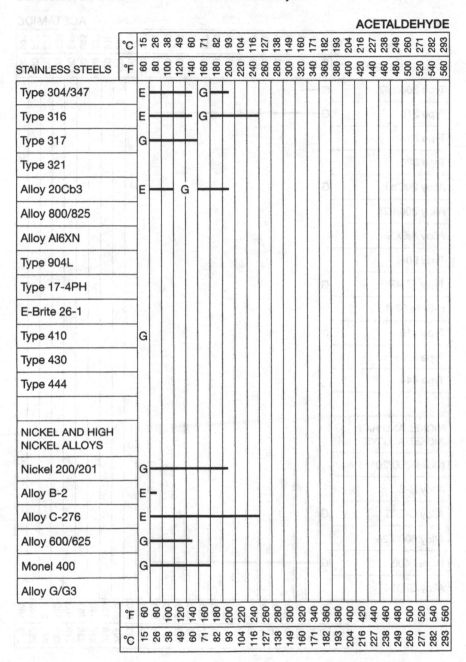

ACETALDEHYDE

STAINLESS STEELS	°C / °F
Type 304/347	E——G—
Type 316	E——G——
Type 317	G——
Type 321	
Alloy 20Cb3	E——G——
Alloy 800/825	
Alloy Al6XN	
Type 904L	
Type 17-4PH	
E-Brite 26-1	
Type 410	G
Type 430	
Type 444	
NICKEL AND HIGH NICKEL ALLOYS	
Nickel 200/201	G——
Alloy B-2	E—
Alloy C-276	E——
Alloy 600/625	G—
Monel 400	G—
Alloy G/G3	

ACETAMIDE

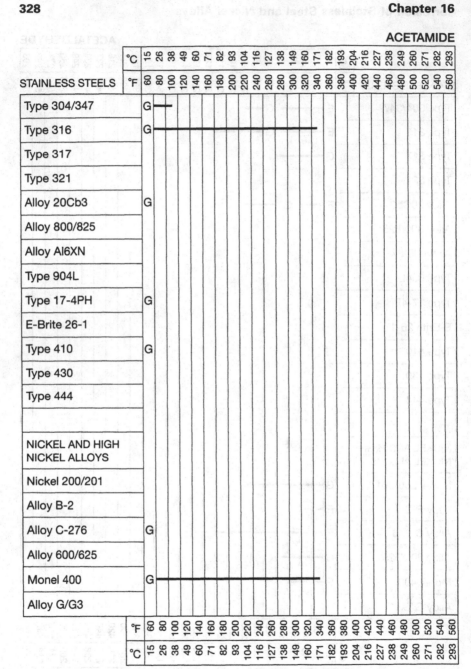

STAINLESS STEELS	°C	15	26	38	49	60	71	82	93	104	116	127	138	149	160	171	182	193	204	216	227	238	249	260	271	282	293
	°F	60	80	100	120	140	160	180	200	220	240	260	280	300	320	340	360	380	400	420	440	460	480	500	520	540	560
Type 304/347	G																										
Type 316	G																										
Type 317																											
Type 321																											
Alloy 20Cb3	G																										
Alloy 800/825																											
Alloy Al6XN																											
Type 904L																											
Type 17-4PH	G																										
E-Brite 26-1																											
Type 410	G																										
Type 430																											
Type 444																											
NICKEL AND HIGH NICKEL ALLOYS																											
Nickel 200/201																											
Alloy B-2																											
Alloy C-276	G																										
Alloy 600/625																											
Monel 400	G																										
Alloy G/G3																											
	°F	60	80	100	120	140	160	180	200	220	240	260	280	300	320	340	360	380	400	420	440	460	480	500	520	540	560
	°C	15	26	38	49	60	71	82	93	104	116	127	138	149	160	171	182	193	204	216	227	238	249	260	271	282	293

ACETIC ACID 10%

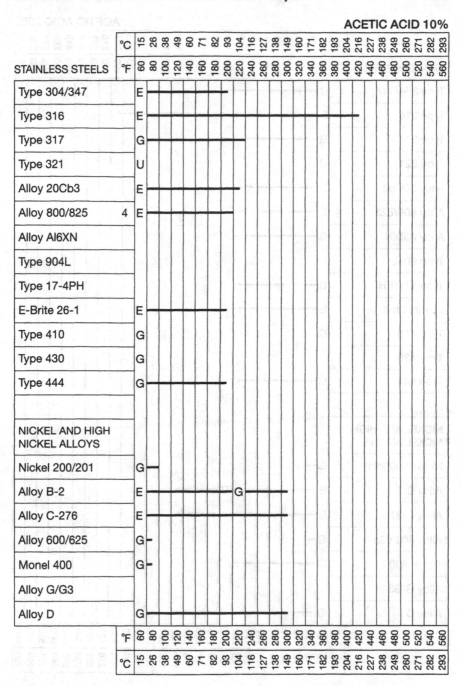

STAINLESS STEELS	°C	15	26	38	49	60	71	82	93	104	116	127	138	149	160	171	182	193	204	216	227	238	249	260	271	282	293
	°F	60	80	100	120	140	160	180	200	220	240	260	280	300	320	340	360	380	400	420	440	460	480	500	520	540	560
Type 304/347	E
Type 316	E
Type 317	G
Type 321	U
Alloy 20Cb3	E
Alloy 800/825 4	E
Alloy AI6XN	
Type 904L	
Type 17-4PH	
E-Brite 26-1	E
Type 410	G
Type 430	G
Type 444	G
NICKEL AND HIGH NICKEL ALLOYS	
Nickel 200/201	G
Alloy B-2	E G
Alloy C-276	E
Alloy 600/625	G
Monel 400	G
Alloy G/G3	
Alloy D	G

ACETIC ACID 20%

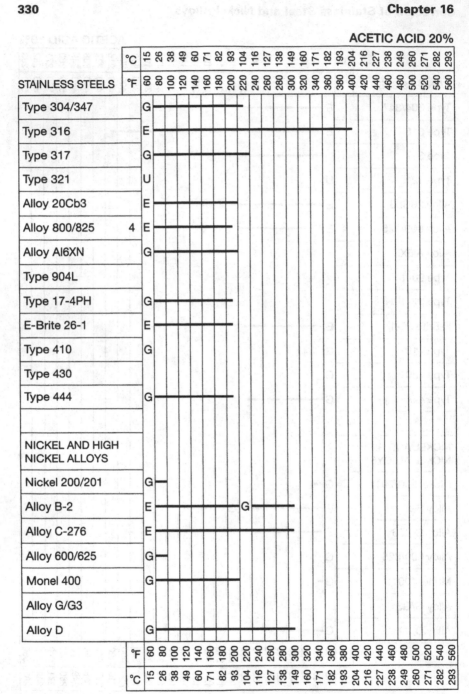

STAINLESS STEELS	°C	15	26	38	49	60	71	82	93	104	116	127	138	149	160	171	182	193	204	216	227	238	249	260	271	282	293
	°F	60	80	100	120	140	160	180	200	220	240	260	280	300	320	340	360	380	400	420	440	460	480	500	520	540	560
Type 304/347	G																										
Type 316	E																										
Type 317	G																										
Type 321	U																										
Alloy 20Cb3	E																										
Alloy 800/825 4	E																										
Alloy Al6XN	G																										
Type 904L																											
Type 17-4PH	G																										
E-Brite 26-1	E																										
Type 410	G																										
Type 430																											
Type 444	G																										
NICKEL AND HIGH NICKEL ALLOYS																											
Nickel 200/201	G																										
Alloy B-2	E						G																				
Alloy C-276	E																										
Alloy 600/625	G																										
Monel 400	G																										
Alloy G/G3																											
Alloy D	G																										

ACETIC ACID 50%

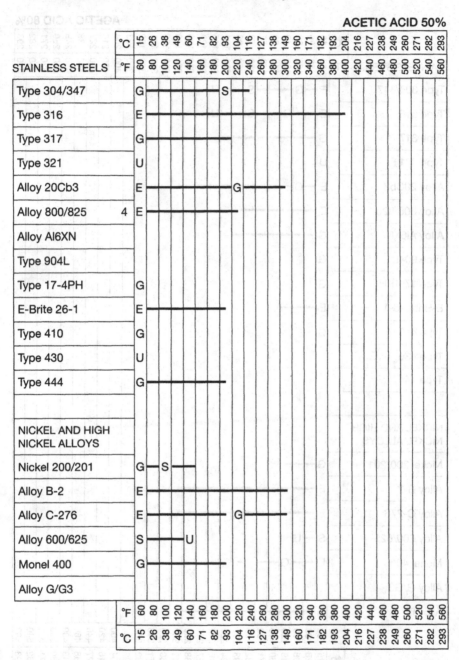

STAINLESS STEELS	°C	15	26	38	49	60	71	82	93	104	116	127	138	149	160	171	182	193	204	216	227	238	249	260	271	282	293
	°F	60	80	100	120	140	160	180	200	220	240	260	280	300	320	340	360	380	400	420	440	460	480	500	520	540	560

ACETIC ACID 80%

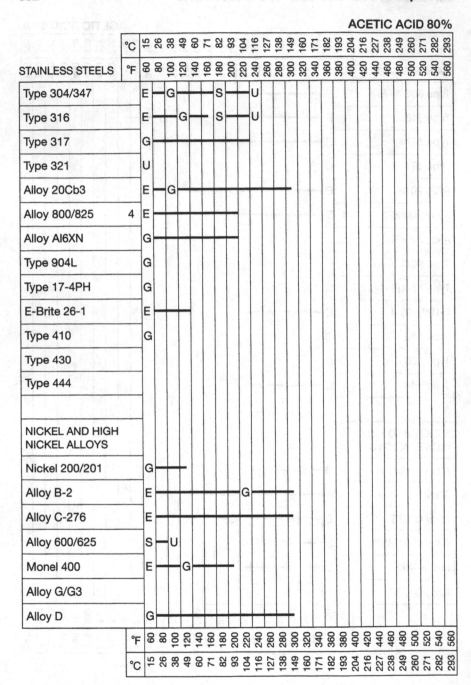

STAINLESS STEELS		°C	15	26	38	49	60	71	82	93	104	116	127	138	149	160	171	182	193	204	216	227	238	249	260	271	282	293
		°F	60	80	100	120	140	160	180	200	220	240	260	280	300	320	340	360	380	400	420	440	460	480	500	520	540	560
Type 304/347		E—G———S—U																										
Type 316		E——G—S—U																										
Type 317		G————————																										
Type 321		U																										
Alloy 20Cb3		E—G——————————																										
Alloy 800/825	4	E———————																										
Alloy Al6XN		G———————																										
Type 904L		G																										
Type 17-4PH		G																										
E-Brite 26-1		E———																										
Type 410		G																										
Type 430																												
Type 444																												
NICKEL AND HIGH NICKEL ALLOYS																												
Nickel 200/201		G——																										
Alloy B-2		E———————G——																										
Alloy C-276		E———————————																										
Alloy 600/625		S—U																										
Monel 400		E——G———																										
Alloy G/G3																												
Alloy D		G———————————																										
		°F	60	80	100	120	140	160	180	200	220	240	260	280	300	320	340	360	380	400	420	440	460	480	500	520	540	560
		°C	15	26	38	49	60	71	82	93	104	116	127	138	149	160	171	182	193	204	216	227	238	249	260	271	282	293

ACETIC ACID, GLACIAL

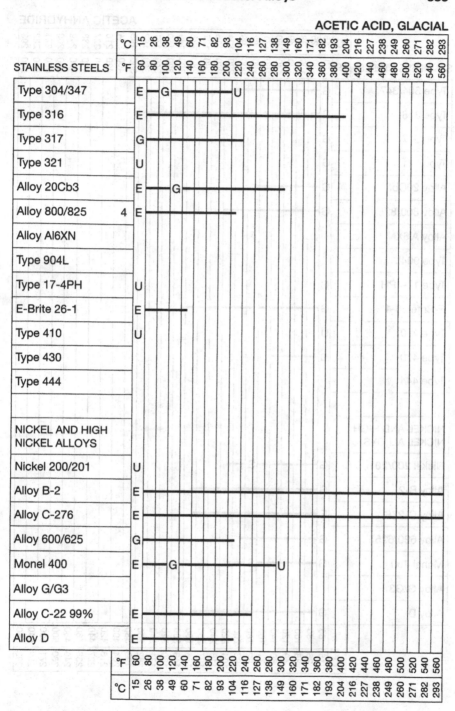

STAINLESS STEELS	°C 15 26 38 49 60 71 82 93 104 116 127 138 149 160 171 182 193 204 216 227 238 249 260 271 282 293 °F 60 80 100 120 140 160 180 200 220 240 260 280 300 320 340 360 380 400 420 440 460 480 500 520 540 560
Type 304/347	E—G————————U
Type 316	E————————————————
Type 317	G——————————
Type 321	U
Alloy 20Cb3	E——G—————————
Alloy 800/825 4	E—————————
Alloy Al6XN	
Type 904L	
Type 17-4PH	U
E-Brite 26-1	E————
Type 410	U
Type 430	
Type 444	
NICKEL AND HIGH NICKEL ALLOYS	
Nickel 200/201	U
Alloy B-2	E—————————————————————————
Alloy C-276	E—————————————————————————
Alloy 600/625	G———————————
Monel 400	E——G——————————U
Alloy G/G3	
Alloy C-22 99%	E————————————
Alloy D	E————————

ACETIC ANHYDRIDE

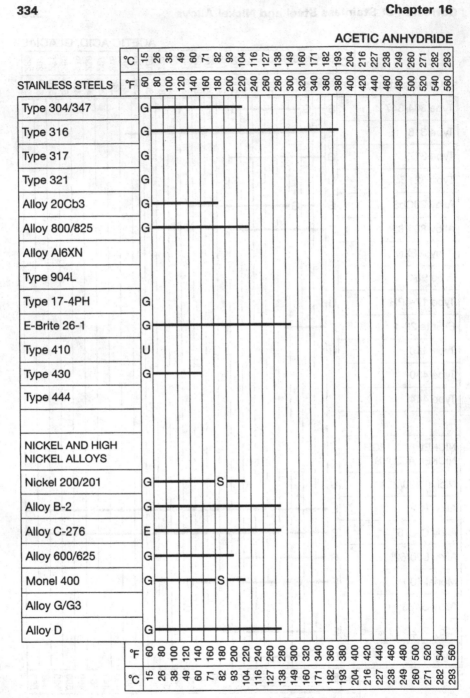

STAINLESS STEELS

| °C | 15 | 26 | 38 | 49 | 60 | 71 | 82 | 93 | 104 | 116 | 127 | 138 | 149 | 160 | 171 | 182 | 193 | 204 | 216 | 227 | 238 | 249 | 260 | 271 | 282 | 293 |
| °F | 60 | 80 | 100 | 120 | 140 | 160 | 180 | 200 | 220 | 240 | 260 | 280 | 300 | 320 | 340 | 360 | 380 | 400 | 420 | 440 | 460 | 480 | 500 | 520 | 540 | 560 |

Type 304/347 — G

Type 316 — G

Type 317 — G

Type 321 — G

Alloy 20Cb3 — G

Alloy 800/825 — G

Alloy Al6XN

Type 904L

Type 17-4PH — G

E-Brite 26-1 — G

Type 410 — U

Type 430 — G

Type 444

NICKEL AND HIGH
NICKEL ALLOYS

Nickel 200/201 — G — S

Alloy B-2 — G

Alloy C-276 — E

Alloy 600/625 — G

Monel 400 — G — S

Alloy G/G3

Alloy D — G

ACETONE

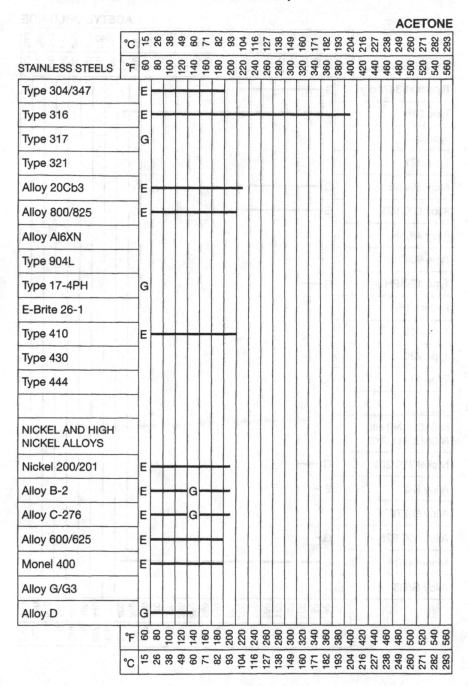

STAINLESS STEELS	°C	15 26 38 49 60 71 82 93 104 116 127 138 149 160 171 182 193 204 216 227 238 249 260 271 282 293
	°F	60 80 100 120 140 160 180 200 220 240 260 280 300 320 340 360 380 400 420 440 460 480 500 520 540 560
Type 304/347	E	
Type 316	E	
Type 317	G	
Type 321		
Alloy 20Cb3	E	
Alloy 800/825	E	
Alloy Al6XN		
Type 904L		
Type 17-4PH	G	
E-Brite 26-1		
Type 410	E	
Type 430		
Type 444		
NICKEL AND HIGH NICKEL ALLOYS		
Nickel 200/201	E	
Alloy B-2	E	G
Alloy C-276	E	G
Alloy 600/625	E	
Monel 400	E	
Alloy G/G3		
Alloy D	G	

ACETYL CHLORIDE

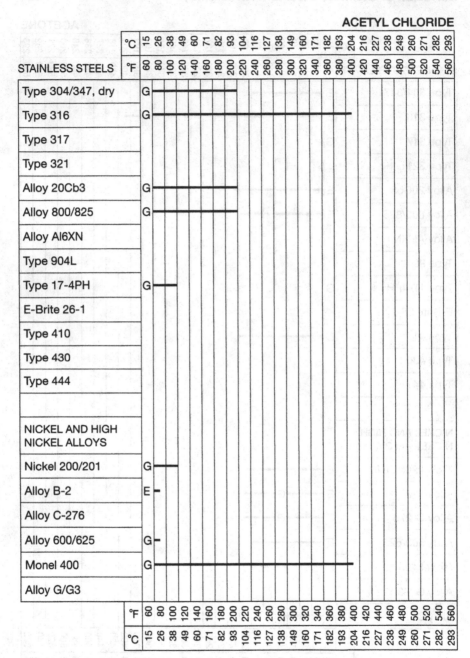

| STAINLESS STEELS | | °C | 15 | 26 | 38 | 49 | 60 | 71 | 82 | 93 | 104 | 116 | 127 | 138 | 149 | 160 | 171 | 182 | 193 | 204 | 216 | 227 | 238 | 249 | 260 | 271 | 282 | 293 |
| | | °F | 60 | 80 | 100 | 120 | 140 | 160 | 180 | 200 | 220 | 240 | 260 | 280 | 300 | 320 | 340 | 360 | 380 | 400 | 420 | 440 | 460 | 480 | 500 | 520 | 540 | 560 |

ACRYLONITRILE

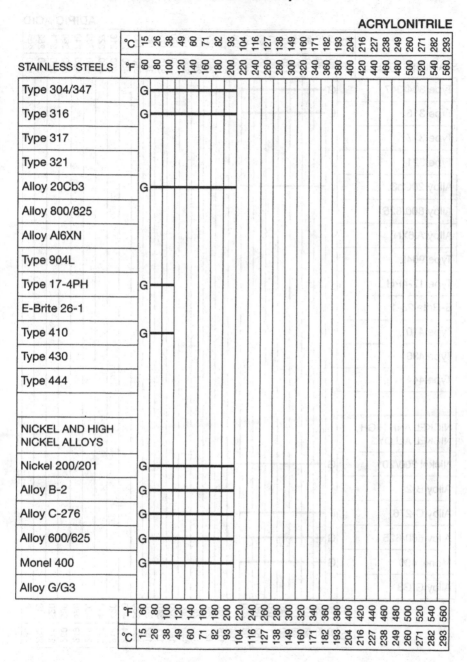

STAINLESS STEELS	°C	15	26	38	49	60	71	82	93	104	116	127	138	149	160	171	182	193	204	216	227	238	249	260	271	282	293
	°F	60	80	100	120	140	160	180	200	220	240	260	280	300	320	340	360	380	400	420	440	460	480	500	520	540	560
Type 304/347	G
Type 316	G
Type 317	
Type 321	
Alloy 20Cb3	G
Alloy 800/825	
Alloy Al6XN	
Type 904L	
Type 17-4PH	G
E-Brite 26-1	
Type 410	G
Type 430	
Type 444	
NICKEL AND HIGH NICKEL ALLOYS	
Nickel 200/201	G
Alloy B-2	G
Alloy C-276	G
Alloy 600/625	G
Monel 400	G
Alloy G/G3	

ADIPIC ACID

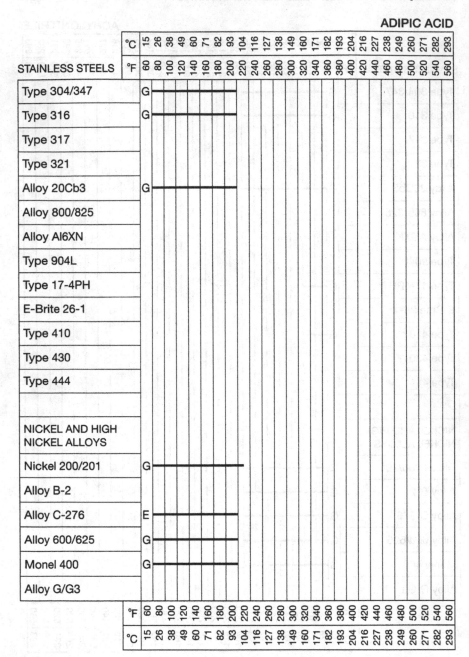

	°C	15	26	38	49	60	71	82	93	104	116	127	138	149	160	171	182	193	204	216	227	238	249	260	271	282	293
STAINLESS STEELS	°F	60	80	100	120	140	160	180	200	220	240	260	280	300	320	340	360	380	400	420	440	460	480	500	520	540	560
Type 304/347	G																										
Type 316	G																										
Type 317																											
Type 321																											
Alloy 20Cb3	G																										
Alloy 800/825																											
Alloy Al6XN																											
Type 904L																											
Type 17-4PH																											
E-Brite 26-1																											
Type 410																											
Type 430																											
Type 444																											
NICKEL AND HIGH NICKEL ALLOYS																											
Nickel 200/201	G																										
Alloy B-2																											
Alloy C-276	E																										
Alloy 600/625	G																										
Monel 400	G																										
Alloy G/G3																											
	°F	60	80	100	120	140	160	180	200	220	240	260	280	300	320	340	360	380	400	420	440	460	480	500	520	540	560
	°C	15	26	38	49	60	71	82	93	104	116	127	138	149	160	171	182	193	204	216	227	238	249	260	271	282	293

ALLYL ALCOHOL

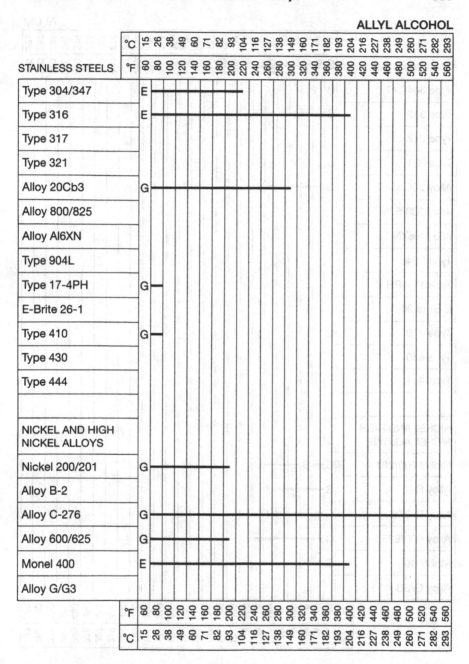

ALUM

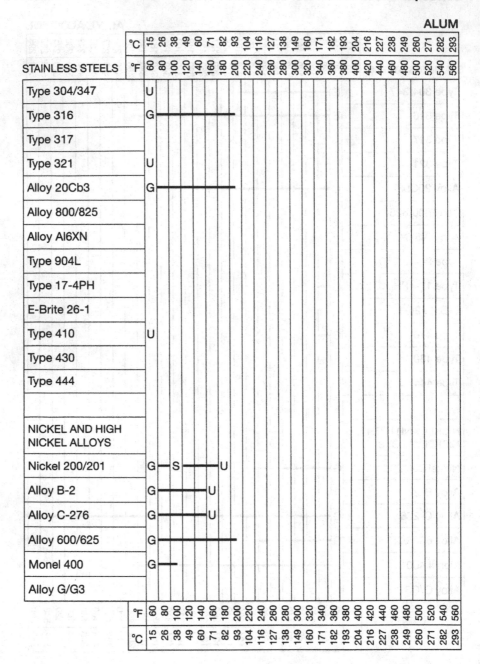

STAINLESS STEELS	°F																										
Type 304/347	U																										
Type 316	G																										
Type 317																											
Type 321	U																										
Alloy 20Cb3	G																										
Alloy 800/825																											
Alloy AI6XN																											
Type 904L																											
Type 17-4PH																											
E-Brite 26-1																											
Type 410	U																										
Type 430																											
Type 444																											
NICKEL AND HIGH NICKEL ALLOYS																											
Nickel 200/201	G	S	U																								
Alloy B-2	G	U																									
Alloy C-276	G	U																									
Alloy 600/625	G																										
Monel 400	G																										
Alloy G/G3																											

ALUMINUM CHLORIDE, AQUEOUS

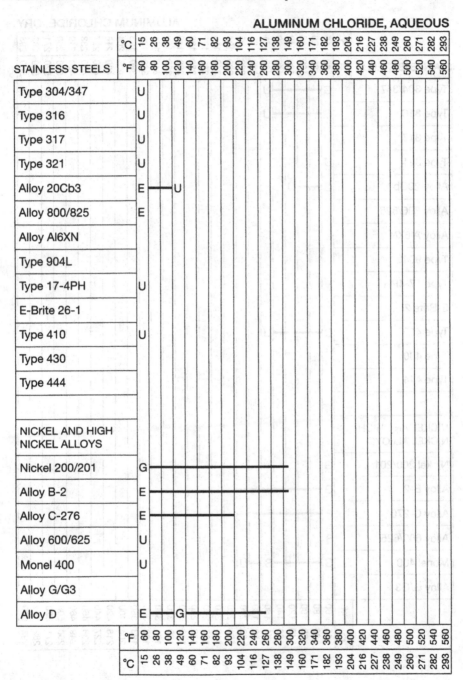

STAINLESS STEELS	Rating
Type 304/347	U
Type 316	U
Type 317	U
Type 321	U
Alloy 20Cb3	E—U
Alloy 800/825	E
Alloy Al6XN	
Type 904L	
Type 17-4PH	U
E-Brite 26-1	
Type 410	U
Type 430	
Type 444	
NICKEL AND HIGH NICKEL ALLOYS	
Nickel 200/201	G—
Alloy B-2	E—
Alloy C-276	E—
Alloy 600/625	U
Monel 400	U
Alloy G/G3	
Alloy D	E—G—

ALUMINUM CHLORIDE, DRY

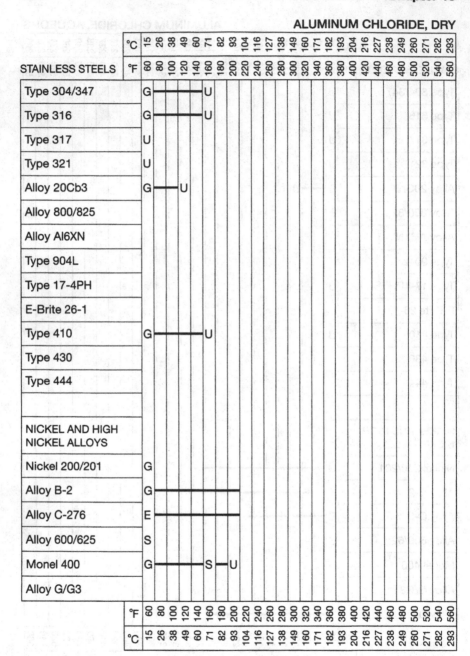

STAINLESS STEELS	°C	15 26 38 49 60 71 82 93 104 116 127 138 149 160 171 182 193 204 216 227 238 249 260 271 282 293
	°F	60 80 100 120 140 160 180 200 220 240 260 280 300 320 340 360 380 400 420 440 460 480 500 520 540 560
Type 304/347		G———————U
Type 316		G———————U
Type 317		U
Type 321		U
Alloy 20Cb3		G———U
Alloy 800/825		
Alloy Al6XN		
Type 904L		
Type 17-4PH		
E-Brite 26-1		
Type 410		G———————U
Type 430		
Type 444		
NICKEL AND HIGH NICKEL ALLOYS		
Nickel 200/201		G
Alloy B-2		G———————
Alloy C-276		E———————
Alloy 600/625		S
Monel 400		G———————S—U
Alloy G/G3		
	°F	60 80 100 120 140 160 180 200 220 240 260 280 300 320 340 360 380 400 420 440 460 480 500 520 540 560
	°C	15 26 38 49 60 71 82 93 104 116 127 138 149 160 171 182 193 204 216 227 238 249 260 271 282 293

ALUMINUM FLUORIDE

STAINLESS STEELS	°C	15	26	38	49	60	71	82	93	104	116	127	138	149	160	171	182	193	204	216	227	238	249	260	271	282	293
	°F	60	80	100	120	140	160	180	200	220	240	260	280	300	320	340	360	380	400	420	440	460	480	500	520	540	560
Type 304/347	U																										
Type 316	G																										
Type 317																											
Type 321																											
Alloy 20Cb3	U																										
Alloy 800/825	G																										
Alloy Al6XN																											
Type 904L																											
Type 17-4PH	U																										
E-Brite 26-1																											
Type 410	U																										
Type 430																											
Type 444																											
NICKEL AND HIGH NICKEL ALLOYS																											
Nickel 200/201	G																										
Alloy B-2 5%	E																										
Alloy C-276 10%	G																										
Alloy 600/625	G																										
Monel 400	G																										
Alloy G/G3																											
	°F	60	80	100	120	140	160	180	200	220	240	260	280	300	320	340	360	380	400	420	440	460	480	500	520	540	560
	°C	15	26	38	49	60	71	82	93	104	116	127	138	149	160	171	182	193	204	216	227	238	249	260	271	282	293

ALUMINUM HYDROXIDE

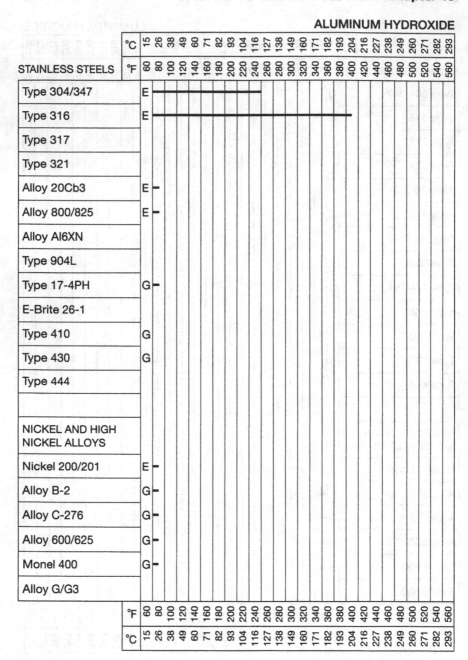

STAINLESS STEELS	°C	15	26	38	49	60	71	82	93	104	116	127	138	149	160	171	182	193	204	216	227	238	249	260	271	282	293
	°F	60	80	100	120	140	160	180	200	220	240	260	280	300	320	340	360	380	400	420	440	460	480	500	520	540	560
Type 304/347	E
Type 316	E
Type 317	
Type 321	
Alloy 20Cb3	E
Alloy 800/825	E
Alloy Al6XN	
Type 904L	
Type 17-4PH	G
E-Brite 26-1	
Type 410	G
Type 430	G
Type 444	
NICKEL AND HIGH NICKEL ALLOYS	
Nickel 200/201	E
Alloy B-2	G
Alloy C-276	G
Alloy 600/625	G
Monel 400	G
Alloy G/G3	
	°F	60	80	100	120	140	160	180	200	220	240	260	280	300	320	340	360	380	400	420	440	460	480	500	520	540	560
	°C	15	26	38	49	60	71	82	93	104	116	127	138	149	160	171	182	193	204	216	227	238	249	260	271	282	293

ALUMINUM NITRATE

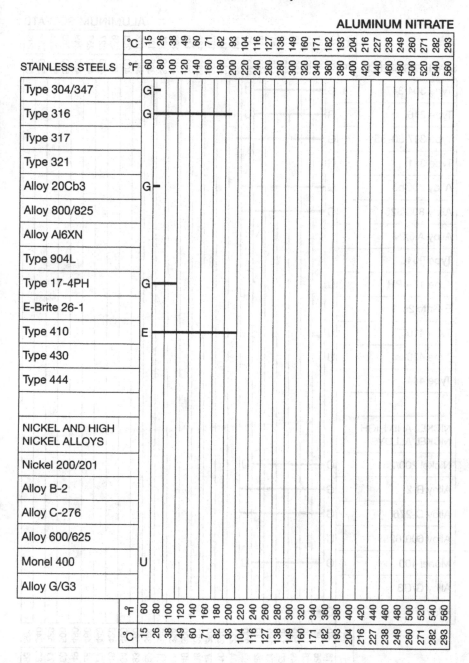

STAINLESS STEELS	°C	15	26	38	49	60	71	82	93	104	116	127	138	149	160	171	182	193	204	216	227	238	249	260	271	282	293
	°F	60	80	100	120	140	160	180	200	220	240	260	280	300	320	340	360	380	400	420	440	460	480	500	520	540	560
Type 304/347	G																										
Type 316	G																										
Type 317																											
Type 321																											
Alloy 20Cb3	G																										
Alloy 800/825																											
Alloy Al6XN																											
Type 904L																											
Type 17-4PH	G																										
E-Brite 26-1																											
Type 410	E																										
Type 430																											
Type 444																											
NICKEL AND HIGH NICKEL ALLOYS																											
Nickel 200/201																											
Alloy B-2																											
Alloy C-276																											
Alloy 600/625																											
Monel 400	U																										
Alloy G/G3																											

ALUMINUM SULFATE

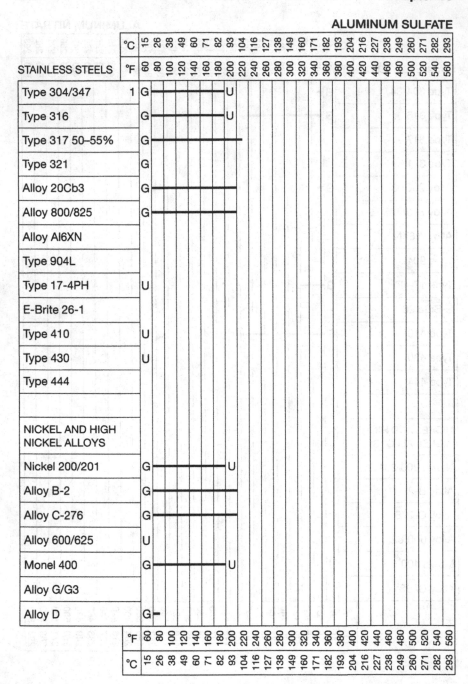

STAINLESS STEELS	°C	15	26	38	49	60	71	82	93	104	116	127	138	149	160	171	182	193	204	216	227	238	249	260	271	282	293
	°F	60	80	100	120	140	160	180	200	220	240	260	280	300	320	340	360	380	400	420	440	460	480	500	520	540	560
Type 304/347		G ──────── U																									
Type 316		G ──────── U																									
Type 317 50–55%		G ────────																									
Type 321		G																									
Alloy 20Cb3		G ────────																									
Alloy 800/825		G ────────																									
Alloy Al6XN																											
Type 904L																											
Type 17-4PH		U																									
E-Brite 26-1																											
Type 410		U																									
Type 430		U																									
Type 444																											
NICKEL AND HIGH NICKEL ALLOYS																											
Nickel 200/201		G ──────── U																									
Alloy B-2		G ────────																									
Alloy C-276		G ────────																									
Alloy 600/625		U																									
Monel 400		G ──────── U																									
Alloy G/G3																											
Alloy D		G ─																									
	°F	60	80	100	120	140	160	180	200	220	240	260	280	300	320	340	360	380	400	420	440	460	480	500	520	540	560
	°C	15	26	38	49	60	71	82	93	104	116	127	138	149	160	171	182	193	204	216	227	238	249	260	271	282	293

AMMONIA, ANHYDROUS

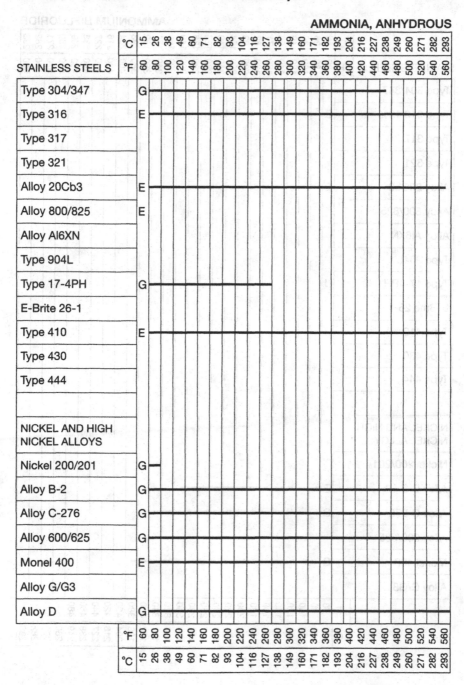

STAINLESS STEELS		°C	15	26	38	49	60	71	82	93	104	116	127	138	149	160	171	182	193	204	216	227	238	249	260	271	282	293
		°F	60	80	100	120	140	160	180	200	220	240	260	280	300	320	340	360	380	400	420	440	460	480	500	520	540	560
Type 304/347	G																											
Type 316	E																											
Type 317																												
Type 321																												
Alloy 20Cb3	E																											
Alloy 800/825	E																											
Alloy Al6XN																												
Type 904L																												
Type 17-4PH	G																											
E-Brite 26-1																												
Type 410	E																											
Type 430																												
Type 444																												
NICKEL AND HIGH NICKEL ALLOYS																												
Nickel 200/201	G																											
Alloy B-2	G																											
Alloy C-276	G																											
Alloy 600/625	G																											
Monel 400	E																											
Alloy G/G3																												
Alloy D	G																											
		°F	60	80	100	120	140	160	180	200	220	240	260	280	300	320	340	360	380	400	420	440	460	480	500	520	540	560
		°C	15	26	38	49	60	71	82	93	104	116	127	138	149	160	171	182	193	204	216	227	238	249	260	271	282	293

AMMONIUM BIFLUORIDE

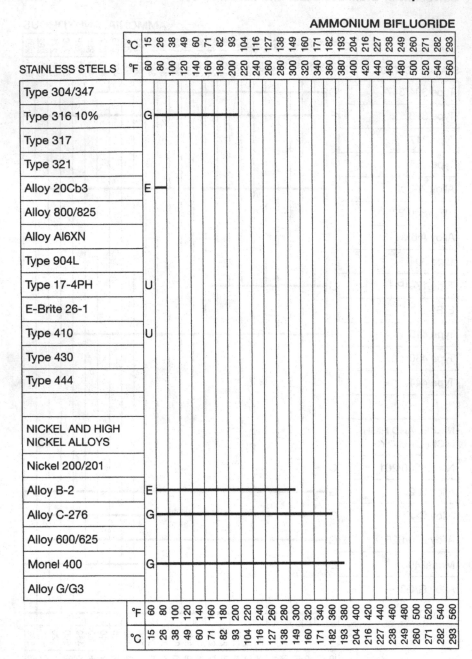

	°C	15	26	38	49	60	71	82	93	104	116	127	138	149	160	171	182	193	204	216	227	238	249	260	271	282	293
STAINLESS STEELS	°F	60	80	100	120	140	160	180	200	220	240	260	280	300	320	340	360	380	400	420	440	460	480	500	520	540	560

STAINLESS STEELS

Type 304/347

Type 316 10% G

Type 317

Type 321

Alloy 20Cb3 E

Alloy 800/825

Alloy Al6XN

Type 904L

Type 17-4PH U

E-Brite 26-1

Type 410 U

Type 430

Type 444

NICKEL AND HIGH NICKEL ALLOYS

Nickel 200/201

Alloy B-2 E

Alloy C-276 G

Alloy 600/625

Monel 400 G

Alloy G/G3

	°F	60	80	100	120	140	160	180	200	220	240	260	280	300	320	340	360	380	400	420	440	460	480	500	520	540	560
	°C	15	26	38	49	60	71	82	93	104	116	127	138	149	160	171	182	193	204	216	227	238	249	260	271	282	293

AMMONIUM CARBONATE

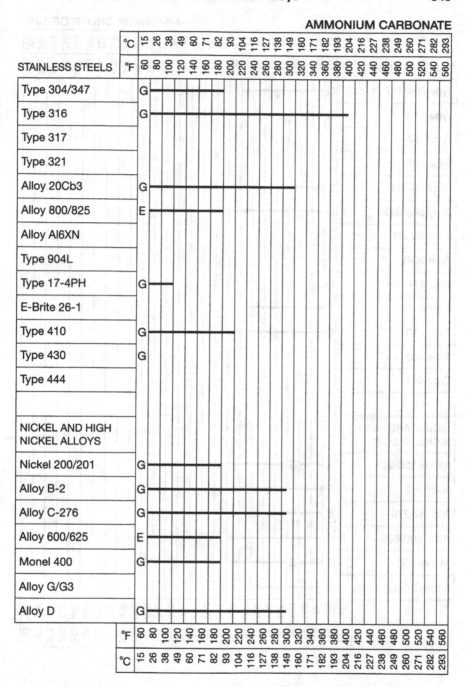

| STAINLESS STEELS | | Type 304/347 | Type 316 | Type 317 | Type 321 | Alloy 20Cb3 | Alloy 800/825 | Alloy Al6XN | Type 904L | Type 17-4PH | E-Brite 26-1 | Type 410 | Type 430 | Type 444 |

AMMONIUM CHLORIDE 10%

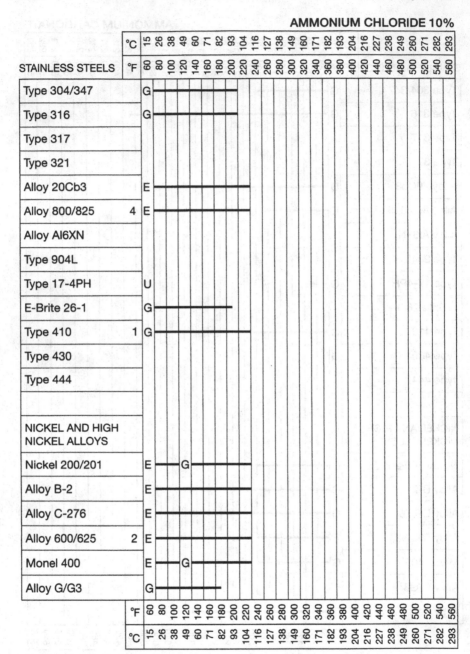

STAINLESS STEELS		°C	15	26	38	49	60	71	82	93	104	116	127	138	149	160	171	182	193	204	216	227	238	249	260	271	282	293
		°F	60	80	100	120	140	160	180	200	220	240	260	280	300	320	340	360	380	400	420	440	460	480	500	520	540	560
Type 304/347		G																										
Type 316		G																										
Type 317																												
Type 321																												
Alloy 20Cb3		E																										
Alloy 800/825	4	E																										
Alloy Al6XN																												
Type 904L																												
Type 17-4PH		U																										
E-Brite 26-1		G																										
Type 410	1	G																										
Type 430																												
Type 444																												
NICKEL AND HIGH NICKEL ALLOYS																												
Nickel 200/201		E		G																								
Alloy B-2		E																										
Alloy C-276		E																										
Alloy 600/625	2	E																										
Monel 400		E		G																								
Alloy G/G3		G																										

	°F	60	80	100	120	140	160	180	200	220	240	260	280	300	320	340	360	380	400	420	440	460	480	500	520	540	560
	°C	15	26	38	49	60	71	82	93	104	116	127	138	149	160	171	182	193	204	216	227	238	249	260	271	282	293

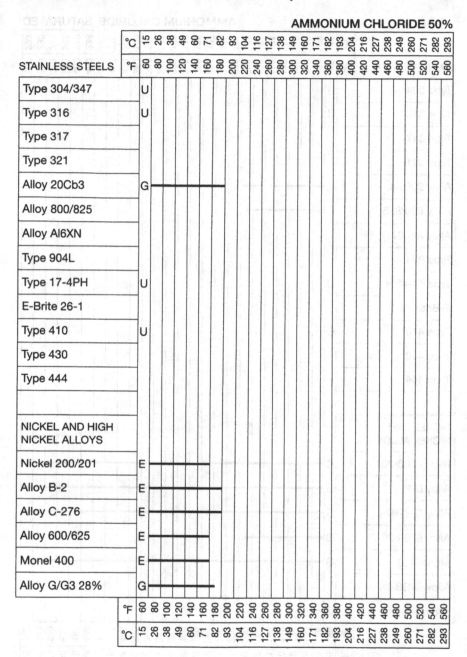

AMMONIUM CHLORIDE 50%

	°C	15	26	38	49	60	71	82	93	104	116	127	138	149	160	171	182	193	204	216	227	238	249	260	271	282	293
STAINLESS STEELS	°F	60	80	100	120	140	160	180	200	220	240	260	280	300	320	340	360	380	400	420	440	460	480	500	520	540	560
Type 304/347	U																										
Type 316	U																										
Type 317																											
Type 321																											
Alloy 20Cb3	G																										
Alloy 800/825																											
Alloy Al6XN																											
Type 904L																											
Type 17-4PH	U																										
E-Brite 26-1																											
Type 410	U																										
Type 430																											
Type 444																											
NICKEL AND HIGH NICKEL ALLOYS																											
Nickel 200/201	E																										
Alloy B-2	E																										
Alloy C-276	E																										
Alloy 600/625	E																										
Monel 400	E																										
Alloy G/G3 28%	G																										
	°F	60	80	100	120	140	160	180	200	220	240	260	280	300	320	340	360	380	400	420	440	460	480	500	520	540	560
	°C	15	26	38	49	60	71	82	93	104	116	127	138	149	160	171	182	193	204	216	227	238	249	260	271	282	293

AMMONIUM CHLORIDE, SATURATED

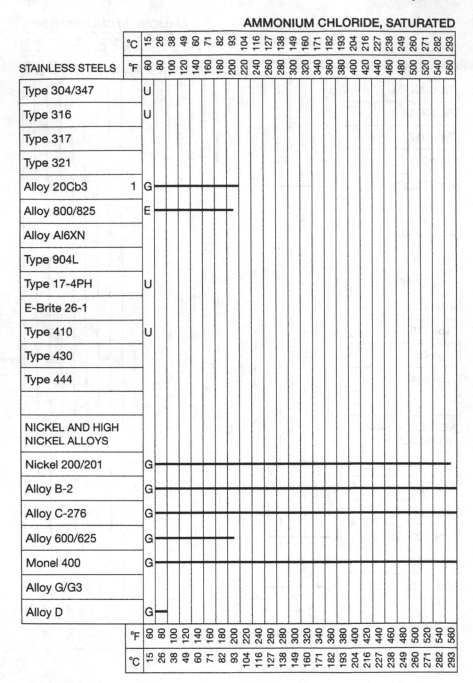

STAINLESS STEELS	°C	15	26	38	49	60	71	82	93	104	116	127	138	149	160	171	182	193	204	216	227	238	249	260	271	282	293
	°F	60	80	100	120	140	160	180	200	220	240	260	280	300	320	340	360	380	400	420	440	460	480	500	520	540	560
Type 304/347	U																										
Type 316	U																										
Type 317																											
Type 321																											
Alloy 20Cb3 1	G																										
Alloy 800/825	E																										
Alloy Al6XN																											
Type 904L																											
Type 17-4PH	U																										
E-Brite 26-1																											
Type 410	U																										
Type 430																											
Type 444																											
NICKEL AND HIGH NICKEL ALLOYS																											
Nickel 200/201	G																										
Alloy B-2	G																										
Alloy C-276	G																										
Alloy 600/625	G																										
Monel 400	G																										
Alloy G/G3																											
Alloy D	G																										
	°F	60	80	100	120	140	160	180	200	220	240	260	280	300	320	340	360	380	400	420	440	460	480	500	520	540	560
	°C	15	26	38	49	60	71	82	93	104	116	127	138	149	160	171	182	193	204	216	227	238	249	260	271	282	293

AMMONIUM HYDROXIDE 10%

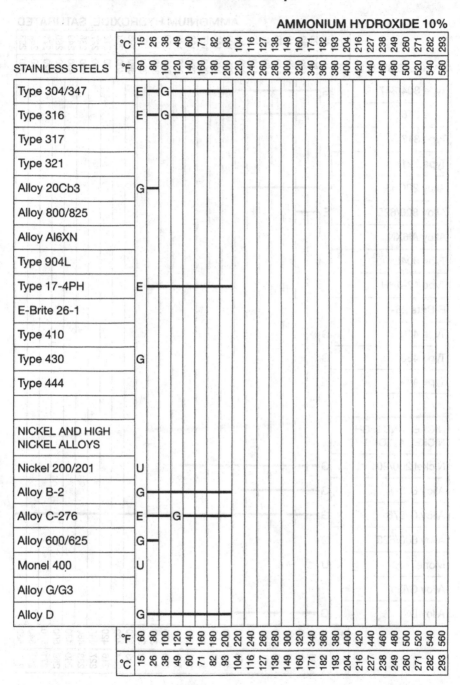

STAINLESS STEELS	°C	15	26	38	49	60	71	82	93	104	116	127	138	149	160	171	182	193	204	216	227	238	249	260	271	282	293
	°F	60	80	100	120	140	160	180	200	220	240	260	280	300	320	340	360	380	400	420	440	460	480	500	520	540	560
Type 304/347		E	—	G																							
Type 316		E	—	G																							
Type 317																											
Type 321																											
Alloy 20Cb3		G																									
Alloy 800/825																											
Alloy AI6XN																											
Type 904L																											
Type 17-4PH		E																									
E-Brite 26-1																											
Type 410																											
Type 430		G																									
Type 444																											
NICKEL AND HIGH NICKEL ALLOYS																											
Nickel 200/201		U																									
Alloy B-2		G																									
Alloy C-276		E	—	G																							
Alloy 600/625		G																									
Monel 400		U																									
Alloy G/G3																											
Alloy D		G																									
	°F	60	80	100	120	140	160	180	200	220	240	260	280	300	320	340	360	380	400	420	440	460	480	500	520	540	560
	°C	15	26	38	49	60	71	82	93	104	116	127	138	149	160	171	182	193	204	216	227	238	249	260	271	282	293

AMMONIUM HYDROXIDE, SATURATED

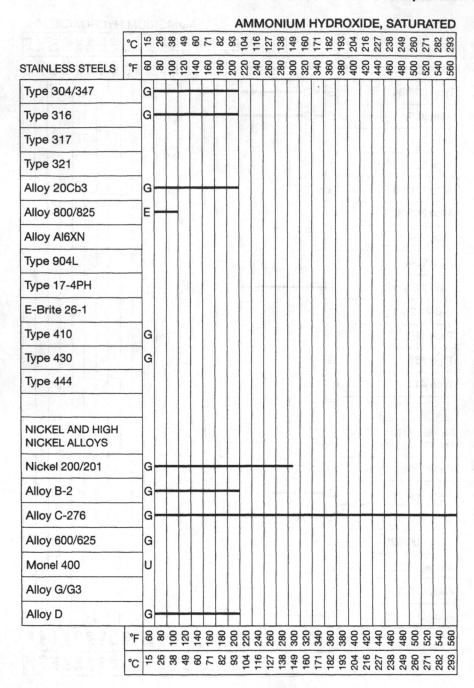

STAINLESS STEELS	Rating	Bar extent (°F)
Type 304/347	G	to ~200
Type 316	G	to ~220
Type 317		
Type 321		
Alloy 20Cb3	G	to ~220
Alloy 800/825	E	to ~100
Alloy AI6XN		
Type 904L		
Type 17-4PH		
E-Brite 26-1		
Type 410	G	
Type 430	G	
Type 444		

NICKEL AND HIGH NICKEL ALLOYS	Rating	Bar extent (°F)
Nickel 200/201	G	to ~300
Alloy B-2	G	to ~220
Alloy C-276	G	to ~560
Alloy 600/625	G	
Monel 400	U	
Alloy G/G3		
Alloy D	G	to ~220

°C: 15 26 38 49 60 71 82 93 104 116 127 138 149 160 171 182 193 204 216 227 238 249 260 271 282 293

°F: 60 80 100 120 140 160 180 200 220 240 260 280 300 320 340 360 380 400 420 440 460 480 500 520 540 560

AMMONIUM NITRATE

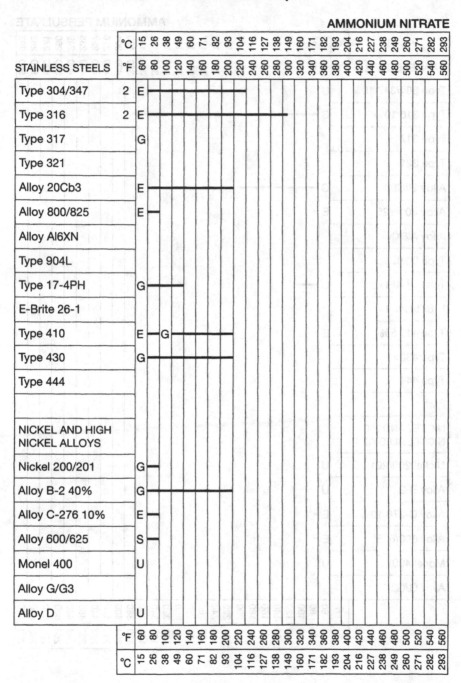

STAINLESS STEELS																												
Type 304/347	2	E																										
Type 316	2	E																										
Type 317		G																										
Type 321																												
Alloy 20Cb3		E																										
Alloy 800/825		E																										
Alloy Al6XN																												
Type 904L																												
Type 17-4PH		G																										
E-Brite 26-1																												
Type 410		E	G																									
Type 430		G																										
Type 444																												
NICKEL AND HIGH NICKEL ALLOYS																												
Nickel 200/201		G																										
Alloy B-2 40%		G																										
Alloy C-276 10%		E																										
Alloy 600/625		S																										
Monel 400		U																										
Alloy G/G3																												
Alloy D		U																										

AMMONIUM PERSULFATE

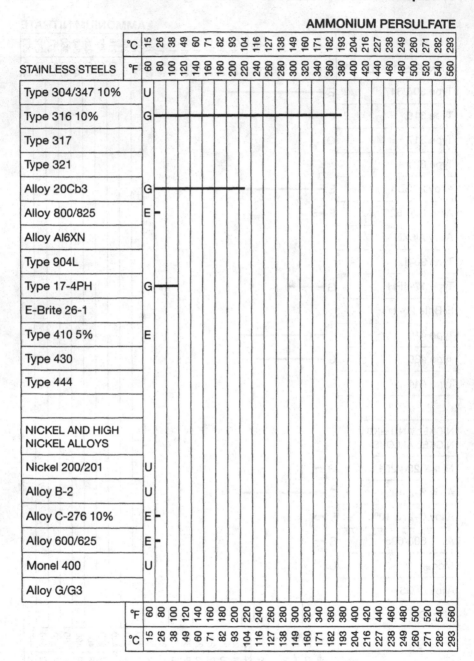

STAINLESS STEELS	°C																											
	°F																											
Type 304/347 10%	U																											
Type 316 10%	G																											
Type 317																												
Type 321																												
Alloy 20Cb3	G																											
Alloy 800/825	E																											
Alloy Al6XN																												
Type 904L																												
Type 17-4PH	G																											
E-Brite 26-1																												
Type 410 5%	E																											
Type 430																												
Type 444																												
NICKEL AND HIGH NICKEL ALLOYS																												
Nickel 200/201	U																											
Alloy B-2	U																											
Alloy C-276 10%	E																											
Alloy 600/625	E																											
Monel 400	U																											
Alloy G/G3																												

°F: 60 80 100 120 140 160 180 200 220 240 260 280 300 320 340 360 380 400 420 440 460 480 500 520 540 560

°C: 15 26 38 49 60 71 82 93 104 116 127 138 149 160 171 182 193 204 216 227 238 249 260 271 282 293

AMMONIUM PHOSPHATE 5%

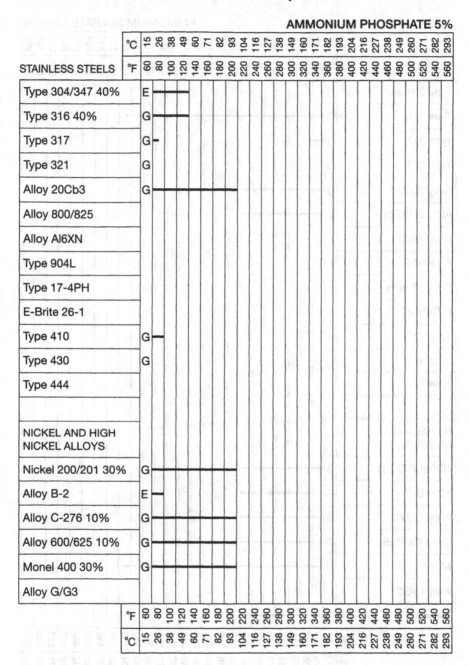

STAINLESS STEELS	°C	15 26 38 49 60 71 82 93 104 116 127 138 149 160 171 182 193 204 216 227 238 249 260 271 282 293
	°F	60 80 100 120 140 160 180 200 220 240 260 280 300 320 340 360 380 400 420 440 460 480 500 520 540 560
Type 304/347 40%	E	
Type 316 40%	G	
Type 317	G	
Type 321	G	
Alloy 20Cb3	G	
Alloy 800/825		
Alloy Al6XN		
Type 904L		
Type 17-4PH		
E-Brite 26-1		
Type 410	G	
Type 430	G	
Type 444		
NICKEL AND HIGH NICKEL ALLOYS		
Nickel 200/201 30%	G	
Alloy B-2	E	
Alloy C-276 10%	G	
Alloy 600/625 10%	G	
Monel 400 30%	G	
Alloy G/G3		
	°F	60 80 100 120 140 160 180 200 220 240 260 280 300 320 340 360 380 400 420 440 460 480 500 520 540 560
	°C	15 26 38 49 60 71 82 93 104 116 127 138 149 160 171 182 193 204 216 227 238 249 260 271 282 293

AMMONIUM SULFATE 10–40%

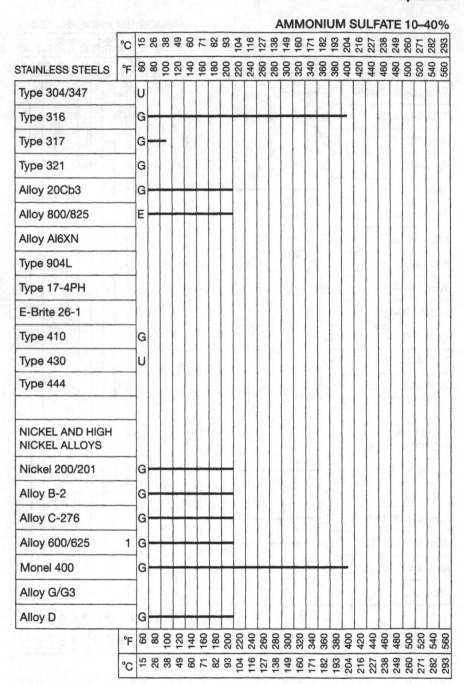

STAINLESS STEELS		°C	15	26	38	49	60	71	82	93	104	116	127	138	149	160	171	182	193	204	216	227	238	249	260	271	282	293
		°F	60	80	100	120	140	160	180	200	220	240	260	280	300	320	340	360	380	400	420	440	460	480	500	520	540	560
Type 304/347		U																										
Type 316		G																										
Type 317		G																										
Type 321		G																										
Alloy 20Cb3		G																										
Alloy 800/825		E																										
Alloy Al6XN																												
Type 904L																												
Type 17-4PH																												
E-Brite 26-1																												
Type 410		G																										
Type 430		U																										
Type 444																												
NICKEL AND HIGH NICKEL ALLOYS																												
Nickel 200/201		G																										
Alloy B-2		G																										
Alloy C-276		G																										
Alloy 600/625	1	G																										
Monel 400		G																										
Alloy G/G3																												
Alloy D		G																										
		°F	60	80	100	120	140	160	180	200	220	240	260	280	300	320	340	360	380	400	420	440	460	480	500	520	540	560
		°C	15	26	38	49	60	71	82	93	104	116	127	138	149	160	171	182	193	204	216	227	238	249	260	271	282	293

AMMONIUM SULFITE

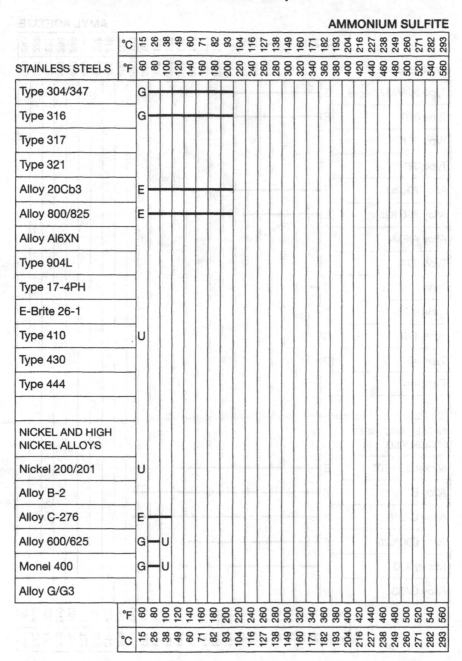

STAINLESS STEELS	°C	15	26	38	49	60	71	82	93	104	116	127	138	149	160	171	182	193	204	216	227	238	249	260	271	282	293
	°F	60	80	100	120	140	160	180	200	220	240	260	280	300	320	340	360	380	400	420	440	460	480	500	520	540	560
Type 304/347		G																									
Type 316		G																									
Type 317																											
Type 321																											
Alloy 20Cb3		E																									
Alloy 800/825		E																									
Alloy Al6XN																											
Type 904L																											
Type 17-4PH																											
E-Brite 26-1																											
Type 410		U																									
Type 430																											
Type 444																											
NICKEL AND HIGH NICKEL ALLOYS																											
Nickel 200/201		U																									
Alloy B-2																											
Alloy C-276		E																									
Alloy 600/625		G	U																								
Monel 400		G	U																								
Alloy G/G3																											

°F	60	80	100	120	140	160	180	200	220	240	260	280	300	320	340	360	380	400	420	440	460	480	500	520	540	560
°C	15	26	38	49	60	71	82	93	104	116	127	138	149	160	171	182	193	204	216	227	238	249	260	271	282	293

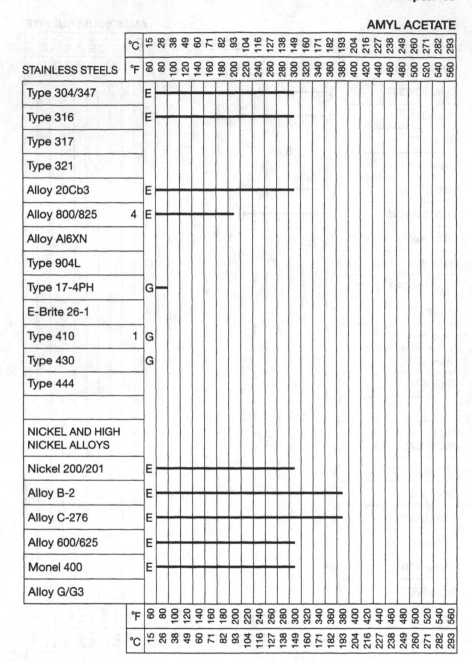

STAINLESS STEELS	°F																										
Type 304/347	E																										
Type 316	E																										
Type 317																											
Type 321																											
Alloy 20Cb3	E																										
Alloy 800/825 4	E																										
Alloy Al6XN																											
Type 904L																											
Type 17-4PH	G																										
E-Brite 26-1																											
Type 410 1	G																										
Type 430	G																										
Type 444																											
NICKEL AND HIGH NICKEL ALLOYS																											
Nickel 200/201	E																										
Alloy B-2	E																										
Alloy C-276	E																										
Alloy 600/625	E																										
Monel 400	E																										
Alloy G/G3																											

AMYL ALCOHOL

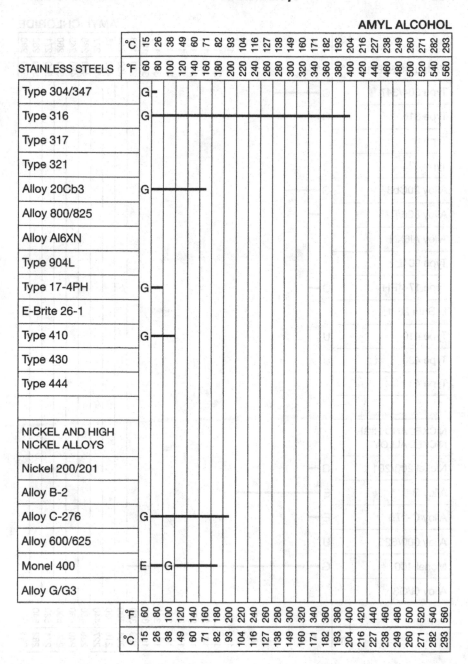

STAINLESS STEELS	°C																							

Chart columns (°C): 15 26 38 49 60 71 82 93 104 116 127 138 149 160 171 182 193 204 216 227 238 249 260 271 282 293
Chart columns (°F): 60 80 100 120 140 160 180 200 220 240 260 280 300 320 340 360 380 400 420 440 460 480 500 520 540 560

STAINLESS STEELS

- Type 304/347 — G
- Type 316 — G
- Type 317
- Type 321
- Alloy 20Cb3 — G
- Alloy 800/825
- Alloy Al6XN
- Type 904L
- Type 17-4PH — G
- E-Brite 26-1
- Type 410 — G
- Type 430
- Type 444

NICKEL AND HIGH NICKEL ALLOYS

- Nickel 200/201
- Alloy B-2
- Alloy C-276 — G
- Alloy 600/625
- Monel 400 — E G
- Alloy G/G3

AMYL CHLORIDE

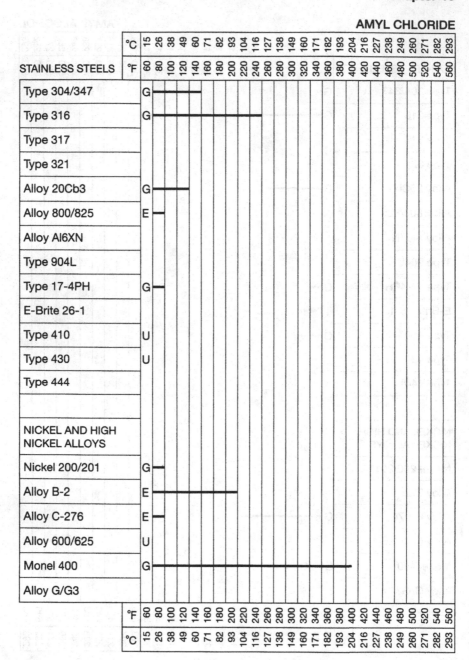

		°C	15	26	38	49	60	71	82	93	104	116	127	138	149	160	171	182	193	204	216	227	238	249	260	271	282	293
STAINLESS STEELS		°F	60	80	100	120	140	160	180	200	220	240	260	280	300	320	340	360	380	400	420	440	460	480	500	520	540	560
Type 304/347	G																											
Type 316	G																											
Type 317																												
Type 321																												
Alloy 20Cb3	G																											
Alloy 800/825	E																											
Alloy Al6XN																												
Type 904L																												
Type 17-4PH	G																											
E-Brite 26-1																												
Type 410	U																											
Type 430	U																											
Type 444																												
NICKEL AND HIGH NICKEL ALLOYS																												
Nickel 200/201	G																											
Alloy B-2	E																											
Alloy C-276	E																											
Alloy 600/625	U																											
Monel 400	G																											
Alloy G/G3																												
		°F	60	80	100	120	140	160	180	200	220	240	260	280	300	320	340	360	380	400	420	440	460	480	500	520	540	560
		°C	15	26	38	49	60	71	82	93	104	116	127	138	149	160	171	182	193	204	216	227	238	249	260	271	282	293

ANILINE

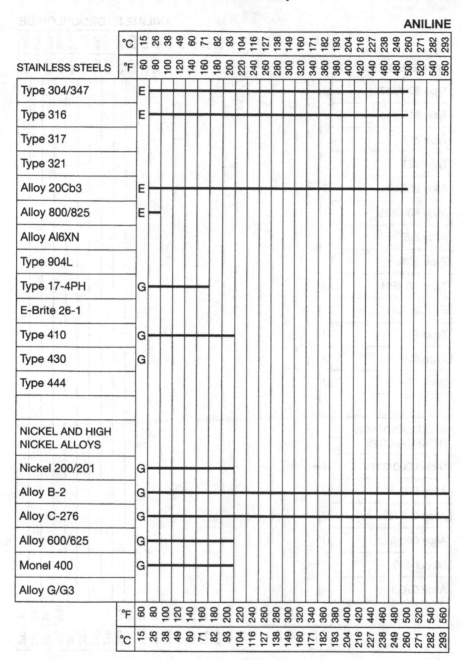

STAINLESS STEELS	°C / °F	
Type 304/347	E	
Type 316	E	
Type 317		
Type 321		
Alloy 20Cb3	E	
Alloy 800/825	E	
Alloy Al6XN		
Type 904L		
Type 17-4PH	G	
E-Brite 26-1		
Type 410	G	
Type 430	G	
Type 444		
NICKEL AND HIGH NICKEL ALLOYS		
Nickel 200/201	G	
Alloy B-2	G	
Alloy C-276	G	
Alloy 600/625	G	
Monel 400	G	
Alloy G/G3		

ANILINE HYDROCHLORIDE

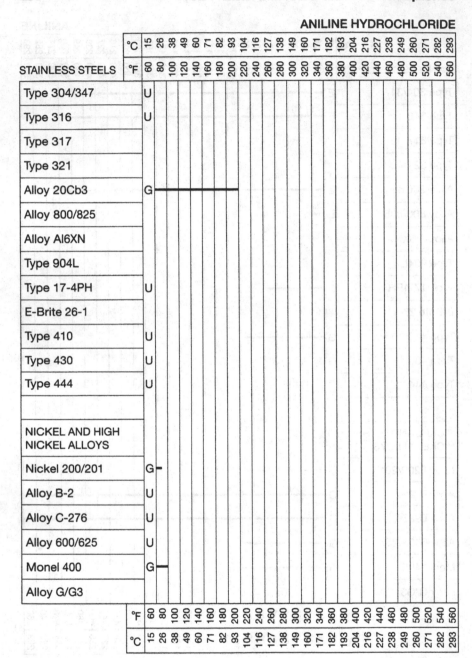

STAINLESS STEELS	°C	15	26	38	49	60	71	82	93	104	116	127	138	149	160	171	182	193	204	216	227	238	249	260	271	282	293
	°F	60	80	100	120	140	160	180	200	220	240	260	280	300	320	340	360	380	400	420	440	460	480	500	520	540	560
Type 304/347	U																										
Type 316	U																										
Type 317																											
Type 321																											
Alloy 20Cb3	G																										
Alloy 800/825																											
Alloy Al6XN																											
Type 904L																											
Type 17-4PH	U																										
E-Brite 26-1																											
Type 410	U																										
Type 430	U																										
Type 444	U																										
NICKEL AND HIGH NICKEL ALLOYS																											
Nickel 200/201	G																										
Alloy B-2	U																										
Alloy C-276	U																										
Alloy 600/625	U																										
Monel 400	G																										
Alloy G/G3																											
	°F	60	80	100	120	140	160	180	200	220	240	260	280	300	320	340	360	380	400	420	440	460	480	500	520	540	560
	°C	15	26	38	49	60	71	82	93	104	116	127	138	149	160	171	182	193	204	216	227	238	249	260	271	282	293

ANTIMONY TRICHLORIDE

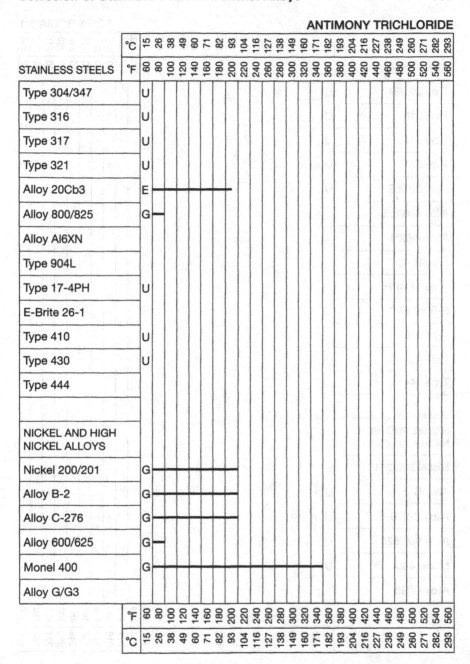

AQUA REGIA 3:1

	°C	15	26	38	49	60	71	82	93	104	116	127	138	149	160	171	182	193	204	216	227	238	249	260	271	282	293
STAINLESS STEELS	°F	60	80	100	120	140	160	180	200	220	240	260	280	300	320	340	360	380	400	420	440	460	480	500	520	540	560
Type 304/347		U																									
Type 316		U																									
Type 317		U																									
Type 321		U																									
Alloy 20Cb3		U																									
Alloy 800/825																											
Alloy Al6XN																											
Type 904L																											
Type 17-4PH																											
E-Brite 26-1		U																									
Type 410																											
Type 430																											
Type 444																											
NICKEL AND HIGH NICKEL ALLOYS																											
Nickel 200/201		U																									
Alloy B-2		U																									
Alloy C-276		U																									
Alloy 600/625		U																									
Monel 400		U																									
Alloy G/G3																											
	°F	60	80	100	120	140	160	180	200	220	240	260	280	300	320	340	360	380	400	420	440	460	480	500	520	540	560
	°C	15	26	38	49	60	71	82	93	104	116	127	138	149	160	171	182	193	204	216	227	238	249	260	271	282	293

ARSENIC ACID

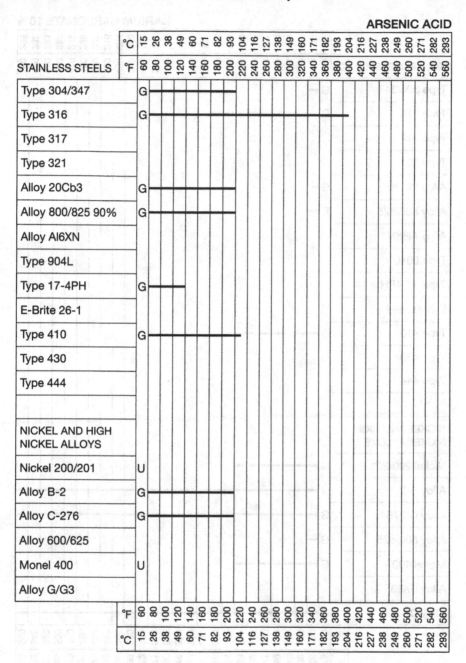

STAINLESS STEELS °C °F				
Type 304/347	G			
Type 316	G			
Type 317				
Type 321				
Alloy 20Cb3	G			
Alloy 800/825 90%	G			
Alloy Al6XN				
Type 904L				
Type 17-4PH	G			
E-Brite 26-1				
Type 410	G			
Type 430				
Type 444				
NICKEL AND HIGH NICKEL ALLOYS				
Nickel 200/201	U			
Alloy B-2	G			
Alloy C-276	G			
Alloy 600/625				
Monel 400	U			
Alloy G/G3				

BARIUM CARBONATE 10%

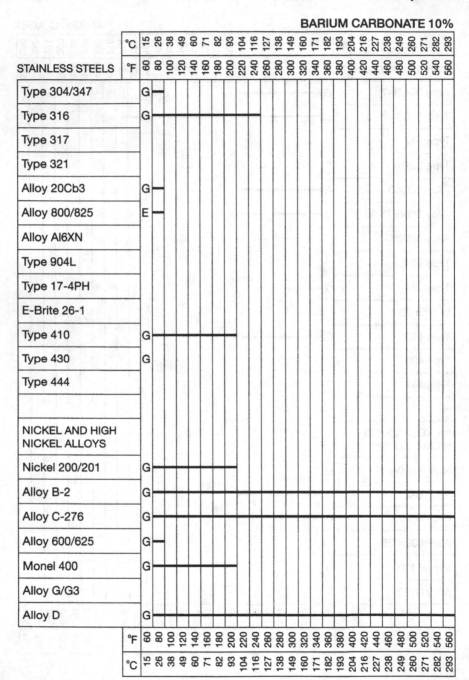

STAINLESS STEELS		
Type 304/347	G	
Type 316	G	
Type 317		
Type 321		
Alloy 20Cb3	G	
Alloy 800/825	E	
Alloy Al6XN		
Type 904L		
Type 17-4PH		
E-Brite 26-1		
Type 410	G	
Type 430	G	
Type 444		
NICKEL AND HIGH NICKEL ALLOYS		
Nickel 200/201	G	
Alloy B-2	G	
Alloy C-276	G	
Alloy 600/625	G	
Monel 400	G	
Alloy G/G3		
Alloy D	G	

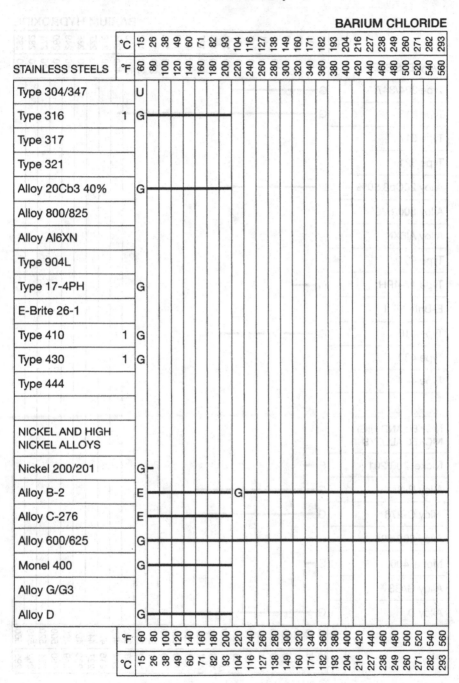

BARIUM CHLORIDE

		°C	15	26	38	49	60	71	82	93	104	116	127	138	149	160	171	182	193	204	216	227	238	249	260	271	282	293
STAINLESS STEELS		°F	60	80	100	120	140	160	180	200	220	240	260	280	300	320	340	360	380	400	420	440	460	480	500	520	540	560
Type 304/347		U																										
Type 316	1	G																										
Type 317																												
Type 321																												
Alloy 20Cb3 40%		G																										
Alloy 800/825																												
Alloy Al6XN																												
Type 904L																												
Type 17-4PH		G																										
E-Brite 26-1																												
Type 410	1	G																										
Type 430	1	G																										
Type 444																												
NICKEL AND HIGH NICKEL ALLOYS																												
Nickel 200/201		G																										
Alloy B-2		E							G																			
Alloy C-276		E																										
Alloy 600/625		G																										
Monel 400		G																										
Alloy G/G3																												
Alloy D		G																										
		°F	60	80	100	120	140	160	180	200	220	240	260	280	300	320	340	360	380	400	420	440	460	480	500	520	540	560
		°C	15	26	38	49	60	71	82	93	104	116	127	138	149	160	171	182	193	204	216	227	238	249	260	271	282	293

BARIUM HYDROXIDE

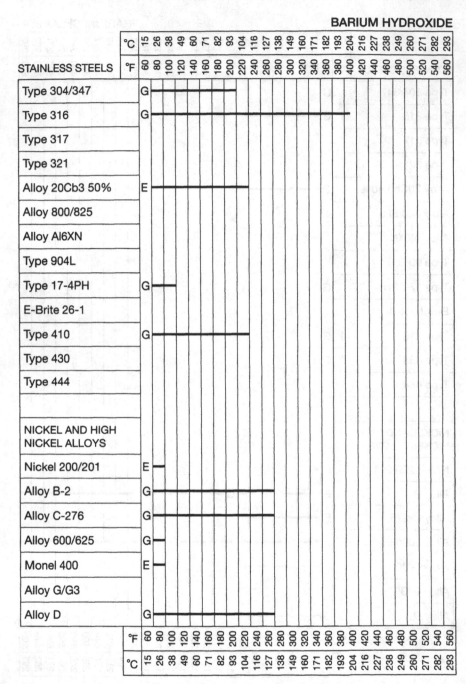

STAINLESS STEELS																															

Type 304/347 — G

Type 316 — G

Type 317

Type 321

Alloy 20Cb3 50% — E

Alloy 800/825

Alloy Al6XN

Type 904L

Type 17-4PH — G

E-Brite 26-1

Type 410 — G

Type 430

Type 444

NICKEL AND HIGH NICKEL ALLOYS

Nickel 200/201 — E

Alloy B-2 — G

Alloy C-276 — G

Alloy 600/625 — G

Monel 400 — E

Alloy G/G3

Alloy D — G

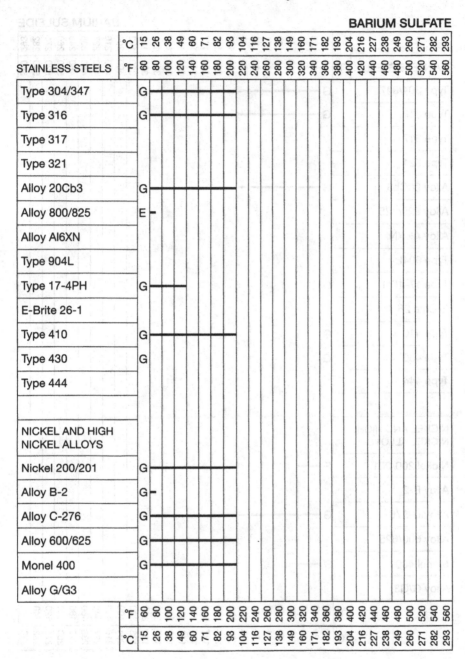

STAINLESS STEELS	°C	15 26 38 49 60 71 82 93 104 116 127 138 149 160 171 182 193 204 216 227 238 249 260 271 282 293
	°F	60 80 100 120 140 160 180 200 220 240 260 280 300 320 340 360 380 400 420 440 460 480 500 520 540 560
Type 304/347	G	
Type 316	G	
Type 317		
Type 321		
Alloy 20Cb3	G	
Alloy 800/825	E	
Alloy Al6XN		
Type 904L		
Type 17-4PH	G	
E-Brite 26-1		
Type 410	G	
Type 430	G	
Type 444		
NICKEL AND HIGH NICKEL ALLOYS		
Nickel 200/201	G	
Alloy B-2	G	
Alloy C-276	G	
Alloy 600/625	G	
Monel 400	G	
Alloy G/G3		
	°F	60 80 100 120 140 160 180 200 220 240 260 280 300 320 340 360 380 400 420 440 460 480 500 520 540 560
	°C	15 26 38 49 60 71 82 93 104 116 127 138 149 160 171 182 193 204 216 227 238 249 260 271 282 293

BARIUM SULFIDE

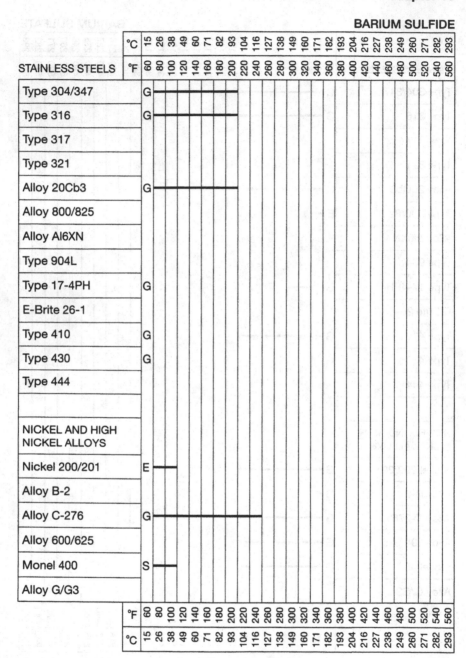

STAINLESS STEELS	°C	°F
Type 304/347	G	
Type 316	G	
Type 317		
Type 321		
Alloy 20Cb3	G	
Alloy 800/825		
Alloy Al6XN		
Type 904L		
Type 17-4PH	G	
E-Brite 26-1		
Type 410	G	
Type 430	G	
Type 444		
NICKEL AND HIGH NICKEL ALLOYS		
Nickel 200/201	E	
Alloy B-2		
Alloy C-276	G	
Alloy 600/625		
Monel 400	S	
Alloy G/G3		

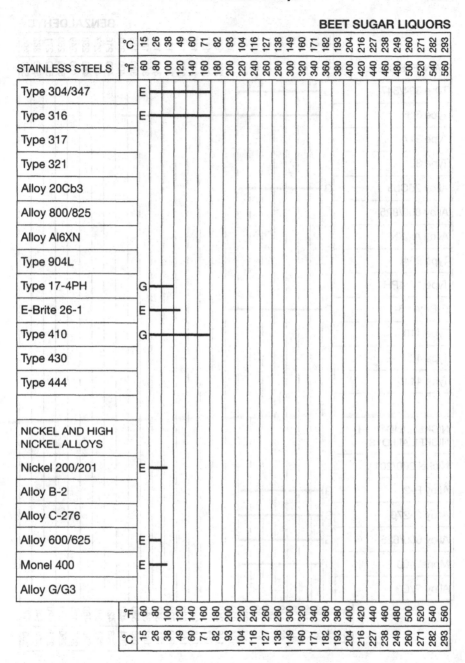

BEET SUGAR LIQUORS

BENZALDEHYDE

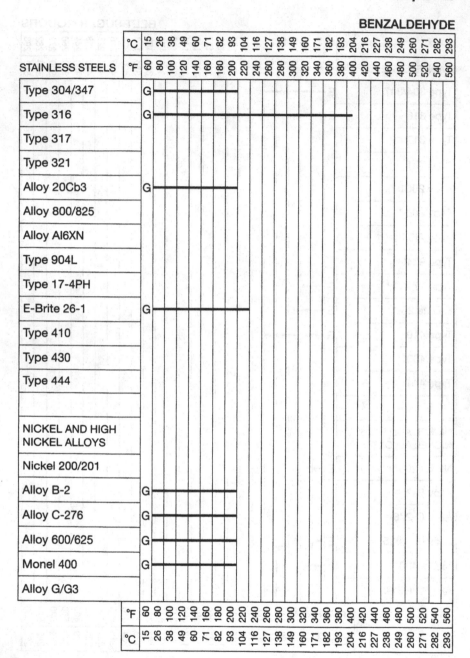

| STAINLESS STEELS | | Type 304/347 | Type 316 | Type 317 | Type 321 | Alloy 20Cb3 | Alloy 800/825 | Alloy AI6XN | Type 904L | Type 17-4PH | E-Brite 26-1 | Type 410 | Type 430 | Type 444 |

BENZENE

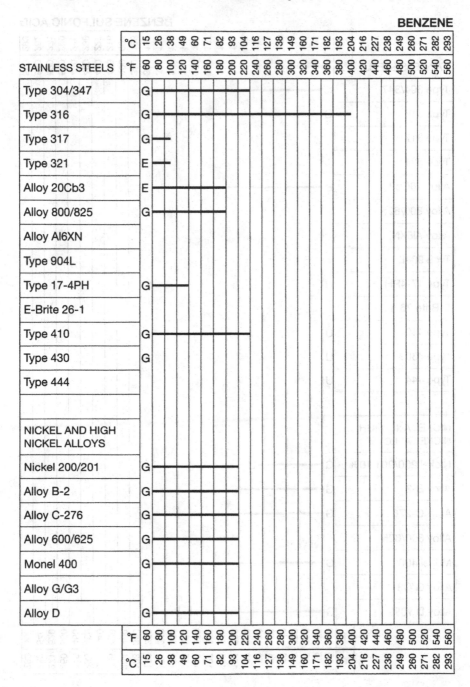

376

BENZENE SULFONIC ACID

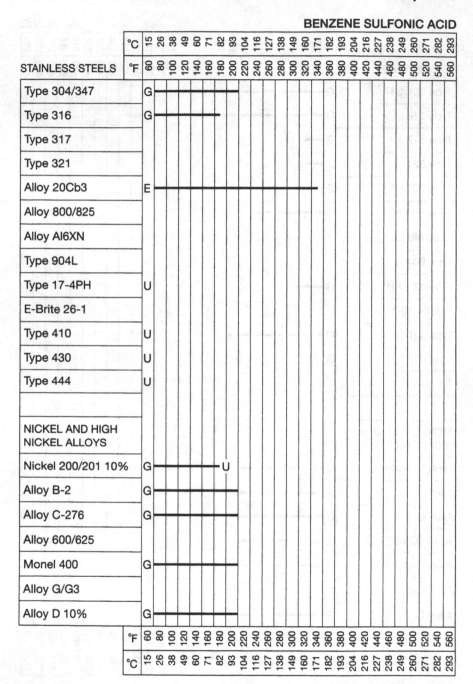

BENZOIC ACID

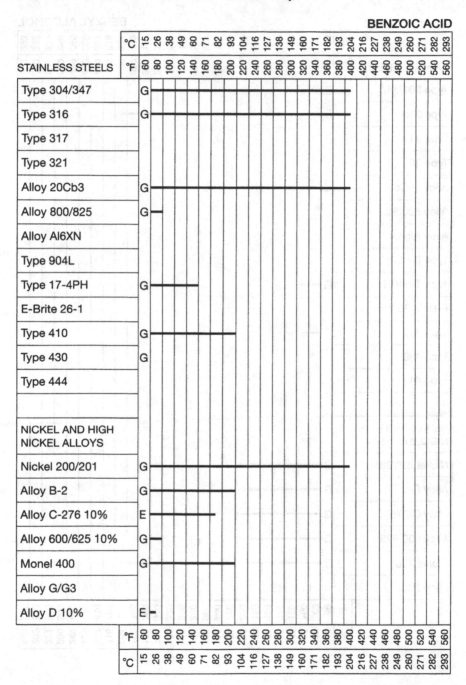

STAINLESS STEELS		°C	15	26	38	49	60	71	82	93	104	116	127	138	149	160	171	182	193	204	216	227	238	249	260	271	282	293
		°F	60	80	100	120	140	160	180	200	220	240	260	280	300	320	340	360	380	400	420	440	460	480	500	520	540	560

BENZYL ALCOHOL

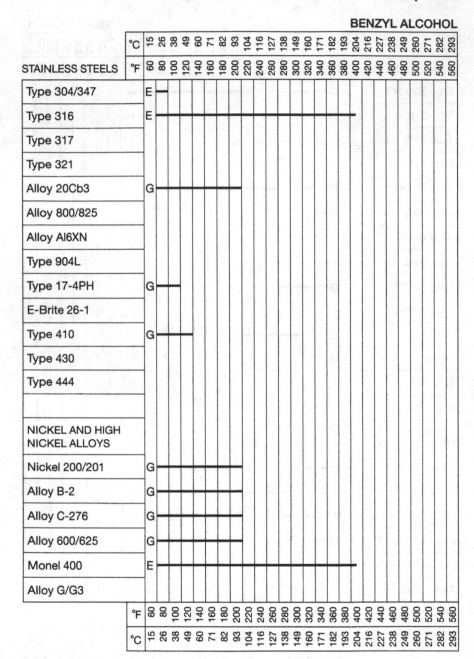

| STAINLESS STEELS | °C | 15 | 26 | 38 | 49 | 60 | 71 | 82 | 93 | 104 | 116 | 127 | 138 | 149 | 160 | 171 | 182 | 193 | 204 | 216 | 227 | 238 | 249 | 260 | 271 | 282 | 293 |
| | °F | 60 | 80 | 100 | 120 | 140 | 160 | 180 | 200 | 220 | 240 | 260 | 280 | 300 | 320 | 340 | 360 | 380 | 400 | 420 | 440 | 460 | 480 | 500 | 520 | 540 | 560 |

Type 304/347 E

Type 316 E

Type 317

Type 321

Alloy 20Cb3 G

Alloy 800/825

Alloy Al6XN

Type 904L

Type 17-4PH G

E-Brite 26-1

Type 410 G

Type 430

Type 444

NICKEL AND HIGH NICKEL ALLOYS

Nickel 200/201 G

Alloy B-2 G

Alloy C-276 G

Alloy 600/625 G

Monel 400 E

Alloy G/G3

| | °F | 60 | 80 | 100 | 120 | 140 | 160 | 180 | 200 | 220 | 240 | 260 | 280 | 300 | 320 | 340 | 360 | 380 | 400 | 420 | 440 | 460 | 480 | 500 | 520 | 540 | 560 |
| | °C | 15 | 26 | 38 | 49 | 60 | 71 | 82 | 93 | 104 | 116 | 127 | 138 | 149 | 160 | 171 | 182 | 193 | 204 | 216 | 227 | 238 | 249 | 260 | 271 | 282 | 293 |

BORAX

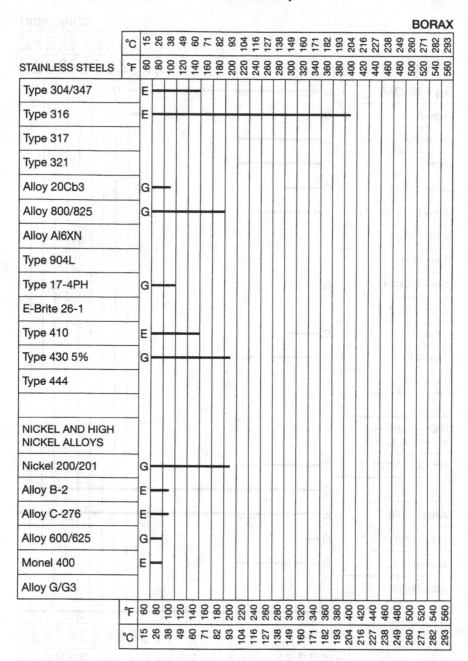

BORIC ACID

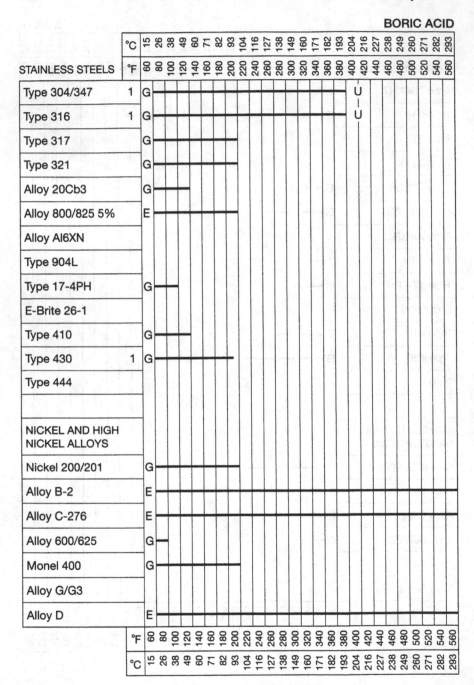

	°C	15	26	38	49	60	71	82	93	104	116	127	138	149	160	171	182	193	204	216	227	238	249	260	271	282	293
STAINLESS STEELS	°F	60	80	100	120	140	160	180	200	220	240	260	280	300	320	340	360	380	400	420	440	460	480	500	520	540	560
Type 304/347	1	G																	U								
Type 316	1	G																	U								
Type 317		G																									
Type 321		G																									
Alloy 20Cb3		G																									
Alloy 800/825 5%		E																									
Alloy AI6XN																											
Type 904L																											
Type 17-4PH		G																									
E-Brite 26-1																											
Type 410		G																									
Type 430	1	G																									
Type 444																											
NICKEL AND HIGH NICKEL ALLOYS																											
Nickel 200/201		G																									
Alloy B-2		E																									
Alloy C-276		E																									
Alloy 600/625		G																									
Monel 400		G																									
Alloy G/G3																											
Alloy D		E																									
	°F	60	80	100	120	140	160	180	200	220	240	260	280	300	320	340	360	380	400	420	440	460	480	500	520	540	560
	°C	15	26	38	49	60	71	82	93	104	116	127	138	149	160	171	182	193	204	216	227	238	249	260	271	282	293

BROMINE GAS, DRY

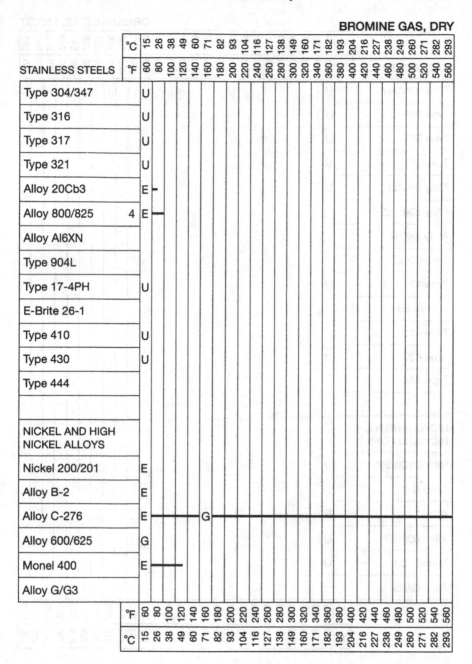

STAINLESS STEELS	°C	15	26	38	49	60	71	82	93	104	116	127	138	149	160	171	182	193	204	216	227	238	249	260	271	282	293
	°F	60	80	100	120	140	160	180	200	220	240	260	280	300	320	340	360	380	400	420	440	460	480	500	520	540	560
Type 304/347	U																										
Type 316	U																										
Type 317	U																										
Type 321	U																										
Alloy 20Cb3	E																										
Alloy 800/825 4	E																										
Alloy Al6XN																											
Type 904L																											
Type 17-4PH	U																										
E-Brite 26-1																											
Type 410	U																										
Type 430	U																										
Type 444																											
NICKEL AND HIGH NICKEL ALLOYS																											
Nickel 200/201	E																										
Alloy B-2	E																										
Alloy C-276	E				G																						
Alloy 600/625	G																										
Monel 400	E																										
Alloy G/G3																											
	°F	60	80	100	120	140	160	180	200	220	240	260	280	300	320	340	360	380	400	420	440	460	480	500	520	540	560
	°C	15	26	38	49	60	71	82	93	104	116	127	138	149	160	171	182	193	204	216	227	238	249	260	271	282	293

BROMINE GAS, MOIST

STAINLESS STEELS °C	15	26	38	49	60	71	82	93	104	116	127	138	149	160	171	182	193	204	216	227	238	249	260	271	282	293
°F	60	80	100	120	140	160	180	200	220	240	260	280	300	320	340	360	380	400	420	440	460	480	500	520	540	560
Type 304/347	U																									
Type 316	U																									
Type 317	U																									
Type 321	U																									
Alloy 20Cb3	U																									
Alloy 800/825																										
Alloy AI6XN																										
Type 904L																										
Type 17-4PH	U																									
E-Brite 26-1																										
Type 410	U																									
Type 430	U																									
Type 444																										
NICKEL AND HIGH NICKEL ALLOYS																										
Nickel 200/201	U																									
Alloy B-2																										
Alloy C-276	E																									
Alloy 600/625	U																									
Monel 400	U																									
Alloy G/G3																										
°F	60	80	100	120	140	160	180	200	220	240	260	280	300	320	340	360	380	400	420	440	460	480	500	520	540	560
°C	15	26	38	49	60	71	82	93	104	116	127	138	149	160	171	182	193	204	216	227	238	249	260	271	282	293

BROMINE LIQUID

	°C	15	26	38	49	60	71	82	93	104	116	127	138	149	160	171	182	193	204	216	227	238	249	260	271	282	293
STAINLESS STEELS	°F	60	80	100	120	140	160	180	200	220	240	260	280	300	320	340	360	380	400	420	440	460	480	500	520	540	560
Type 304/347	U																										
Type 316	U																										
Type 317	U																										
Type 321	U																										
Alloy 20Cb3																											
Alloy 800/825																											
Alloy Al6XN																											
Type 904L																											
Type 17-4PH	U																										
E-Brite 26-1																											
Type 410	U																										
Type 430	U																										
Type 444																											
NICKEL AND HIGH NICKEL ALLOYS																											
Nickel 200/201																											
Alloy B-2																											
Alloy C-276																											
Alloy 600/625																											
Monel 400																											
Alloy G/G3																											
	°F	60	80	100	120	140	160	180	200	220	240	260	280	300	320	340	360	380	400	420	440	460	480	500	520	540	560
	°C	15	26	38	49	60	71	82	93	104	116	127	138	149	160	171	182	193	204	216	227	238	249	260	271	282	293

BUTADIENE

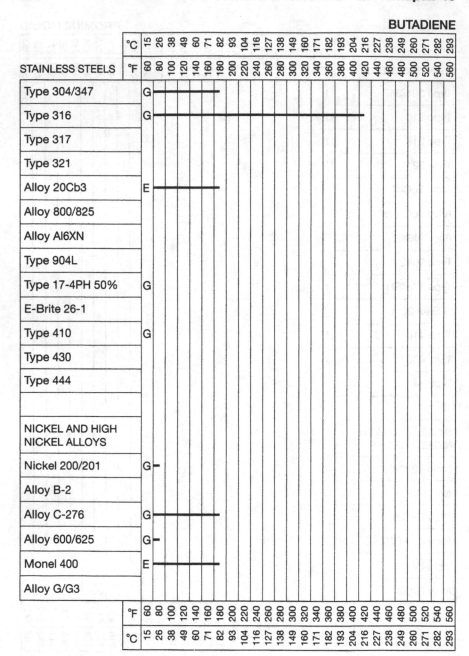

STAINLESS STEELS	°C	15	26	38	49	60	71	82	93	104	116	127	138	149	160	171	182	193	204	216	227	238	249	260	271	282	293
	°F	60	80	100	120	140	160	180	200	220	240	260	280	300	320	340	360	380	400	420	440	460	480	500	520	540	560
Type 304/347	G																										
Type 316	G																										
Type 317																											
Type 321																											
Alloy 20Cb3	E																										
Alloy 800/825																											
Alloy Al6XN																											
Type 904L																											
Type 17-4PH 50%	G																										
E-Brite 26-1																											
Type 410	G																										
Type 430																											
Type 444																											
NICKEL AND HIGH NICKEL ALLOYS																											
Nickel 200/201	G																										
Alloy B-2																											
Alloy C-276	G																										
Alloy 600/625	G																										
Monel 400	E																										
Alloy G/G3																											

BUTYL ACETATE

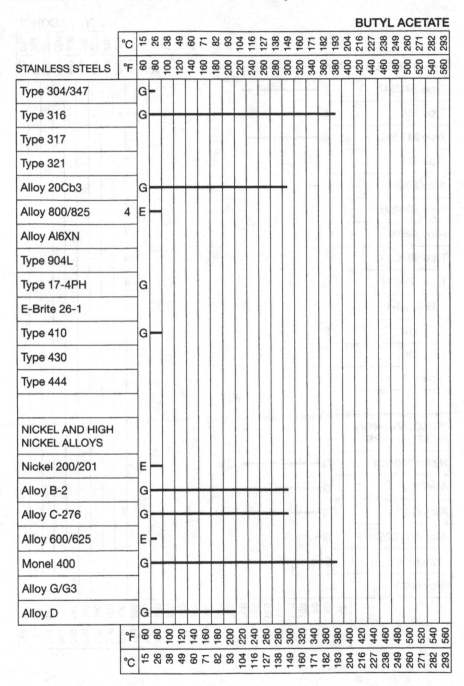

STAINLESS STEELS	°C	15	26	38	49	60	71	82	93	104	116	127	138	149	160	171	182	193	204	216	227	238	249	260	271	282	293
	°F	60	80	100	120	140	160	180	200	220	240	260	280	300	320	340	360	380	400	420	440	460	480	500	520	540	560
Type 304/347	G																										
Type 316	G																										
Type 317																											
Type 321																											
Alloy 20Cb3	G																										
Alloy 800/825 4	E																										
Alloy Al6XN																											
Type 904L																											
Type 17-4PH	G																										
E-Brite 26-1																											
Type 410	G																										
Type 430																											
Type 444																											
NICKEL AND HIGH NICKEL ALLOYS																											
Nickel 200/201	E																										
Alloy B-2	G																										
Alloy C-276	G																										
Alloy 600/625	E																										
Monel 400	G																										
Alloy G/G3																											
Alloy D	G																										

BUTYL ALCOHOL

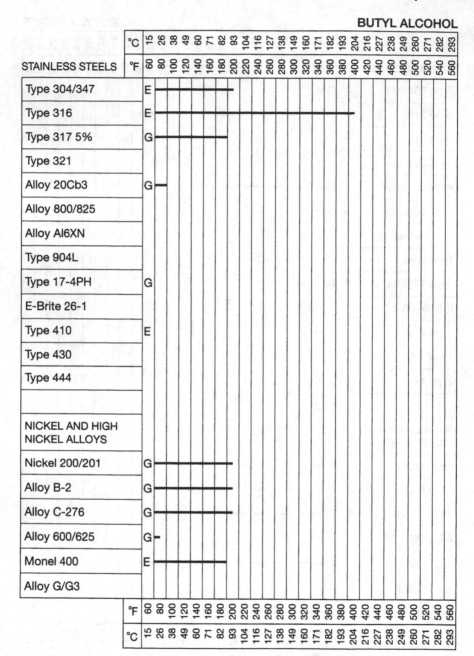

BUTYRIC ACID

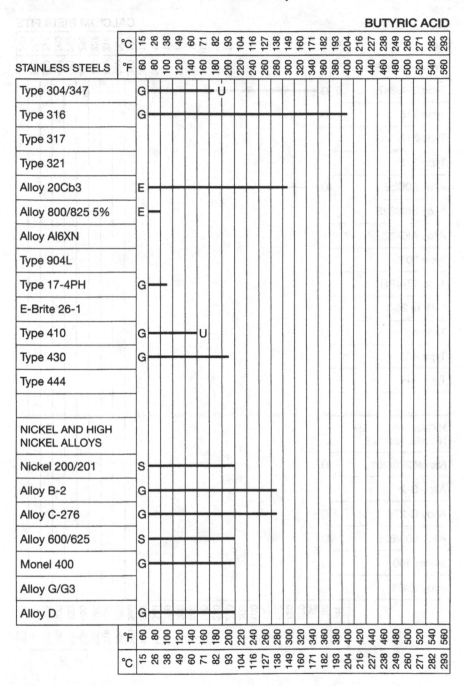

CALCIUM BISULFITE

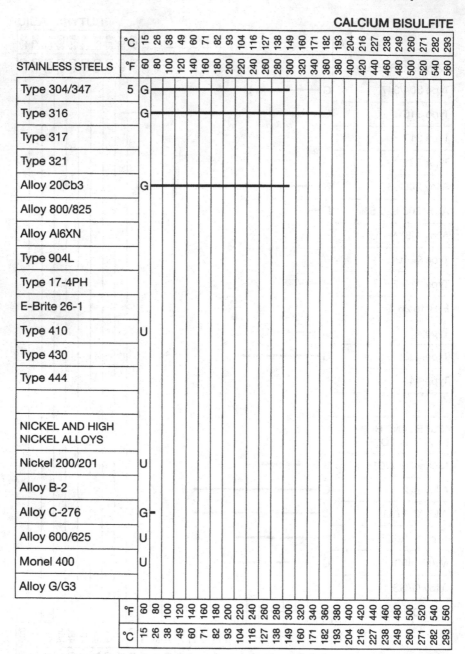

STAINLESS STEELS	°C	15	26	38	49	60	71	82	93	104	116	127	138	149	160	171	182	193	204	216	227	238	249	260	271	282	293
	°F	60	80	100	120	140	160	180	200	220	240	260	280	300	320	340	360	380	400	420	440	460	480	500	520	540	560
Type 304/347	5 G																										
Type 316	G																										
Type 317																											
Type 321																											
Alloy 20Cb3	G																										
Alloy 800/825																											
Alloy Al6XN																											
Type 904L																											
Type 17-4PH																											
E-Brite 26-1																											
Type 410	U																										
Type 430																											
Type 444																											
NICKEL AND HIGH NICKEL ALLOYS																											
Nickel 200/201	U																										
Alloy B-2																											
Alloy C-276	G																										
Alloy 600/625	U																										
Monel 400	U																										
Alloy G/G3																											
	°F	60	80	100	120	140	160	180	200	220	240	260	280	300	320	340	360	380	400	420	440	460	480	500	520	540	560
	°C	15	26	38	49	60	71	82	93	104	116	127	138	149	160	171	182	193	204	216	227	238	249	260	271	282	293

CALCIUM CARBONATE

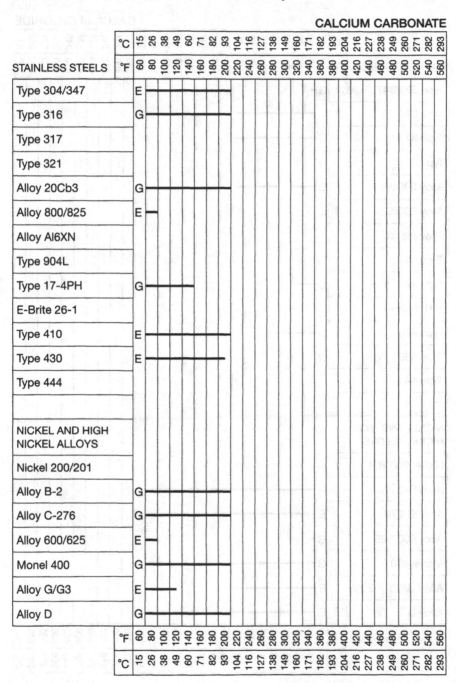

| STAINLESS STEELS | °C | 15 | 26 | 38 | 49 | 60 | 71 | 82 | 93 | 104 | 116 | 127 | 138 | 149 | 160 | 171 | 182 | 193 | 204 | 216 | 227 | 238 | 249 | 260 | 271 | 282 | 293 |
| | °F | 60 | 80 | 100 | 120 | 140 | 160 | 180 | 200 | 220 | 240 | 260 | 280 | 300 | 320 | 340 | 360 | 380 | 400 | 420 | 440 | 460 | 480 | 500 | 520 | 540 | 560 |

Type 304/347 — E
Type 316 — G
Type 317
Type 321
Alloy 20Cb3 — G
Alloy 800/825 — E
Alloy Al6XN
Type 904L
Type 17-4PH — G
E-Brite 26-1
Type 410 — E
Type 430 — E
Type 444

NICKEL AND HIGH NICKEL ALLOYS
Nickel 200/201
Alloy B-2 — G
Alloy C-276 — G
Alloy 600/625 — E
Monel 400 — G
Alloy G/G3 — E
Alloy D — G

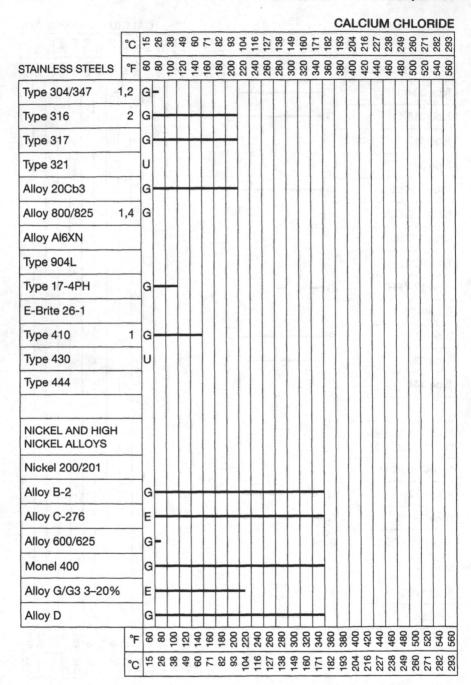

CALCIUM CHLORIDE

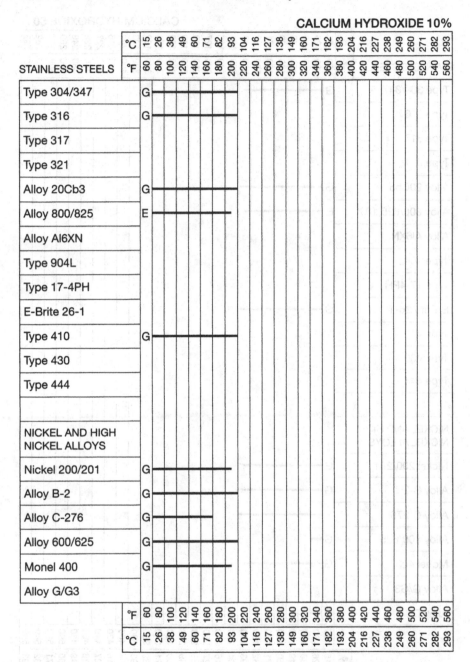

CALCIUM HYDROXIDE 10%

STAINLESS STEELS	°C / °F	
Type 304/347	G	
Type 316	G	
Type 317		
Type 321		
Alloy 20Cb3	G	
Alloy 800/825	E	
Alloy Al6XN		
Type 904L		
Type 17-4PH		
E-Brite 26-1		
Type 410	G	
Type 430		
Type 444		
NICKEL AND HIGH NICKEL ALLOYS		
Nickel 200/201	G	
Alloy B-2	G	
Alloy C-276	G	
Alloy 600/625	G	
Monel 400	G	
Alloy G/G3		

CALCIUM HYDROXIDE 50%

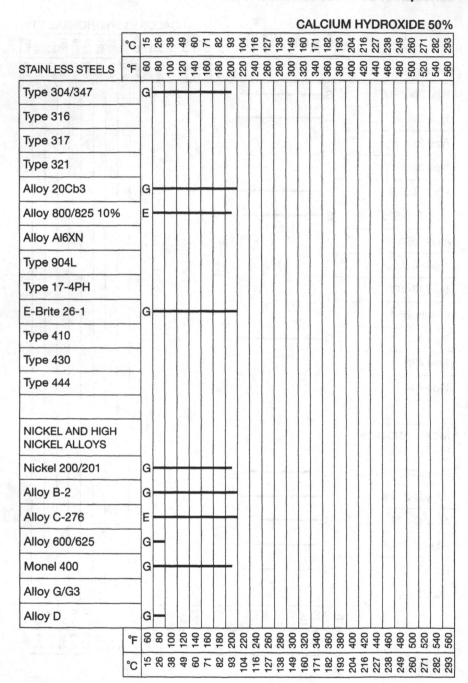

STAINLESS STEELS	°C	15	26	38	49	60	71	82	93	104	116	127	138	149	160	171	182	193	204	216	227	238	249	260	271	282	293
	°F	60	80	100	120	140	160	180	200	220	240	260	280	300	320	340	360	380	400	420	440	460	480	500	520	540	560
Type 304/347	G																										
Type 316																											
Type 317																											
Type 321																											
Alloy 20Cb3	G																										
Alloy 800/825 10%	E																										
Alloy Al6XN																											
Type 904L																											
Type 17-4PH																											
E-Brite 26-1	G																										
Type 410																											
Type 430																											
Type 444																											
NICKEL AND HIGH NICKEL ALLOYS																											
Nickel 200/201	G																										
Alloy B-2	G																										
Alloy C-276	E																										
Alloy 600/625	G																										
Monel 400	G																										
Alloy G/G3																											
Alloy D	G																										
	°F	60	80	100	120	140	160	180	200	220	240	260	280	300	320	340	360	380	400	420	440	460	480	500	520	540	560
	°C	15	26	38	49	60	71	82	93	104	116	127	138	149	160	171	182	193	204	216	227	238	249	260	271	282	293

CALCIUM HYPOCHLORITE

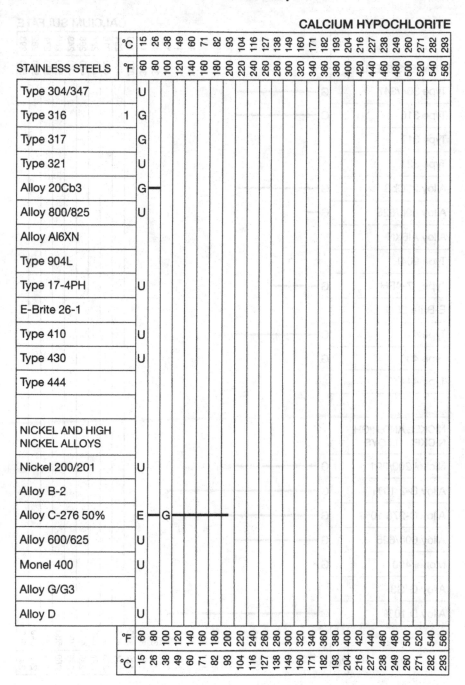

	°C	15	26	38	49	60	71	82	93	104	116	127	138	149	160	171	182	193	204	216	227	238	249	260	271	282	293
STAINLESS STEELS	°F	60	80	100	120	140	160	180	200	220	240	260	280	300	320	340	360	380	400	420	440	460	480	500	520	540	560
Type 304/347		U																									
Type 316	1	G																									
Type 317		G																									
Type 321		U																									
Alloy 20Cb3		G	—																								
Alloy 800/825		U																									
Alloy Al6XN																											
Type 904L																											
Type 17-4PH		U																									
E-Brite 26-1																											
Type 410		U																									
Type 430		U																									
Type 444																											
NICKEL AND HIGH NICKEL ALLOYS																											
Nickel 200/201		U																									
Alloy B-2																											
Alloy C-276 50%		E	—	G	—	—	—	—	—																		
Alloy 600/625		U																									
Monel 400		U																									
Alloy G/G3																											
Alloy D		U																									
	°F	60	80	100	120	140	160	180	200	220	240	260	280	300	320	340	360	380	400	420	440	460	480	500	520	540	560
	°C	15	26	38	49	60	71	82	93	104	116	127	138	149	160	171	182	193	204	216	227	238	249	260	271	282	293

CALCIUM SULFATE

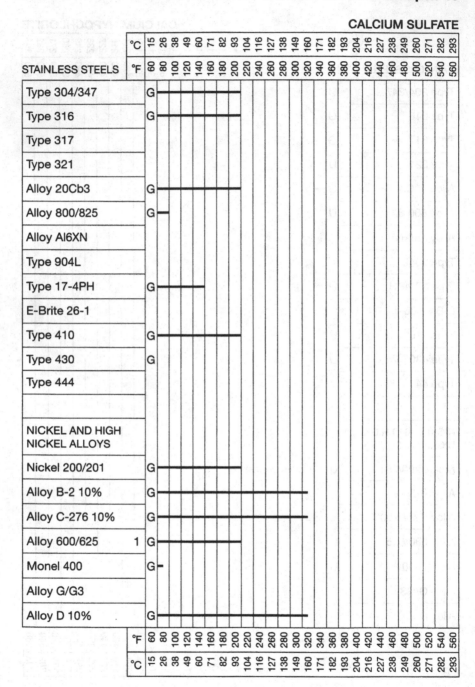

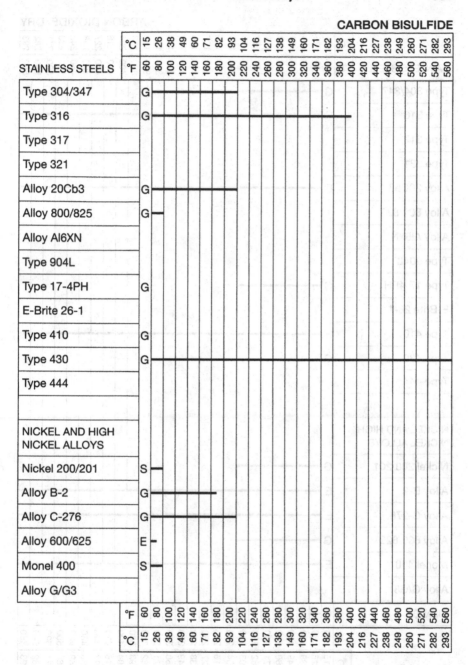

CARBON BISULFIDE

STAINLESS STEELS	°F/°C		
Type 304/347	G		
Type 316	G		
Type 317			
Type 321			
Alloy 20Cb3	G		
Alloy 800/825	G		
Alloy Al6XN			
Type 904L			
Type 17-4PH	G		
E-Brite 26-1			
Type 410	G		
Type 430	G		
Type 444			
NICKEL AND HIGH NICKEL ALLOYS			
Nickel 200/201	S		
Alloy B-2	G		
Alloy C-276	G		
Alloy 600/625	E		
Monel 400	S		
Alloy G/G3			

CARBON DIOXIDE, DRY

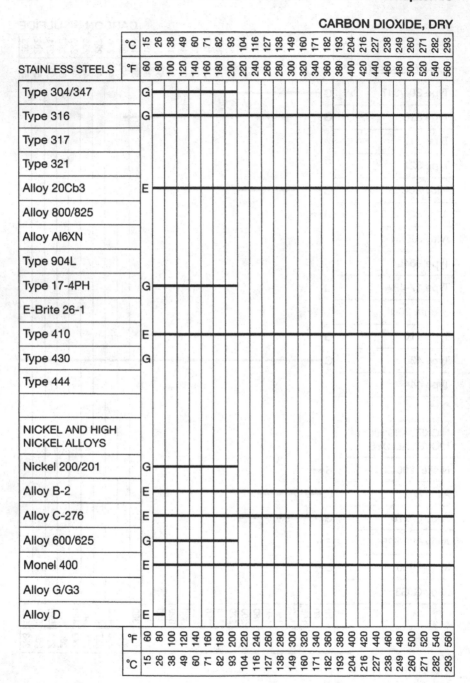

| STAINLESS STEELS | °C | 15 | 26 | 38 | 49 | 60 | 71 | 82 | 93 | 104 | 116 | 127 | 138 | 149 | 160 | 171 | 182 | 193 | 204 | 216 | 227 | 238 | 249 | 260 | 271 | 282 | 293 |
| | °F | 60 | 80 | 100 | 120 | 140 | 160 | 180 | 200 | 220 | 240 | 260 | 280 | 300 | 320 | 340 | 360 | 380 | 400 | 420 | 440 | 460 | 480 | 500 | 520 | 540 | 560 |

Type 304/347 — G

Type 316 — G

Type 317

Type 321

Alloy 20Cb3 — E

Alloy 800/825

Alloy Al6XN

Type 904L

Type 17-4PH — G

E-Brite 26-1

Type 410 — E

Type 430 — G

Type 444

NICKEL AND HIGH NICKEL ALLOYS

Nickel 200/201 — G

Alloy B-2 — E

Alloy C-276 — E

Alloy 600/625 — G

Monel 400 — E

Alloy G/G3

Alloy D — E

CARBON DIOXIDE, WET

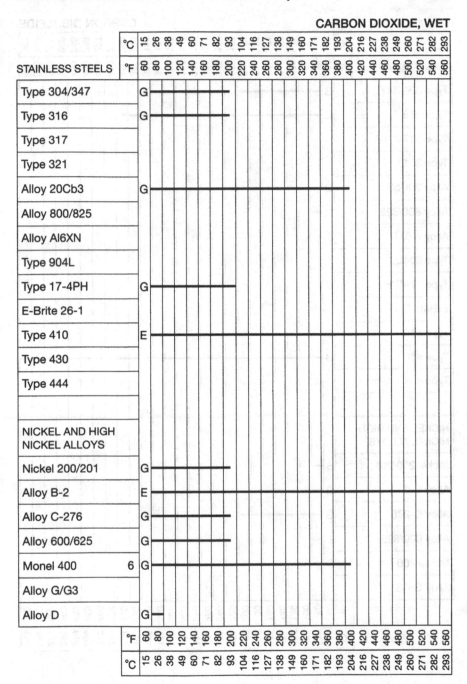

STAINLESS STEELS		
Type 304/347	G	
Type 316	G	
Type 317		
Type 321		
Alloy 20Cb3	G	
Alloy 800/825		
Alloy Al6XN		
Type 904L		
Type 17-4PH	G	
E-Brite 26-1		
Type 410	E	
Type 430		
Type 444		
NICKEL AND HIGH NICKEL ALLOYS		
Nickel 200/201	G	
Alloy B-2	E	
Alloy C-276	G	
Alloy 600/625	G	
Monel 400	6	G
Alloy G/G3		
Alloy D	G	

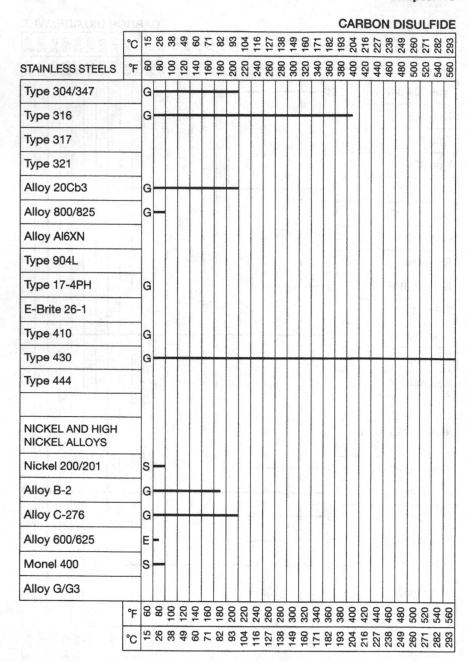

CARBON DISULFIDE

| STAINLESS STEELS | °C | 15 | 26 | 38 | 49 | 60 | 71 | 82 | 93 | 104 | 116 | 127 | 138 | 149 | 160 | 171 | 182 | 193 | 204 | 216 | 227 | 238 | 249 | 260 | 271 | 282 | 293 |
| | °F | 60 | 80 | 100 | 120 | 140 | 160 | 180 | 200 | 220 | 240 | 260 | 280 | 300 | 320 | 340 | 360 | 380 | 400 | 420 | 440 | 460 | 480 | 500 | 520 | 540 | 560 |

CARBON MONOXIDE

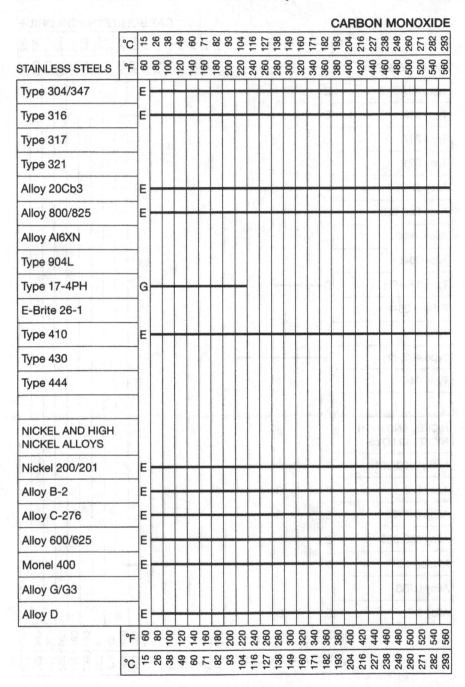

STAINLESS STEELS	°C	15	26	38	49	60	71	82	93	104	116	127	138	149	160	171	182	193	204	216	227	238	249	260	271	282	293
	°F	60	80	100	120	140	160	180	200	220	240	260	280	300	320	340	360	380	400	420	440	460	480	500	520	540	560
Type 304/347	E																										
Type 316	E																										
Type 317																											
Type 321																											
Alloy 20Cb3	E																										
Alloy 800/825	E																										
Alloy AI6XN																											
Type 904L																											
Type 17-4PH	G																										
E-Brite 26-1																											
Type 410	E																										
Type 430																											
Type 444																											
NICKEL AND HIGH NICKEL ALLOYS																											
Nickel 200/201	E																										
Alloy B-2	E																										
Alloy C-276	E																										
Alloy 600/625	E																										
Monel 400	E																										
Alloy G/G3																											
Alloy D	E																										
	°F	60	80	100	120	140	160	180	200	220	240	260	280	300	320	340	360	380	400	420	440	460	480	500	520	540	560
	°C	15	26	38	49	60	71	82	93	104	116	127	138	149	160	171	182	193	204	216	227	238	249	260	271	282	293

CARBON TETRACHLORIDE

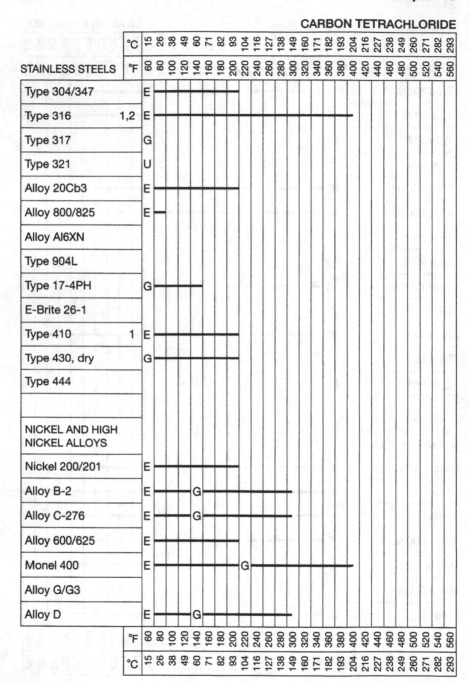

STAINLESS STEELS

Type 304/347
Type 316 1,2
Type 317
Type 321
Alloy 20Cb3
Alloy 800/825
Alloy AI6XN
Type 904L
Type 17-4PH
E-Brite 26-1
Type 410 1
Type 430, dry
Type 444

NICKEL AND HIGH NICKEL ALLOYS

Nickel 200/201
Alloy B-2
Alloy C-276
Alloy 600/625
Monel 400
Alloy G/G3
Alloy D

CARBONIC ACID

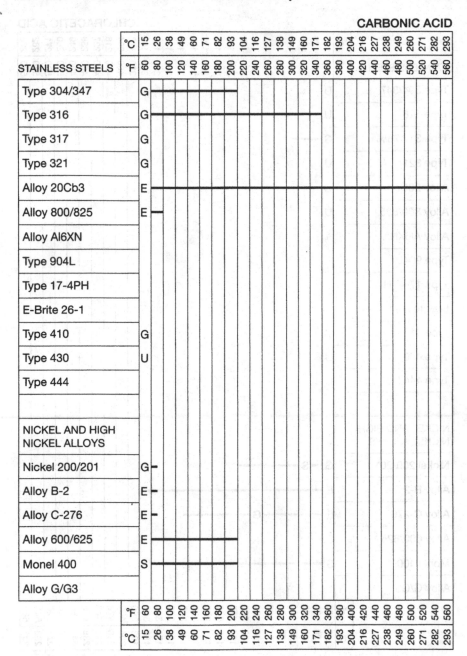

STAINLESS STEELS		°C / °F
Type 304/347	G	
Type 316	G	
Type 317	G	
Type 321	G	
Alloy 20Cb3	E	
Alloy 800/825	E	
Alloy Al6XN		
Type 904L		
Type 17-4PH		
E-Brite 26-1		
Type 410	G	
Type 430	U	
Type 444		
NICKEL AND HIGH NICKEL ALLOYS		
Nickel 200/201	G	
Alloy B-2	E	
Alloy C-276	E	
Alloy 600/625	E	
Monel 400	S	
Alloy G/G3		

°F: 60 80 100 120 140 160 180 200 220 240 260 280 300 320 340 360 380 400 420 440 460 480 500 520 540 560

°C: 15 26 38 49 60 71 82 93 104 116 127 138 149 160 171 182 193 204 216 227 238 249 260 271 282 293

CHLORACETIC ACID

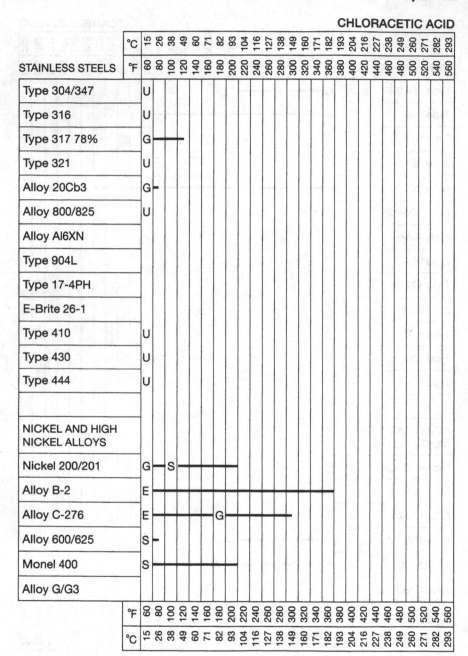

STAINLESS STEELS		°C	15	26	38	49	60	71	82	93	104	116	127	138	149	160	171	182	193	204	216	227	238	249	260	271	282	293
		°F	60	80	100	120	140	160	180	200	220	240	260	280	300	320	340	360	380	400	420	440	460	480	500	520	540	560
Type 304/347	U																											
Type 316	U																											
Type 317 78%	G																											
Type 321	U																											
Alloy 20Cb3	G																											
Alloy 800/825	U																											
Alloy Al6XN																												
Type 904L																												
Type 17-4PH																												
E-Brite 26-1																												
Type 410	U																											
Type 430	U																											
Type 444	U																											
NICKEL AND HIGH NICKEL ALLOYS																												
Nickel 200/201	G		S																									
Alloy B-2	E																											
Alloy C-276	E						G																					
Alloy 600/625	S																											
Monel 400	S																											
Alloy G/G3																												
		°F	60	80	100	120	140	160	180	200	220	240	260	280	300	320	340	360	380	400	420	440	460	480	500	520	540	560
		°C	15	26	38	49	60	71	82	93	104	116	127	138	149	160	171	182	193	204	216	227	238	249	260	271	282	293

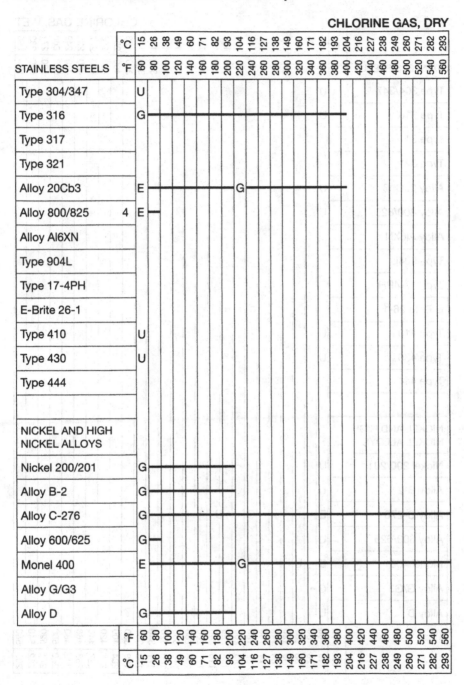

CHLORINE GAS, WET

STAINLESS STEELS	°C	15	26	38	49	60	71	82	93	104	116	127	138	149	160	171	182	193	204	216	227	238	249	260	271	282	293
	°F	60	80	100	120	140	160	180	200	220	240	260	280	300	320	340	360	380	400	420	440	460	480	500	520	540	560
Type 304/347	U																										
Type 316	U																										
Type 317																											
Type 321																											
Alloy 20Cb3	U																										
Alloy 800/825	U																										
Alloy Al6XN																											
Type 904L																											
Type 17-4PH																											
E-Brite 26-1																											
Type 410	U																										
Type 430	U																										
Type 444																											
NICKEL AND HIGH NICKEL ALLOYS																											
Nickel 200/201	U																										
Alloy B-2	U																										
Alloy C-276	E							G	U																		
Alloy 600/625	U																										
Monel 400	S	U																									
Alloy G/G3	G																										
Alloy D	U																										
	°F	60	80	100	120	140	160	180	200	220	240	260	280	300	320	340	360	380	400	420	440	460	480	500	520	540	560
	°C	15	26	38	49	60	71	82	93	104	116	127	138	149	160	171	182	193	204	216	227	238	249	260	271	282	293

CHLORINE LIQUID

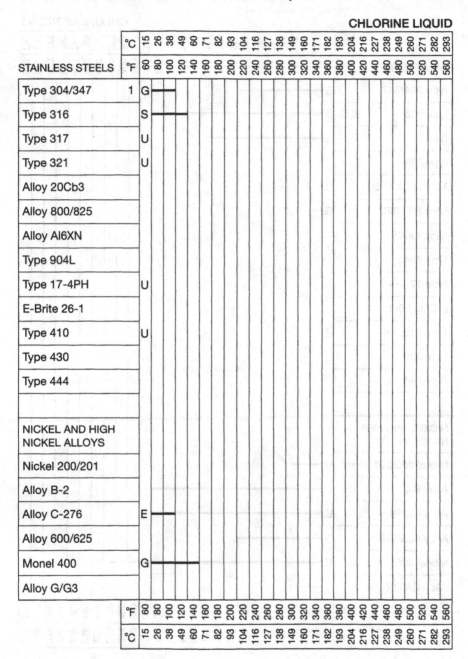

STAINLESS STEELS	°C																										
	°F																										
Type 304/347	1	G																									
Type 316		S																									
Type 317		U																									
Type 321		U																									
Alloy 20Cb3																											
Alloy 800/825																											
Alloy Al6XN																											
Type 904L																											
Type 17-4PH		U																									
E-Brite 26-1																											
Type 410		U																									
Type 430																											
Type 444																											
NICKEL AND HIGH NICKEL ALLOYS																											
Nickel 200/201																											
Alloy B-2																											
Alloy C-276		E																									
Alloy 600/625																											
Monel 400		G																									
Alloy G/G3																											

CHLOROBENZENE

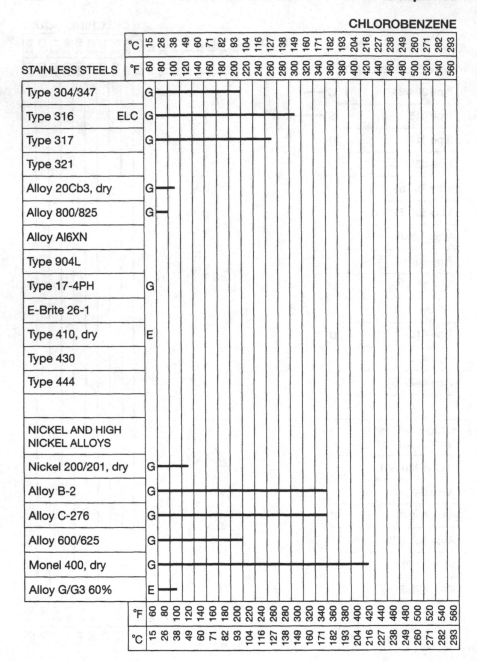

CHLOROFORM

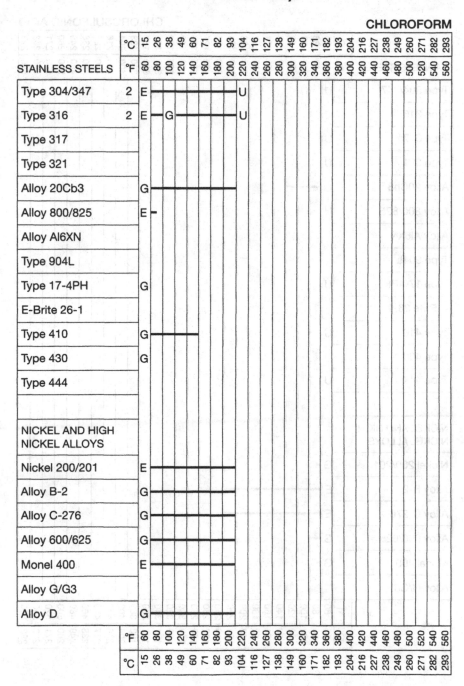

STAINLESS STEELS	°C °F	15 60	26 80	38 100	49 120	60 140	71 160	82 180	93 200	104 220	116 240	127 260	138 280	149 300	160 320	171 340	182 360	193 380	204 400	216 420	227 440	238 460	249 480	260 500	271 520	282 540	293 560
Type 304/347	2	E							U																		
Type 316	2	E	G						U																		
Type 317																											
Type 321																											
Alloy 20Cb3		G																									
Alloy 800/825		E																									
Alloy Al6XN																											
Type 904L																											
Type 17-4PH		G																									
E-Brite 26-1																											
Type 410		G																									
Type 430		G																									
Type 444																											
NICKEL AND HIGH NICKEL ALLOYS																											
Nickel 200/201		E																									
Alloy B-2		G																									
Alloy C-276		G																									
Alloy 600/625		G																									
Monel 400		E																									
Alloy G/G3																											
Alloy D		G																									
	°F °C	60 15	80 26	100 38	120 49	140 60	160 71	180 82	200 93	220 104	240 116	260 127	280 138	300 149	320 160	340 171	360 182	380 193	400 204	420 216	440 227	460 238	480 249	500 260	520 271	540 282	560 293

CHLOROSULFONIC ACID

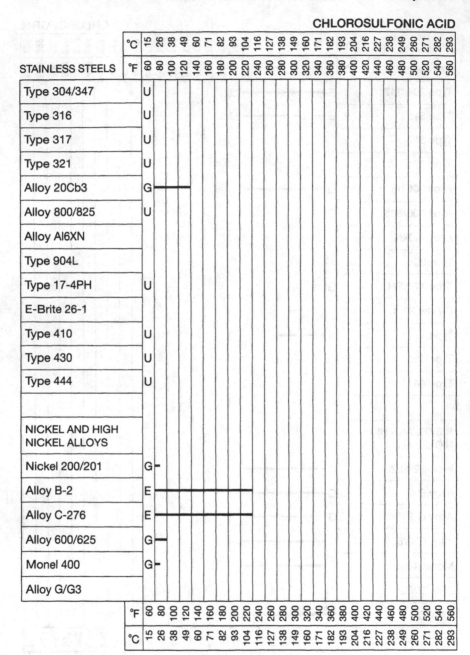

STAINLESS STEELS	°C	15	26	38	49	60	71	82	93	104	116	127	138	149	160	171	182	193	204	216	227	238	249	260	271	282	293
	°F	60	80	100	120	140	160	180	200	220	240	260	280	300	320	340	360	380	400	420	440	460	480	500	520	540	560
Type 304/347	U																										
Type 316	U																										
Type 317	U																										
Type 321	U																										
Alloy 20Cb3	G																										
Alloy 800/825	U																										
Alloy Al6XN																											
Type 904L																											
Type 17-4PH	U																										
E-Brite 26-1																											
Type 410	U																										
Type 430	U																										
Type 444	U																										
NICKEL AND HIGH NICKEL ALLOYS																											
Nickel 200/201	G																										
Alloy B-2	E																										
Alloy C-276	E																										
Alloy 600/625	G																										
Monel 400	G																										
Alloy G/G3																											
	°F	60	80	100	120	140	160	180	200	220	240	260	280	300	320	340	360	380	400	420	440	460	480	500	520	540	560
	°C	15	26	38	49	60	71	82	93	104	116	127	138	149	160	171	182	193	204	216	227	238	249	260	271	282	293

CHROMIC ACID 10%

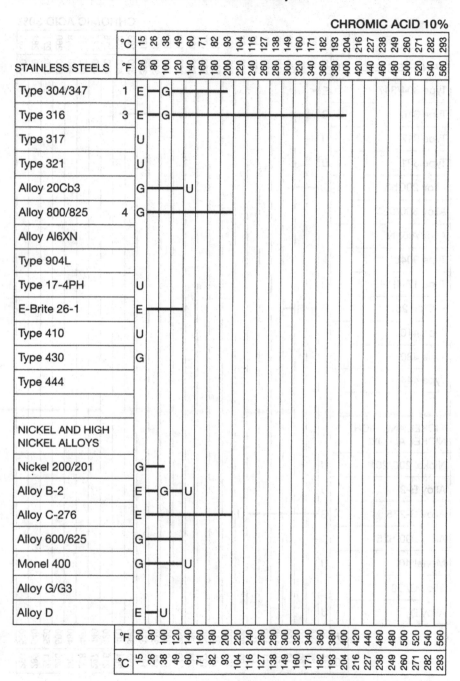

CHROMIC ACID 30%

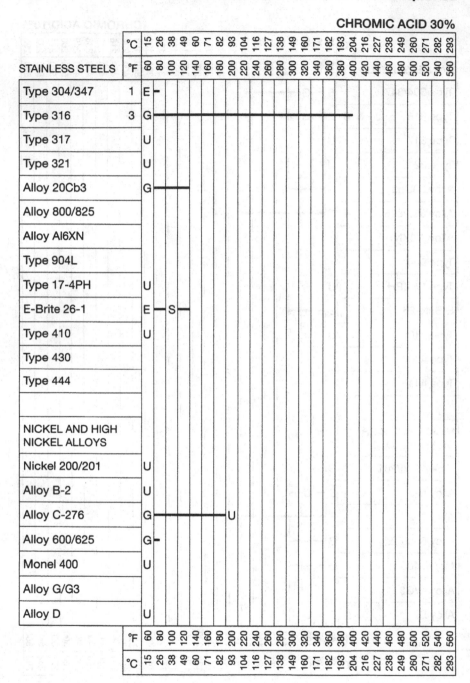

STAINLESS STEELS	°C	15	26	38	49	60	71	82	93	104	116	127	138	149	160	171	182	193	204	216	227	238	249	260	271	282	293
	°F	60	80	100	120	140	160	180	200	220	240	260	280	300	320	340	360	380	400	420	440	460	480	500	520	540	560
Type 304/347	1	E–																									
Type 316	3	G————————————————																									
Type 317		U																									
Type 321		U																									
Alloy 20Cb3		G———																									
Alloy 800/825																											
Alloy Al6XN																											
Type 904L																											
Type 17-4PH		U																									
E-Brite 26-1		E—S—																									
Type 410		U																									
Type 430																											
Type 444																											
NICKEL AND HIGH NICKEL ALLOYS																											
Nickel 200/201		U																									
Alloy B-2		U																									
Alloy C-276		G————————U																									
Alloy 600/625		G–																									
Monel 400		U																									
Alloy G/G3																											
Alloy D		U																									
	°F	60	80	100	120	140	160	180	200	220	240	260	280	300	320	340	360	380	400	420	440	460	480	500	520	540	560
	°C	15	26	38	49	60	71	82	93	104	116	127	138	149	160	171	182	193	204	216	227	238	249	260	271	282	293

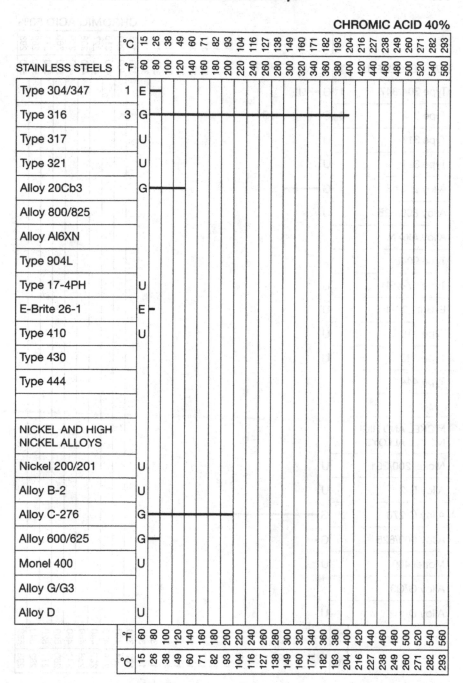

CHROMIC ACID 40%

STAINLESS STEELS		°C	15	26	38	49	60	71	82	93	104	116	127	138	149	160	171	182	193	204	216	227	238	249	260	271	282	293
		°F	60	80	100	120	140	160	180	200	220	240	260	280	300	320	340	360	380	400	420	440	460	480	500	520	540	560
Type 304/347	1	E
Type 316	3	G
Type 317		U
Type 321		U
Alloy 20Cb3		G
Alloy 800/825		
Alloy Al6XN		
Type 904L		
Type 17-4PH		U
E-Brite 26-1		E
Type 410		U
Type 430		
Type 444		
NICKEL AND HIGH NICKEL ALLOYS		
Nickel 200/201		U
Alloy B-2		U
Alloy C-276		G
Alloy 600/625		G
Monel 400		U
Alloy G/G3		
Alloy D		U

CHROMIC ACID 50%

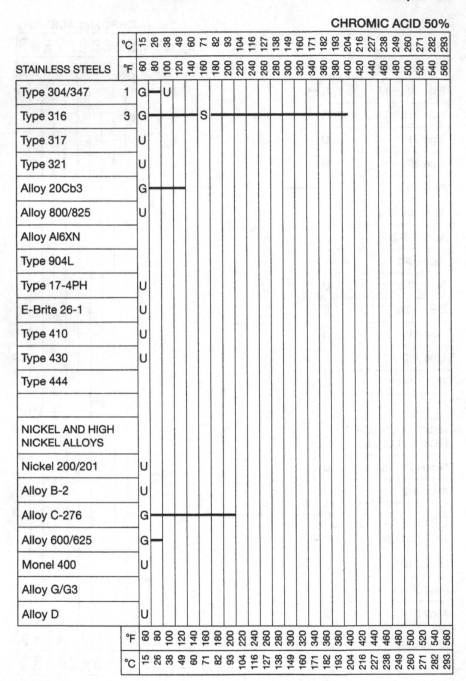

STAINLESS STEELS	°C		15	26	38	49	60	71	82	93	104	116	127	138	149	160	171	182	193	204	216	227	238	249	260	271	282	293
	°F		60	80	100	120	140	160	180	200	220	240	260	280	300	320	340	360	380	400	420	440	460	480	500	520	540	560
Type 304/347	1	G—U
Type 316	3	G——S————
Type 317		U
Type 321		U
Alloy 20Cb3		G——
Alloy 800/825		U
Alloy Al6XN		
Type 904L		
Type 17-4PH		U
E-Brite 26-1		U
Type 410		U
Type 430		U
Type 444		
NICKEL AND HIGH NICKEL ALLOYS		
Nickel 200/201		U
Alloy B-2		U
Alloy C-276		G———
Alloy 600/625		G—
Monel 400		U
Alloy G/G3		
Alloy D		U
	°F		60	80	100	120	140	160	180	200	220	240	260	280	300	320	340	360	380	400	420	440	460	480	500	520	540	560
	°C		15	26	38	49	60	71	82	93	104	116	127	138	149	160	171	182	193	204	216	227	238	249	260	271	282	293

CITRIC ACID 10%

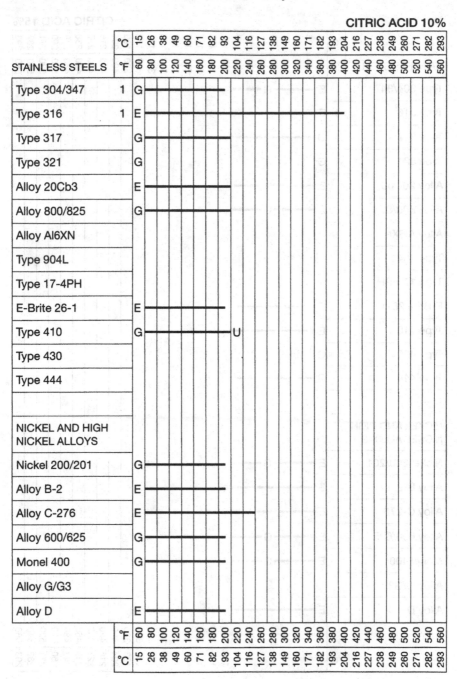

STAINLESS STEELS	°C																													
	°F																													
Type 304/347	1	G																												
Type 316	1	E																												
Type 317		G																												
Type 321		G																												
Alloy 20Cb3		E																												
Alloy 800/825		G																												
Alloy AI6XN																														
Type 904L																														
Type 17-4PH																														
E-Brite 26-1		E																												
Type 410		G						U																						
Type 430																														
Type 444																														
NICKEL AND HIGH NICKEL ALLOYS																														
Nickel 200/201		G																												
Alloy B-2		E																												
Alloy C-276		E																												
Alloy 600/625		G																												
Monel 400		G																												
Alloy G/G3																														
Alloy D		E																												

CITRIC ACID 15%

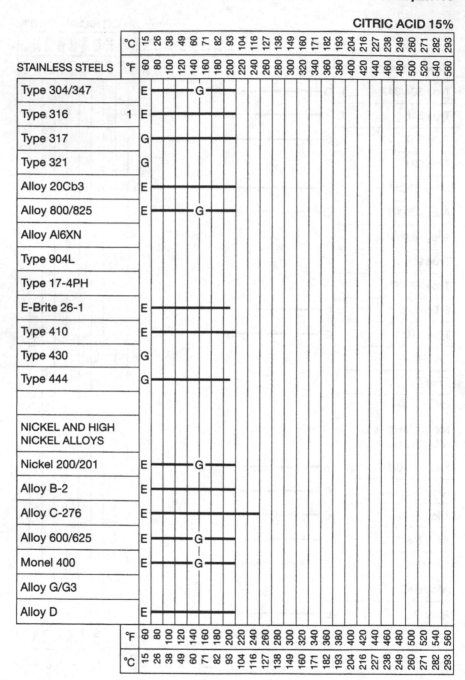

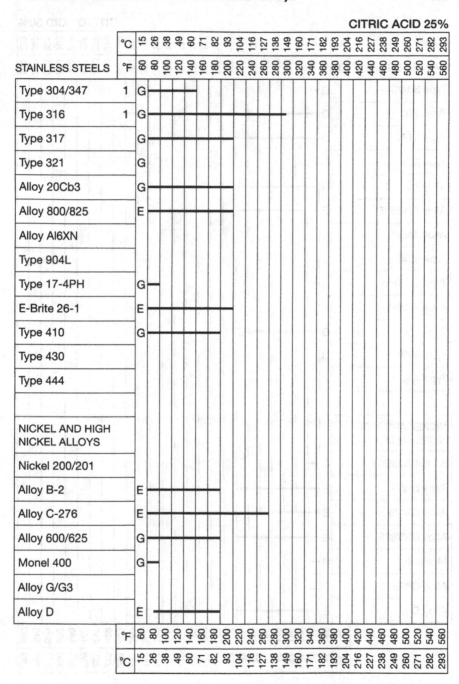

CITRIC ACID 25%

STAINLESS STEELS		304/347 data
Type 304/347	1	G
Type 316	1	G
Type 317		G
Type 321		G
Alloy 20Cb3		G
Alloy 800/825		E
Alloy Al6XN		
Type 904L		
Type 17-4PH		G
E-Brite 26-1		E
Type 410		G
Type 430		
Type 444		
NICKEL AND HIGH NICKEL ALLOYS		
Nickel 200/201		
Alloy B-2		E
Alloy C-276		E
Alloy 600/625		G
Monel 400		G
Alloy G/G3		
Alloy D		E

CITRIC ACID 50%

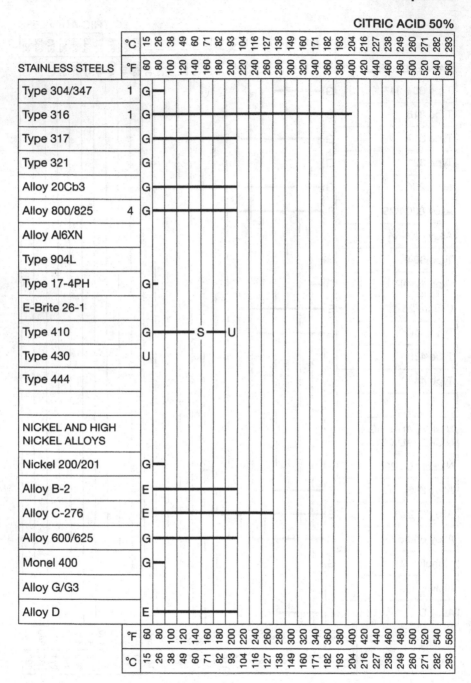

STAINLESS STEELS		°C	15	26	38	49	60	71	82	93	104	116	127	138	149	160	171	182	193	204	216	227	238	249	260	271	282	293
		°F	60	80	100	120	140	160	180	200	220	240	260	280	300	320	340	360	380	400	420	440	460	480	500	520	540	560
Type 304/347	1	G—																										
Type 316	1	G—————————————————————————																										
Type 317		G—————————																										
Type 321		G																										
Alloy 20Cb3		G—————————																										
Alloy 800/825	4	G—————————																										
Alloy Al6XN																												
Type 904L																												
Type 17-4PH		G—																										
E-Brite 26-1																												
Type 410		G———————S———U																										
Type 430		U																										
Type 444																												
NICKEL AND HIGH NICKEL ALLOYS																												
Nickel 200/201		G—																										
Alloy B-2		E—————————																										
Alloy C-276		E—————————————																										
Alloy 600/625		G—————————																										
Monel 400		G—																										
Alloy G/G3																												
Alloy D		E—————————																										
		°F	60	80	100	120	140	160	180	200	220	240	260	280	300	320	340	360	380	400	420	440	460	480	500	520	540	560
		°C	15	26	38	49	60	71	82	93	104	116	127	138	149	160	171	182	193	204	216	227	238	249	260	271	282	293

COPPER ACETATE

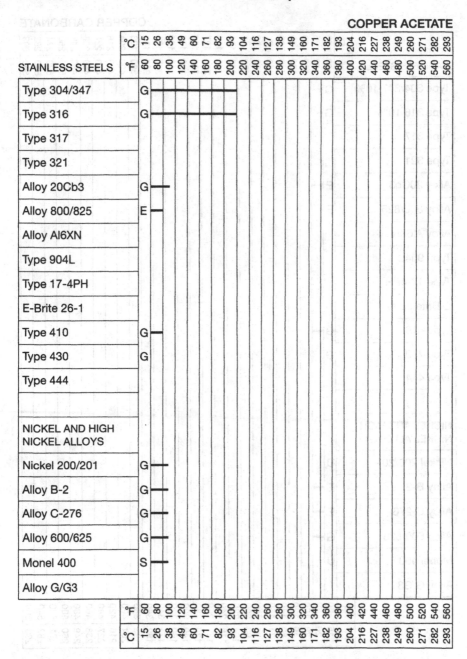

STAINLESS STEELS	°C	15	26	38	49	60	71	82	93	104	116	127	138	149	160	171	182	193	204	216	227	238	249	260	271	282	293
	°F	60	80	100	120	140	160	180	200	220	240	260	280	300	320	340	360	380	400	420	440	460	480	500	520	540	560
Type 304/347	G																										
Type 316	G																										
Type 317																											
Type 321																											
Alloy 20Cb3	G																										
Alloy 800/825	E																										
Alloy Al6XN																											
Type 904L																											
Type 17-4PH																											
E-Brite 26-1																											
Type 410	G																										
Type 430	G																										
Type 444																											
NICKEL AND HIGH NICKEL ALLOYS																											
Nickel 200/201	G																										
Alloy B-2	G																										
Alloy C-276	G																										
Alloy 600/625	G																										
Monel 400	S																										
Alloy G/G3																											
	°F	60	80	100	120	140	160	180	200	220	240	260	280	300	320	340	360	380	400	420	440	460	480	500	520	540	560
	°C	15	26	38	49	60	71	82	93	104	116	127	138	149	160	171	182	193	204	216	227	238	249	260	271	282	293

COPPER CARBONATE

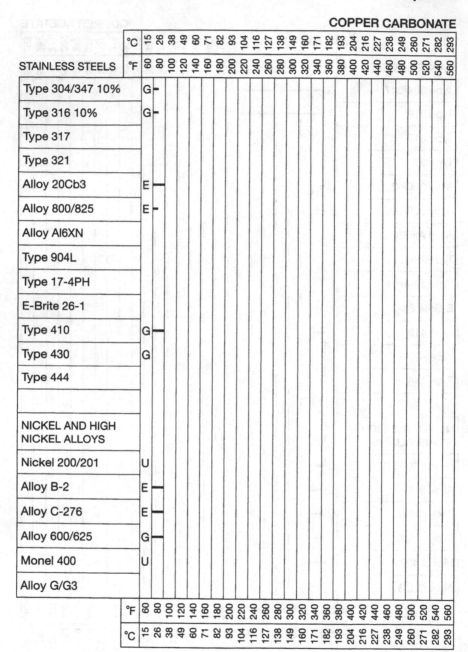

STAINLESS STEELS	°C	15	26	38	49	60	71	82	93	104	116	127	138	149	160	171	182	193	204	216	227	238	249	260	271	282	293
	°F	60	80	100	120	140	160	180	200	220	240	260	280	300	320	340	360	380	400	420	440	460	480	500	520	540	560
Type 304/347 10%	G																										
Type 316 10%	G																										
Type 317																											
Type 321																											
Alloy 20Cb3	E																										
Alloy 800/825	E																										
Alloy Al6XN																											
Type 904L																											
Type 17-4PH																											
E-Brite 26-1																											
Type 410	G																										
Type 430	G																										
Type 444																											
NICKEL AND HIGH NICKEL ALLOYS																											
Nickel 200/201	U																										
Alloy B-2	E																										
Alloy C-276	E																										
Alloy 600/625	G																										
Monel 400	U																										
Alloy G/G3																											
	°F	60	80	100	120	140	160	180	200	220	240	260	280	300	320	340	360	380	400	420	440	460	480	500	520	540	560
	°C	15	26	38	49	60	71	82	93	104	116	127	138	149	160	171	182	193	204	216	227	238	249	260	271	282	293

COPPER CHLORIDE

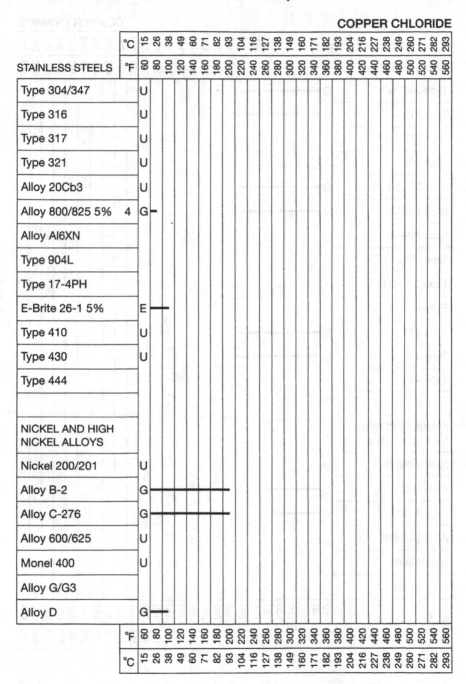

STAINLESS STEELS	°C	15	26	38	49	60	71	82	93	104	116	127	138	149	160	171	182	193	204	216	227	238	249	260	271	282	293
	°F	60	80	100	120	140	160	180	200	220	240	260	280	300	320	340	360	380	400	420	440	460	480	500	520	540	560
Type 304/347		U																									
Type 316		U																									
Type 317		U																									
Type 321		U																									
Alloy 20Cb3		U																									
Alloy 800/825 5% 4		G—																									
Alloy Al6XN																											
Type 904L																											
Type 17-4PH																											
E-Brite 26-1 5%		E——																									
Type 410		U																									
Type 430		U																									
Type 444																											
NICKEL AND HIGH NICKEL ALLOYS																											
Nickel 200/201		U																									
Alloy B-2		G———————																									
Alloy C-276		G———————																									
Alloy 600/625		U																									
Monel 400		U																									
Alloy G/G3																											
Alloy D		G—																									
	°F	60	80	100	120	140	160	180	200	220	240	260	280	300	320	340	360	380	400	420	440	460	480	500	520	540	560
	°C	15	26	38	49	60	71	82	93	104	116	127	138	149	160	171	182	193	204	216	227	238	249	260	271	282	293

COPPER CYANIDE

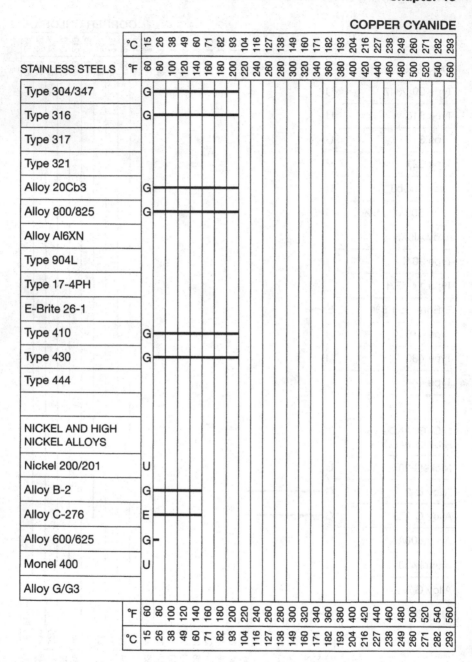

STAINLESS STEELS		
Type 304/347	G	
Type 316	G	
Type 317		
Type 321		
Alloy 20Cb3	G	
Alloy 800/825	G	
Alloy Al6XN		
Type 904L		
Type 17-4PH		
E-Brite 26-1		
Type 410	G	
Type 430	G	
Type 444		
NICKEL AND HIGH NICKEL ALLOYS		
Nickel 200/201	U	
Alloy B-2	G	
Alloy C-276	E	
Alloy 600/625	G	
Monel 400	U	
Alloy G/G3		

COPPER SULFATE

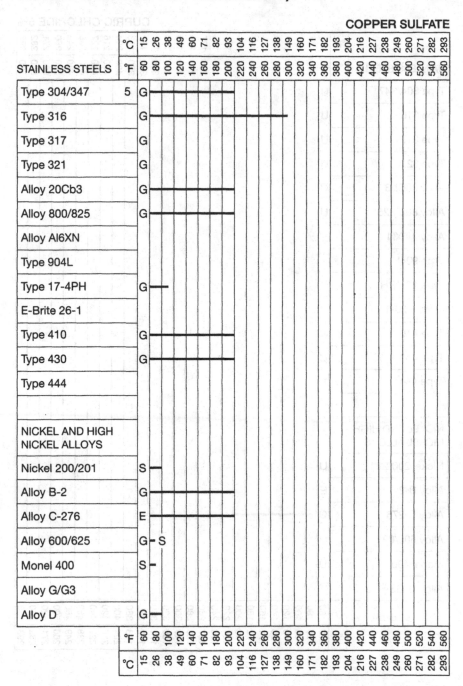

STAINLESS STEELS	°C	15	26	38	49	60	71	82	93	104	116	127	138	149	160	171	182	193	204	216	227	238	249	260	271	282	293
	°F	60	80	100	120	140	160	180	200	220	240	260	280	300	320	340	360	380	400	420	440	460	480	500	520	540	560
Type 304/347	5 G																										
Type 316	G																										
Type 317	G																										
Type 321	G																										
Alloy 20Cb3	G																										
Alloy 800/825	G																										
Alloy Al6XN																											
Type 904L																											
Type 17-4PH	G																										
E-Brite 26-1																											
Type 410	G																										
Type 430	G																										
Type 444																											
NICKEL AND HIGH NICKEL ALLOYS																											
Nickel 200/201	S																										
Alloy B-2	G																										
Alloy C-276	E																										
Alloy 600/625	G S																										
Monel 400	S																										
Alloy G/G3																											
Alloy D	G																										
	°F	60	80	100	120	140	160	180	200	220	240	260	280	300	320	340	360	380	400	420	440	460	480	500	520	540	560
	°C	15	26	38	49	60	71	82	93	104	116	127	138	149	160	171	182	193	204	216	227	238	249	260	271	282	293

CUPRIC CHLORIDE 5%

STAINLESS STEELS	°C	15	26	38	49	60	71	82	93	104	116	127	138	149	160	171	182	193	204	216	227	238	249	260	271	282	293
	°F	60	80	100	120	140	160	180	200	220	240	260	280	300	320	340	360	380	400	420	440	460	480	500	520	540	560
Type 304/347	U																										
Type 316	U																										
Type 317	U																										
Type 321	U																										
Alloy 20Cb3	G																										
Alloy 800/825	U																										
Alloy Al6XN																											
Type 904L																											
Type 17-4PH																											
E-Brite 26-1																											
Type 410	U																										
Type 430	U																										
Type 444																											
NICKEL AND HIGH NICKEL ALLOYS																											
Nickel 200/201	U																										
Alloy B-2	E																										
Alloy C-276	G	▬	▬	▬	▬	▬	▬	▬																			
Alloy 600/625	U																										
Monel 400	U																										
Alloy G/G3																											
	°F	60	80	100	120	140	160	180	200	220	240	260	280	300	320	340	360	380	400	420	440	460	480	500	520	540	560
	°C	15	26	38	49	60	71	82	93	104	116	127	138	149	160	171	182	193	204	216	227	238	249	260	271	282	293

CUPRIC CHLORIDE 50%

STAINLESS STEELS	°C / °F (start 15°C/60°F)	Rating
Type 304/347	15°C / 60°F	U
Type 316	15°C / 60°F	U
Type 317	15°C / 60°F	U
Type 321	15°C / 60°F	U
Alloy 20Cb3	15°C / 60°F	U
Alloy 800/825	15°C / 60°F	U
Alloy Al6XN		
Type 904L		
Type 17-4PH		
E-Brite 26-1		
Type 410	15°C / 60°F	U
Type 430	15°C / 60°F	U
Type 444		

NICKEL AND HIGH NICKEL ALLOYS

Alloy	Rating
Nickel 200/201	U
Alloy B-2	G (to ~93°C / 200°F)
Alloy C-276	G (to ~93°C / 200°F)
Alloy 600/625	U
Monel 400	U
Alloy G/G3	

Temperature headers °C: 15, 26, 38, 49, 60, 71, 82, 93, 104, 116, 127, 138, 149, 160, 171, 182, 193, 204, 216, 227, 238, 249, 260, 271, 282, 293

Temperature headers °F: 60, 80, 100, 120, 140, 160, 180, 200, 220, 240, 260, 280, 300, 320, 340, 360, 380, 400, 420, 440, 460, 480, 500, 520, 540, 560

CYCLOHEXANE

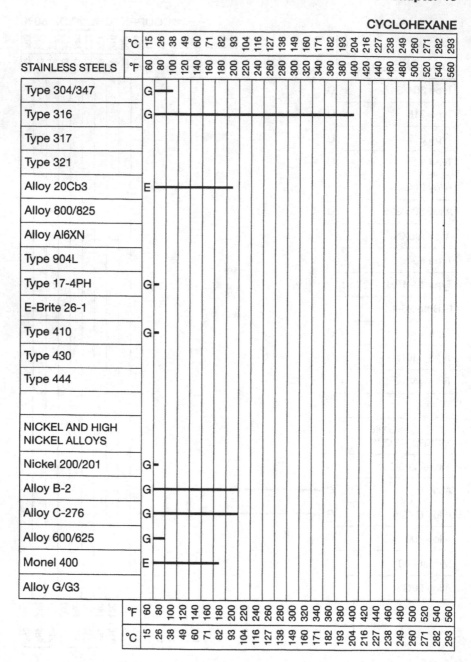

STAINLESS STEELS		Rating
Type 304/347	G	to ~38°C (100°F)
Type 316	G	to ~204°C (400°F)
Type 317		
Type 321		
Alloy 20Cb3	E	to ~82°C (180°F)
Alloy 800/825		
Alloy Al6XN		
Type 904L		
Type 17-4PH	G	
E-Brite 26-1		
Type 410	G	
Type 430		
Type 444		
NICKEL AND HIGH NICKEL ALLOYS		
Nickel 200/201	G	
Alloy B-2	G	to ~93°C (200°F)
Alloy C-276	G	to ~93°C (200°F)
Alloy 600/625	G	
Monel 400	E	to ~82°C (180°F)
Alloy G/G3		

CYCLOHEXANOL

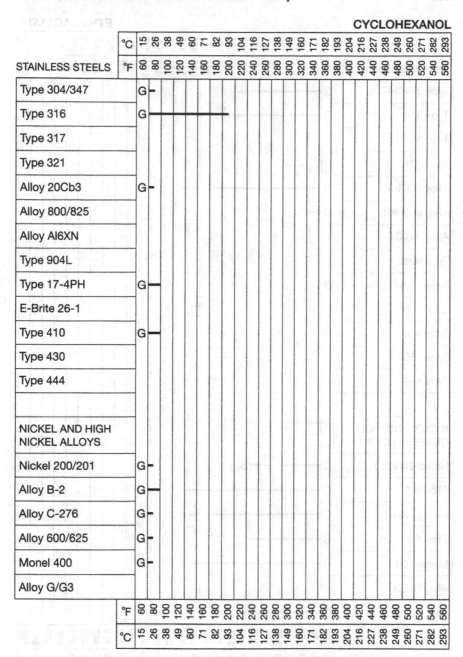

ETHYL ACETATE

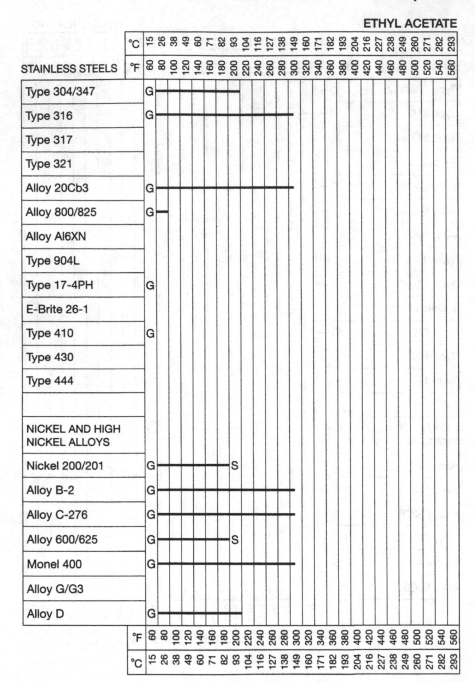

STAINLESS STEELS	°C	15	26	38	49	60	71	82	93	104	116	127	138	149	160	171	182	193	204	216	227	238	249	260	271	282	293
	°F	60	80	100	120	140	160	180	200	220	240	260	280	300	320	340	360	380	400	420	440	460	480	500	520	540	560
Type 304/347	G
Type 316	G
Type 317		
Type 321		
Alloy 20Cb3	G
Alloy 800/825	G
Alloy Al6XN		
Type 904L		
Type 17-4PH	G
E-Brite 26-1		
Type 410	G
Type 430		
Type 444		
NICKEL AND HIGH NICKEL ALLOYS		
Nickel 200/201	G								S
Alloy B-2	G
Alloy C-276	G
Alloy 600/625	G								S
Monel 400	G
Alloy G/G3		
Alloy D	G

	°F	60	80	100	120	140	160	180	200	220	240	260	280	300	320	340	360	380	400	420	440	460	480	500	520	540	560
	°C	15	26	38	49	60	71	82	93	104	116	127	138	149	160	171	182	193	204	216	227	238	249	260	271	282	293

ETHYL ALCOHOL

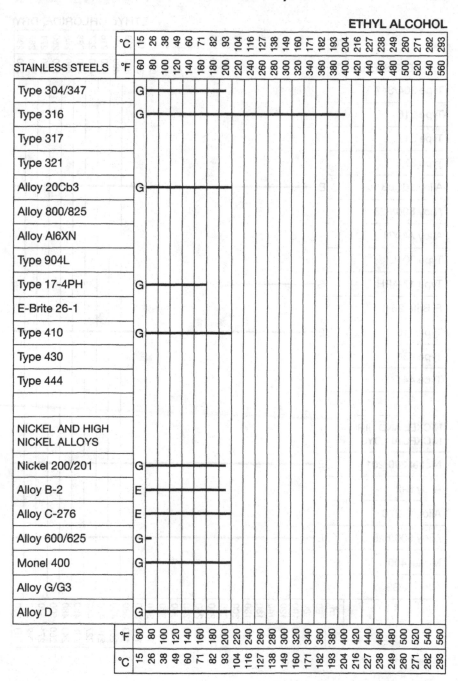

STAINLESS STEELS	°C	15 26 38 49 60 71 82 93 104 116 127 138 149 160 171 182 193 204 216 227 238 249 260 271 282 293
	°F	60 80 100 120 140 160 180 200 220 240 260 280 300 320 340 360 380 400 420 440 460 480 500 520 540 560
Type 304/347	G	
Type 316	G	
Type 317		
Type 321		
Alloy 20Cb3	G	
Alloy 800/825		
Alloy AI6XN		
Type 904L		
Type 17-4PH	G	
E-Brite 26-1		
Type 410	G	
Type 430		
Type 444		
NICKEL AND HIGH NICKEL ALLOYS		
Nickel 200/201	G	
Alloy B-2	E	
Alloy C-276	E	
Alloy 600/625	G	
Monel 400	G	
Alloy G/G3		
Alloy D	G	
	°F	60 80 100 120 140 160 180 200 220 240 260 280 300 320 340 360 380 400 420 440 460 480 500 520 540 560
	°C	15 26 38 49 60 71 82 93 104 116 127 138 149 160 171 182 193 204 216 227 238 249 260 271 282 293

ETHYL CHLORIDE, DRY

	°C	15	26	38	49	60	71	82	93	104	116	127	138	149	160	171	182	193	204	216	227	238	249	260	271	282	293
STAINLESS STEELS	°F	60	80	100	120	140	160	180	200	220	240	260	280	300	320	340	360	380	400	420	440	460	480	500	520	540	560
Type 304/347																											
Type 316																											
Type 317																											
Type 321																											
Alloy 20Cb3	E																										
Alloy 800/825																											
Alloy Al6XN																											
Type 904L																											
Type 17-4PH	G																										
E-Brite 26-1																											
Type 410	E																										
Type 430																											
Type 444																											
NICKEL AND HIGH NICKEL ALLOYS																											
Nickel 200/201	E																										
Alloy B-2																											
Alloy C-276																											
Alloy 600/625	E																										
Monel 400																											
Alloy G/G3																											
	°F	60	80	100	120	140	160	180	200	220	240	260	280	300	320	340	360	380	400	420	440	460	480	500	520	540	560
	°C	15	26	38	49	60	71	82	93	104	116	127	138	149	160	171	182	193	204	216	227	238	249	260	271	282	293

ETHYLENE GLYCOL

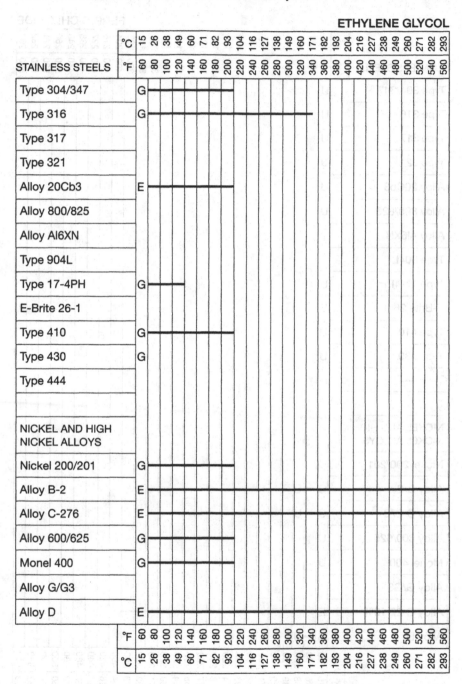

STAINLESS STEELS		°C	15	26	38	49	60	71	82	93	104	116	127	138	149	160	171	182	193	204	216	227	238	249	260	271	282	293
		°F	60	80	100	120	140	160	180	200	220	240	260	280	300	320	340	360	380	400	420	440	460	480	500	520	540	560
Type 304/347	G																											
Type 316	G																											
Type 317																												
Type 321																												
Alloy 20Cb3	E																											
Alloy 800/825																												
Alloy Al6XN																												
Type 904L																												
Type 17-4PH	G																											
E-Brite 26-1																												
Type 410	G																											
Type 430	G																											
Type 444																												
NICKEL AND HIGH NICKEL ALLOYS																												
Nickel 200/201	G																											
Alloy B-2	E																											
Alloy C-276	E																											
Alloy 600/625	G																											
Monel 400	G																											
Alloy G/G3																												
Alloy D	E																											

FERRIC CHLORIDE

STAINLESS STEELS	°C 15	26	38	49	60	71	82	93	104	116	127	138	149	160	171	182	193	204	216	227	238	249	260	271	282	293
	°F 60	80	100	120	140	160	180	200	220	240	260	280	300	320	340	360	380	400	420	440	460	480	500	520	540	560
Type 304/347	U																									
Type 316	U																										
Type 317	U																										
Type 321	U																										
Alloy 20Cb3	U																										
Alloy 800/825	U																										
Alloy Al6XN																											
Type 904L																											
Type 17-4PH	U																										
E-Brite 26-1	E—																										
Type 410	U																										
Type 430	U																										
Type 444																											
NICKEL AND HIGH NICKEL ALLOYS																											
Nickel 200/201	U																										
Alloy B-2	G—	U																									
Alloy C-276	G—	U																										
Alloy 600/625	U																										
Monel 400	U																										
Alloy G/G3																											
Alloy C-22 10%	E—																|—									
Alloy D	G—	U																									

°F	60	80	100	120	140	160	180	200	220	240	260	280	300	320	340	360	380	400	420	440	460	480	500	520	540	560
°C	15	26	38	49	60	71	82	93	104	116	127	138	149	160	171	182	193	204	216	227	238	249	260	271	282	293

FERRIC CHLORIDE 50%, IN WATER

STAINLESS STEELS	°C → 15 26 38 49 60 71 82 93 104 116 127 138 149 160 171 182 193 204 216 227 238 249 260 271 282 293
	°F → 60 80 100 120 140 160 180 200 220 240 260 280 300 320 340 360 380 400 420 440 460 480 500 520 540 560
Type 304/347	U
Type 316	U
Type 317	U
Type 321	U
Alloy 20Cb3	U
Alloy 800/825	U
Alloy Al6XN	
Type 904L 10%	G–
Type 17-4PH	
E-Brite 26-1	
Type 410	U
Type 430	
Type 444	
NICKEL AND HIGH NICKEL ALLOYS	
Nickel 200/201	U
Alloy B-2	U
Alloy C-276 10%	G ———————— (to ~200°F)
Alloy 600/625	S–
Monel 400	U
Alloy G/G3	

°F: 60 80 100 120 140 160 180 200 220 240 260 280 300 320 340 360 380 400 420 440 460 480 500 520 540 560

°C: 15 26 38 49 60 71 82 93 104 116 127 138 149 160 171 182 193 204 216 227 238 249 260 271 282 293

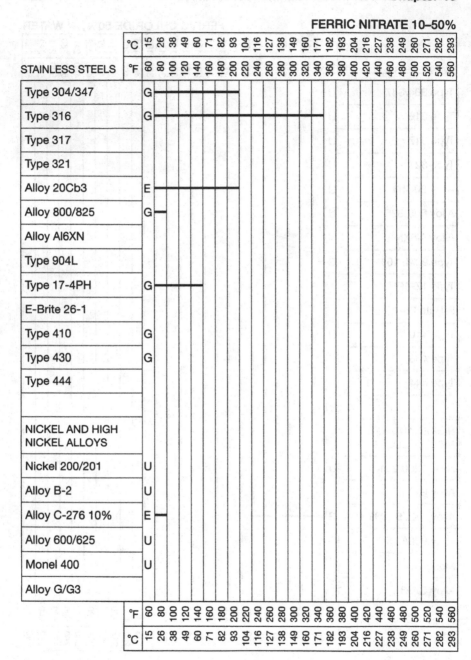

FERRIC NITRATE 10–50%

FERROUS CHLORIDE

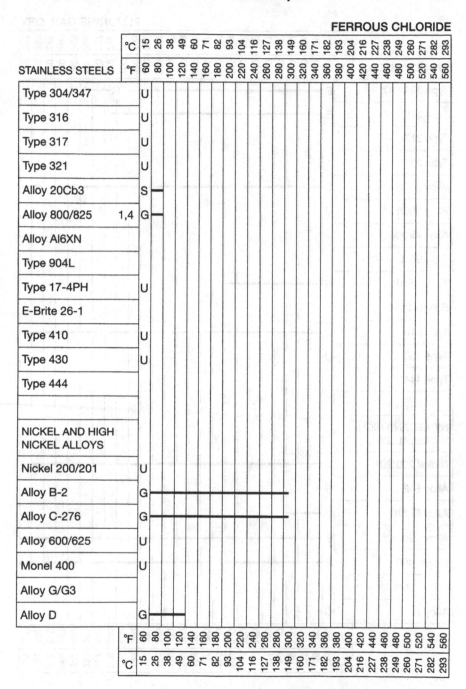

STAINLESS STEELS	°F	60 80 100 120 140 160 180 200 220 240 260 280 300 320 340 360 380 400 420 440 460 480 500 520 540 560
	°C	15 26 38 49 60 71 82 93 104 116 127 138 149 160 171 182 193 204 216 227 238 249 260 271 282 293
Type 304/347	U	
Type 316	U	
Type 317	U	
Type 321	U	
Alloy 20Cb3	S	
Alloy 800/825 1,4	G	
Alloy Al6XN		
Type 904L		
Type 17-4PH	U	
E-Brite 26-1		
Type 410	U	
Type 430	U	
Type 444		
NICKEL AND HIGH NICKEL ALLOYS		
Nickel 200/201	U	
Alloy B-2	G	
Alloy C-276	G	
Alloy 600/625	U	
Monel 400	U	
Alloy G/G3		
Alloy D	G	

FLUORINE GAS, DRY

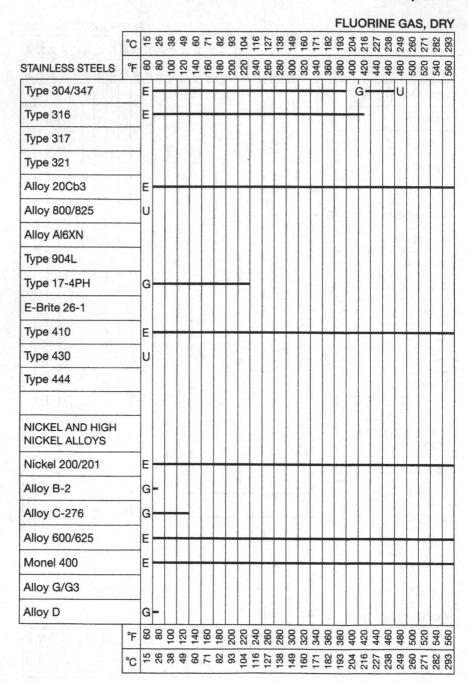

STAINLESS STEELS	°C	15	26	38	49	60	71	82	93	104	116	127	138	149	160	171	182	193	204	216	227	238	249	260	271	282	293
	°F	60	80	100	120	140	160	180	200	220	240	260	280	300	320	340	360	380	400	420	440	460	480	500	520	540	560
Type 304/347	E																			G	U						
Type 316	E																										
Type 317																											
Type 321																											
Alloy 20Cb3	E																										
Alloy 800/825	U																										
Alloy Al6XN																											
Type 904L																											
Type 17-4PH	G																										
E-Brite 26-1																											
Type 410	E																										
Type 430	U																										
Type 444																											
NICKEL AND HIGH NICKEL ALLOYS																											
Nickel 200/201	E																										
Alloy B-2	G																										
Alloy C-276	G																										
Alloy 600/625	E																										
Monel 400	E																										
Alloy G/G3																											
Alloy D	G																										

FLUORINE GAS, MOIST

STAINLESS STEELS	°C	15	26	38	49	60	71	82	93	104	116	127	138	149	160	171	182	193	204	216	227	238	249	260	271	282	293
	°F	60	80	100	120	140	160	180	200	220	240	260	280	300	320	340	360	380	400	420	440	460	480	500	520	540	560
Type 304/347	U																										
Type 316	U																										
Type 317	U																										
Type 321	U																										
Alloy 20Cb3	U																										
Alloy 800/825	U																										
Alloy Al6XN																											
Type 904L																											
Type 17-4PH																											
E-Brite 26-1																											
Type 410	U																										
Type 430	U																										
Type 444																											
NICKEL AND HIGH NICKEL ALLOYS																											
Nickel 200/201	G–																										
Alloy B-2																											
Alloy C-276	E————————————————————————																									
Alloy 600/625	G–																										
Monel 400	U																										
Alloy G/G3																											
	°F	60	80	100	120	140	160	180	200	220	240	260	280	300	320	340	360	380	400	420	440	460	480	500	520	540	560
	°C	15	26	38	49	60	71	82	93	104	116	127	138	149	160	171	182	193	204	216	227	238	249	260	271	282	293

FORMIC ACID 50%

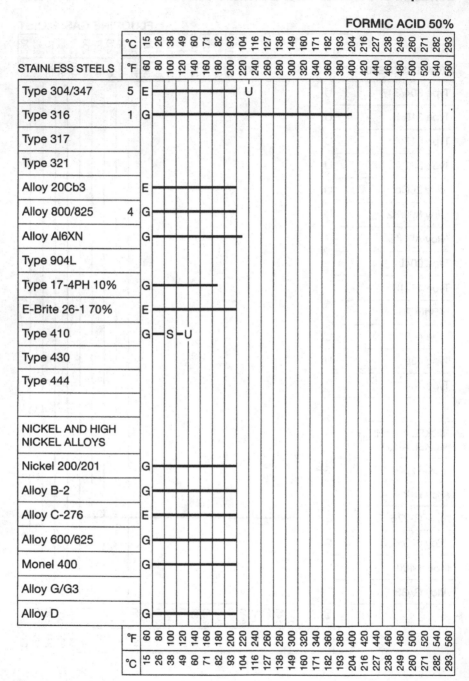

STAINLESS STEELS	°C	15	26	38	49	60	71	82	93	104	116	127	138	149	160	171	182	193	204	216	227	238	249	260	271	282	293
	°F	60	80	100	120	140	160	180	200	220	240	260	280	300	320	340	360	380	400	420	440	460	480	500	520	540	560
Type 304/347	5	E								U																	
Type 316	1	G																									
Type 317																											
Type 321																											
Alloy 20Cb3		E																									
Alloy 800/825	4	G																									
Alloy Al6XN		G																									
Type 904L																											
Type 17-4PH 10%		G																									
E-Brite 26-1 70%		E																									
Type 410		G	S	U																							
Type 430																											
Type 444																											
NICKEL AND HIGH NICKEL ALLOYS																											
Nickel 200/201		G																									
Alloy B-2		G																									
Alloy C-276		E																									
Alloy 600/625		G																									
Monel 400		G																									
Alloy G/G3																											
Alloy D		G																									
	°F	60	80	100	120	140	160	180	200	220	240	260	280	300	320	340	360	380	400	420	440	460	480	500	520	540	560
	°C	15	26	38	49	60	71	82	93	104	116	127	138	149	160	171	182	193	204	216	227	238	249	260	271	282	293

FORMIC ACID 80%

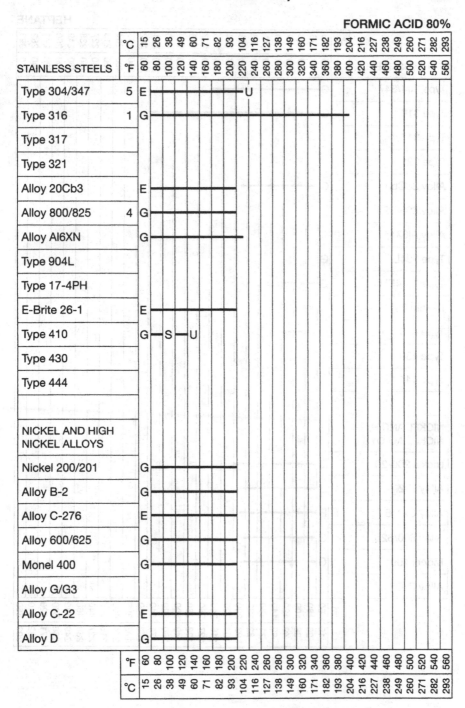

		°C	15	26	38	49	60	71	82	93	104	116	127	138	149	160	171	182	193	204	216	227	238	249	260	271	282	293
STAINLESS STEELS		°F	60	80	100	120	140	160	180	200	220	240	260	280	300	320	340	360	380	400	420	440	460	480	500	520	540	560
Type 304/347	5	E							U																			
Type 316	1	G																										
Type 317																												
Type 321																												
Alloy 20Cb3		E																										
Alloy 800/825	4	G																										
Alloy Al6XN		G																										
Type 904L																												
Type 17-4PH																												
E-Brite 26-1		E																										
Type 410		G	S	U																							
Type 430																												
Type 444																												
NICKEL AND HIGH NICKEL ALLOYS																												
Nickel 200/201		G																										
Alloy B-2		G																										
Alloy C-276		E																										
Alloy 600/625		G																										
Monel 400		G																										
Alloy G/G3																												
Alloy C-22		E																										
Alloy D		G																										
		°F	60	80	100	120	140	160	180	200	220	240	260	280	300	320	340	360	380	400	420	440	460	480	500	520	540	560
		°C	15	26	38	49	60	71	82	93	104	116	127	138	149	160	171	182	193	204	216	227	238	249	260	271	282	293

HEPTANE

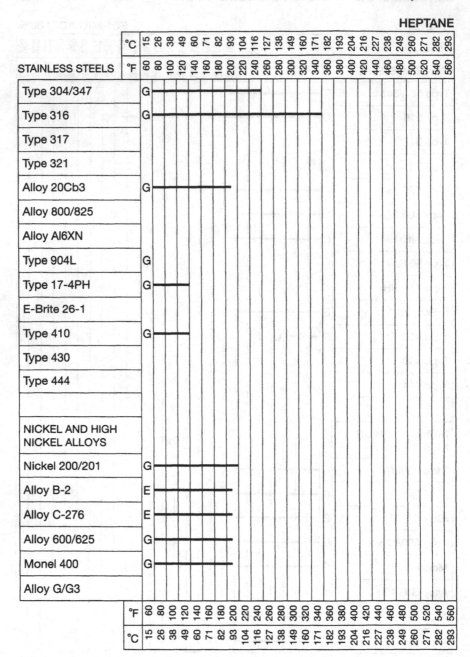

	°F	60	80	100	120	140	160	180	200	220	240	260	280	300	320	340	360	380	400	420	440	460	480	500	520	540	560

STAINLESS STEELS — °C scale: 15 26 38 49 60 71 82 93 104 116 127 138 149 160 171 182 193 204 216 227 238 249 260 271 282 293

- Type 304/347 — G
- Type 316 — G
- Type 317
- Type 321
- Alloy 20Cb3 — G
- Alloy 800/825
- Alloy Al6XN
- Type 904L — G
- Type 17-4PH — G
- E-Brite 26-1
- Type 410 — G
- Type 430
- Type 444

NICKEL AND HIGH NICKEL ALLOYS

- Nickel 200/201 — G
- Alloy B-2 — E
- Alloy C-276 — E
- Alloy 600/625 — G
- Monel 400 — G
- Alloy G/G3

HYDROBROMIC ACID, DILUTE

STAINLESS STEELS	°C	15	26	38	49	60	71	82	93	104	116	127	138	149	160	171	182	193	204	216	227	238	249	260	271	282	293
	°F	60	80	100	120	140	160	180	200	220	240	260	280	300	320	340	360	380	400	420	440	460	480	500	520	540	560
Type 304/347	U																										
Type 316	U																										
Type 317	U																										
Type 321	U																										
Alloy 20Cb3	U																										
Alloy 800/825	U																										
Alloy Al6XN																											
Type 904L																											
Type 17-4PH	U																										
E-Brite 26-1																											
Type 410	U																										
Type 430	U																										
Type 444																											
NICKEL AND HIGH NICKEL ALLOYS																											
Nickel 200/201	U																										
Alloy B-2	G	———————— (to 200°F)																									
Alloy C-276																											
Alloy 600/625	G	— (to 80°F)																									
Monel 400	U																										
Alloy G/G3																											

	°F	60	80	100	120	140	160	180	200	220	240	260	280	300	320	340	360	380	400	420	440	460	480	500	520	540	560
	°C	15	26	38	49	60	71	82	93	104	116	127	138	149	160	171	182	193	204	216	227	238	249	260	271	282	293

HYDROBROMIC ACID 20%

STAINLESS STEELS	°F / °C	60 / 15	80 / 26	100 / 38	120 / 49	140 / 60	160 / 71	180 / 82	200 / 93	220 / 104	240 / 116	260 / 127	280 / 138	300 / 149	320 / 160	340 / 171	360 / 182	380 / 193	400 / 204	420 / 216	440 / 227	460 / 238	480 / 249	500 / 260	520 / 271	540 / 282	560 / 293
Type 304/347		U																									
Type 316		U																									
Type 317		U																									
Type 321		U																									
Alloy 20Cb3		U																									
Alloy 800/825		U																									
Alloy Al6XN																											
Type 904L																											
Type 17-4PH		U																									
E-Brite 26-1																											
Type 410		U																									
Type 430		U																									
Type 444																											
NICKEL AND HIGH NICKEL ALLOYS																											
Nickel 200/201		U																									
Alloy B-2		G																									
Alloy C-276		E																									
Alloy 600/625		G–U																									
Monel 400		U																									
Alloy G/G3																											

HYDROBROMIC ACID 50%

STAINLESS STEELS	°C	15	26	38	49	60	71	82	93	104	116	127	138	149	160	171	182	193	204	216	227	238	249	260	271	282	293
	°F	60	80	100	120	140	160	180	200	220	240	260	280	300	320	340	360	380	400	420	440	460	480	500	520	540	560
Type 304/347	U																										
Type 316	U																										
Type 317	U																										
Type 321	U																										
Alloy 20Cb3	U																										
Alloy 800/825	U																										
Alloy Al6XN																											
Type 904L																											
Type 17-4PH	U																										
E-Brite 26-1																											
Type 410	U																										
Type 430	U																										
Type 444																											
NICKEL AND HIGH NICKEL ALLOYS																											
Nickel 200/201	U																										
Alloy B-2	G																										
Alloy C-276	G																										
Alloy 600/625	U																										
Monel 400	U																										
Alloy G/G3																											
	°F	60	80	100	120	140	160	180	200	220	240	260	280	300	320	340	360	380	400	420	440	460	480	500	520	540	560
	°C	15	26	38	49	60	71	82	93	104	116	127	138	149	160	171	182	193	204	216	227	238	249	260	271	282	293

HYDROCHLORIC ACID 20%

STAINLESS STEELS	°C 15 / °F 60	26/80	38/100	49/120	60/140	71/160	82/180	93/200	104/220	116/240	127/260	138/280	149/300	160/320	171/340	182/360	193/380	204/400	216/420	227/440	238/460	249/480	260/500	271/520	282/540	293/560
Type 304/347	U																									
Type 316	U																									
Type 317	U																									
Type 321	U																									
Alloy 20Cb3	U																									
Alloy 800/825 4	G	——	U																							
Alloy Al6XN																										
Type 904L																										
Type 17-4PH	U																									
E-Brite 26-1	U																									
Type 410	U																									
Type 430	U																									
Type 444																										
NICKEL AND HIGH NICKEL ALLOYS																										
Nickel 200/201	G	——	U																							
Alloy B-2	G	——	——	——	——	——	S	——	——																	
Alloy C-276	E	——	G	——	——	——	U																			
Alloy 600/625	G	——	U																							
Monel 400	G	——	U																							
Alloy G/G3																										
Alloy C-22	U																									
Alloy D	S	——	U																							

HYDROCHLORIC ACID 38%

	°C	15	26	38	49	60	71	82	93	104	116	127	138	149	160	171	182	193	204	216	227	238	249	260	271	282	293
STAINLESS STEELS	°F	60	80	100	120	140	160	180	200	220	240	260	280	300	320	340	360	380	400	420	440	460	480	500	520	540	560
Type 304/347	U																										
Type 316	U																										
Type 317	U																										
Type 321	U																										
Alloy 20Cb3	U																										
Alloy 800/825	U																										
Alloy Al6XN																											
Type 904L																											
Type 17–4PH	U																										
E-Brite 26-1	U																										
Type 410	U																										
Type 430	U																										
Type 444																											
NICKEL AND HIGH NICKEL ALLOYS																											
Nickel 200/201	U																										
Alloy B-2	E		—	—	U																						
Alloy C-276	E	—	S	—	U																						
Alloy 600/625	U																										
Monel 400	U																										
Alloy G/G3																											
	°F	60	80	100	120	140	160	180	200	220	240	260	280	300	320	340	360	380	400	420	440	460	480	500	520	540	560
	°C	15	26	38	49	60	71	82	93	104	116	127	138	149	160	171	182	193	204	216	227	238	249	260	271	282	293

HYDROCYANIC ACID 10%

STAINLESS STEELS	°C	15	26	38	49	60	71	82	93	104	116	127	138	149	160	171	182	193	204	216	227	238	249	260	271	282	293
	°F	60	80	100	120	140	160	180	200	220	240	260	280	300	320	340	360	380	400	420	440	460	480	500	520	540	560
Type 304/347	G																										
Type 316	G																										
Type 317																											
Type 321																											
Alloy 20Cb3	G																										
Alloy 800/825	E																										
Alloy Al6XN																											
Type 904L																											
Type 17-4PH	U																										
E-Brite 26-1																											
Type 410	G																										
Type 430	U																										
Type 444																											
NICKEL AND HIGH NICKEL ALLOYS																											
Nickel 200/201																											
Alloy B-2																											
Alloy C-276																											
Alloy 600/625																											
Monel 400	G																										
Alloy G/G3																											
	°F	60	80	100	120	140	160	180	200	220	240	260	280	300	320	340	360	380	400	420	440	460	480	500	520	540	560
	°C	15	26	38	49	60	71	82	93	104	116	127	138	149	160	171	182	193	204	216	227	238	249	260	271	282	293

HYDROFLUORIC ACID 30%

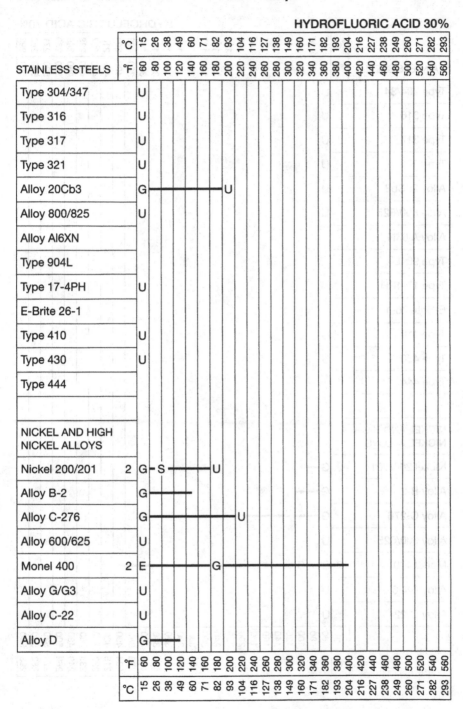

STAINLESS STEELS	°C	15 26 38 49 60 71 82 93 104 116 127 138 149 160 171 182 193 204 216 227 238 249 260 271 282 293
	°F	60 80 100 120 140 160 180 200 220 240 260 280 300 320 340 360 380 400 420 440 460 480 500 520 540 560
Type 304/347		U
Type 316		U
Type 317		U
Type 321		U
Alloy 20Cb3		G———————U
Alloy 800/825		U
Alloy Al6XN		
Type 904L		
Type 17-4PH		U
E-Brite 26-1		
Type 410		U
Type 430		U
Type 444		
NICKEL AND HIGH NICKEL ALLOYS		
Nickel 200/201	2	G–S————U
Alloy B-2		G———
Alloy C-276		G————————U
Alloy 600/625		U
Monel 400	2	E————————G—————————
Alloy G/G3		U
Alloy C-22		U
Alloy D		G——
	°F	60 80 100 120 140 160 180 200 220 240 260 280 300 320 340 360 380 400 420 440 460 480 500 520 540 560
	°C	15 26 38 49 60 71 82 93 104 116 127 138 149 160 171 182 193 204 216 227 238 249 260 271 282 293

HYDROFLUORIC ACID 70%

STAINLESS STEELS	°C	15	26	38	49	60	71	82	93	104	116	127	138	149	160	171	182	193	204	216	227	238	249	260	271	282	293
	°F	60	80	100	120	140	160	180	200	220	240	260	280	300	320	340	360	380	400	420	440	460	480	500	520	540	560
Type 304/347	U																										
Type 316	U																										
Type 317	U																										
Type 321	U																										
Alloy 20Cb3	U																										
Alloy 800/825	U																										
Alloy Al6XN																											
Type 904L																											
Type 17-4PH																											
E-Brite 26-1																											
Type 410	U																										
Type 430	U																										
Type 444																											
NICKEL AND HIGH NICKEL ALLOYS																											
Nickel 200/201 2	G	——																									
Alloy B-2	G	————																									
Alloy C-276	G	—————————																								
Alloy 600/625	U																										
Monel 400 2	E	—G—————————————————————————————																									
Alloy G/G3	U																										
Alloy C-22	U																										

	°F	60	80	100	120	140	160	180	200	220	240	260	280	300	320	340	360	380	400	420	440	460	480	500	520	540	560
	°C	15	26	38	49	60	71	82	93	104	116	127	138	149	160	171	182	193	204	216	227	238	249	260	271	282	293

HYDROFLUORIC ACID 100%

	°C	15	26	38	49	60	71	82	93	104	116	127	138	149	160	171	182	193	204	216	227	238	249	260	271	282	293
STAINLESS STEELS	°F	60	80	100	120	140	160	180	200	220	240	260	280	300	320	340	360	380	400	420	440	460	480	500	520	540	560
Type 304/347	U																										
Type 316	G—																										
Type 317	U																										
Type 321	U																										
Alloy 20Cb3	G—																										
Alloy 800/825	U																										
Alloy Al6XN																											
Type 904L																											
Type 17-4PH																											
E-Brite 26-1																											
Type 410	U																										
Type 430	U																										
Type 444																											
NICKEL AND HIGH NICKEL ALLOYS																											
Nickel 200/201	2 G————																										
Alloy B-2	G—																										
Alloy C-276	G————————————																										
Alloy 600/625	G———																										
Monel 400	2 E—G—————————																										
Alloy G/G3	U																										
Alloy C-22	U																										
Alloy D	G—																										
	°F	60	80	100	120	140	160	180	200	220	240	260	280	300	320	340	360	380	400	420	440	460	480	500	520	540	560
	°C	15	26	38	49	60	71	82	93	104	116	127	138	149	160	171	182	193	204	216	227	238	249	260	271	282	293

HYDROGEN PEROXIDE 30%

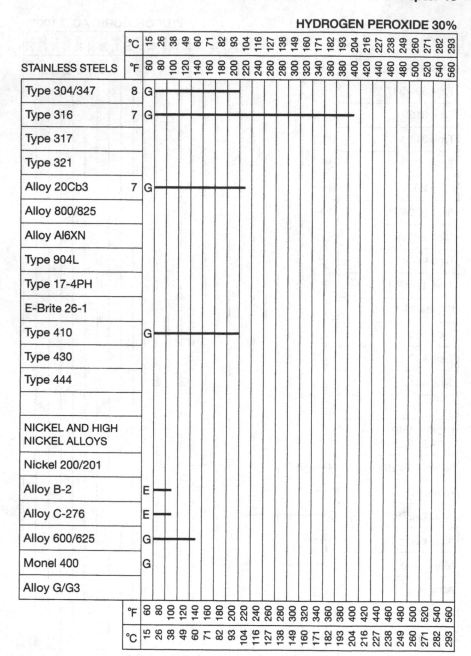

		°C	15	26	38	49	60	71	82	93	104	116	127	138	149	160	171	182	193	204	216	227	238	249	260	271	282	293
STAINLESS STEELS		°F	60	80	100	120	140	160	180	200	220	240	260	280	300	320	340	360	380	400	420	440	460	480	500	520	540	560
Type 304/347	8	G																										
Type 316	7	G																										
Type 317																												
Type 321																												
Alloy 20Cb3	7	G																										
Alloy 800/825																												
Alloy AI6XN																												
Type 904L																												
Type 17-4PH																												
E-Brite 26-1																												
Type 410		G																										
Type 430																												
Type 444																												
NICKEL AND HIGH NICKEL ALLOYS																												
Nickel 200/201																												
Alloy B-2		E																										
Alloy C-276		E																										
Alloy 600/625		G																										
Monel 400		G																										
Alloy G/G3																												
		°F	60	80	100	120	140	160	180	200	220	240	260	280	300	320	340	360	380	400	420	440	460	480	500	520	540	560
		°C	15	26	38	49	60	71	82	93	104	116	127	138	149	160	171	182	193	204	216	227	238	249	260	271	282	293

HYDROGEN PEROXIDE 90%

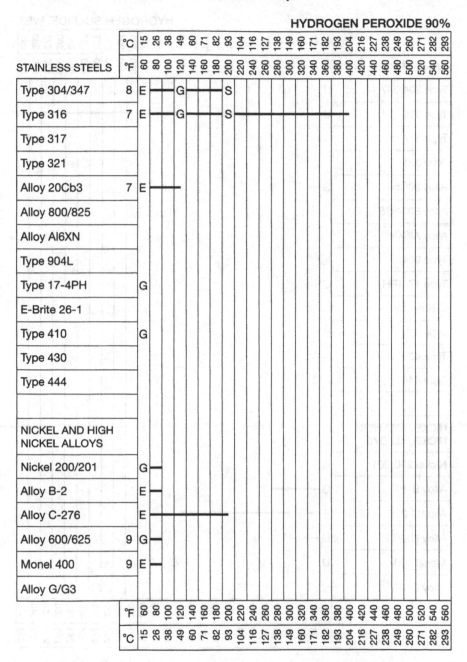

		°C	15	26	38	49	60	71	82	93	104	116	127	138	149	160	171	182	193	204	216	227	238	249	260	271	282	293
STAINLESS STEELS		°F	60	80	100	120	140	160	180	200	220	240	260	280	300	320	340	360	380	400	420	440	460	480	500	520	540	560
Type 304/347	8	E		G			S																					
Type 316	7	E		G			S																					
Type 317																												
Type 321																												
Alloy 20Cb3	7	E																										
Alloy 800/825																												
Alloy Al6XN																												
Type 904L																												
Type 17-4PH		G																										
E-Brite 26-1																												
Type 410		G																										
Type 430																												
Type 444																												
NICKEL AND HIGH NICKEL ALLOYS																												
Nickel 200/201		G																										
Alloy B-2		E																										
Alloy C-276		E																										
Alloy 600/625	9	G																										
Monel 400	9	E																										
Alloy G/G3																												
		°F	60	80	100	120	140	160	180	200	220	240	260	280	300	320	340	360	380	400	420	440	460	480	500	520	540	560
		°C	15	26	38	49	60	71	82	93	104	116	127	138	149	160	171	182	193	204	216	227	238	249	260	271	282	293

HYDROGEN SULFIDE, WET

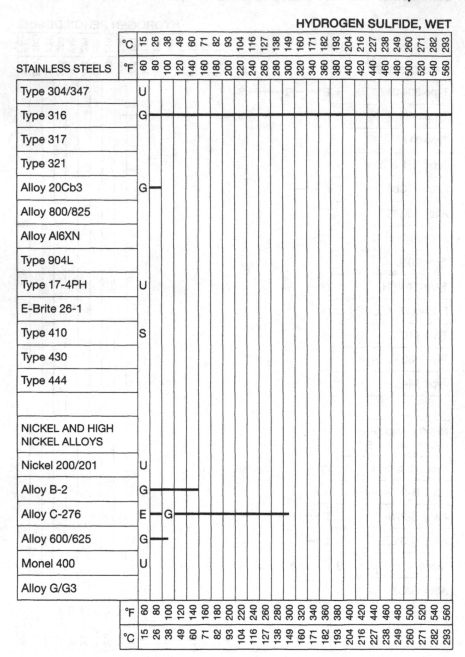

STAINLESS STEELS	°C 15 26 38 49 60 71 82 93 104 116 127 138 149 160 171 182 193 204 216 227 238 249 260 271 282 293	°F 60 80 100 120 140 160 180 200 220 240 260 280 300 320 340 360 380 400 420 440 460 480 500 520 540 560
Type 304/347	U	
Type 316	G	
Type 317		
Type 321		
Alloy 20Cb3	G	
Alloy 800/825		
Alloy Al6XN		
Type 904L		
Type 17-4PH	U	
E-Brite 26-1		
Type 410	S	
Type 430		
Type 444		
NICKEL AND HIGH NICKEL ALLOYS		
Nickel 200/201	U	
Alloy B-2	G	
Alloy C-276	E G	
Alloy 600/625	G	
Monel 400	U	
Alloy G/G3		

IODINE SOLUTION 10%

STAINLESS STEELS	°C	15	26	38	49	60	71	82	93	104	116	127	138	149	160	171	182	193	204	216	227	238	249	260	271	282	293
	°F	60	80	100	120	140	160	180	200	220	240	260	280	300	320	340	360	380	400	420	440	460	480	500	520	540	560
Type 304/347	U																										
Type 316	U																										
Type 317	G																										
Type 321	U																										
Alloy 20Cb3	U																										
Alloy 800/825																											
Alloy Al6XN																											
Type 904L																											
Type 17-4PH	U																										
E-Brite 26-1																											
Type 410																											
Type 430	U																										
Type 444																											
NICKEL AND HIGH NICKEL ALLOYS																											
Nickel 200/201																											
Alloy B-2																											
Alloy C-276	G	▬	▬	▬	▬																						
Alloy 600/625																											
Monel 400	U																										
Alloy G/G3																											
	°F	60	80	100	120	140	160	180	200	220	240	260	280	300	320	340	360	380	400	420	440	460	480	500	520	540	560
	°C	15	26	38	49	60	71	82	93	104	116	127	138	149	160	171	182	193	204	216	227	238	249	260	271	282	293

KETONES, GENERAL

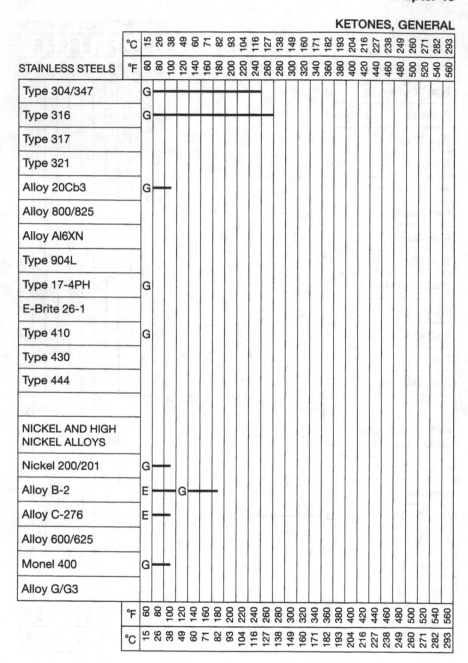

STAINLESS STEELS	°C	°F
Type 304/347	G	
Type 316	G	
Type 317		
Type 321		
Alloy 20Cb3	G	
Alloy 800/825		
Alloy Al6XN		
Type 904L		
Type 17-4PH	G	
E-Brite 26-1		
Type 410	G	
Type 430		
Type 444		
NICKEL AND HIGH NICKEL ALLOYS		
Nickel 200/201	G	
Alloy B-2	E	G
Alloy C-276	E	
Alloy 600/625		
Monel 400	G	
Alloy G/G3		

LACTIC ACID 25%

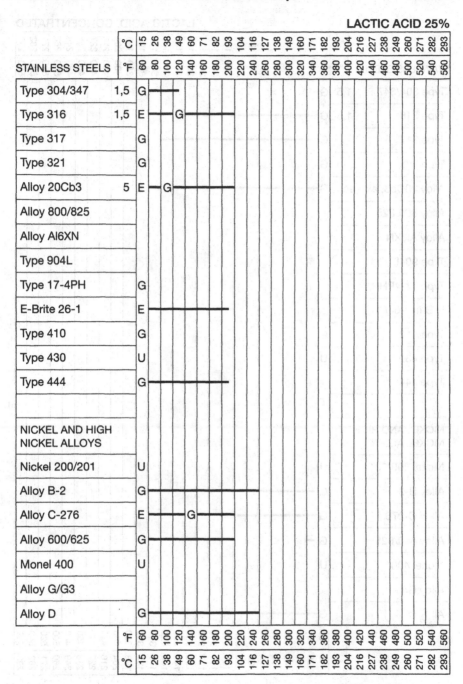

STAINLESS STEELS		°C	15	26	38	49	60	71	82	93	104	116	127	138	149	160	171	182	193	204	216	227	238	249	260	271	282	293
		°F	60	80	100	120	140	160	180	200	220	240	260	280	300	320	340	360	380	400	420	440	460	480	500	520	540	560
Type 304/347	1,5	G																										
Type 316	1,5	E			G																							
Type 317		G																										
Type 321		G																										
Alloy 20Cb3	5	E		G																								
Alloy 800/825																												
Alloy Al6XN																												
Type 904L																												
Type 17-4PH		G																										
E-Brite 26-1		E																										
Type 410		G																										
Type 430		U																										
Type 444		G																										
NICKEL AND HIGH NICKEL ALLOYS																											
Nickel 200/201		U																										
Alloy B-2		G																										
Alloy C-276		E				G																						
Alloy 600/625		G																										
Monel 400		U																										
Alloy G/G3																												
Alloy D		G																										
		°F	60	80	100	120	140	160	180	200	220	240	260	280	300	320	340	360	380	400	420	440	460	480	500	520	540	560
		°C	15	26	38	49	60	71	82	93	104	116	127	138	149	160	171	182	193	204	216	227	238	249	260	271	282	293

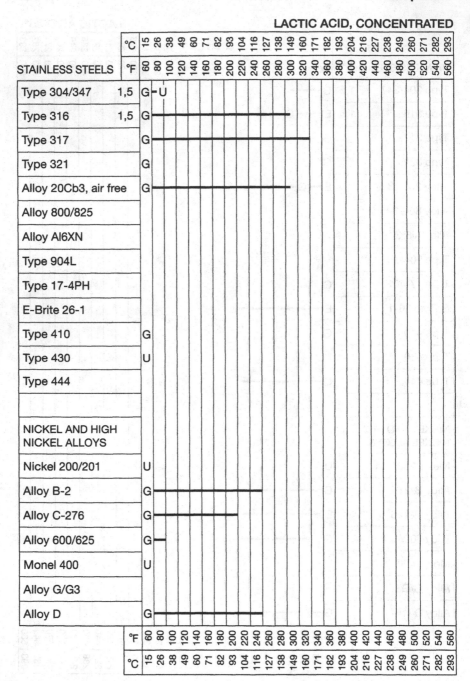

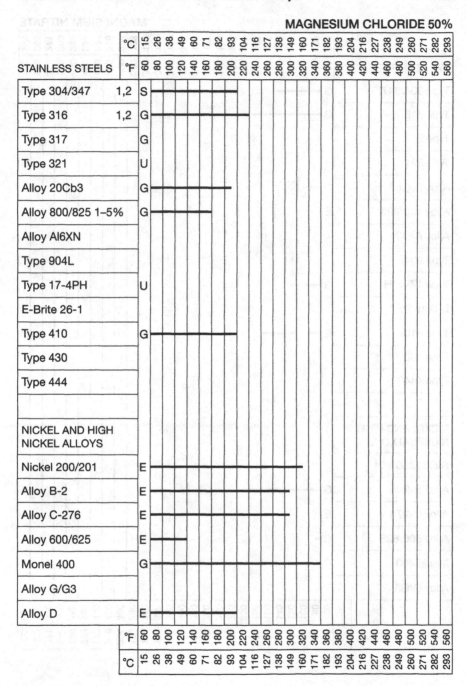

MAGNESIUM CHLORIDE 50%

| STAINLESS STEELS | | | °C | 15 | 26 | 38 | 49 | 60 | 71 | 82 | 93 | 104 | 116 | 127 | 138 | 149 | 160 | 171 | 182 | 193 | 204 | 216 | 227 | 238 | 249 | 260 | 271 | 282 | 293 |

Type 304/347 1,2 S

Type 316 1,2 G

Type 317 G

Type 321 U

Alloy 20Cb3 G

Alloy 800/825 1–5% G

Alloy Al6XN

Type 904L

Type 17-4PH U

E-Brite 26-1

Type 410 G

Type 430

Type 444

NICKEL AND HIGH NICKEL ALLOYS

Nickel 200/201 E

Alloy B-2 E

Alloy C-276 E

Alloy 600/625 E

Monel 400 G

Alloy G/G3

Alloy D E

MAGNESIUM NITRATE

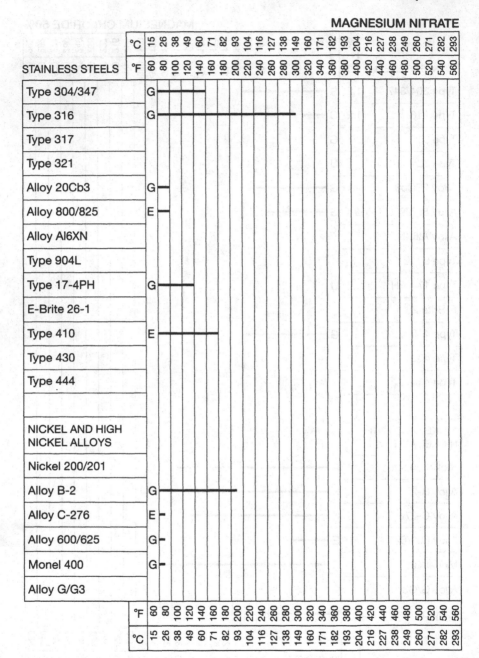

MAGNESIUM SULFATE

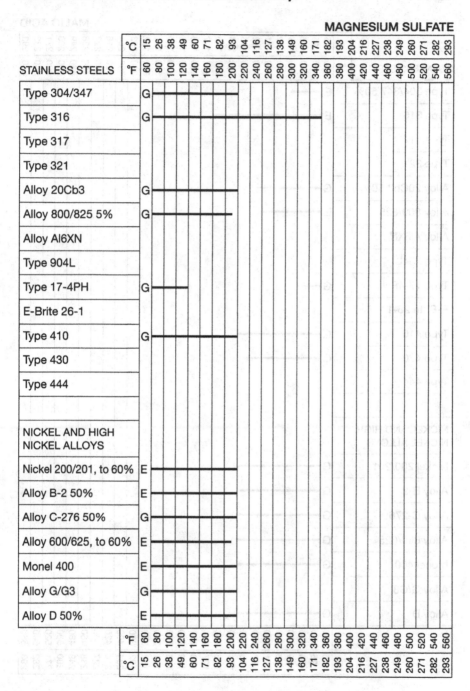

STAINLESS STEELS		°C	15	26	38	49	60	71	82	93	104	116	127	138	149	160	171	182	193	204	216	227	238	249	260	271	282	293
		°F	60	80	100	120	140	160	180	200	220	240	260	280	300	320	340	360	380	400	420	440	460	480	500	520	540	560
Type 304/347	G																											
Type 316	G																											
Type 317																												
Type 321																												
Alloy 20Cb3	G																											
Alloy 800/825 5%	G																											
Alloy Al6XN																												
Type 904L																												
Type 17-4PH	G																											
E-Brite 26-1																												
Type 410	G																											
Type 430																												
Type 444																												
NICKEL AND HIGH NICKEL ALLOYS																												
Nickel 200/201, to 60%	E																											
Alloy B-2 50%	E																											
Alloy C-276 50%	G																											
Alloy 600/625, to 60%	E																											
Monel 400	E																											
Alloy G/G3	G																											
Alloy D 50%	E																											
		°F	60	80	100	120	140	160	180	200	220	240	260	280	300	320	340	360	380	400	420	440	460	480	500	520	540	560
		°C	15	26	38	49	60	71	82	93	104	116	127	138	149	160	171	182	193	204	216	227	238	249	260	271	282	293

MALIC ACID

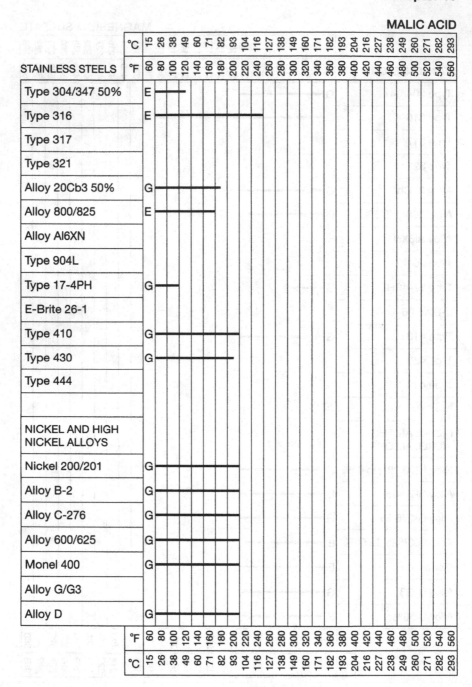

METHYL CHLORIDE

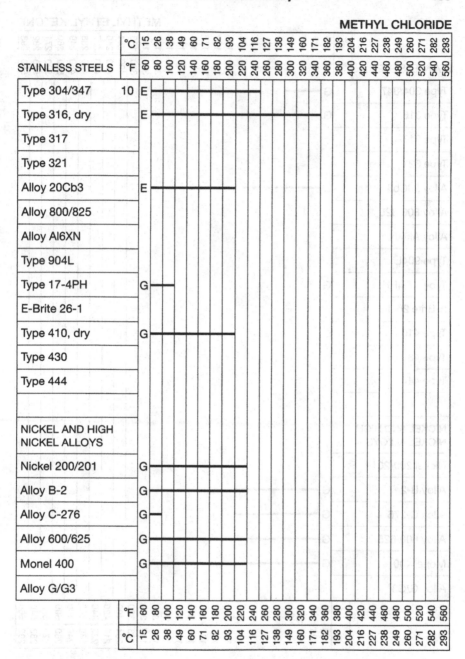

		°C	15	26	38	49	60	71	82	93	104	116	127	138	149	160	171	182	193	204	216	227	238	249	260	271	282	293
STAINLESS STEELS		°F	60	80	100	120	140	160	180	200	220	240	260	280	300	320	340	360	380	400	420	440	460	480	500	520	540	560
Type 304/347	10	E																										
Type 316, dry		E																										
Type 317																												
Type 321																												
Alloy 20Cb3		E																										
Alloy 800/825																												
Alloy Al6XN																												
Type 904L																												
Type 17-4PH		G																										
E-Brite 26-1																												
Type 410, dry		G																										
Type 430																												
Type 444																												
NICKEL AND HIGH NICKEL ALLOYS																												
Nickel 200/201		G																										
Alloy B-2		G																										
Alloy C-276		G																										
Alloy 600/625		G																										
Monel 400		G																										
Alloy G/G3																												
		°F	60	80	100	120	140	160	180	200	220	240	260	280	300	320	340	360	380	400	420	440	460	480	500	520	540	560
		°C	15	26	38	49	60	71	82	93	104	116	127	138	149	160	171	182	193	204	216	227	238	249	260	271	282	293

METHYL ETHYL KETONE

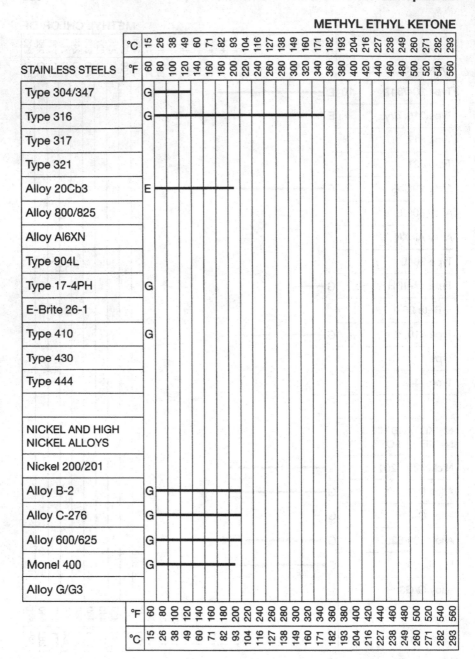

STAINLESS STEELS	°C	15	26	38	49	60	71	82	93	104	116	127	138	149	160	171	182	193	204	216	227	238	249	260	271	282	293
	°F	60	80	100	120	140	160	180	200	220	240	260	280	300	320	340	360	380	400	420	440	460	480	500	520	540	560
Type 304/347	G																										
Type 316	G																										
Type 317																											
Type 321																											
Alloy 20Cb3	E																										
Alloy 800/825																											
Alloy Al6XN																											
Type 904L																											
Type 17-4PH	G																										
E-Brite 26-1																											
Type 410	G																										
Type 430																											
Type 444																											
NICKEL AND HIGH NICKEL ALLOYS																											
Nickel 200/201																											
Alloy B-2	G																										
Alloy C-276	G																										
Alloy 600/625	G																										
Monel 400	G																										
Alloy G/G3																											
	°F	60	80	100	120	140	160	180	200	220	240	260	280	300	320	340	360	380	400	420	440	460	480	500	520	540	560
	°C	15	26	38	49	60	71	82	93	104	116	127	138	149	160	171	182	193	204	216	227	238	249	260	271	282	293

METHYLENE CHLORIDE

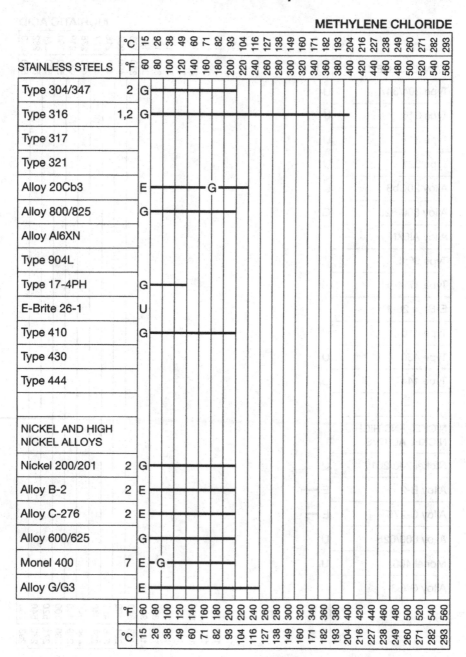

STAINLESS STEELS		°C	15	26	38	49	60	71	82	93	104	116	127	138	149	160	171	182	193	204	216	227	238	249	260	271	282	293
		°F	60	80	100	120	140	160	180	200	220	240	260	280	300	320	340	360	380	400	420	440	460	480	500	520	540	560
Type 304/347	2	G																										
Type 316	1,2	G																										
Type 317																												
Type 321																												
Alloy 20Cb3		E				G																						
Alloy 800/825		G																										
Alloy Al6XN																												
Type 904L																												
Type 17-4PH		G																										
E-Brite 26-1		U																										
Type 410		G																										
Type 430																												
Type 444																												
NICKEL AND HIGH NICKEL ALLOYS																												
Nickel 200/201	2	G																										
Alloy B-2	2	E																										
Alloy C-276	2	E																										
Alloy 600/625		G																										
Monel 400	7	E	G																									
Alloy G/G3		E																										
		°F	60	80	100	120	140	160	180	200	220	240	260	280	300	320	340	360	380	400	420	440	460	480	500	520	540	560
		°C	15	26	38	49	60	71	82	93	104	116	127	138	149	160	171	182	193	204	216	227	238	249	260	271	282	293

MURIATIC ACID

STAINLESS STEELS		°C	15	26	38	49	60	71	82	93	104	116	127	138	149	160	171	182	193	204	216	227	238	249	260	271	282	293
		°F	60	80	100	120	140	160	180	200	220	240	260	280	300	320	340	360	380	400	420	440	460	480	500	520	540	560
Type 304/347		U																										
Type 316		U																										
Type 317		U																										
Type 321		U																										
Alloy 20Cb3		U																										
Alloy 800/825	4	G																										
Alloy Al6XN																												
Type 904L																												
Type 17-4PH																												
E-Brite 26-1																												
Type 410		U																										
Type 430		U																										
Type 444		U																										
NICKEL AND HIGH NICKEL ALLOYS																												
Nickel 200/201		U																										
Alloy B-2		E																										
Alloy C-276		E																										
Alloy 600/625		U																										
Monel 400		U																										
Alloy G/G3																												
		°F	60	80	100	120	140	160	180	200	220	240	260	280	300	320	340	360	380	400	420	440	460	480	500	520	540	560
		°C	15	26	38	49	60	71	82	93	104	116	127	138	149	160	171	182	193	204	216	227	238	249	260	271	282	293

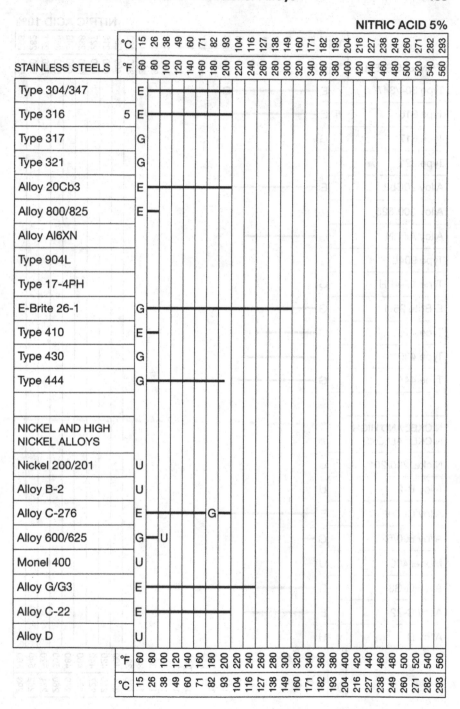

NITRIC ACID 5%

STAINLESS STEELS	°C	15	26	38	49	60	71	82	93	104	116	127	138	149	160	171	182	193	204	216	227	238	249	260	271	282	293
	°F	60	80	100	120	140	160	180	200	220	240	260	280	300	320	340	360	380	400	420	440	460	480	500	520	540	560
Type 304/347	E ———————————																										
Type 316 5	E ———————————																										
Type 317	G																										
Type 321	G																										
Alloy 20Cb3	E ———————————																										
Alloy 800/825	E —																										
Alloy AI6XN																											
Type 904L																											
Type 17-4PH																											
E-Brite 26-1	G —————————————————————																										
Type 410	E —																										
Type 430	G																										
Type 444	G —————————————																										
NICKEL AND HIGH NICKEL ALLOYS																											
Nickel 200/201	U																										
Alloy B-2	U																										
Alloy C-276	E ——————————— G ——																										
Alloy 600/625	G — U																										
Monel 400	U																										
Alloy G/G3	E —————————————————																										
Alloy C-22	E —————————————————																										
Alloy D	U																										
	°F	60	80	100	120	140	160	180	200	220	240	260	280	300	320	340	360	380	400	420	440	460	480	500	520	540	560
	°C	15	26	38	49	60	71	82	93	104	116	127	138	149	160	171	182	193	204	216	227	238	249	260	271	282	293

NITRIC ACID 10%

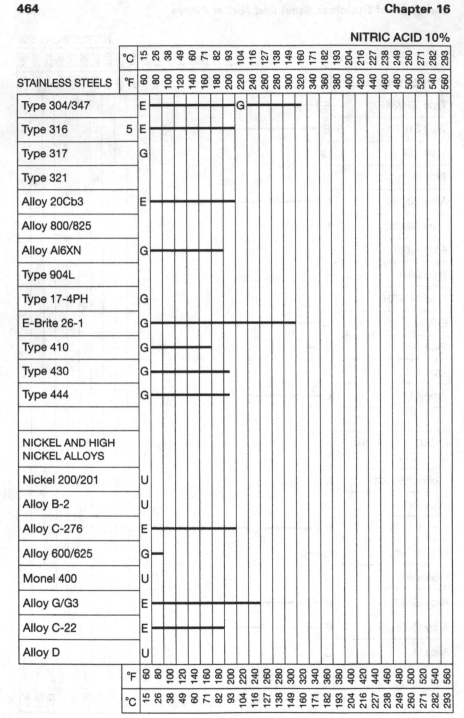

STAINLESS STEELS	°C	15	26	38	49	60	71	82	93	104	116	127	138	149	160	171	182	193	204	216	227	238	249	260	271	282	293
	°F	60	80	100	120	140	160	180	200	220	240	260	280	300	320	340	360	380	400	420	440	460	480	500	520	540	560
Type 304/347	E								G																	
Type 316 5	E																									
Type 317	G																									
Type 321																										
Alloy 20Cb3	E																									
Alloy 800/825																										
Alloy Al6XN	G																									
Type 904L																										
Type 17-4PH	G																									
E-Brite 26-1	G																									
Type 410	G																									
Type 430	G																									
Type 444	G																									
NICKEL AND HIGH NICKEL ALLOYS																										
Nickel 200/201	U																									
Alloy B-2	U																									
Alloy C-276	E																									
Alloy 600/625	G																									
Monel 400	U																									
Alloy G/G3	E																									
Alloy C-22	E																									
Alloy D	U																									

NITRIC ACID 20%

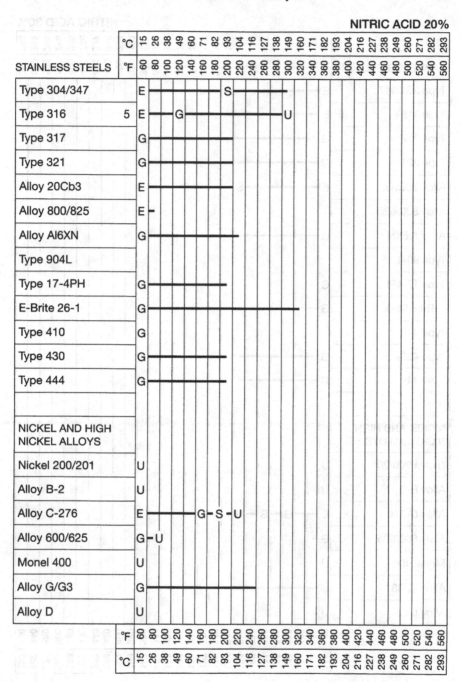

NITRIC ACID 30%

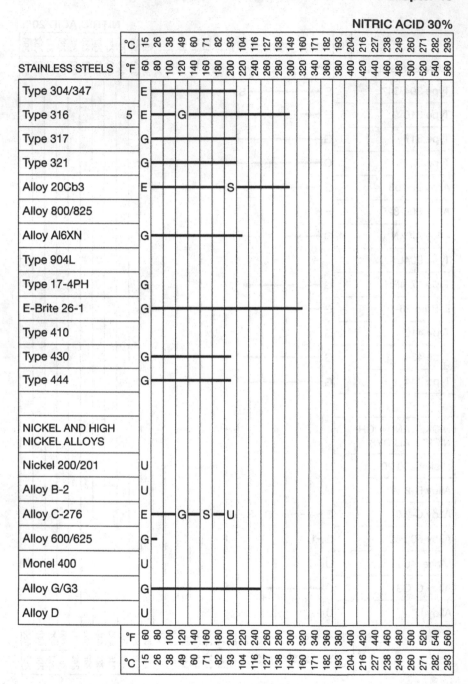

NITRIC ACID 40%

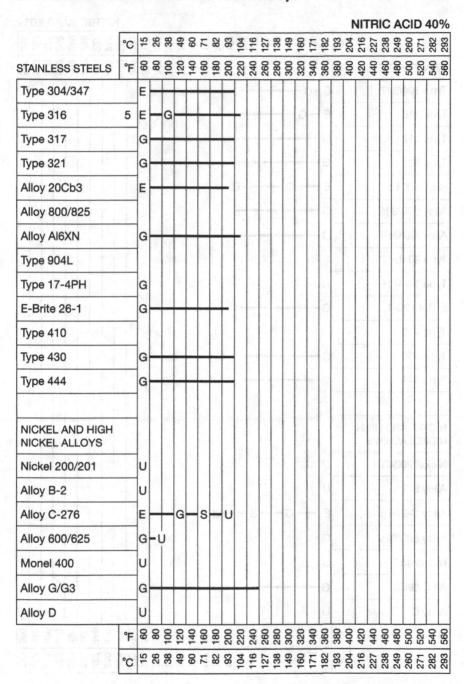

STAINLESS STEELS	°F
Type 304/347	E
Type 316	E —G
Type 317	G
Type 321	G
Alloy 20Cb3	E
Alloy 800/825	
Alloy Al6XN	G
Type 904L	
Type 17-4PH	G
E-Brite 26-1	G
Type 410	
Type 430	G
Type 444	G
NICKEL AND HIGH NICKEL ALLOYS	
Nickel 200/201	U
Alloy B-2	U
Alloy C-276	E —G —S—U
Alloy 600/625	G–U
Monel 400	U
Alloy G/G3	G
Alloy D	U

NITRIC ACID 50%

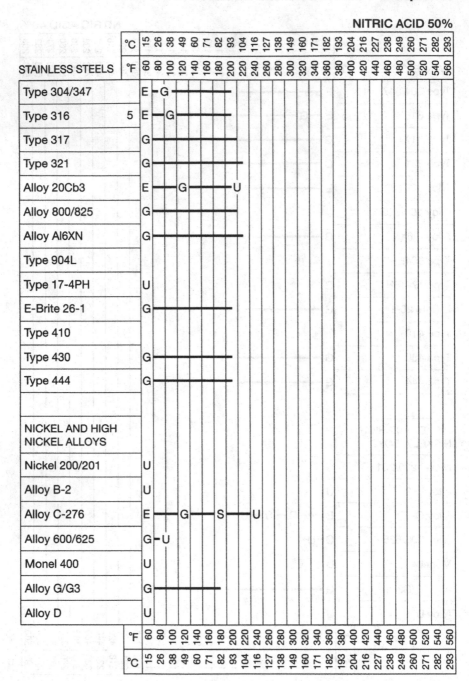

STAINLESS STEELS	°C	15	26	38	49	60	71	82	93	104	116	127	138	149	160	171	182	193	204	216	227	238	249	260	271	282	293
	°F	60	80	100	120	140	160	180	200	220	240	260	280	300	320	340	360	380	400	420	440	460	480	500	520	540	560
Type 304/347		E	G																								
Type 316	5	E	G																								
Type 317		G																									
Type 321		G																									
Alloy 20Cb3		E		G			U																				
Alloy 800/825		G																									
Alloy Al6XN		G																									
Type 904L																											
Type 17-4PH		U																									
E-Brite 26-1		G																									
Type 410																											
Type 430		G																									
Type 444		G																									
NICKEL AND HIGH NICKEL ALLOYS																											
Nickel 200/201		U																									
Alloy B-2		U																									
Alloy C-276		E			G		S		U																		
Alloy 600/625		G	U																								
Monel 400		U																									
Alloy G/G3		G																									
Alloy D		U																									
	°F	60	80	100	120	140	160	180	200	220	240	260	280	300	320	340	360	380	400	420	440	460	480	500	520	540	560
	°C	15	26	38	49	60	71	82	93	104	116	127	138	149	160	171	182	193	204	216	227	238	249	260	271	282	293

NITRIC ACID 70%

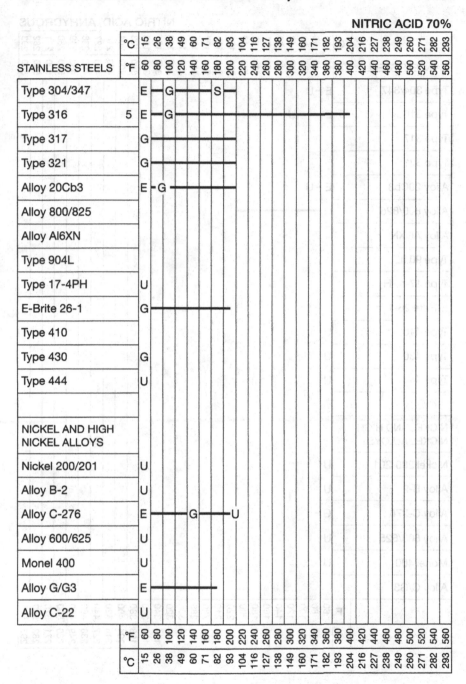

STAINLESS STEELS		°C	15	26	38	49	60	71	82	93	104	116	127	138	149	160	171	182	193	204	216	227	238	249	260	271	282	293
		°F	60	80	100	120	140	160	180	200	220	240	260	280	300	320	340	360	380	400	420	440	460	480	500	520	540	560
Type 304/347		E	—G——————S—																									
Type 316	5	E	—G																									
Type 317		G																										
Type 321		G																										
Alloy 20Cb3		E	⊢G																									
Alloy 800/825																												
Alloy Al6XN																												
Type 904L																												
Type 17-4PH		U																										
E-Brite 26-1		G																										
Type 410																												
Type 430		G																										
Type 444		U																										
NICKEL AND HIGH NICKEL ALLOYS																												
Nickel 200/201		U																										
Alloy B-2		U																										
Alloy C-276		E	———G———U																								
Alloy 600/625		U																										
Monel 400		U																										
Alloy G/G3		E	——————																								
Alloy C-22		U																										
		°F	60	80	100	120	140	160	180	200	220	240	260	280	300	320	340	360	380	400	420	440	460	480	500	520	540	560
		°C	15	26	38	49	60	71	82	93	104	116	127	138	149	160	171	182	193	204	216	227	238	249	260	271	282	293

NITRIC ACID, ANHYDROUS

STAINLESS STEELS		°C	15	26	38	49	60	71	82	93	104	116	127	138	149	160	171	182	193	204	216	227	238	249	260	271	282	293
		°F	60	80	100	120	140	160	180	200	220	240	260	280	300	320	340	360	380	400	420	440	460	480	500	520	540	560
Type 304/347			E	U																								
Type 316	5		E		U																							
Type 317																												
Type 321																												
Alloy 20Cb3			E	U																								
Alloy 800/825			G							U																		
Alloy Al6XN																												
Type 904L																												
Type 17-4PH																												
E-Brite 26-1																												
Type 410			U																									
Type 430			U																									
Type 444			U																									
NICKEL AND HIGH NICKEL ALLOYS																												
Nickel 200/201			U																									
Alloy B-2			U																									
Alloy C-276			G																									
Alloy 600/625			U																									
Monel 400			U																									
Alloy G/G3																												
		°F	60	80	100	120	140	160	180	200	220	240	260	280	300	320	340	360	380	400	420	440	460	480	500	520	540	560
		°C	15	26	38	49	60	71	82	93	104	116	127	138	149	160	171	182	193	204	216	227	238	249	260	271	282	293

NITROUS ACID, CONCENTRATED

STAINLESS STEELS °C	15	26	38	49	60	71	82	93	104	116	127	138	149	160	171	182	193	204	216	227	238	249	260	271	282	293
°F	60	80	100	120	140	160	180	200	220	240	260	280	300	320	340	360	380	400	420	440	460	480	500	520	540	560
Type 304/347	G–																									
Type 316	G–																									
Type 317																										
Type 321																										
Alloy 20Cb3	G–																									
Alloy 800/825																										
Alloy Al6XN																										
Type 904L																										
Type 17-4PH																										
E-Brite 26-1																										
Type 410	G																									
Type 430 5%	G																									
Type 444																										
NICKEL AND HIGH NICKEL ALLOYS																										
Nickel 200/201	U																									
Alloy B-2	U																									
Alloy C-276	U																									
Alloy 600/625	U																									
Monel 400	U																									
Alloy G/G3																										
°F	60	80	100	120	140	160	180	200	220	240	260	280	300	320	340	360	380	400	420	440	460	480	500	520	540	560
°C	15	26	38	49	60	71	82	93	104	116	127	138	149	160	171	182	193	204	216	227	238	249	260	271	282	293

OLEUM

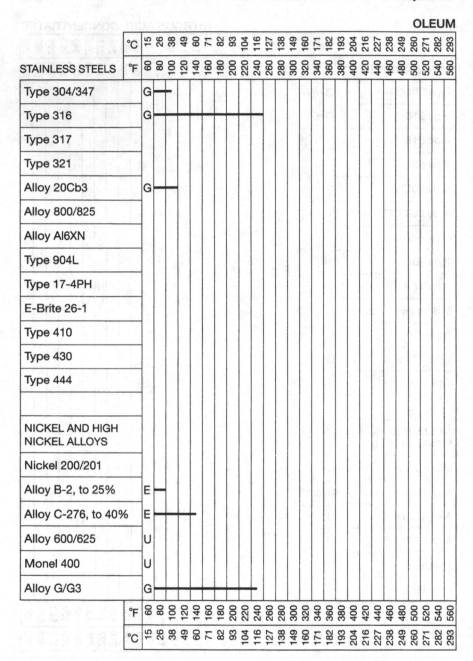

STAINLESS STEELS	°C	15	26	38	49	60	71	82	93	104	116	127	138	149	160	171	182	193	204	216	227	238	249	260	271	282	293
	°F	60	80	100	120	140	160	180	200	220	240	260	280	300	320	340	360	380	400	420	440	460	480	500	520	540	560
Type 304/347	G
Type 316	G
Type 317	
Type 321	
Alloy 20Cb3	G
Alloy 800/825	
Alloy Al6XN	
Type 904L	
Type 17-4PH	
E-Brite 26-1	
Type 410	
Type 430	
Type 444	
NICKEL AND HIGH NICKEL ALLOYS	
Nickel 200/201	
Alloy B-2, to 25%	E
Alloy C-276, to 40%	E
Alloy 600/625	U
Monel 400	U
Alloy G/G3	G
	°F	60	80	100	120	140	160	180	200	220	240	260	280	300	320	340	360	380	400	420	440	460	480	500	520	540	560
	°C	15	26	38	49	60	71	82	93	104	116	127	138	149	160	171	182	193	204	216	227	238	249	260	271	282	293

OXALIC ACID 10%

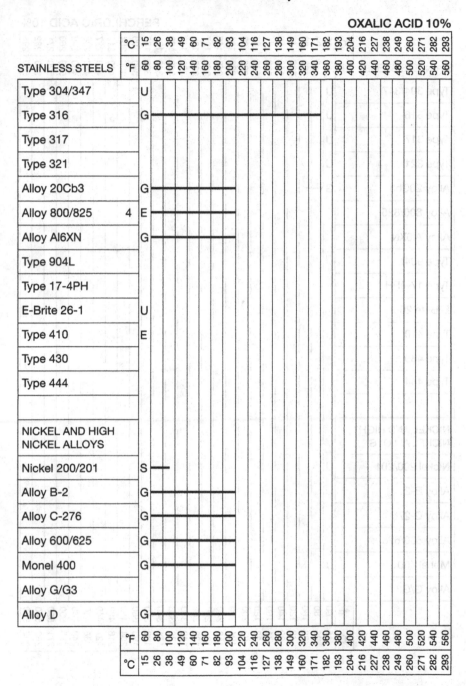

STAINLESS STEELS		
Type 304/347	U	
Type 316	G	
Type 317		
Type 321		
Alloy 20Cb3	G	
Alloy 800/825	4	E
Alloy Al6XN	G	
Type 904L		
Type 17-4PH		
E-Brite 26-1	U	
Type 410	E	
Type 430		
Type 444		
NICKEL AND HIGH NICKEL ALLOYS		
Nickel 200/201	S	
Alloy B-2	G	
Alloy C-276	G	
Alloy 600/625	G	
Monel 400	G	
Alloy G/G3		
Alloy D	G	

PERCHLORIC ACID 10%

STAINLESS STEELS °C	15	26	38	49	60	71	82	93	104	116	127	138	149	160	171	182	193	204	216	227	238	249	260	271	282	293
°F	60	80	100	120	140	160	180	200	220	240	260	280	300	320	340	360	380	400	420	440	460	480	500	520	540	560
Type 304/347	U																									
Type 316	U																									
Type 317	U																									
Type 321	U																									
Alloy 20Cb3	G																									
Alloy 800/825																										
Alloy Al6XN																										
Type 904L																										
Type 17-4PH																										
E-Brite 26-1																										
Type 410	U																									
Type 430																										
Type 444																										
NICKEL AND HIGH NICKEL ALLOYS																										
Nickel 200/201	U																									
Alloy B-2																										
Alloy C-276																										
Alloy 600/625																										
Monel 400	U																									
Alloy G/G3																										
°F	60	80	100	120	140	160	180	200	220	240	260	280	300	320	340	360	380	400	420	440	460	480	500	520	540	560
°C	15	26	38	49	60	71	82	93	104	116	127	138	149	160	171	182	193	204	216	227	238	249	260	271	282	293

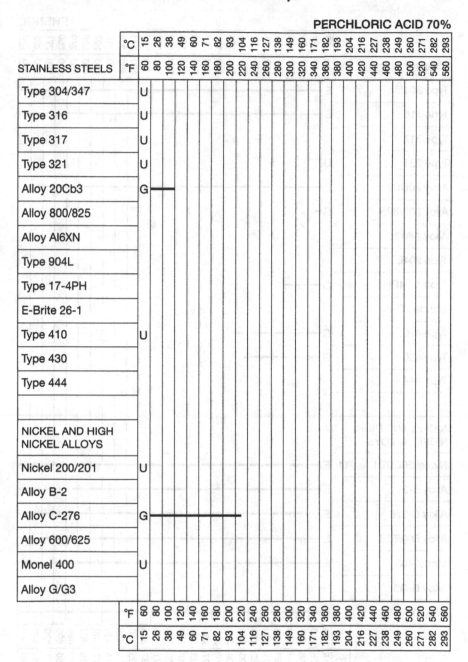

PERCHLORIC ACID 70%

STAINLESS STEELS	°C	15	26	38	49	60	71	82	93	104	116	127	138	149	160	171	182	193	204	216	227	238	249	260	271	282	293
	°F	60	80	100	120	140	160	180	200	220	240	260	280	300	320	340	360	380	400	420	440	460	480	500	520	540	560
Type 304/347	U																										
Type 316	U																										
Type 317	U																										
Type 321	U																										
Alloy 20Cb3	G																										
Alloy 800/825																											
Alloy Al6XN																											
Type 904L																											
Type 17-4PH																											
E-Brite 26-1																											
Type 410	U																										
Type 430																											
Type 444																											
NICKEL AND HIGH NICKEL ALLOYS																											
Nickel 200/201	U																										
Alloy B-2																											
Alloy C-276	G																										
Alloy 600/625																											
Monel 400	U																										
Alloy G/G3																											

PHENOL

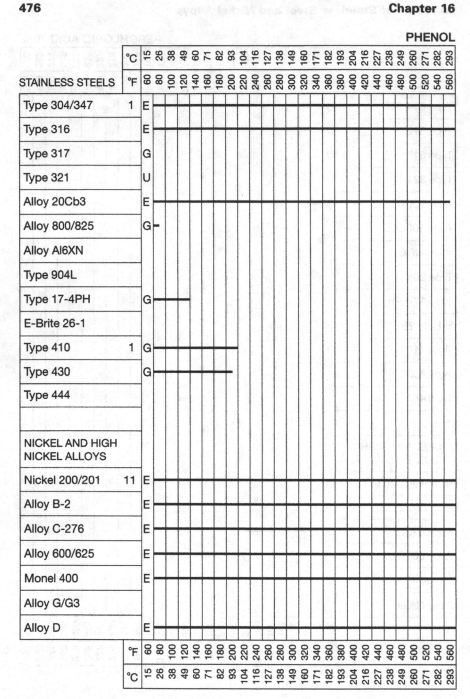

STAINLESS STEELS		°C	15	26	38	49	60	71	82	93	104	116	127	138	149	160	171	182	193	204	216	227	238	249	260	271	282	293
		°F	60	80	100	120	140	160	180	200	220	240	260	280	300	320	340	360	380	400	420	440	460	480	500	520	540	560
Type 304/347	1	E																										
Type 316		E																										
Type 317		G																										
Type 321		U																										
Alloy 20Cb3		E																										
Alloy 800/825		G																										
Alloy Al6XN																												
Type 904L																												
Type 17-4PH		G																										
E-Brite 26-1																												
Type 410	1	G																										
Type 430		G																										
Type 444																												
NICKEL AND HIGH NICKEL ALLOYS																												
Nickel 200/201	11	E																										
Alloy B-2		E																										
Alloy C-276		E																										
Alloy 600/625		E																										
Monel 400		E																										
Alloy G/G3																												
Alloy D		E																										
		°F	60	80	100	120	140	160	180	200	220	240	260	280	300	320	340	360	380	400	420	440	460	480	500	520	540	560
		°C	15	26	38	49	60	71	82	93	104	116	127	138	149	160	171	182	193	204	216	227	238	249	260	271	282	293

PHOSPHORIC ACID 5%

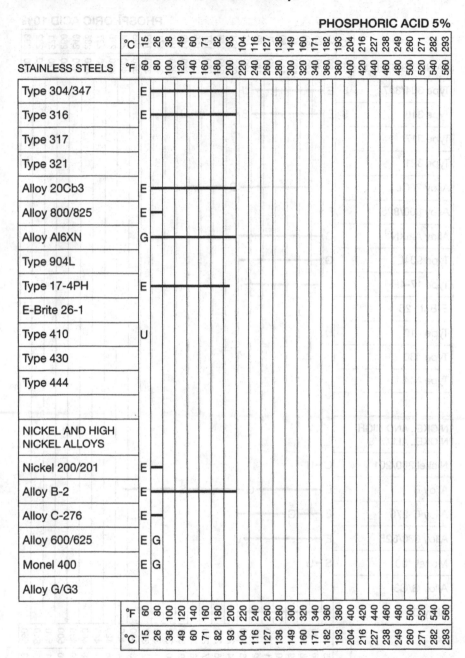

STAINLESS STEELS	°C	15 26 38 49 60 71 82 93 104 116 127 138 149 160 171 182 193 204 216 227 238 249 260 271 282 293
	°F	60 80 100 120 140 160 180 200 220 240 260 280 300 320 340 360 380 400 420 440 460 480 500 520 540 560
Type 304/347	E	
Type 316	E	
Type 317		
Type 321		
Alloy 20Cb3	E	
Alloy 800/825	E	
Alloy AI6XN	G	
Type 904L		
Type 17-4PH	E	
E-Brite 26-1		
Type 410	U	
Type 430		
Type 444		
NICKEL AND HIGH NICKEL ALLOYS		
Nickel 200/201	E	
Alloy B-2	E	
Alloy C-276	E	
Alloy 600/625	E G	
Monel 400	E G	
Alloy G/G3		

PHOSPHORIC ACID 10%

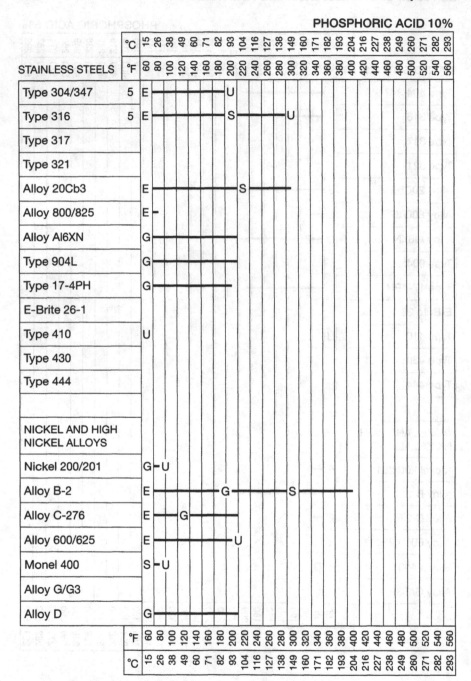

STAINLESS STEELS	°C	°F		
Type 304/347	5	E —————— U		
Type 316	5	E —————— S ——— U		
Type 317				
Type 321				
Alloy 20Cb3		E —————— S ———		
Alloy 800/825		E ⊢		
Alloy Al6XN		G ——————		
Type 904L		G ——————		
Type 17-4PH		G ————		
E-Brite 26-1				
Type 410		U		
Type 430				
Type 444				
NICKEL AND HIGH NICKEL ALLOYS				
Nickel 200/201		G ⊢ U		
Alloy B-2		E —————— G ——— S ———		
Alloy C-276		E ⊢ G ———		
Alloy 600/625		E —————— U		
Monel 400		S ⊢ U		
Alloy G/G3				
Alloy D		G ——————		

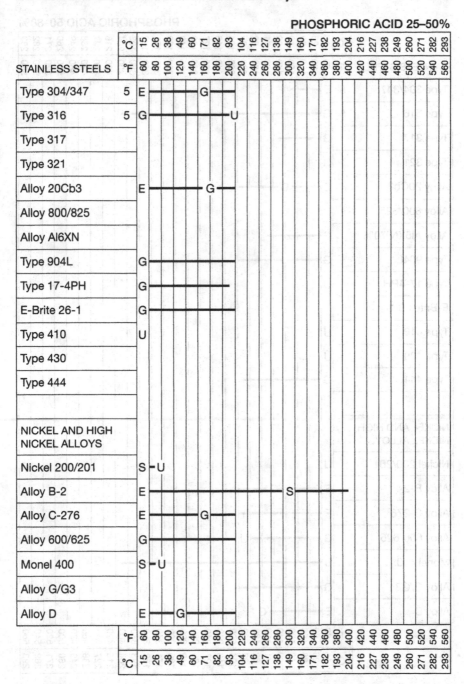

PHOSPHORIC ACID 25–50%

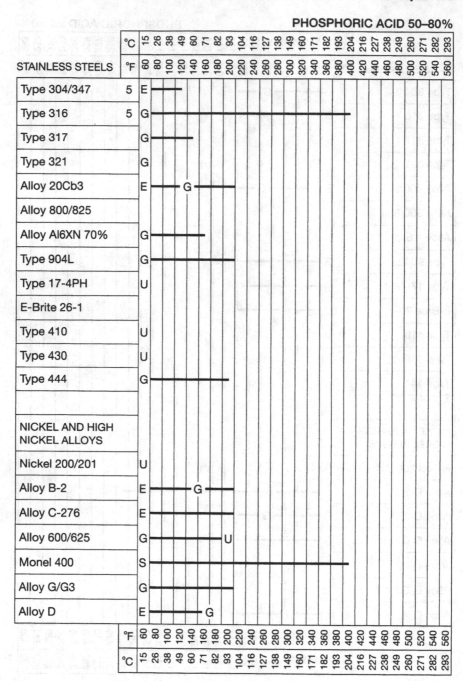

PHOSPHORIC ACID 50–80%

STAINLESS STEELS		°C	15	26	38	49	60	71	82	93	104	116	127	138	149	160	171	182	193	204	216	227	238	249	260	271	282	293
		°F	60	80	100	120	140	160	180	200	220	240	260	280	300	320	340	360	380	400	420	440	460	480	500	520	540	560
Type 304/347	5	E																										
Type 316	5	G																										
Type 317		G																										
Type 321		G																										
Alloy 20Cb3		E		G																								
Alloy 800/825																												
Alloy Al6XN 70%		G																										
Type 904L		G																										
Type 17-4PH		U																										
E-Brite 26-1																												
Type 410		U																										
Type 430		U																										
Type 444		G																										
NICKEL AND HIGH NICKEL ALLOYS																												
Nickel 200/201		U																										
Alloy B-2		E			G																							
Alloy C-276		E																										
Alloy 600/625		G					U																					
Monel 400		S																										
Alloy G/G3		G																										
Alloy D		E			G																							
		°F	60	80	100	120	140	160	180	200	220	240	260	280	300	320	340	360	380	400	420	440	460	480	500	520	540	560
		°C	15	26	38	49	60	71	82	93	104	116	127	138	149	160	171	182	193	204	216	227	238	249	260	271	282	293

PHTHALIC ACID

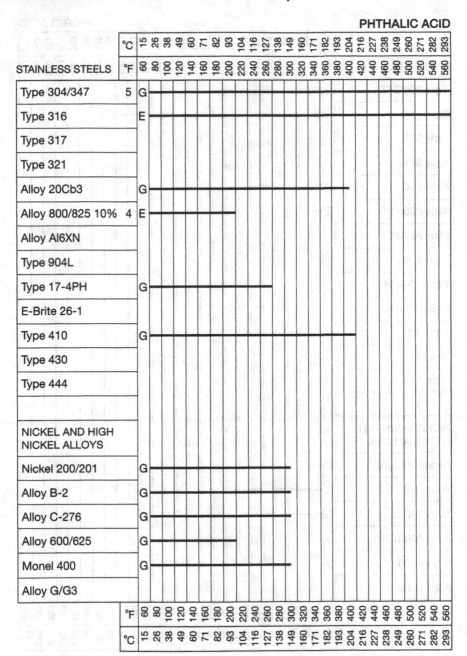

STAINLESS STEELS	°F	
Type 304/347	5	G
Type 316		E
Type 317		
Type 321		
Alloy 20Cb3		G
Alloy 800/825 10%	4	E
Alloy Al6XN		
Type 904L		
Type 17-4PH		G
E-Brite 26-1		
Type 410		G
Type 430		
Type 444		
NICKEL AND HIGH NICKEL ALLOYS		
Nickel 200/201		G
Alloy B-2		G
Alloy C-276		G
Alloy 600/625		G
Monel 400		G
Alloy G/G3		

PICRIC ACID

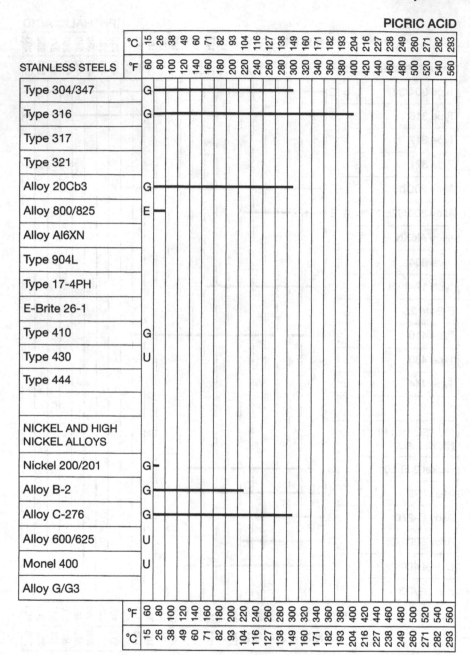

STAINLESS STEELS	°C	°F																								
	15	60	26	38	49	60	71	82	93	104	116	127	138	149	160	171	182	193	204	216	227	238	249	260	271	282	293

Type 304/347 G

Type 316 G

Type 317

Type 321

Alloy 20Cb3 G

Alloy 800/825 E

Alloy Al6XN

Type 904L

Type 17-4PH

E-Brite 26-1

Type 410 G

Type 430 U

Type 444

NICKEL AND HIGH NICKEL ALLOYS

Nickel 200/201 G

Alloy B-2 G

Alloy C-276 G

Alloy 600/625 U

Monel 400 U

Alloy G/G3

°F 60 80 100 120 140 160 180 200 220 240 260 280 300 320 340 360 380 400 420 440 460 480 500 520 540 560

°C 15 26 38 49 60 71 82 93 104 116 127 138 149 160 171 182 193 204 216 227 238 249 260 271 282 293

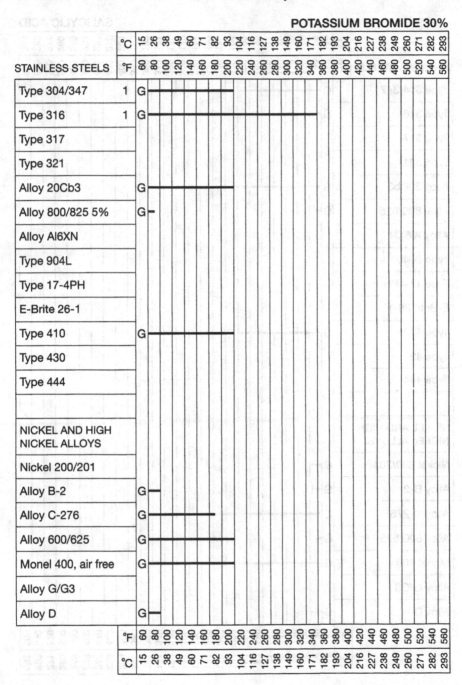

POTASSIUM BROMIDE 30%

STAINLESS STEELS	
Type 304/347	1
Type 316	1
Type 317	
Type 321	
Alloy 20Cb3	
Alloy 800/825 5%	
Alloy Al6XN	
Type 904L	
Type 17-4PH	
E-Brite 26-1	
Type 410	
Type 430	
Type 444	
NICKEL AND HIGH NICKEL ALLOYS	
Nickel 200/201	
Alloy B-2	
Alloy C-276	
Alloy 600/625	
Monel 400, air free	
Alloy G/G3	
Alloy D	

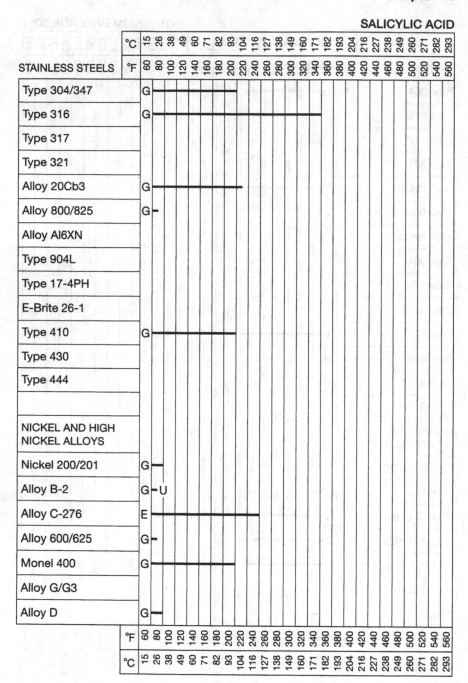

SALICYLIC ACID

| STAINLESS STEELS | °C | 15 | 26 | 38 | 49 | 60 | 71 | 82 | 93 | 104 | 116 | 127 | 138 | 149 | 160 | 171 | 182 | 193 | 204 | 216 | 227 | 238 | 249 | 260 | 271 | 282 | 293 |
| | °F | 60 | 80 | 100 | 120 | 140 | 160 | 180 | 200 | 220 | 240 | 260 | 280 | 300 | 320 | 340 | 360 | 380 | 400 | 420 | 440 | 460 | 480 | 500 | 520 | 540 | 560 |

Type 304/347 — G
Type 316 — G
Type 317
Type 321
Alloy 20Cb3 — G
Alloy 800/825 — G
Alloy Al6XN
Type 904L
Type 17-4PH
E-Brite 26-1
Type 410 — G
Type 430
Type 444

NICKEL AND HIGH NICKEL ALLOYS

Nickel 200/201 — G
Alloy B-2 — G-U
Alloy C-276 — E
Alloy 600/625 — G
Monel 400 — G
Alloy G/G3
Alloy D — G

SILVER BROMIDE 10%

STAINLESS STEELS	°C	15	26	38	49	60	71	82	93	104	116	127	138	149	160	171	182	193	204	216	227	238	249	260	271	282	293
	°F	60	80	100	120	140	160	180	200	220	240	260	280	300	320	340	360	380	400	420	440	460	480	500	520	540	560
Type 304/347	U																										
Type 316	U																										
Type 317	U																										
Type 321	U																										
Alloy 20Cb3	E —																										
Alloy 800/825 4	E —																										
Alloy Al6XN																											
Type 904L																											
Type 17-4PH																											
E-Brite 26-1																											
Type 410	U																										
Type 430	U																										
Type 444																											
NICKEL AND HIGH NICKEL ALLOYS																											
Nickel 200/201																											
Alloy B-2	E —																										
Alloy C-276	E —																										
Alloy 600/625																											
Monel 400	E	G																									
Alloy G/G3																											
	°F	60	80	100	120	140	160	180	200	220	240	260	280	300	320	340	360	380	400	420	440	460	480	500	520	540	560
	°C	15	26	38	49	60	71	82	93	104	116	127	138	149	160	171	182	193	204	216	227	238	249	260	271	282	293

SODIUM CARBONATE 10–30%

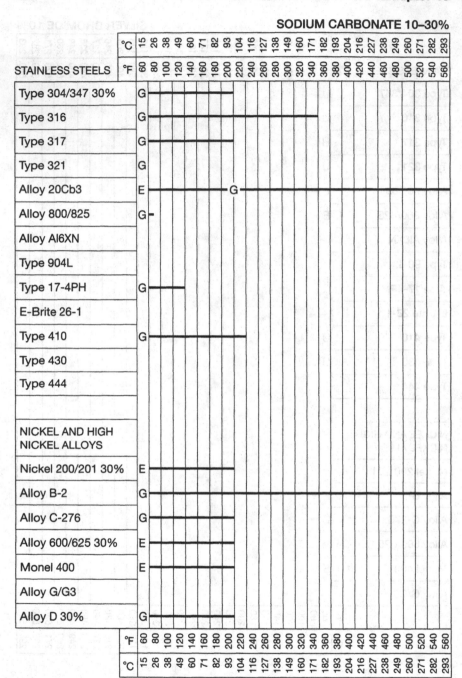

STAINLESS STEELS	°C	15 26 38 49 60 71 82 93 104 116 127 138 149 160 171 182 193 204 216 227 238 249 260 271 282 293
	°F	60 80 100 120 140 160 180 200 220 240 260 280 300 320 340 360 380 400 420 440 460 480 500 520 540 560
Type 304/347 30%	G	
Type 316	G	
Type 317	G	
Type 321	G	
Alloy 20Cb3	E	G
Alloy 800/825	G	
Alloy Al6XN		
Type 904L		
Type 17-4PH	G	
E-Brite 26-1		
Type 410	G	
Type 430		
Type 444		
NICKEL AND HIGH NICKEL ALLOYS		
Nickel 200/201 30%	E	
Alloy B-2	G	
Alloy C-276	G	
Alloy 600/625 30%	E	
Monel 400	E	
Alloy G/G3		
Alloy D 30%	G	
	°F	60 80 100 120 140 160 180 200 220 240 260 280 300 320 340 360 380 400 420 440 460 480 500 520 540 560
	°C	15 26 38 49 60 71 82 93 104 116 127 138 149 160 171 182 193 204 216 227 238 249 260 271 282 293

SODIUM CHLORIDE

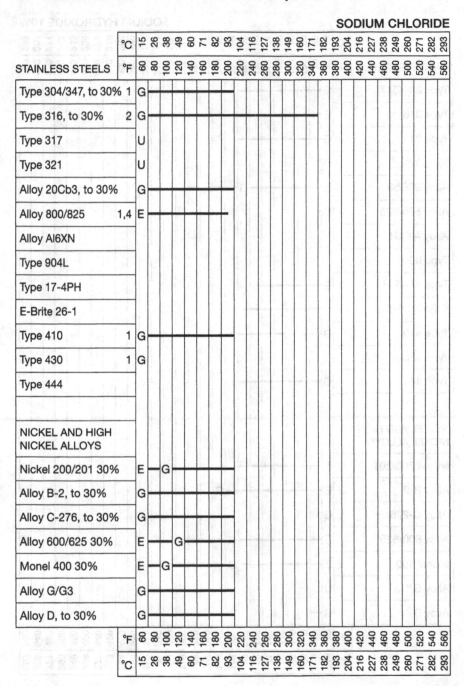

Temperature scale — °C: 15, 26, 38, 49, 60, 71, 82, 93, 104, 116, 127, 138, 149, 160, 171, 182, 193, 204, 216, 227, 238, 249, 260, 271, 282, 293

Temperature scale — °F: 60, 80, 100, 120, 140, 160, 180, 200, 220, 240, 260, 280, 300, 320, 340, 360, 380, 400, 420, 440, 460, 480, 500, 520, 540, 560

STAINLESS STEELS	Note	Rating
Type 304/347, to 30%	1	G
Type 316, to 30%	2	G
Type 317		U
Type 321		U
Alloy 20Cb3, to 30%		G
Alloy 800/825	1,4	E
Alloy Al6XN		
Type 904L		
Type 17-4PH		
E-Brite 26-1		
Type 410	1	G
Type 430	1	G
Type 444		

NICKEL AND HIGH NICKEL ALLOYS	Rating
Nickel 200/201 30%	E —G
Alloy B-2, to 30%	G
Alloy C-276, to 30%	G
Alloy 600/625 30%	E —G
Monel 400 30%	E —G
Alloy G/G3	G
Alloy D, to 30%	G

SODIUM HYDROXIDE 10%

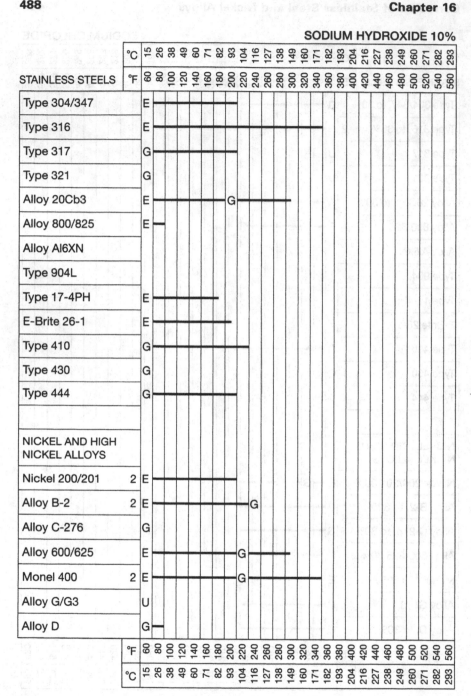

| STAINLESS STEELS | | °C | 15 | 26 | 38 | 49 | 60 | 71 | 82 | 93 | 104 | 116 | 127 | 138 | 149 | 160 | 171 | 182 | 193 | 204 | 216 | 227 | 238 | 249 | 260 | 271 | 282 | 293 |
| | | °F | 60 | 80 | 100 | 120 | 140 | 160 | 180 | 200 | 220 | 240 | 260 | 280 | 300 | 320 | 340 | 360 | 380 | 400 | 420 | 440 | 460 | 480 | 500 | 520 | 540 | 560 |

SODIUM HYDROXIDE 15%

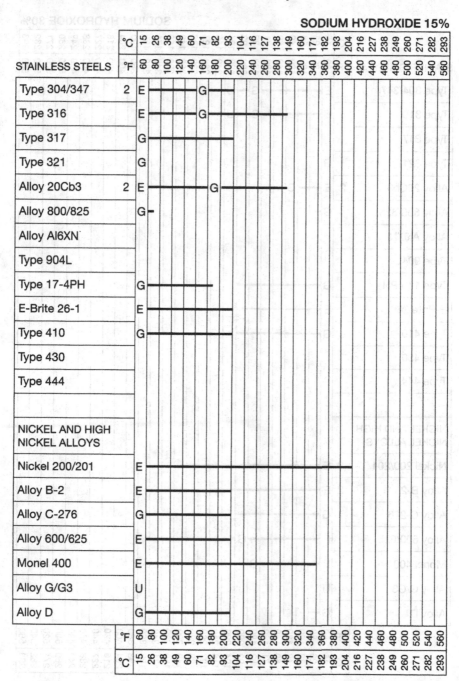

STAINLESS STEELS	°C	15	26	38	49	60	71	82	93	104	116	127	138	149	160	171	182	193	204	216	227	238	249	260	271	282	293
	°F	60	80	100	120	140	160	180	200	220	240	260	280	300	320	340	360	380	400	420	440	460	480	500	520	540	560
Type 304/347	2	E			G																						
Type 316		E			G																						
Type 317		G																									
Type 321		G																									
Alloy 20Cb3	2	E				G																					
Alloy 800/825		G																									
Alloy Al6XN																											
Type 904L																											
Type 17-4PH		G																									
E-Brite 26-1		E																									
Type 410		G																									
Type 430																											
Type 444																											
NICKEL AND HIGH NICKEL ALLOYS																											
Nickel 200/201		E																									
Alloy B-2		E																									
Alloy C-276		G																									
Alloy 600/625		E																									
Monel 400		E																									
Alloy G/G3		U																									
Alloy D		G																									
	°F	60	80	100	120	140	160	180	200	220	240	260	280	300	320	340	360	380	400	420	440	460	480	500	520	540	560
	°C	15	26	38	49	60	71	82	93	104	116	127	138	149	160	171	182	193	204	216	227	238	249	260	271	282	293

SODIUM HYDROXIDE 30%

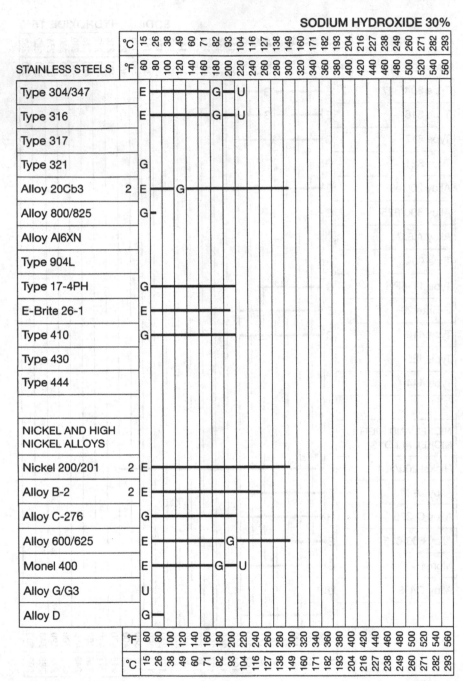

STAINLESS STEELS	°C	15 26 38 49 60 71 82 93 104 116 127 138 149 160 171 182 193 204 216 227 238 249 260 271 282 293
	°F	60 80 100 120 140 160 180 200 220 240 260 280 300 320 340 360 380 400 420 440 460 480 500 520 540 560
Type 304/347		E————G—U
Type 316		E————G—U
Type 317		
Type 321		G
Alloy 20Cb3	2	E—G———————
Alloy 800/825		G—
Alloy Al6XN		
Type 904L		
Type 17-4PH		G———————
E-Brite 26-1		E———————
Type 410		G———————
Type 430		
Type 444		
NICKEL AND HIGH NICKEL ALLOYS		
Nickel 200/201	2	E——————————
Alloy B-2	2	E—————————
Alloy C-276		G—————
Alloy 600/625		E————G————
Monel 400		E————G—U
Alloy G/G3		U
Alloy D		G—
	°F	60 80 100 120 140 160 180 200 220 240 260 280 300 320 340 360 380 400 420 440 460 480 500 520 540 560
	°C	15 26 38 49 60 71 82 93 104 116 127 138 149 160 171 182 193 204 216 227 238 249 260 271 282 293

SODIUM HYDROXIDE 50%

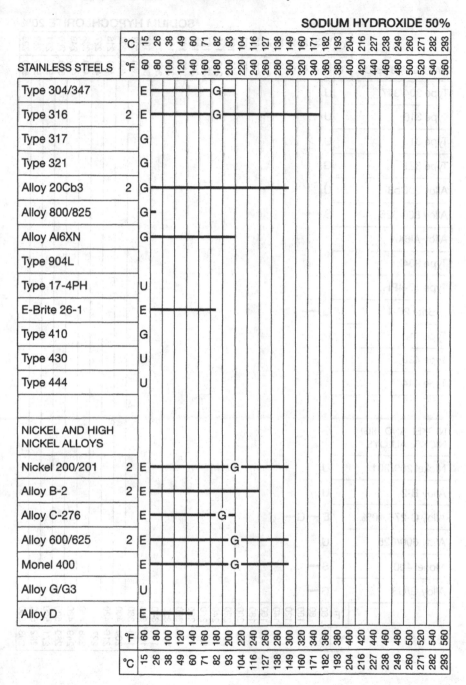

STAINLESS STEELS		°C	15 26 38 49 60 71 82 93 104 116 127 138 149 160 171 182 193 204 216 227 238 249 260 271 282 293
		°F	60 80 100 120 140 160 180 200 220 240 260 280 300 320 340 360 380 400 420 440 460 480 500 520 540 560
Type 304/347		E	—————G—
Type 316	2	E	————————G————————
Type 317		G	
Type 321		G	
Alloy 20Cb3	2	G	————————————
Alloy 800/825		G	-
Alloy Al6XN		G	————————
Type 904L			
Type 17-4PH		U	
E-Brite 26-1		E	——————
Type 410		G	
Type 430		U	
Type 444		U	
NICKEL AND HIGH NICKEL ALLOYS			
Nickel 200/201	2	E	——————G———————
Alloy B-2	2	E	———————————
Alloy C-276		E	————G—
Alloy 600/625	2	E	——————G———————
Monel 400		E	——————G———————
Alloy G/G3		U	
Alloy D		E	————

	°F	60 80 100 120 140 160 180 200 220 240 260 280 300 320 340 360 380 400 420 440 460 480 500 520 540 560
	°C	15 26 38 49 60 71 82 93 104 116 127 138 149 160 171 182 193 204 216 227 238 249 260 271 282 293

SODIUM HYPOCHLORITE 20%

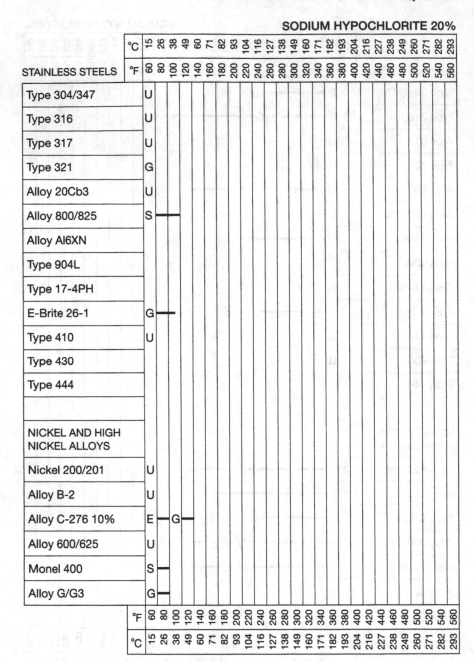

STAINLESS STEELS		°C	15	26	38	49	60	71	82	93	104	116	127	138	149	160	171	182	193	204	216	227	238	249	260	271	282	293
		°F	60	80	100	120	140	160	180	200	220	240	260	280	300	320	340	360	380	400	420	440	460	480	500	520	540	560
Type 304/347	U																											
Type 316	U																											
Type 317	U																											
Type 321	G																											
Alloy 20Cb3	U																											
Alloy 800/825	S																											
Alloy Al6XN																												
Type 904L																												
Type 17-4PH																												
E-Brite 26-1	G																											
Type 410	U																											
Type 430																												
Type 444																												
NICKEL AND HIGH NICKEL ALLOYS																												
Nickel 200/201	U																											
Alloy B-2	U																											
Alloy C-276 10%	E	G																										
Alloy 600/625	U																											
Monel 400	S																											
Alloy G/G3	G																											

	°F	60	80	100	120	140	160	180	200	220	240	260	280	300	320	340	360	380	400	420	440	460	480	500	520	540	560
	°C	15	26	38	49	60	71	82	93	104	116	127	138	149	160	171	182	193	204	216	227	238	249	260	271	282	293

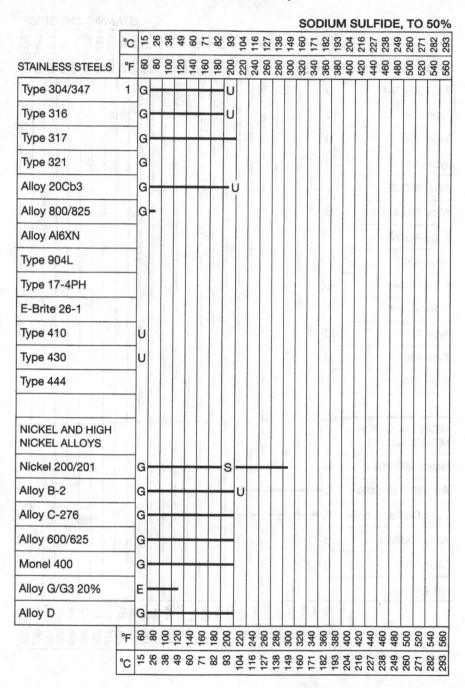

SODIUM SULFIDE, TO 50%

STANNIC CHLORIDE

STAINLESS STEELS	°C	15	26	38	49	60	71	82	93	104	116	127	138	149	160	171	182	193	204	216	227	238	249	260	271	282	293
	°F	60	80	100	120	140	160	180	200	220	240	260	280	300	320	340	360	380	400	420	440	460	480	500	520	540	560
Type 304/347	U																										
Type 316	U																										
Type 317	U																										
Type 321	U																										
Alloy 20Cb3	U																										
Alloy 800/825	U																										
Alloy Al6XN																											
Type 904L																											
Type 17-4PH	U																										
E–Brite 26-1																											
Type 410	U																										
Type 430	U																										
Type 444																											
NICKEL AND HIGH NICKEL ALLOYS																											
Nickel 200/201	U																										
Alloy B-2 10–50%	G																										
Alloy C-276, to 50%	G																										
Alloy 600/625	U																										
Monel 400	U																										
Alloy G/G3																											
	°F	60	80	100	120	140	160	180	200	220	240	260	280	300	320	340	360	380	400	420	440	460	480	500	520	540	560
	°C	15	26	38	49	60	71	82	93	104	116	127	138	149	160	171	182	193	204	216	227	238	249	260	271	282	293

STANNOUS CHLORIDE

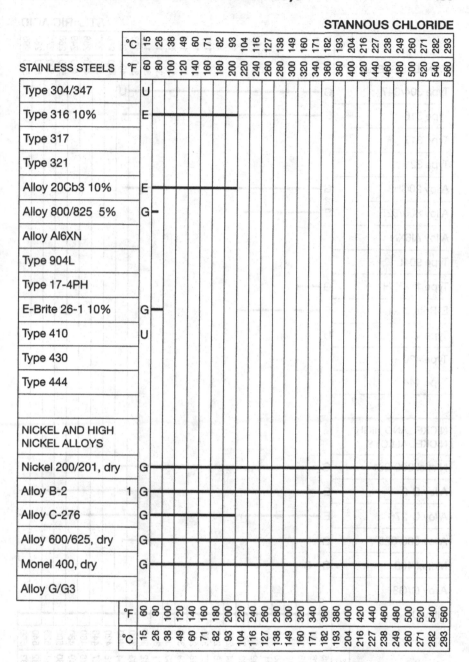

STAINLESS STEELS	°C 15 26 38 49 60 71 82 93 104 116 127 138 149 160 171 182 193 204 216 227 238 249 260 271 282 293 / °F 60 80 100 120 140 160 180 200 220 240 260 280 300 320 340 360 380 400 420 440 460 480 500 520 540 560
Type 304/347	U
Type 316 10%	E————————————
Type 317	
Type 321	
Alloy 20Cb3 10%	E————————————
Alloy 800/825 5%	G—
Alloy Al6XN	
Type 904L	
Type 17-4PH	
E-Brite 26-1 10%	G—
Type 410	U
Type 430	
Type 444	
NICKEL AND HIGH NICKEL ALLOYS	
Nickel 200/201, dry	G————————————————————————————
Alloy B-2 1	G————————————————————————————
Alloy C-276	G————————————
Alloy 600/625, dry	G————————————————————————————
Monel 400, dry	G————————————————————————————
Alloy G/G3	

STEARIC ACID

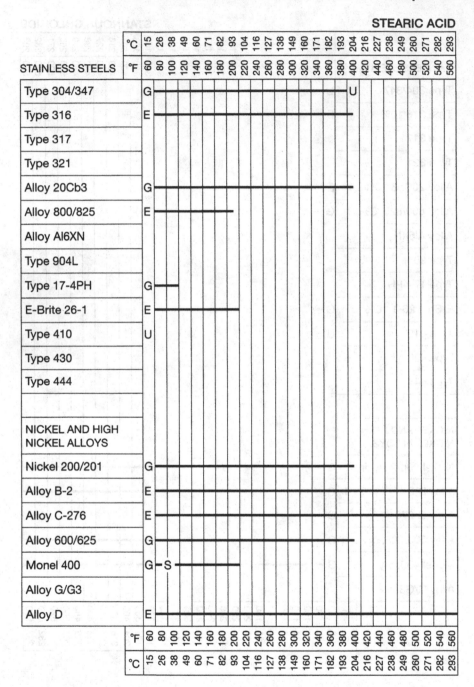

	°C	15	26	38	49	60	71	82	93	104	116	127	138	149	160	171	182	193	204	216	227	238	249	260	271	282	293
STAINLESS STEELS	°F	60	80	100	120	140	160	180	200	220	240	260	280	300	320	340	360	380	400	420	440	460	480	500	520	540	560
Type 304/347	G																		U								
Type 316	E																										
Type 317																											
Type 321																											
Alloy 20Cb3	G																										
Alloy 800/825	E																										
Alloy AI6XN																											
Type 904L																											
Type 17-4PH	G																										
E-Brite 26-1	E																										
Type 410	U																										
Type 430																											
Type 444																											
NICKEL AND HIGH NICKEL ALLOYS																											
Nickel 200/201	G																										
Alloy B-2	E																										
Alloy C-276	E																										
Alloy 600/625	G																										
Monel 400	G–S																										
Alloy G/G3																											
Alloy D	E																										
	°F	60	80	100	120	140	160	180	200	220	240	260	280	300	320	340	360	380	400	420	440	460	480	500	520	540	560
	°C	15	26	38	49	60	71	82	93	104	116	127	138	149	160	171	182	193	204	216	227	238	249	260	271	282	293

SULFAMIC ACID

STAINLESS STEELS	°C	15	26	38	49	60	71	82	93	104	116	127	138	149	160	171	182	193	204	216	227	238	249	260	271	282	293
	°F	60	80	100	120	140	160	180	200	220	240	260	280	300	320	340	360	380	400	420	440	460	480	500	520	540	560
Type 304/347																											
Type 316 20%		G——————																									
Type 317																											
Type 321																											
Alloy 20Cb3 20%		G——————																									
Alloy 800/825																											
Alloy Al6XN 10%		G————————————																									
Type 904L																											
Type 17-4PH																											
E-Brite 26-1		E———																									
Type 410																											
Type 430																											
Type 444																											
NICKEL AND HIGH NICKEL ALLOYS																											
Nickel 200/201																											
Alloy B-2																											
Alloy C-276																											
Alloy 600/625																											
Monel 400																											
Alloy G/G3																											
	°F	60	80	100	120	140	160	180	200	220	240	260	280	300	320	340	360	380	400	420	440	460	480	500	520	540	560
	°C	15	26	38	49	60	71	82	93	104	116	127	138	149	160	171	182	193	204	216	227	238	249	260	271	282	293

SULFUR DIOXIDE, WET

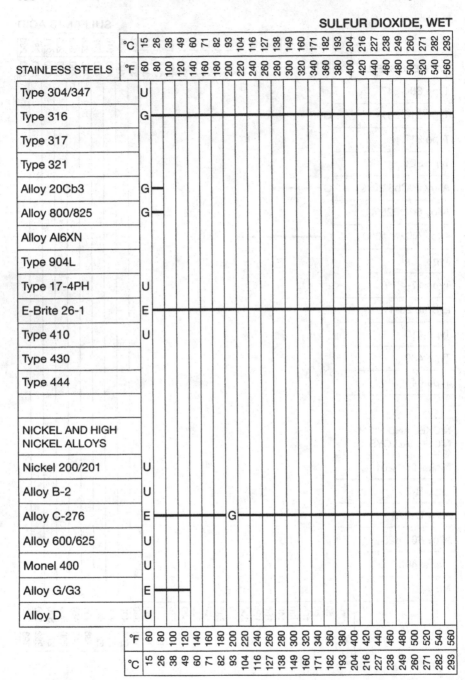

SULFURIC ACID 10%

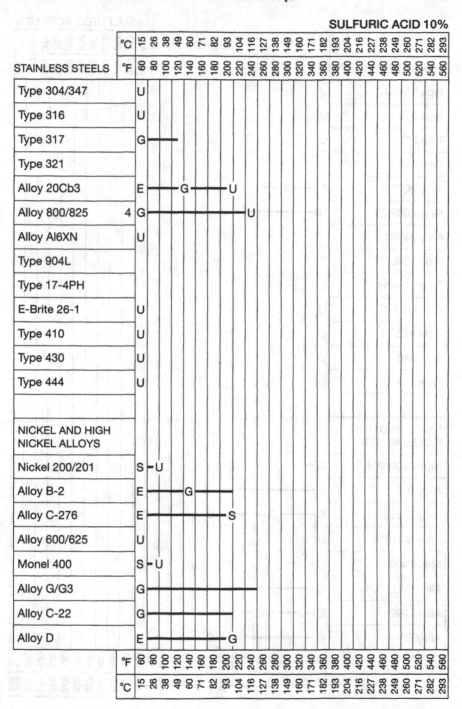

STAINLESS STEELS	°C	15	26	38	49	60	71	82	93	104	116	127	138	149	160	171	182	193	204	216	227	238	249	260	271	282	293
	°F	60	80	100	120	140	160	180	200	220	240	260	280	300	320	340	360	380	400	420	440	460	480	500	520	540	560
Type 304/347		U																									
Type 316		U																									
Type 317		G	——																								
Type 321																											
Alloy 20Cb3		E	——	G	——	U																					
Alloy 800/825	4	G	——————					U																			
Alloy AI6XN		U																									
Type 904L																											
Type 17-4PH																											
E-Brite 26-1		U																									
Type 410		U																									
Type 430		U																									
Type 444		U																									
NICKEL AND HIGH NICKEL ALLOYS																											
Nickel 200/201		S	U																								
Alloy B-2		E	——	G	——																						
Alloy C-276		E	——————	S																						
Alloy 600/625		U																									
Monel 400		S	U																								
Alloy G/G3		G	————————																							
Alloy C-22		G	——————																								
Alloy D		E	——————	G																						
	°F	60	80	100	120	140	160	180	200	220	240	260	280	300	320	340	360	380	400	420	440	460	480	500	520	540	560
	°C	15	26	38	49	60	71	82	93	104	116	127	138	149	160	171	182	193	204	216	227	238	249	260	271	282	293

SULFURIC ACID 50%

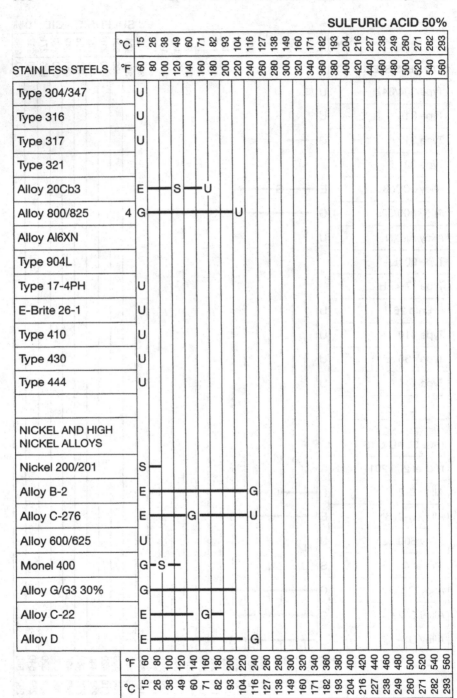

STAINLESS STEELS	°C	15 26 38 49 60 71 82 93 104 116 127 138 149 160 171 182 193 204 216 227 238 249 260 271 282 293
	°F	60 80 100 120 140 160 180 200 220 240 260 280 300 320 340 360 380 400 420 440 460 480 500 520 540 560
Type 304/347		U
Type 316		U
Type 317		U
Type 321		
Alloy 20Cb3		E—S—U
Alloy 800/825	4	G———————U
Alloy Al6XN		
Type 904L		
Type 17-4PH		U
E-Brite 26-1		U
Type 410		U
Type 430		U
Type 444		U
NICKEL AND HIGH NICKEL ALLOYS		
Nickel 200/201		S—
Alloy B-2		E——————G
Alloy C-276		E———G————U
Alloy 600/625		U
Monel 400		G—S—
Alloy G/G3 30%		G——————
Alloy C-22		E————G—
Alloy D		E—————G
	°F	60 80 100 120 140 160 180 200 220 240 260 280 300 320 340 360 380 400 420 440 460 480 500 520 540 560
	°C	15 26 38 49 60 71 82 93 104 116 127 138 149 160 171 182 193 204 216 227 238 249 260 271 282 293

SULFURIC ACID 70%

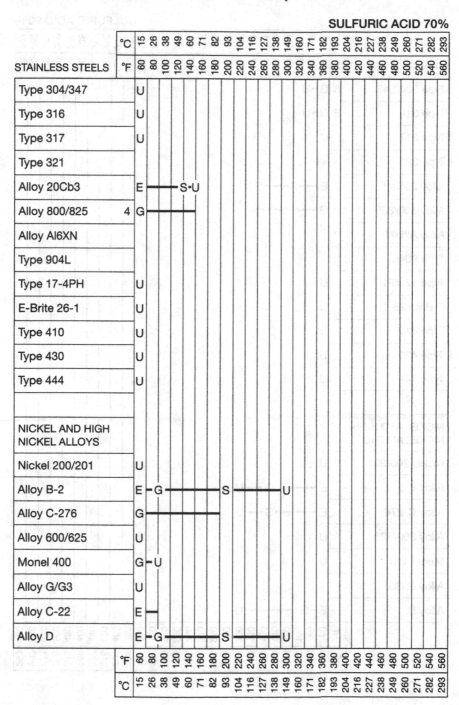

STAINLESS STEELS	°C	15	26	38	49	60	71	82	93	104	116	127	138	149	160	171	182	193	204	216	227	238	249	260	271	282	293
	°F	60	80	100	120	140	160	180	200	220	240	260	280	300	320	340	360	380	400	420	440	460	480	500	520	540	560
Type 304/347	U																										
Type 316	U																										
Type 317	U																										
Type 321																											
Alloy 20Cb3	E ——— S-U																										
Alloy 800/825 4	G ———																										
Alloy Al6XN																											
Type 904L																											
Type 17-4PH	U																										
E-Brite 26-1	U																										
Type 410	U																										
Type 430	U																										
Type 444	U																										
NICKEL AND HIGH NICKEL ALLOYS																											
Nickel 200/201	U																										
Alloy B-2	E -G ————— S ————— U																										
Alloy C-276	G —————																										
Alloy 600/625	U																										
Monel 400	G -U																										
Alloy G/G3	U																										
Alloy C-22	E —																										
Alloy D	E -G ————— S ————— U																										
	°F	60	80	100	120	140	160	180	200	220	240	260	280	300	320	340	360	380	400	420	440	460	480	500	520	540	560
	°C	15	26	38	49	60	71	82	93	104	116	127	138	149	160	171	182	193	204	216	227	238	249	260	271	282	293

SULFURIC ACID 90%

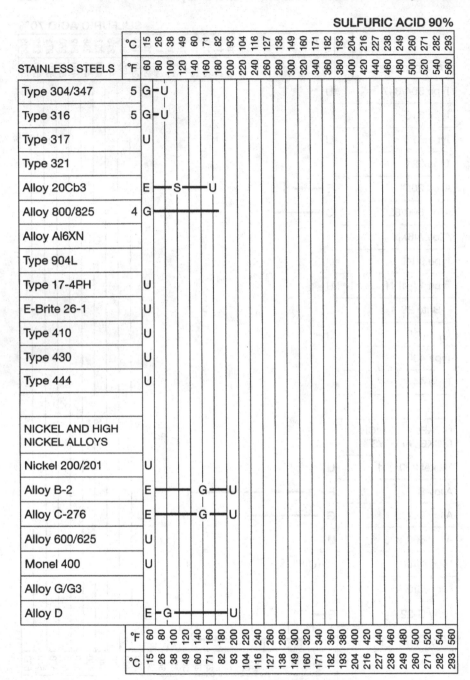

STAINLESS STEELS	°C	15	26	38	49	60	71	82	93	104	116	127	138	149	160	171	182	193	204	216	227	238	249	260	271	282	293
	°F	60	80	100	120	140	160	180	200	220	240	260	280	300	320	340	360	380	400	420	440	460	480	500	520	540	560
Type 304/347	5	G–U																									
Type 316	5	G–U																									
Type 317		U																									
Type 321																											
Alloy 20Cb3		E — S — U																									
Alloy 800/825	4	G ———————																									
Alloy AI6XN																											
Type 904L																											
Type 17-4PH		U																									
E-Brite 26-1		U																									
Type 410		U																									
Type 430		U																									
Type 444		U																									
NICKEL AND HIGH NICKEL ALLOYS																											
Nickel 200/201		U																									
Alloy B-2		E ——— G — U																									
Alloy C-276		E ——— G — U																									
Alloy 600/625		U																									
Monel 400		U																									
Alloy G/G3																											
Alloy D		E – G ——— U																									
	°F	60	80	100	120	140	160	180	200	220	240	260	280	300	320	340	360	380	400	420	440	460	480	500	520	540	560
	°C	15	26	38	49	60	71	82	93	104	116	127	138	149	160	171	182	193	204	216	227	238	249	260	271	282	293

SULPHURIC ACID 95%

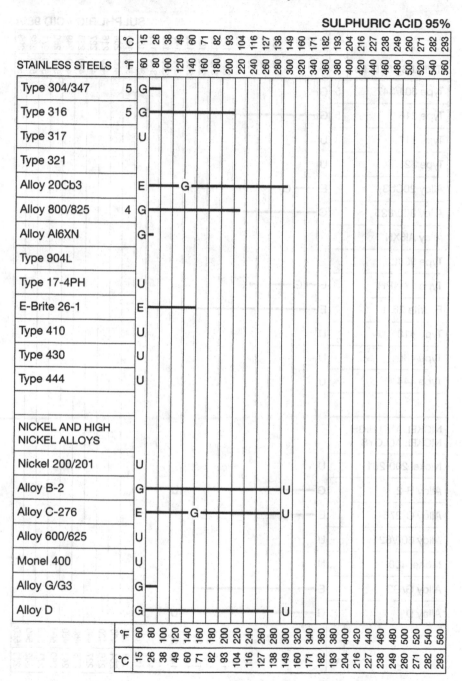

STAINLESS STEELS		°C	15	26	38	49	60	71	82	93	104	116	127	138	149	160	171	182	193	204	216	227	238	249	260	271	282	293
		°F	60	80	100	120	140	160	180	200	220	240	260	280	300	320	340	360	380	400	420	440	460	480	500	520	540	560
Type 304/347	5	G																										
Type 316	5	G																										
Type 317		U																										
Type 321																												
Alloy 20Cb3		E			G																							
Alloy 800/825	4	G																										
Alloy Al6XN		G																										
Type 904L																												
Type 17-4PH		U																										
E-Brite 26-1		E																										
Type 410		U																										
Type 430		U																										
Type 444		U																										
NICKEL AND HIGH NICKEL ALLOYS																												
Nickel 200/201		U																										
Alloy B-2		G											U															
Alloy C-276		E			G								U															
Alloy 600/625		U																										
Monel 400		U																										
Alloy G/G3		G																										
Alloy D		G											U															

	°F	60	80	100	120	140	160	180	200	220	240	260	280	300	320	340	360	380	400	420	440	460	480	500	520	540	560
	°C	15	26	38	49	60	71	82	93	104	116	127	138	149	160	171	182	193	204	216	227	238	249	260	271	282	293

SULPHURIC ACID 98%

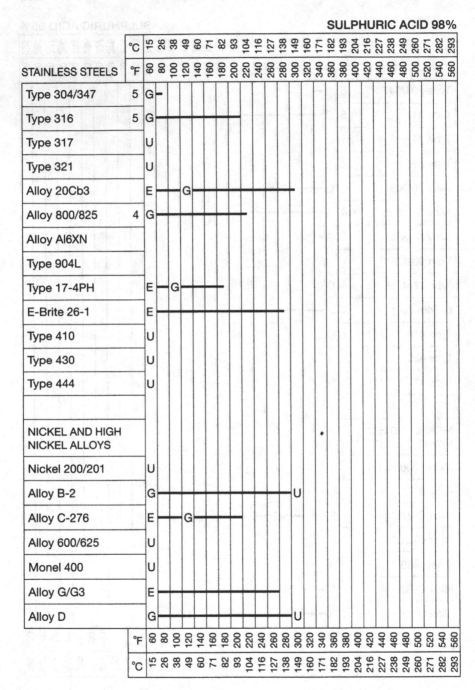

STAINLESS STEELS	°C	15	26	38	49	60	71	82	93	104	116	127	138	149	160	171	182	193	204	216	227	238	249	260	271	282	293
	°F	60	80	100	120	140	160	180	200	220	240	260	280	300	320	340	360	380	400	420	440	460	480	500	520	540	560
Type 304/347	5	G
Type 316	5	G
Type 317	U
Type 321	U
Alloy 20Cb3	E
Alloy 800/825	4	G
Alloy Al6XN
Type 904L
Type 17-4PH	E
E-Brite 26-1	E
Type 410	U
Type 430	U
Type 444	U
NICKEL AND HIGH NICKEL ALLOYS
Nickel 200/201	U
Alloy B-2	G
Alloy C-276	E
Alloy 600/625	U
Monel 400	U
Alloy G/G3	E
Alloy D	G

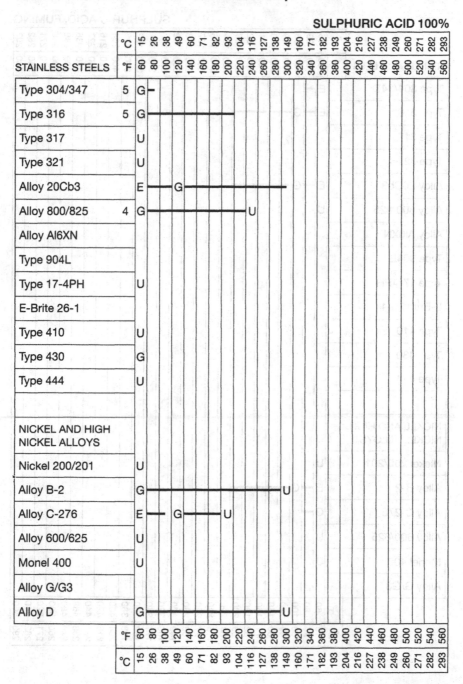

SULPHURIC ACID 100%

SULPHURIC ACID, FUMING

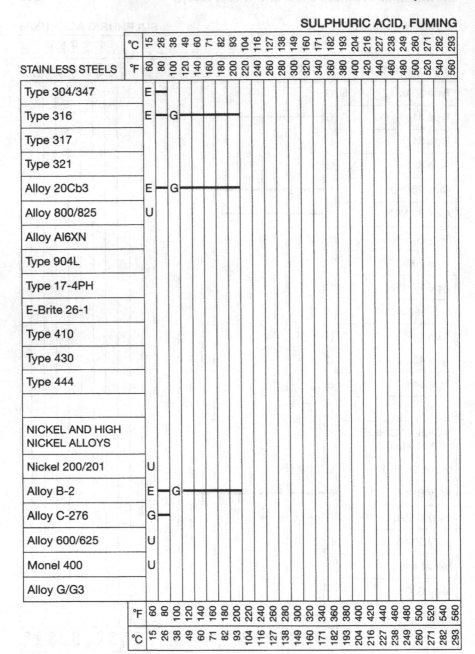

STAINLESS STEELS	°F	60	80	100	120	140	160	180	200	220	240	260	280	300	320	340	360	380	400	420	440	460	480	500	520	540	560
	°C	15	26	38	49	60	71	82	93	104	116	127	138	149	160	171	182	193	204	216	227	238	249	260	271	282	293
Type 304/347		E																									
Type 316		E		G																							
Type 317																											
Type 321																											
Alloy 20Cb3		E		G																							
Alloy 800/825		U																									
Alloy Al6XN																											
Type 904L																											
Type 17-4PH																											
E-Brite 26-1																											
Type 410																											
Type 430																											
Type 444																											
NICKEL AND HIGH NICKEL ALLOYS																											
Nickel 200/201		U																									
Alloy B-2		E		G																							
Alloy C-276		G																									
Alloy 600/625		U																									
Monel 400		U																									
Alloy G/G3																											

	°F	60	80	100	120	140	160	180	200	220	240	260	280	300	320	340	360	380	400	420	440	460	480	500	520	540	560
	°C	15	26	38	49	60	71	82	93	104	116	127	138	149	160	171	182	193	204	216	227	238	249	260	271	282	293

TARTARIC ACID

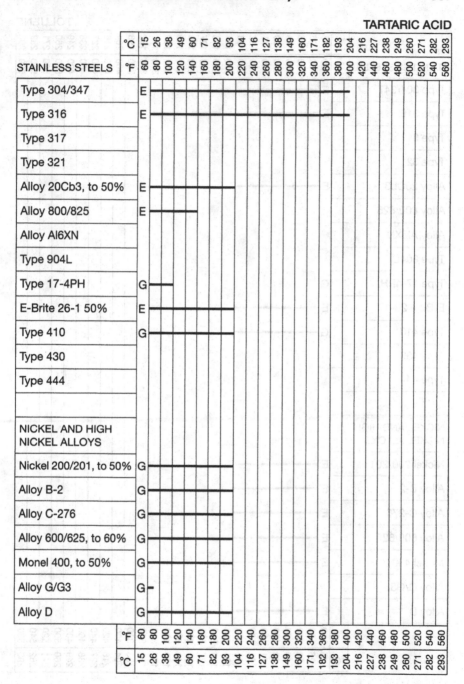

STAINLESS STEELS	
Type 304/347	E
Type 316	E
Type 317	
Type 321	
Alloy 20Cb3, to 50%	E
Alloy 800/825	E
Alloy Al6XN	
Type 904L	
Type 17-4PH	G
E-Brite 26-1 50%	E
Type 410	G
Type 430	
Type 444	
NICKEL AND HIGH NICKEL ALLOYS	
Nickel 200/201, to 50%	G
Alloy B-2	G
Alloy C-276	G
Alloy 600/625, to 60%	G
Monel 400, to 50%	G
Alloy G/G3	G
Alloy D	G

TOLUENE

STAINLESS STEELS	°C	15	26	38	49	60	71	82	93	104	116	127	138	149	160	171	182	193	204	216	227	238	249	260	271	282	293
	°F	60	80	100	120	140	160	180	200	220	240	260	280	300	320	340	360	380	400	420	440	460	480	500	520	540	560

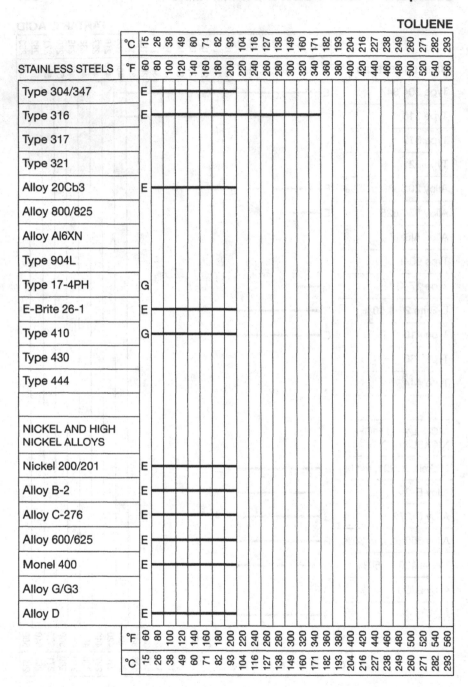

| | | °C | 15 | 26 | 38 | 49 | 60 | 71 | 82 | 93 | 104 | 116 | 127 | 138 | 149 | 160 | 171 | 182 | 193 | 204 | 216 | 227 | 238 | 249 | 260 | 271 | 282 | 293 |

Rows (left labels):
- Type 304/347 — E
- Type 316 — E
- Type 317
- Type 321
- Alloy 20Cb3 — E
- Alloy 800/825
- Alloy Al6XN
- Type 904L
- Type 17-4PH — G
- E-Brite 26-1 — E
- Type 410 — G
- Type 430
- Type 444

NICKEL AND HIGH NICKEL ALLOYS
- Nickel 200/201 — E
- Alloy B-2 — E
- Alloy C-276 — E
- Alloy 600/625 — E
- Monel 400 — E
- Alloy G/G3
- Alloy D — E

TRICHLORACETIC ACID

STAINLESS STEELS	°C: 15 26 38 49 60 71 82 93 104 116 127 138 149 160 171 182 193 204 216 227 238 249 260 271 282 293 / °F: 60 80 100 120 140 160 180 200 220 240 260 280 300 320 340 360 380 400 420 440 460 480 500 520 540 560
Type 304/347	U
Type 316	U
Type 317	U
Type 321	U
Alloy 20Cb3	
Alloy 800/825	
Alloy Al6XN	
Type 904L	
Type 17-4PH	
E-Brite 26-1	
Type 410	U
Type 430	U
Type 444	
NICKEL AND HIGH NICKEL ALLOYS	
Nickel 200/201	G—
Alloy B-2	G———————————
Alloy C-276	G———————————
Alloy 600/625	G—
Monel 400	G—————
Alloy G/G3	
Alloy D	G—————

ZINC CHLORIDE

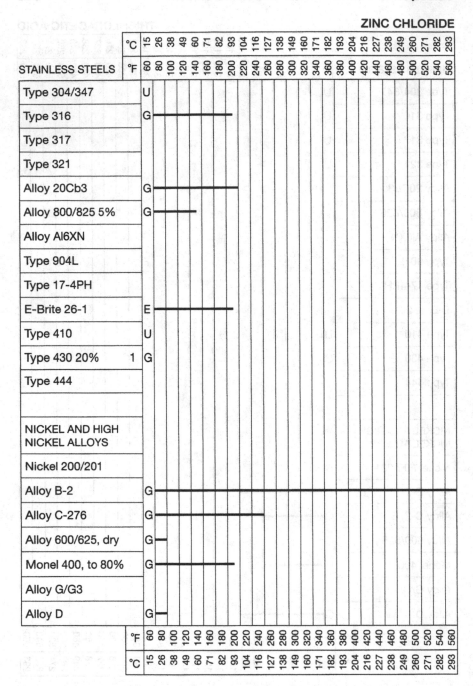

STAINLESS STEELS		°C	15	26	38	49	60	71	82	93	104	116	127	138	149	160	171	182	193	204	216	227	238	249	260	271	282	293
		°F	60	80	100	120	140	160	180	200	220	240	260	280	300	320	340	360	380	400	420	440	460	480	500	520	540	560
Type 304/347		U																										
Type 316		G																										
Type 317																												
Type 321																												
Alloy 20Cb3		G																										
Alloy 800/825 5%		G																										
Alloy Al6XN																												
Type 904L																												
Type 17-4PH																												
E-Brite 26-1		E																										
Type 410		U																										
Type 430 20%	1	G																										
Type 444																												
NICKEL AND HIGH NICKEL ALLOYS																												
Nickel 200/201																												
Alloy B-2		G																										
Alloy C-276		G																										
Alloy 600/625, dry		G																										
Monel 400, to 80%		G																										
Alloy G/G3																												
Alloy D		G																										
		°F	60	80	100	120	140	160	180	200	220	240	260	280	300	320	340	360	380	400	420	440	460	480	500	520	540	560
		°C	15	26	38	49	60	71	82	93	104	116	127	138	149	160	171	182	193	204	216	227	238	249	260	271	282	293

17

Cast Stainless Steel and Nickel Base Alloys

Most wrought alloy compositions are also available in an equivalent grade casting. In addition there are many alloy castings available in compositions that are not available as wrought materials. This is the result of the design freedom associated with the various casting processes. The compositions of many cast grades are modified relative to their wrought equivalents to take advantage of the casting process because little or no mechanical working of the cast component will be necessary. Because of this ability to modify the alloy compositions of the cast grades, improved and unique properties may be imparted to some cast alloys compared to their wrought equivalents. Various alloy designation systems have been developed to separate the wrought alloys from the cast alloys because of the variations in compositions and the resulting physical and mechanical properties. Three designation systems are presently used for the identification of alloys: Unified Numbering System (UNS), Alloy Casting Institute (ACI), and American Society for Testing and Materials (ASTM).

The unified numbering system was developed by the Society of Automotive Engineers (SAE) and ASTM. Under the UNS system metals and alloys are divided into 18 series. The designations start with a single letter followed by five numerical digits. As much as possible the letter is suggestive of the family of metals it identifies, for example, A is for aluminum alloys, C is for copper alloys, N is for nickel alloys, and S is for stainless alloys. A complete listing of the letters used will be found in Table 17.1. If possible, common designations are used within the five numerical digits for user convenience, for example, A92024 is for 2024 aluminum, C36000 for copper alloy 360, and S31600 for type 316 stainless steel.

TABLE 17.1 Letter Prefixes Used in the Unified Numbering
System

Prefix	Alloy series
A	Aluminum and aluminum alloys
C	Copper and copper alloys
D	Steels with specified mechanical properties
E	Rare earths and rare earth-like metals and alloys
F	Cast irons
G	Carbon and alloy steels
H	AISI H-steels (hardenability controlled)
J	Cast steels (except tool steels)
K	Miscellaneous steels and ferrous alloys
L	Low-melting metals and alloys
M	Miscellaneous nonferrous metals and alloys
N	Nickel and nickel alloys
P	Precious metals and alloys
R	Reactive and refractory metals and alloys
S	Heat and corrosion resistant (stainless) steels
T	Tool steels
W	Welding filler materials
Z	Zinc and zinc alloys

The UNS system is more commonly used for wrought materials. ACI designations are more appropriate for cast alloys since the designations are more indicative of the compositions. Most ACI designations begin with two letters followed by two or three numerical digits. Some may also end with additional letters and/or numerical digits. In general the designations begin with either a C for corrosion resistant materials or an H for heat resistant materials. The second letter in the designation ranges from A to Z, depending upon the nickel and, to a lesser degree, the chromium content. For example, a corrosion resistant material with 12% chromium and no nickel begins with CA; an alloy with 100% nickel begins with CZ; alloys inbetween have intermediate letters. The maximum carbon content is indicated by the numerical digits (percent × 100). Additional letters following the numerical digits indicate the presence of other alloying ingredients. Table 17.2 gives examples.

There are two groups of materials whose designations do not follow the scheme shown in Table 17.2. Nickel-copper materials use M as the first letter (examples are M35-1 and M25S). Nickel-molybdenum alloys begin with the letter N, such as N7M and N12MV. Although the Alloy Casting

TABLE 17.2 Examples of ACI Designations

| Designation | Alloying elements (wt%) | | | |
	Chromium, nominal	Nickel, nominal	Carbon, max.	Other alloying elements, nominal
CA 15	12	—	0.15	—
CD 4MCu	25	6	0.04	Mo 3, Cu 3
CF 8M	19	10	0.08	Mo 2.5
CF 3M	19	10	0.03	Mo 2.5
CN 7M	21	29	0.07	Mo 2.5
CW 2M	16	68	0.02	Mo 16
CZ 100	0	100	1.0	—
HK 40	25	20	0.40	—

Institute is no longer in existance, the system has been adopted by ASTM and appropriate ASTM committees assign designations for new cast alloys.

In addition to the UNS designations previously described there is also a series of UNS designations specifically for cast materials. Table 17.3 lists several alloys giving both ACI and UNS designations.

TABLE 17.3 ACI and UNS Designations for Stainless Steel Castings (nominal weight percent)

ACI	UNS	Cr	Ni (max.)	Other
CA[a]	—	12	1	C
CB[a]	—	19	2	C
CD 4MCu	J93370	26	6	C 0.04 max., Mo 2, Cu 3
CF 8	J92600	19	11	C 0.08 max.
CF 3	J92500	18	12	C 0.03 max.
CF 8M	J92900	19	12	C 0.08 max., Mo 2
CF 3M	J92800	18	13	C 0.03 max., Mo 2
CH 20	J93402	23	15	C 0.20 max.
CK 20	J94202	24	22	C 0.20 max.
CN 7M	N08007	20	21	C 0.07 max., Mo 2, Cu 3
HK[b]	—	24	20	C 0.60 max.

[a]Maximum carbon.
[b]±0.05% carbon.

ASTM designations are used for many special carbon and alloy steel products and for cast iron.

Castings have several advantages over wrought materials. Among the advantages are

1. Unlimited freedom on design configuration
2. Minimization or elimination of machining and material waste
3. Wide range of alloy choice
4. Mechanical property isotropy
5. Production economies

However, there are limitations—the most serious of which is variations in quality from casting to casting and foundry to foundry. Potential quality shortcomings involve

1. Surface finish
2. Compositional purity
3. Internal integrity
4. Dimensional control

These limitations can be overcome through application of sound foundry practices.

It should be kept in mind that specifications for castings should be based on ACI designations. A specification such as cast type 316 stainless steel should never be used because a foundry might pour just that from bar stock (to meet your specification) without regard to the proper balance of constituents.

I. STAINLESS STEELS

Iron-based alloys containing at least 11.5% chromium are referred to as stainless steels. This level of chromium is necessary to produce passivity. Cast stainless steels may be had in all grades comparable with the wrought grades plus many additional grades for special end-use applications. Cast alloys can be produced with improvement in specific properties, but the composition cannot be produced in the wrought form. Some cast alloys have high silicon and/or carbon contents for superior corrosion or abrasion resistance, but the low ductility and high strength may make rolling and/or forging impossible.

The cast stainless steel alloys will be discussed by microstructure.

A. Martensitic Alloys

The chemical composition of typical cast martensitic stainless steel alloys will be found in Table 17.4 and their mechanical properties in Table 17.5.

TABLE 17.4 Chemical Composition of Cast Martensitic Stainless Alloys

	Alloy (wt%)				
Chemical	CA 6NM	CA 15	CA 15M	CA 28MWV	CA 40
Carbon	0.06	0.05	0.15	0.2–0.28	0.20–0.40
Manganese	1.00	1.00	1.00	—	1.00
Silicon	1.00	1.50	0.65	—	1.50
Phosphorus	0.04	0.04	0.04	—	0.04
Sulfur	0.03	0.04	0.04	—	0.04
Chromium	11.5–14.0	11.5–14.0	11.5–14.0	11.0–12.5	11.5–14.0
Nickel	3.5–4.5	1.00	1.00	—	1.0
Molybdenum	0.40–1.0	0.50	0.15–1.0	0.9–1.25	0.5
Tungsten	—	—	—	0.9–1.25	—
Vanadium	—	—	—	0.2–0.3	—
Iron	Balance	Balance	Balance	Balance	Balance

Maximum unless otherwise indicated.

TABLE 17.5 Mechanical Properties of Cast Martensitic Stainless
Steel Alloys

	Alloy				
Property	Ca 6NM	CA 15	CA 15M	CA 28MWV	CA 40
Wrought equivalent	410, modified	410	—	422	420
ASTM designation	A743	A743	A743, A217	A743	A743
Tensile strength (ksi)	110	90	90	140	100
Yield strength (ksi)	80	65	65	110	70
Brinell hardness	240	210	220	302–352	240
Elongation (%)	15	18	18	—	15
Impact toughness (ft-lb)	70	20	—	—	2

Alloy CA-15 contains the minimum amount of chromium required to make it a rust-proof alloy. It exhibits good resistance to atmospheric corrosion and finds applications in mildly corrosive organic services. Because the alloy is martensitic it is used in some abrasive applications. Specific areas of application include alkaline liquids, ammonia water, boiler feedwater, pulp, steam, and food products.

1. Mechanical Properties

Alloy CA-40 is a higher carbon version of CA-15. The higher carbon content permits heat treatment to higher strength and hardness levels. Corrosion resistance is comparable to that of alloy CA-15, and this alloy finds use in similar applications. The addition of molybdenum forms alloy CA-15M, which has improved elevated-temperature resistance over alloy CA-15.

CA-40 is a higher carbon version of alloy CA-15. Because of the higher carbon content this alloy may be heat treated to provide higher strength and hardness levels. Corrosion resistance is comparable to that of Alloy CA-15 and the alloy CA-40 finds use in similar applications.

CA-6NM is an iron-chromium-nickel-molybdenum alloy that is hardenable by heat treatment. The major advantages of CA-6NM over CA-15 are its increased toughness and improved foundry characteristics. Its corrosion resistance is comparable to that of CA-15, with improved corrosion resistance in seawater as a result of the molybdenum content. Typical applications include seawater, boiler feedwater, and other waters up to a temperature of 400°F (204°C). A particular application is for large turbine runners for power generation.

CA-28MWV is a modified version of the wrought alloy type 410 with improved high temperature strength.

The primary advantages of the martensitic grades are their low cost and the ability to be hardened for wear resistance. They can be heat treated similar to carbon steels to produce hardnesses varying by grade as high as Rockwell C-60.

The martenisitic grades are resistant to corrosion in mild atmospheres, water, steam, and other nonsevere environments. They will rust quickly in marine and humid industrial atmospheres and are attacked by most inorganic acids. When used at high hardness levels they are susceptible to several forms of stress corrosion cracking. Hardened martensitic grades have poor resistance to sour environments and may crack in humid industrial atmospheres. Resistance is greatly improved in the quenched and fully tempered condition (generally below Rockwell C-25), especially for CA-6NM. In general the martensitic grades are less corrosion resistant than the austenitic grades.

TABLE 17.6 Chemical Composition of Cast Ferritic Stainless Steels

Chemical	Alloy (wt%)	
	CB 30	CC 50
Carbon	0.3	0.5
Manganese	1.00	1.00
Silicon	1.50	1.50
Phosphorus	0.04	0.04
Sulfur	0.04	0.04
Chromium	18.0–21.0	26.0–30.0
Nickel	2.00	4.00
Iron	Balance	Balance

Maximum unless otherwise noted.

B. Ferritic Alloys

The ferritic stainless castings have properties much different from those of the austenitic stainless castings, some of which can be very advantageous in certain applications. The two most common cast ferritic stainless steels are CB-30 and CC-50. Their chemical composition will be found in Table 17.6.

Compared to the austenitic grades these alloys have very poor impact resistance. At elevated temperatures a brittle σ phase is formed which limits the use of most to below 650°F (343°C). The ferritic castings also have poor weldability. Mechanical properties of these alloys will be found in Table 17.7.

TABLE 17.7 Mechanical Properties of Cast Ferritic Stainless Steels

Property	Alloy	
	CB 30	CC 50
Wrought equivalent	442	446
ASTM designation	A743	A743
Tensile strength (ksi)	65	55
Yield strength (ksi)	30	—
Brinell hardness	180	200
Impact toughness (ft-lb)	2	—

CB-30 is essentially all ferritic and therefore is nonhardenable. The chromium content of CB-30 is sufficient to give this alloy much better corrosion resistance in oxidizing environments. This alloy has found application in food products, nitric acid, steam, sulfur atmospheres, and other oxidizing atmospheres at temperatures up to 400°F (204°C). It is also resistant to alkaline solutions and many inorganic chemicals.

CC-50 has a higher chromium content than CB-30, which gives it improved corrosion resistance in oxidizing media. In addition at least 2% nickel and 0.15% nitrogen are usually added to CC-50 giving it improved toughness. Applications for CC-50 include acid mine waters, sulfuric and nitric mixtures, alkaline liquors, and sulfurous liquors.

Because of the low nickel content these alloys have better resistance to stress corrosion cracking than the austenitic alloys.

C. Austenitic Alloys

The austenitic cast alloys represent the largest group of cast stainless steels both in terms of the number of compositions and the quantity of material produced. This group of alloys also illustrates the differences that can exist between the so-called cast and wrought grades.

The austenitic cast alloys are the equivalents of the wrought 300 series stainless steels. Wrought 300 series stainless steels are fully austenitic. This structure is necessary to permit the hot and cold forming operations used to produce the various wrought shapes. Since castings are produced essentially to the finished shape, it is not necessary for the cast alloys to be fully austenitic. Even though these alloys are referred to as the cast austenitic alloys, the cast compositions can be balanced such that the microstructure contains from 5 to 40% ferrite. Using Fig. 17.1 as a guide, the amount of ferrite present can be estimated from the composition. The cast equivalents of the 300 series alloys can display a magnetic response from none to quite strong. The wrought 300 series alloys contain no ferrite and in the annealed condition are nonmagnetic.

The cast alloys, because of the presence of ferrite, offer a number of advantages. The mechanical strength increases in proportion to the amount of ferrite present. Refer to Fig. 17.2. Weldability is improved, particularly in relation to resistance of fissuring. The presence of ferrite also increases the resistance of the cast alloys to stress corrosion cracking compared to the fully austenitic wrought equivalents. It is possible to specify castings with specific ferrite levels.

The high temperature service of these alloys is limited because of the presence of a continuous phase of ferrite. Avove 600°F (315°C) chromium precipitates in the ferrite phase, which embrittles the ferrite. The impact

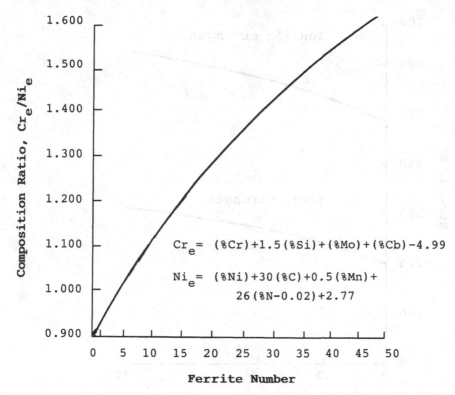

$$Cr_e = (\%Cr)+1.5(\%Si)+(\%Mo)+(\%Cb)-4.99$$

$$Ni_e = (\%Ni)+30(\%C)+0.5(\%Mn)+26(\%N-0.02)+2.77$$

FIGURE 17.1 Diagram for estimating ferrite content in cast stainless steels. (From Ref. 1.)

properties (ductility) of a casting with a continuous ferrite phase will be greatly reduced. A noncontinuous ferrite phase can be provided as long as the ferrite number is maintained below 10%.

Table 17.8 provides the chemical composition of the cast austenitic stainless steels. The CF series of cast alloys makes up the majority of the corrosion resistant casting alloys. These are 19% chromium/9% nickel materials which have particularly good mechanical properties and processing characteristics. The alloys are generally ferrite in austenite, with the composition balanced to provide 5 to 25% ferrite. Fully austenitic castings can be provided, particularly for applications requiring nonmagnetic properties. However, because of the many benefits derived from the presence of ferrite in the structure, it is usually present in the alloy. Refer to Table 17.9 for the mechanical properties of cast austenitic stainless steels.

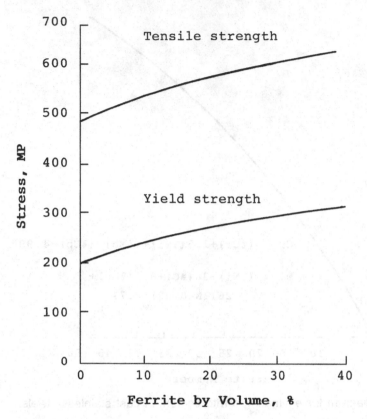

FIGURE 17.2 Relationship of yield strength and tensile strength to ferrite content for CF-8 and CF-8M. (From Ref. 1.)

These alloys are furnished in the solution-annealed condition in order to provide the maximum corrosion resistance, which is at least equal to its wrought equivalent. The CF alloys are not hardenable by heat treatment.

CF-8 is the base composition for the CF alloy group, and its wrought equivalent is type 304. The alloy has good strength and ductility and maintains its impact toughness to below −400°F (−240°C). As with wrought 304 alloy, CF-8 is resistant to strongly oxidizing media, such as boiling nitric acid. Other typical applications include adipic acid, copper sulfate, fatty acids, organic acids and liquids, sewage, sodium sulfite, sodium carbonate, vinegars, and white liquor.

CF-3 is the low carbon version of CF-8, equivalent to wrought 304L. When welding is required and postweld heat treatment is not possible CF-3 is usually specified. For the optimum corrosion resistance and mechanical

TABLE 17.8 Chemical Composition of Cast Austenitic Stainless Steels

Alloy	Chemical (wt%)								
	C	Mn	Si	P	S	Cr	Ni	Mo	Other
CE 30	0.30	1.50	2.00	0.04	0.04	26.0–30.0	8.0–11.0	—	—
CF 3	0.03	1.50	2.00	0.04	0.04	12.0–21.0	8.0–12.0	0.5	—
CF 3A	0.03	1.50	2.00	0.04	0.04	17.0–21.0	8.0–12.0	0.5	—
CF 3M	0.03	1.50	1.50	0.04	0.04	17.0–21.0	9.0–13.0	2.0–3.0	—
CF 8	0.08	1.50	2.00	0.04	0.04	18.0–21.0	8.0–11.0	0.5	—
CF 8A	0.08	1.50	2.00	0.04	0.04	18.0–21.0	8.0–11.0	0.5	—
CF 20	0.20	1.50	2.00	0.04	0.04	18.0–21.0	8.0–11.0	—	—
CF 3MA	0.03	1.50	1.50	0.04	0.04	12.0–21.0	9.0–13.0	2.0–3.0	—
CF 8M	0.08	1.50	2.00	0.04	0.04	18.0–21.0	9.0–12.0	2.0–3.0	—
CF 8C	0.08	1.50	2.00	0.04	0.04	18.0–21.0	9.0–12.0	0.5	8 × C Cb, 1.0 Cb
CF 10MC	0.10	1.50	1.50	0.04	0.04	15.0–18.0	13.0–16.0	1.75–2.25	10 × C Cb,[a] 1.2 Cb
CF 10SMnN	0.1	7–9	3.5–4.5	—	—	16.0–18.0	8.0–9.0	—	0.08–0.18 N
CF 16F	0.16	1.50	2.00	0.17	0.04	18.0–21.0	9.0–12.0	1.50	0.20–0.35 Se
CG 6MMn	0.06	4–6	—	—	—	20.5–23.5	11.5–13.5	1.5–3	0.1–0.3 Cb, 0.1–0.3 V, 0.2–0.4 N
CG 8M	0.08	1.50	1.50	0.04	0.04	18.0–21.0	9.0–13.0	3.0–4.0	—
CG 12	0.12	1.50	2.00	0.04	0.04	20.0–23.0	10.0–13.0	—	—
CH 20	0.20	1.50	2.00	0.04	0.04	22.0–26.0	12.0–15.0	0.05	—
CK 20	0.20	2.00	2.00	0.04	0.04	23.0–27.0	19.0–22.0	0.05	—

Maximum unless otherwise specified; iron balance in all cases.   a = minimum.

TABLE 17.9   Mechanical Properties of Cast Austenitic Stainless Steel

Alloy	Wrought equivalent	ASTM designation	Tensile strength (ksi)	Yield strength (ksi)	Elongation (%)	Brinell hardness	Impact toughness (ft-lb)
CE 30	312	A743	80	40	10	190	7
CF 3	304L	A351, A743, A744	70	30	35	150	110
CF 3A	—	A351	77	35	35	160	100
CF 3M	316L	A351, A743, A744	70	30	30	150	120
CF 8	304	A351, A743, A744	70	30	35	150	70
CF 8A	—	A351	77	35	35	160	70
CF 20	302	A743	70	30	30	150	60
CF 3MA	—	A351	80	37	30	170	100
CF 8M	316	A351, A743, A744	70	30	30	160	70
CF 8C	347	A351, A743, A744	70	30	30	150	30
CF 10MC	—	A351	70	30	20	150	—
CF 10SMnN	Nitronic 60	A351, A743	85	42	—	—	—
CF 16F	303	A743	70	30	25	150	75
CG 6MMN	Nitronic 50	A351, A743	85	42	—	—	—
CG 8M	317	A743, A744	75	35	25	160	80
CG 12	—	A743	70	28	35	160	—
CH 20	309	A351, A743	70	30	30	160	30
CK 20	310	A351, A743	65	28	30	140	50

properties the casting should be solution annealed, although because of the low carbon content some castings have been used in the as-cast condition. The overall corrosion resistance of CF-3 is somewhat better than that of CF-8 but in general they are used in the same applications.

CF-8C is the stabilized grade of CF-8 and is equivalent to wrought 347. Carbon in the alloy is tied up to prevent the formation of chromium carbides by the addition of columbium (niobium) or columbium plus tantalum. This alloy finds application in the same areas as alloy CF-3 and has the equivalent corrosion resistance of CF-8.

CF-8A and *CF-3A* are controlled ferrite grades. By controlling the composition and thus the precentage of ferrite present, it is possible to receive a 5 ksi increase in the minimum yield strength and a 7 ksi increase in the tensile strength. In addition an increase in resistance to stress corrosion cracking is also achieved, without any loss of corrosion resistance. CF-3A finds application in nuclear power plant construction. Thermal instability of the higher ferrite microstructures limits the upper temperature of CF-3A to 650°F (349°C) and CF-8A to 800°F (425°C).

CF-20 is a high carbon version of CF-8 and is equivalent to wrought type 302. This alloy is resistant to moderately oxidizing environments. The alloy is fully austenitic and is nonmagnetic. Applications include caustic salts, food products, sulfite liquor, and sulfurous acid.

CF-8M is the cast equivalent of wrought type 316. It contains slightly more nickel than CF-8 to offset the ferritizing influence of the molybdenum and thus maintain a comparable ferrite level in the microstructure. This alloy can be made fully austenitic and thus nonmagnetic. Addition of the molybdenum improves the general corrosion resistance, provides greater elevated-temperature strength, and particularly improves pitting resistance in chloride environments. By adding the molybdenum some resistance to strongly oxidizing environments is sacrificed, such as in boiling nitric acid. However, passivity is increased in weakly oxidizing conditions compared to CF-8. This alloy has good resistance in the presence of reducing media. Overall corrosion resistance is equal to or better than that of wrought 316. Typical services include acetic acid, acetone, black liquor, chloride solution, hot dyes, fatty acids, phosphoric acid, sulfurous acid, and vinyl alcohol. The alloy is supplied in the solution-annealed condition.

CF-8M has excellent corrosion resistance in normal atmospheric conditions including seacoast exposure. It also resists most water and brines at ambient temperature. Under low flow or stagnant conditions or at elevated temperatures seawater may cause pitting. One application of CF-8M is the handling of 80–100% sulfuric acid at ambient temperature. Good resistance is also exhibited to phosphoric acid at all concentrations up to 170°F (77°C). It is also used for nitric acid up to boiling at all concentrations to 65%.

Although CF-8M is not attacked by organic solvents, chlorinated organics may attack CF-8M, particularly under condensing conditions as when water is present.

CF-8M resists many alkaline solutions and alkaline salts: ammonium hydroxide at all concentrations to boiling and sodium hydroxide at all concentrations up to 150°F (65°C), above which stress corrosion cracking may occur.

Metallic chloride salts, such as ferric chloride and cupric chloride, can be very corrosive to CF-8M. Chloride can cause stress corrosion cracking above 160°F (71°C). The combination of chlorides, oxygen, water, and surface tensile stress can result in cracking at stresses far below the tensile strength of all austenitic stainless steels. Whenever a few hundred parts per million chlorides are present and the temperature exceeds 160°F (71°C), there is the possibility of stress corrosion cracking developing.

CF-3M is the cast equivalent of 316L and is intended for use where postweld heat treatment is not possible. The areas of application for CF-3M are essentially the same as for CF-8M.

CF-3MA is a controlled ferrite grade. Increases in minimum yield strength of 7 ksi and tensile strength of 10 ksi can be achieved. CF-3MA has an upper temperature limit of 800°F (425°C).

CF-10MC is the stabilized grade of CF-8M for field welding applications.

CF-16F is the cast equivalent of wrought type 303. It is a free machining stainless steel. Small additions of phosphorus and selenium form inclusions that act as chip breakers during machining. In addition to selenium inclusions the microstructure also contains about 15% ferrite. This alloy and CF-20 are used in similar applications, although the corrosion resistance of CF-16F is inferior to that of CF-20.

CG-8M is the cast version of wrought type 317. This alloy is resistant to reducing media and is resistant to sulfuric and sulfurous acids. It also resists the pitting action of halogen compounds. Strongly oxidizing environments will attack CG-8M. As a result of the high ferrite content this alloy exhibits very good stress corrosion cracking resistance but also has an upper temperature limit of 800°F (425°C). Applications are found in the pulp and paper industry where it resists attack from pulping liquors and bleach-containing water.

CE-30 is a high carbon cast stainless steel. The alloy has a microstructure of ferrite in austenite with carbide precipitates present in the as-cast condition. Resistance to intergranular corrosion is not seriously impaired since there is sufficient chromium present. Since the alloy does retain good corrosion resistance as cast, it is useful where heat treatment is not possible or where heat treatment following welding cannot be performed. By solution annealing of the casting, corrosion resistance and ductility can be greatly improved.

The alloy is resistant to sulfurous acid, sulfites, mixtures of sulfurous and sulfuric acids, and sulfuric and nitric acids.

CF-30A, which is a controlled ferrite grade, is resistant to stress corrosion cracking in polythionic acid and chlorides. Other applications are found in pulp and paper manufacture, caustic soda, organic acids, and acid mine water.

CH-20 is similar to CE-30 but with a composition containing a greater amount of nickel and a lesser amount of chromium. This alloy is stronger than CF-8 and more ductile than CE-30. It is also considerably more corrosion resistant than CF-8 and less susceptible to intergranular corrosion than CF-8 after exposure to sensitizing temperatures. To achieve maximum mechanical properties and corrosion resistance the alloy must be solution annealed.

CK-20 is used in the same applications as CH-20 but at higher temperatures.

CG-6MMN is the cast equivalent of Nitronic 50™ (Armco, Inc.). This alloy is used in place of CF-8M when higher strength and/or better corrosion resistance is required.

CF-10SMMN is the cast equivalent of Nitronic 60™ (Armco, Inc.). The corrosion resistance is similar to CF-8 but not as good in hot nitric acid. This alloy does have the advantage of better galling resistance than the other CF grades.

D. Duplex Alloys

By the proper adjustment of the chemistry of stainless steel it is possible to have both ferrite and austenite present at room temperature. Stainless steels with approximately 50% ferrite and 50% austenite are known as duplex stainless steels. These alloys have superior corrosion resistance and higher yield strengths than the austenitics with lower alloy content. Refer to Table 17.10 for the chemical composition of cast duplex stainless steels.

These alloys are limited to a maximum operating temperature of 500°F (260°C) as the result of the formation of a sigma phase at elevated temperatures. Above 600°F (315°C) the ferrite becomes embrittled by the "855°F embrittlement" that occurs in ferritic stainless steels. Continuous operation above 600°F (315°C) could result in catastrophic brittle failure of cast components.

Both toughness and corrosion resistance are adversely affected by the formation of a sigma phase. Welding of duplex alloys is somewhat difficult because of the potential of the formation of a sigma phase. When the casting

TABLE 17.10 Chemical Composition of Cast Duplex Stainless Steel

Chemical	Alloy (wt%)			
	CD 4MCu	CD 3MN	CD 3MWN	Z6CNDU20.08M
Carbon	0.04	0.03	0.03	0.08
Manganese	1.00	—	—	—
Silicon	1.00	—	—	—
Phosphorus	0.04	—	—	—
Sulfur	0.04	—	—	—
Chromium	24.5–26.5	21–23.5	24–26	19–23
Nickel	4.75–6.00	4.5–6.5	6.5–8.5	7–9
Molybdenum	1.75–2.25	2.5–3.5	3–4	2.3
Copper	2.75–3.25	—	0.0–1	1.2
Nitrogen	—	0.1–0.3	0.2–0.3	—
Tungsten	—	—	0.5–1	—
Iron	Balance	Balance	Balance	Balance

Maximum unless otherwise noted.

is to be re–solution heat treated after welding, a welding filler material containing about 1–2% more nickel than the casting is normally used. When the casting is not to be re–solution heat treated a filler material with 3% additional nickel is used.

Duplex alloys are strongly magnetic as a result of the high percentage of ferrite present. These alloys develop yield strengths and hardness levels about twice those of the cast 304 and 316 equivalents. Refer to Table 17.11 for the mechanical properties of the cast duplex stainless steels. The duplex alloys exhibit improved resistance to erosion and velocity conditions as a result of the increased hardness. They also exhibit exceptional resistance to chloride stress cracking.

The duplex alloys are completely resistant to corrosion from atmospheric and marine environments, fresh water, brine, boiler feedwater, and steam. They are especially suitable for high-temperature, chloride-containing environments where stress corrosion cracking and pitting are common causes of failure of other stainless steels. Because of the two phases of the duplex alloys they have inherently better stress corrosion cracking resistance than single-phase alloys. Usually at least one of the phases is resistant to cracking in a given environment. These alloys are also highly resistant to acetic, formic, and other organic acids and compounds.

CD-4MCu is the cast equivalent of Ferralium 255. Its high chromium level makes it particularly useful in oxidizing environments such as nitric

TABLE 17.11 Mechanical Properties of Cast Duplex Stainless Steels

Property	Alloy			
	CD 4MCu	CD 3MN	CD 3MWN	Z6CNDU20.08M
Wrought equivalent	Ferralium 255	S31803, 2205	Zeron 100	Uranus 50M
ASTM designation	A351, A743, A744	A890 Grade 5A	A890 Grade 5A	NF A32-055
Tensile strength (ksi)	100	90	100	87
Yield strength (ksi)	70	60	65	46
Elongation (%)	16	—	—	—
Brinell hardness	240	—	—	—
Impact toughness (ft-lb)	35	—	—	—

acid. Since the alloy possesses good hardness, low concentrations of solids present little problems. The alloy can also be used in reducing environments. CD-4MCu has been widely used in dilute sulfuric acid services up to fairly high temperatures. It has also performed well in fertilizer production and the wet process method for producing phosphoric acid. This alloy also performs well in sodium hydroxide even though it is low in nickel content. Other services for which this alloy is suitable include concentrated brines, fatty acids, seawater, hot oils, pulp liquors, scrubber solutions containing alumina and hydrofluoric acid, and dye slurries.

CD-3MN is the cast version of wrought UNS S31803 or 2205. Compared to the other duplex grades this has a lower alloy content. Consequently its cost is lower, but some corrosion resistance is sacrificed.

CD-3MWN is the cast version of wrought Zeron 100. It has higher alloy content than the other duplex grades and a corrosion resistance nearly as good as that of the superaustenitic alloys.

Z6CNDU20.08M is the cast version of Uranus 50M. This alloy is a borderline duplex alloy since it is limited to 25–40% ferrite. Its corrosion resistance is slightly better than CF-8M but inferior to the other duplex stainless steels.

E. Superaustenitic Alloys

Superaustenitic alloys are those austenitic stainless steels having alloying elements, particularly nickel and/or molybdenum, in higher percentages than the conventional 300 series stainless steels. Table 17.12 lists the chemical

TABLE 17.12   Chemical Composition of Cast Superaustenitic Stainless Steel

Chemical	Alloy (wt%)				
	CD 7M	CN 7MS	CK 3MCuN	CE 3MN	CUSMCuC
Carbon	0.07	0.07	0.025	0.03	0.05
Manganese	1.50	—	—	—	—
Silicon	1.50	—	—	—	—
Phosphorus	0.04	—	—	—	—
Sulfur	0.04	—	—	—	—
Chromium	19.0–22.0	18.0–20.0	19.5–20.5	20.0–22.0	19.5–23.5
Nickel	27.5–30.5	22.0–25.0	17.5–19.5	23.5–25.5	38.0–46.0
Molybdenum	2.0–3.0	2.5–3.0	6.0–7.0	6.0–7.0	2.50–3.50
Copper	3.0–4.0	1.5–2.0	0.5–1.0	—	1.50–3.50
Nitrogen	—	—	0.18–0.24	0.18–0.26	—
Columbium	—	—	—	—	0.6–1.2
Iron	Balance	Balance	Balance	—	—

Maximum unless otherwise noted.

compositions of the cast superaustenitic stainless steels. In some instances these alloys have been classified as nickel alloys. As can be seen in Table 17.12 these alloys contain 16–25% chromium; 30–35% nickel, molybdenum, and nitrogen; and some also contain copper. No single element exceeds 50%.

Added resistance to reducing environments is provided by the additional nickel, while the extra molybdenum, copper, and nitrogen increase the resistance to pitting in chlorides. These cast alloys are fully austenitic, which makes them more difficult to cast than the ferrite-containing grades.

In order to preserve the special corrosion resistant properties of the superaustenitics weld procedures must be carefully developed. Heat input must be kept to a minimum, and interpass temperatures must be in the 250–350°F (121–177°C) range. Overmatching weld filler materials are generally used for weld repairs and fabrication welds. The most commonly used grades are AWS filler metal grades NiCrMo-3, NiCrMo-7, NiCrMo-10, and NiCrMo-12. Welding with matching filler requires re-solution heat treatment after all welding.

Superaustenitics are used for high-temperature, chloride-containing environments where pitting and stress corrosion cracking are common causes of failure with other stainless steels. These alloys resist chloride stress corrosion cracking above 250°F (121°C). They also exhibit excellent resistance to sulfide stress cracking.

CN-7M is the cast equivalent of wrought alloy 20Cb3. The mechanical properties of cast superaustenitic stainless steels will be found in Table 17.13. This alloy resists sulfuric acid in all concentrations at temperatures up to 150°F (65°C) and higher for most concentrations. The high nickel content of alloy CN-7M imparts excellent resistance to alkaline environments, such as sodium hydroxide where it can be used up to 73% and temperatures to 300°F (149°C). The chromium content of this alloy makes it superior to the CF grades in nitric acid—even better than CF-3, which is generally considered the best alloy for this service. Hydrochloric acid, certain chlorides, and strong reducing agents such as hydrogen sulfide, carbon disulfide, and sulfur dioxide will accelerate corrosion. Applications for CN-7M have also included hot acetic acid, dilute hydrofluoric and hydrofluosilicic acids, nitric-hydrofluoric pickling solutions, phosphoric acid, and some plating solutions.

Alloy *CN-7MS* is a modified version of alloy CN-7M.

Alloys *CK-3MCuN* and *CE-3MN* are superior for chloride environments and are the cast equivalents of wrought alloys 254SMO and AL6XN, respectively.

Cu-5MCuC is the cast version of wrought alloy 825, although columbium is substituted for titanium. Titanium will oxidize rapidly during air melting, while columbium will not. This alloy is similar in properties, corrosion resistance, and weldability to CN-7M. It has equal corrosion resistance in sulfuric, nitric, and phosphoric acids; seawater; and other environments.

F. Precipitation Hardening Alloys

Table 17.14 lists the cast precipitation hardening stainless steels and gives their chemical compositions.

CB-7Cu is the cast equivalent of wrought alloy 326. The alloy is martensitic with minor amounts of retained austenite present in the microstructure. When the alloy is aged at 925°F (496°C) after soltuon annealing, submicroscopic copper particles precipitate and strengthen the alloy. Typical mechanical properties are

Tensile	130 ksi
Yield strength	110 ksi
Elongation in 2 in.	10%
Brinell hardness	320
Impact toughness	25 ft-lb

TABLE 17.13 Mechanical Properties of Cast Duplex Stainless Steel

Property	Alloy						
	CN-7M	CN-7MS	CK-3MCuN	CE-3MN	CU-SMCuC		
Wrought equivalent	20Cb3	20Cb3 modified	254SMO	A16XN	Alloy 825		
ASTM designation	A351, A743, A744	A743, A764	A351, A743, A744	A351, A744	A494		
Tensile strength (ksi)	62	70	80	80	75		
Yield strength (ksi)	25	30	38	38	35		
Elongation (%)	35	—	—	—	—		
Brinell hardness	130	—	—	—	—		
Impact toughness (ft-lb)	70	—	—	—	—		

TABLE 17.14 Chemical Composition of Cast Precipitation Hardening Stainless Steels

Chemical	Alloy (wt%)		
	CB-7CU	CB-7Cu-1	CB-7Cu-2
Carbon	0.07	0.07	0.07
Manganese	1.00	—	—
Silicon	1.00	—	—
Phosphorus	0.04	—	—
Sulfur	0.04	—	—
Chromium	15.5–17.0	15.5–17.7	14.0–15.5
Nickel	3.6–4.6	3.6–4.6	4.5–5.5
Copper	2.3–3.3	2.5–3.2	2.5–3.2
Columbium	—	0.15–0.35	0.15–0.35
Iron	Balance	Balance	Balance

Maximum unless otherwise noted.

In the age-hardened condition the alloy exhibits corrosion resistance superior to the straight martensitic and ferretic grades. This alloy is used where moderate corrosion resistance and high strength are required. Typical applications include aircraft parts, pump shafting, and food processing equipment.

CB-7Cu-1 and CB-7Cu-2 are cast versions of wrought alloys 17-4PH and 15-5PH. These are high-strength, precipitation hardening, martensitic stainless steels, which are cast, solution heat treated, and aged. CB-7Cu-1 is more commonly cast than CB-7Cu2.

The higher hardness conditions of these alloys are susceptible to stress corrosion cracking. As the aging temperature is increased and the strength and hardness decreased, the resistance to stress cracking improves. Table 17.15 compares the strength and hardness values for various aging conditions.

These alloys are similar in corrosion resistance to alloys CF-8 and wrought type 304, and better than the 400 series of stainless steels. They will resist atmospheric attack in all but the most severe environments. When in contact with seawater the alloys will pit, but they are resistant to natural water. Applications include use in steam, boiler feedwater, condensate, and dry gases.

II. NICKEL BASE ALLOYS

The nickel base alloys are more difficult to cast than the austenitics. A wrought trade name should never be used when purchasing a nickel alloy

TABLE 17.15 Minimum Strength of Cast Precipitation Hardening
Stainless Steels

Condition	Minimum tensile strength (ksi)	Yield strength (ksi)	Brinell hardness
H-900	170	145	375 min.
H-1075	145	115	277 min.
H-1150	125	97	269 min.
H-1150 DBL	—	—	310 max.

casting. Foundry selection is critical in obtaining high-quality, corrosion-resistant castings. Expertise is obtained only by pouring nickel alloys on a daily basis.

Because of the high cost of these alloys they are generally used only in specialty areas and very severe service. As with the stainless steels ACI designations have been adopted for these alloys since their compositions and properties in many cases vary significantly from the wrought equivalents. ASTM standard A494 covers cast nickel base alloys. The chemical composition and mechanical properties of the cast nickel base alloys will be found in Table 17.16.

CZ-100 is the cast equivalent of wrought nickel 200. In order to provide adequate castability the carbon and silicon levels are higher in the cast grade than in the wrought grade. While in the molten state the alloy is treated with magnesium, which causes the carbon to nodularize leading to an increase in the mechanical properties, much as with ductile iron. The alloy is generally supplied in the as-cast condition since the properties of CZ-100 cannot be improved by heat treatment.

This alloy is used for dry halogen gases and liquids and ambient temperature hydrofluoric acid, but its widest use is in alkaline services. It has excellent resistance to all bases except ammonium hydroxide, which will cause rapid attack at any concentration above 1%. CZ-100 is resistant to all concentrations and temperatures of sodium and potassium hydroxide. If chlorates or oxidizable sulfur compounds are present in caustic, the corrosion rate will be accelerated. CZ-100 also finds application in food processing where product purity is important.

M-35-1, M-35-2, M-30-G, and *M-25-S* cast alloys are the equivalent of wrought Monel 400. The most common cast grade is M-35-1. This grade is the one most preferred for welding because of its lower silicon level. The lower level of silicon also makes alloy M-35-1 suitable for the handling of air-free hydrofluoric acid. Since alloy M-35-1 exhibits good resistance to

TABLE 17.16 Cast Nickel Base Alloys

Specification and grade	Wrought equivalent	C max.	Cr	Ni	Fe	Mo	Others	Minimum strength (ksi/MPa)	
								Tensile	Yield
ASTM A494 Grade CZ100	Nickel 200	1	—	95[a]	3[a]	—	—	50/345	18/125
ASTM A494 Grade M35-1	Monel 400	0.35	—	Balance	3.5[a]	—	Si 1.25[a]	65/450	25/170
ASTM A494 Grade M35-2	Monel 400	0.35	—	Balance	3.5[a]	—	Si 2[a]	65/450	30/205
ASTM A494 Grade M30C	Monel 400	0.3	—	Balance	3.5[a]	—	Si 1–2, Cb 1–3	65/450	32.5/225
ASTM A494 Grade M25S	S-Monel	0.25	—	Balance	3.5[a]	—	Si 3.5–4.5	300 HB min. aged condition	—
ASTM A494 Grade CY40	Inconel 600	0.4	14–17	Balance	11[a]	—	—	70/485	28/195
ASTM A494 Grade CW6MC	Inconel 625	0.06	20–23	Balance	5[a]	8–10	Cb 3.15–4.5	70/495	40/275
ASTM A494 Grade CW2M	Hastelloy C	0.02	15–17.5	Balance	2[a]	15–17.5	—	72/495	40/275
ASTM A494 Grade CX2MW	Hastelloy C22	0.02	20–22.5	Balance	2–6	12.5–14.5	W 2.5–3.6	80/550	45/280
ASTM A494 Grade CW6M	Chlorimet 3	0.07	17–20	Balance	3[a]	17–20	—	72/495	40/275
ASTM A494 Grade N7M	Hastelloy B2	0.07	1[a]	Balance	3[a]	30–33	—	76/525	40/275
ASTM A494 Grade CY5SnBiM	Waukesha 88	0.05	11–14	Balance	2[a]	2–3.5	Bi 3–5, Sn 3–5	—	—

[a]Maximum content.

fluorides it is finding widespread use in uranium enrichment. The higher silicon grade alloy M-30-H is used for rotating parts and wear rings since it combines corrosion resistance with high strength and wear resistance.

When exceptional resistance to galling is required alloy M-25-S is the alloy of choice. At higher silicon levels the ductility will decrease sharply and the tensile and yield strengths will increase. The only mechanical property recorded for M-25-S is hardness because the strength and ductility cannot be controlled readily. The higher silicon grades M-25-S and M-30-C are considered unweldable. Grade M-30-C obtains its high strength from the addition of columbium instead of higher silicon levels. However the columbium-carbide phase which forms decreases the corrosion resistance of this alloy in hydrofluoric acid.

In general the cast Monel alloys exhibit excellent resistance to mineral acids, organic acids, and salt solutions. These alloys are also used in sulfuric acid services where reducing conditions are present and in chlorinated solvents. Oxidizing conditions accelerate the corrosion rate in all services.

CY-40 is the cast equivalent of wrought Inconel alloy 600. It is a nickel-chromium alloy without the molybdenum content of most nickel-chromium alloys. In order to provide adequate castability the carbon and silicon are higher than in the wrought alloy. In order to maximize corrosion resistance and mechanical properties this alloy is solution annealed.

Applications for this alloy are found where oxidation resistance and strength retention at high temperatures are required. This alloy resists stress corrosion cracking in chloride environments and at times is substituted for CZ-100 in caustic soda containing halogens. CY-40 is widely used in nuclear reactor services because of its resistance to chloride stress corrosion cracking and corrosion by high purity water. It also finds application in steam, boiler feedwater, and alkaline solutions including ammonium hydroxide.

CW-12MW is the original cast equivalent of alloy C-276. Because of segregation problems with the alloy the corrosion resistance is inferior to wrought C-276.

CW-2M is essentially a low carbon version of CW-12MW, having improved ductility and high temperature service and is the presently used cast version of wrought alloy 276. This alloy may be used in corrosive environments in the welded condition without postweld heat treatment since it resists the formation of grain-boundary precipitation. It has excellent corrosion resistance in hydrochloric and sulfuric acids at temperatures below 129°F (52°C), with a much higher temperature range at low concentrations. Excellent resistance is also exhibited in organic acids. Contamination by strong oxidizing species such as cupric and ferric ions will not cause accelerated attack of CW-2M such as is experienced with other alloys. It is also resistant to most forms of stress corrosion cracking including chloride, caustic, and hydrogen sulfide. Good as-welded corrosion resistance can be maintained with the use of AWS filler materials NiCrMo-7 or NiCrMo-10.

CW-6M is the cast version of Chlorimet 3™ (Duriron Co.) and is intended primarily for corrosive services. The tungsten and vanadium have been removed and the chromium, molybdenum, and nickel levels raised. These modifications in the composition result in improved corrosion resistance and increased mechanical properties, particularly ductility.

CW-6MC is the cast equivalent of wrought Inconel 625. In order to maximize corrosion resistance and mechanical properties the alloy is solution annealed. The alloy is used primarily for oxidation resistance and strength retention at high temperatures. Alloy CW-6MC has superior corrosion resistance compared with alloy CY-40.

CX-2MW is the cast version of wrought alloy C-22.

CY-5SnBiM is a proprietary alloy of Waukesha Foundry and is known as Waukesha 88™. In order to improve galling resistance against stainless steel, tin and bismuth are added as solid metal lubricants. Weld repairs are prohibited. The alloy is not as corrosion resistant as other nickel base alloys but performs well in the food industry.

N-7M is the cast equivalent of wrought alloy B-2 and is a nickel-molybdenum alloy. To insure maximum corrosion resistance, solution annealing, heat treatment, and alloy purity are essential to produce a suitable microstructure. Impurities such as carbon and silicon must be kept to as low levels as possible to prevent the formation of a secondary phase in the microstructure that will adversely affect corrosion resistance. This alloy is particularly recommended for handling hydrochloric acid at all concentrations and temperatures including boiling. Oxidizing contaminents or conditions can lead to rapid failure. Accelerated corrosion will result when cupric or ferric chloride, hypochlorites, nitric acid, or even aeration are present. In 10% hydrochloric acid the maximum allowable ferric ion concentration is 5000 ppm at 78°F (26°C), while at 150°F (66°C) the maximum allowable concentration is less than 1000 ppm and boiling less than 75 ppm. This alloy is also resistant to hot sulfuric acid as long as no oxidizing contaminents are present. Phosphoric acid in all concentrations up to 300°F (148°C) can also be handled.

N-12MV is also a nickel-molybdenum alloy. This alloy is similar to N-7M but with considerably less ductility. Its corrosion resistance is basically the same as that of N-7M.

REFERENCES

1. GW George, PG Brey. Cast alloys. In: PA Schweitzer, ed. Corrosion and Corrosion Protection Handbook, 2nd ed. New York: Marcel Dekker, 1988.
2. JL Gossett. Corrosion resistance of cast alloys. In: PA Schweitzer, ed. Corrosion Engineering Handbook. New York: Marcel Dekker, 1996.
3. CP Dillon. Corrosion Control in the Chemical Process Industries, 2nd ed. St. Louis, MO: Materials Technology Institute of the Chemical Process Industries Inc., 1986.

18

Copper and Copper Alloys

Since before the dawn of history, when primative people first discovered the red metal, copper has been serving us. The craftsmen who built the great pyramid for the Egyptian pharaoh Cheops used copper pipe to convey water to the royal bath. A remnant of this pipe was unearthed some years ago, still in usable condition, a testimonial to copper's durability and resistance to corrosion. Today, nearly 5000 years after Cheops, copper is still used to convey water and is a prime material for this purpose.

Copper ores found in the United States contain approximately 1% copper in the form of copper sulfide. Copper sulfide concentrates are smelted to yield a matte that is a mixture of copper and iron sulfides. These mattes are remelted in a converter where air is blown through the mattes to oxidize the remaining sulfur. The remaining copper is approximately 99% pure. This is referred to as blister copper and is further refined to remove other impurities. The final "tough-pitch" copper has a purity of about 99.5%, which is suitable for many alloys. A higher purity of 99.9% can be obtained by electrolytic refinement of the tough-pitch copper.

Copper is a very useful material. It has excellent electrical and thermal conductivity properties, is malleable and machinable, but has low mechanical properties. In order to obtain strength the metal must be cold worked or alloyed. As a result there are hundreds of copper alloys. The Copper Development Association, together with the American Society of Testing and Materials and the Society of Automotive Engineers, has developed a five-digit system to identify these alloys. The system is part of the Unified Numbering System for Metals and Alloys. The numbers C-10000 through C79999 denote the wrought alloys, while the cast copper and copper alloys are numbered C80000 through C99999. This designation system is used throughout North America, Australia, and Brazil. Each number refers to a specific alloy composition. The UNS numbers for specific groups of wrought alloys are given in Table 18.1 along with applications for each specific group.

There are more than 100 temper designations for copper and copper alloys. These may be found in ASTM B601 "Standard Practice for Temper

TABLE 18.1   Unified Numbering System for Wrought Copper and Copper Alloys

UNS no.	Application
Coppers	
C10100 to C15999	General for group; high electrical conductivity requirements
C10100 to C10700	Highest conductivity coppers
C11000	Electrical wire and cable
C12200	Household water tube
C12800	Refrigerators, air conditioners, and gasoline
C14200, C14300	Contacts, terminals; resists softening when soldered
C15215 to C16200	Electrical components, lead frames, integrated circuits
High copper alloys	
C16200 to C19199	General for group; electrical and electronic connectors and contacts
C17000 to C17300	Highest strength copper-beryllium alloys; bellows, diaphragms, fasteners, relay parts
C18000	Switches, circuit breakers, contacts
C18200 to C18300	Cable connectors, electrodes, arcing and bridging parts
C19400	Terminals, flexible hose, gaskets, fuse clips
Copper-zinc brasses	
C21000 to C28000	General; sheet for stamping, springs, electrical switches and sockets, plumbing
C23000	Red brass, condenser and heat exchanger tubes, plumbing, architectural trim
C26000 to C26200	Cartridge brass, radiator cores, hardware, ammunition, plumbing accessories
C26800	Muntz metal, architectural sheet and trim, large nuts and bolts, brazing rod
Copper-zinc-lead brasses	
C31200 to C38500	General for group; leaded brasses, high machinability requirements
C34500	Clock parts, gears, wheels
C36000	Free-cutting brass, screw machine materials, gears, pinions
C37700	Forging brass

TABLE 18.1 Continued

UNS no.	Application
Copper-zinc-tin brasses	
C40400 to C48600	General; corrosion resistance and higher strength requirements
C42500	Supplied as strip for fabricating electrical connectors, springs, terminals
C46400 to C46700	Naval brass, marine hardware, propeller shafts, structural uses
C48200 to C48500	Free machining, marine hardware, valve stems, screw machine products
Copper-tin-phosphorus bronzes	
C50100 to C54200	General; superb spring qualities, good formability and solderability, high fatigue resistance, high corrosion resistance
C50500	Flexible hose, electrical contacts
C51100	Fasteners, bellows, fuse clips, switch parts
Copper-tin-lead-phosphorus bronzes	
C53400 to C53500	Leaded phosphor bronzes; general; combine high strength and fatigue resistance with good machinability and wear resistance
Copper-aluminum bronzes	
C60800 to C64210	General; combine high strength and excellent corrosion resistance
C61000	Marine hardware, pumps, valves, nuts, bolts, shaft tie rods, machine parts, condenser tubing
C63000	Nuts, bolts, marine shafts, aircraft parts, forgings
Copper-silicone bronzes	
C64700 to C66100	General; properties similar to aluminum bronzes; excellent weldability; hydraulic fluid lines, high strength fasteners, wear plates, marine hardware
Copper-nickel	
C70100 to C72950	General; excellent corrosion resistance, strength retention at high temperatures, condenser tubes, marine products
Copper-zinc-nickel	
C73500 to C79800	Good corrosion resistance with moderately high strength, silver luster; food and beverage handling equipment, decorative hardware, hollow-ware

Designations for Copper and Copper Alloys—Wrought and Cast." Table 18.2 lists the more common temper designations.

Wrought copper and copper alloys can be divided into the following groups:

Coppers	Metal which has a minimum copper content of 99.3% copper.
High copper alloys	Alloys which contain 96.0 to 99.3% copper.
Brasses	Alloys which have zinc as the major alloying element. This group contains the copper-zinc alloys, the copper-zinc-lead alloys (leaded brass), and the copper-zinc-tin alloys (tin brasses).
Bronzes	Alloys in which the major alloying element is neither zinc nor nickel. Four major families of bronze exist: copper-tin-phosphorus (phosphor bronzes), copper-tin-lead-phosphorus (leaded phosphor bronzes), copper-tin-nickel alloys (nickel-tin bronzes), and copper-aluminum alloys (aluminum bronzes).
Miscellaneous copper-zinc alloys	This group contains alloys formerly known as manganese or nickel bronzes. However, since zinc is the principal alloying element they are really brasses. The two major subgroups in this category are the copper-nickel alloys called cupronickels and the copper-zinc-nickel alloys commonly called nickel-silvers.

For the most part copper alloys to a large extent attain their mechanical properties through different amounts of cold work. The main exceptions are

TABLE 18.2 Temper Designations of Wrought Copper and Copper Alloys

Tempers	Description
H00 to H14	Degree of cold work: 1/8 hard to super spring
H50 to H90	Cold-work tempers based on manufacturing process
HR01 to HR50	Cold-worked and stress-relieved tempers
HT04 to HT08	Cold-rolled and thermal-strengthened tempers
M01 to M45	As manufactured tempers
010 to 082	Annealed tempers to meet specific mechanical properties
0S005 to 0S200	Annealed tempers to meet prescribed grain sizes
TD00 to TD04	Solution-treated and cold-worked tempers
TH01 to TH04	Cold-worked and precipitation-hardened tempers

the copper-beryllium alloys (the highest strength copper alloys), the aluminum bronzes, and some copper-chromium alloys. The aluminum bronzes are the second highest strength copper alloy group. Their high strengths are achieved by means of heat treatments that produce second phase particles.

The physical properties of copper which are of the most interest are the electrical and thermal conductivities. The recognized standard for metal electrical conductivity is the International Annealed Copper Standard (IACS). Conductivities of 100% are exhibited by high-purity, oxygen-free coppers (1.7241 $\mu\Omega$-cm at 20°C). The conductivity is reduced by alloying. For example, alloy C71500 has a nickel content of 30% which reduces the conductivity to only 4% IACS.

Applications such as heat exchangers, condensers, and other heat transfer devices take advantage of the high thermal conductivity of copper and its alloys. As temperatures increase, the thermal conductivity of many copper alloys increases, unlike most metals. Alloying reduces the thermal conductivity. Physical and mechanical properties of copper and each of the alloys will be given as each is discussed in detail.

Copper and its alloys are very easily fabricated; they can be shaped by the common forming processes. They can be cold-rolled, stamped, drawn, bent, and headed as well as being extruded, forged, and rolled at elevated temperatures. Annealing is required for some copper alloys that work-harden during cold working more rapidly than others. The annealing must be done intermittently in order for the alloy to be cold-reduced extensively. Bending and stamping operations do not normally require an annealing step. Forming of the leaded copper alloys is somewhat more difficult because of their low ductility, particularly the free-cutting brasses. Age-hardenable copper-beryllium alloys must be formed in the solution-annealed (unhardened) condition and subsequently aged to the desired strength level.

The design engineer should be familiar with the various tempers. In general the higher the temper numbers, the harder the metal. The higher the temper number of cold-worked metal, the more prior cold work the material has received. Cold-worked metal resists further deformation, which makes it more difficult to shape. Fracture of the part can result from extensive additional forming.

All copper alloys are machinable using standard tooling. All but the hardest alloys can be machined using high-speed tool steels. Free-cutting copper alloys should be used for screw machine production.

A wide choice of means of joining copper alloys is available. They can be joined by soldering, brazing, welding, and mechanical means such as crimping, riveting, and bolting. During welding and brazing operations it is necessary to take into account the high thermal conductivities of copper alloys. Their high thermal expansion coefficient must also be considered when fitting pieces together.

Before copper alloys can be soldered it is necessary to remove the protective film. Although mechanical cleaning is helpful it is usually necessary to use fluxes to remove and prevent re-formation of tenaceous films. Cleaning with 15–20% aqueous solution of nitric acid is necessary with copper-beryllium and aluminum bronzes since they tend to resist fluxing. It is also possible that the properties of the age-hardening alloys may change during application of heat.

Copper alloys are used primarily for room temperature applications. Heat transfer apparatus applications are the exception. Most copper alloys can be used to −328°F (−200°C). The approximate maximum temperature limits for a selection of alloys are listed in Table 18.3.

I. COPPERS

To be classified as a copper the alloy must contain a minimum of 99.3% copper. Elements such as silver, arsenic, lead, phosphorus, antimony, tellurium, nickel, cadmium, sulfur, zirconium, manganese, boron, and bismuth

TABLE 18.3 Maximum Allowable Operating Temperature of Wrought Copper Alloys

| Alloy | Maximum operating temp. | |
	°F	°C
Coppers	176	80
Phosphorus deoxidized copper	248	120
Copper-beryllium	482	250
Copper-chromium	662	350
Silicon bronze	392	200
Aluminum bronze (7–9% Aluminum)	572	300
Cartridge red brass	392	200
Naval brass	302	150
Muntz metal	356	180
Leaded brass	212	100
Manganese brass	356	180
Phosphor bronze (9% tin)	320	160
Cupronickel (5–10% nickel)	302	150
Cupronickel (20–30% nickel)	392	200
Nickel-silvers	392	200
Cu-Cd bearing alloys	392	200

may be present, singly or in combination. Because of the good electrical properties of copper, it is used primarily in electrical or electronic applications such as bus bars, waveguides, wire, switches, and transfer components. Since copper is a noble metal it also finds many applications in corrosive environments. Table 18.4 gives the chemical composition of some of the coppers used in corrosion applications.

The mechanical and physical properties of copper are shown in Table 18.5. Large quantities of copper are used to produce copper pipe and tubing. Table 18.6 lists the mechanical and physical properties of copper pipe, and Table 18.7 the allowable design stress for copper pipe.

Copper itself is inherently corrosion resistant. It is noble to hydrogen in the emf series and thermodynamically stable with no tendency to corrode in water and in nonoxidizing acids free of dissolved oxygen. With copper and its alloys the predominent cathode reaction is the reduction of oxygen to form hydroxide ions. Therefore oxygen or other oxidizing agents are necessary for corrosion to take place. In oxidizing acids or in aerated solutions of ions that form copper complexes, e.g., CN, MH_A, corrosion can be severe. Copper is also subject to attack by turbulently flowing solutions,

TABLE 18.4 Chemical Composition of Coppers: Maximum[a] Composition (%)

Copper UNS no.	Cu	Ag min.	P	As	Sb	Te	Other
C10200	99.95						
C10300	99.95		0.001–0.005				
C10400	99.95	0.027					
C10800	99.95		0.005–0.012				
C11000	99.90						
C11300	99.90	0.027					
C12000	99.90		0.004–0.012				
C12200	99.90		0.015–0.040				
C12500	99.88			0.012	0.003	0.025	0.050 Ni, 0.003 Bi, 0.004 Pb
C13000	99.88	0.085		0.012	0.003	0.025	0.050 Ni, 0.003 Bi, 0.004 Pb
C14200	99.40		0.015–0.040	0.15–0.50			

[a]Except for Ag and when shown as a range.
Source: Ref. 3.

TABLE 18.5 Mechanical and Physical Properties of Wrought Copper

Property	C11000 spring H08	C10800 hard rod H04	C10800 tube H55
		Alloy	
Modulus of elasticity $\times 10^6$ (psi)	17	17	17
Yield strength 0.5% ext. (ksi)	500	44	32
Elongation in 2 in. (%)	4	16	25
Rockwell hardness	B-60	B-47	B-35
Field strength 10^8 cycles (ksi)	14	—	—
Shear strength (ksi)	—	27	26
Density (lb/in.^{3})	0.323	0.323	0.323
Coefficient of thermal expansion at 70–572°F, (10^{-6}/°F)	9.8	9.8	9.8
Thermal conductivity (Btu/ft^2/ft/hr/°F)	226	202	202
Specific heat	0.092	0.092	0.092
Electrical conductivity (% IACS)	101	92	92

even though the metal may be resistant to the solution in a stagnant condition. Most of the corrosion products formed on copper and copper alloys produce adherent, relatively impervious films with low solubility that provide the corrosion protection.

TABLE 18.6 Mechanical and Physical Properties of Copper Pipe

Property	Annealed	Hard-drawn
Modulus of elasticity $\times 10^6$ (psi)	17	17
Tensile strength $\times 10^3$ (psi)	33	45
Yield strength 0.2% offset $\times 10^3$ (psi)	10	40
Elongation in 2 in. (%)	45	10
Rockwell hardness	F-45	B-40
Density (lb/in.^{3})	0.323	0.323
Specific gravity	8.91	8.91
Specific heat (Btu/hr °F)	.092	.092
Thermal conductivity at 68°F (Btu/hr ft^2 °F)	2364	2364
Coefficient of thermal expansion at 77–572°F $\times 10^{-6}$(in./in.°F)	9.8	9.8

TABLE 18.7 Allowable Design Stress for Copper Pipe

Operating temp. (°F/°C)	Allowable design stress (psi)		
	Annealed	Hard-drawn	Light-drawn
≤150/66	6000	11,300	9000
250/121	5800	10,500	8300
300/149	5000	8000	8000
350/177	3800	5000	5000
400/204	2500	2500	2500
450/332	1500	15,000	1500
500/260	750	750	750

Copper finds many applications in the handling of seawater and/or fresh water. The corrosion resistance of copper, when in contact with fresh water or seawater, is dependent upon the surface oxide film that forms. In order for corrosion to continue, oxygen must diffuse through this film. High velocity water will disturb the film, while carbonic acid or organic acids, which are present in some fresh waters and soils, will dissolve the film. Either situation leads to an appreciably high corrosion rate. If the water velocity is limited to 4–5 ft/s, the film will not be disturbed.

Sodium and potassium hydroxide solutions can be handled at room temperature by copper in all concentrations. Copper is not corroded by perfectly dry ammonia, but it may be rapidly corroded by moist ammonia and ammonium hydroxide solutions. Alkaline salts, such as sodium carbonate, sodium phosphate, or sodium silicate act like the hydroxides but are less corrosive.

When exposed to the atmosphere over long periods of time the protective film that forms is initially dark in color, gradually turning green. This corrosion product is known as patina. Since the coloration is given by copper hydroxide products, the length of time required to form this coloration is dependent upon the atmosphere. In marine atmospheres the compound is a mixture of copper/hydroxide/chloride and in an urban or industrial atmosphere a mixture of copper/hydroxide/sulfate.

Pure copper is immune to stress corrosion cracking. However, alloys of copper containing more than 15% zinc are particularly subject to this type of corrosion.

The coppers are resistant to urban, marine, and industrial atmospheres. For this reason copper is used in many architectural applications such as building fronts, downspouts, flashing, gutters, roofing, and screening. In addition to their corrosion resistance, their good thermal conductivity properties make the coppers ideal for use in solar panels and related tubing and piping used in solar energy conversion. These same properties plus their resistance to engine coolants have made the coppers suitable for use as radiators.

Large amounts of copper are also used in the beverage industry, particularly in the brewing and distilling operations.

In general the coppers are generally resistant to

1. Seawater
2. Fresh waters, hot or cold
3. Deaerated, hot or cold,, dilute sulfuric acid, phosphoric acid, acetic acid, and other nonoxidizing acids
4. Atmospheric exposure

The coppers are not resistant to

1. Oxidizing acids such as nitric and, hot concentrated sulfuric acid, and aerated nonoxidizing acids (including carbonic acid).
2. Ammonium hydroxide (plus oxygen), a complex ion $Cu(NH_3)_4^{2+}$ forms. Substituted ammonia compounds (amines) are also corrosive.
3. High velocity aerated waters and aqueous solutions.
4. Oxidizing heavy metal salts (ferric chloride, ferric sulfate, etc.).
5. Hydrogen sulfide and some sulfur compounds.

The compatibility of copper with selected corrodents is shown in Table 18.8. Reference 6 provides a more detailed listing.

II. HIGH COPPER ALLOYS

Wrought high copper alloys contain a minimum of 96% copper. Table 18.9 lists the chemical composition of some of the high copper alloys. High copper alloys are used primarily for electrical and electronic applications. The physical and mechanical properties of three of the high copper alloys are shown in Table 18.10.

The corrosion resistance of the high copper alloys is approximately the same as that of the coppers. These alloys are used in corrosion service when mechanical strength is needed as well as corrosion resistance. Alloy C-19400 is basically copper that has about 2.4% iron added to improve corrosion resistance. It is used in seam-welded condenser tubing in desalting services.

TABLE 18.8 Compatibility of Copper, Aluminum Bronze, and Red Brass with Selected Corrodents

	Maximum temp. (°F/°C)		
Chemical	Copper	Aluminum bronze	Red brass
Acetaldehyde	X	X	X
Acetamide		60/16	
Acetic acid, 10%	100/38		X
Acetic acid, 50%	X	X	X
Acetic acid, 80%	X	X	X
Acetic acid, glacial	X	X	X
Acetic anhydride	80/27	90/32	X
Acetone	140/60	90/32	220/104
Acetyl chloride	X	60/16	X
Acrylonitrile	80/27	90/32	210/99
Adipic acid	80/27		
Allyl alcohol	90/32	90/32	90/32
Alum	90/32	60/16	80/27
Aluminum acetate	60/16		
Aluminum chloride, aqueous	X	X	X
Aluminum chloride, dry	60/16		
Aluminum fluoride	X	90/32	
Aluminum hydroxide	90/32	X	80/27
Aluminum nitrate		X	X
Aluminum oxychloride			
Aluminum sulfate	80/27	X	X
Ammonia, gas	X	90/32	X
Ammonium bifluoride	X		X
Ammonium carbonate	X		X
Ammonium chloride, 10%	X	X	X
Ammonium chloride, 50%	X	X	X
Ammonium chloride, sat.	X	X	X
Ammonium fluoride, 10%	X		X
Ammonium fluoride, 25%	X	X	X
Ammonium hydroxide, 25%	X	X	X
Ammonium hydroxide, sat.	X	X	X
Ammonium nitrate	X	X	X
Ammonium persulfate	90/32	X	X
Ammonium phosphate	X	90/32	X
Ammonium sulfate, 10–40%	X	X	X
Ammonium sulfide	X	X	X
Ammonium sulfite	X		X
Amyl acetate	90/32	X	400/204
Amyl alcohol	80/27	90/32	90/32

TABLE 18.8   Continued

Chemical	Copper	Aluminum bronze	Red brass
		Maximum temp. (°F/°C)	
Amyl chloride	80/27	90/32	80/27
Aniline	X	90/32	X
Antimony trichloride	80/27	X	X
Aqua regia, 3:1	X	X	X
Barium carbonate	80/27	90/32	90/32
Barium chloride	80/27	80/27	80/27
Barium hydroxide	80/27	X	80/27
Barium sulfate	80/27	60/16	210/99
Barium sulfide	X	X	X
Benzaldehyde	80/27	90/32	210/99
Benzene	100/38	80/27	210/99
Benzene sulfonic acid, 10%			90/32
Benzoic acid, 10%	80/27	90/32	210/99
Benzyl alcohol	80/27	90/32	210/99
Benzyl chloride	X		X
Borax	80/27	90/32	80/27
Boric acid	100/38	90/32	X
Bromine gas, dry	60/16	X	
Bromine gas, moist	X	X	
Bromine, liquid		X	
Butadiene	80/27		80/27
Butyl acetate	80/27		300/149
Butyl alcohol	80/27	90/32	90/32
Butyl phthalate	80/27		210/99
Butyric acid	60/16	90/32	80/27
Calcium bisulfite	80/27	X	X
Calcium carbonate	80/27	X	80/27
Calcium chlorate	X	X	X
Calcium chloride	210/99	X	80/27
Calcium hydroxide, 10%	210/99	80/27	
Calcium hydroxide, sat.	210/99	60/16	210/99
Calcium hypochlorite	X	X	X
Calcium nitrate		X	X
Calcium sulfate	80/27	X	80/27
Caprylic acid	X		X
Carbon bisulfide	80/27		X
Carbon dioxide, dry	90/32	90/32	570/299
Carbon dioxide, wet	90/32	90/32	X
Carbon disulfide	80/27		X
Carbon monoxide		60/16	570/299

TABLE 18.8 Continued

Chemical	Maximum temp. (°F/°C)		
	Copper	Aluminum bronze	Red brass
Carbon tetrachloride	210/99	90/32	180/82
Carbonic acid	80/27	X	210/99
Cellosolve	80/27	60/16	210/99
Chloracetic acid, 50% water	X		
Chloracetic acid	X	80/27	X
Chlorine gas, dry	210/99	90/32	570/299
Chlorine gas, wet	X	X	X
Chlorine, liquid			
Chlorobenzene	90/32	60/16	210/99
Chloroform	80/27	90/32	80/27
Chlorosulfonic acid	X	X	X
Chromic acid, 10%	X	X	X
Chromic acid, 50%	X	X	X
Citric acid, 15%	210/99	90/32	X
Citric acid, conc.	X	X	X
Copper acetate	90/32	X	X
Copper carbonate	90/32		
Copper chloride	X	X	X
Copper cyanide	X	X	X
Copper sulfate	X	X	X
Cupric chloride, 5%	X		
Cupric chloride, 50%			
Cyclohexane	80/27	80/27	80/27
Cyclohexanol	80/27		80/27
Dichloroethane			210/99
Ethylene glycol	100/38	80/27	80/27
Ferric chloride	80/27	X	X
Ferric chloride, 50% in water	X	X	X
Ferric nitrate, 10–50%	X	X	X
Ferrous chloride		X	X
Ferrous nitrate			
Fluorine gas, dry	X	X	X
Fluorine gas, moist	X		
Hydrobromic acid, dilute	X	X	X
Hydrobromic acid, 20%	X	X	X
Hydrobromic acid, 50%	X	X	X
Hydrochloric acid, 20%	X	X	X
Hydrochloric acid, 38%	X	X	X
Hydrocyanic acid, 10%	X	X	X
Hydrofluoric acid, 30%	X	X	X

TABLE 18.8   Continued

Chemical	Maximum temp. (°F/°C)		
	Copper	Aluminum bronze	Red brass
Hydrofluoric acid, 70%	X	X	X
Hydrofluoric acid, 100%	X	X	X
Hypochlorous acid	X		X
Iodine solution, 10%			
Ketones, general		90/32	100/38
Lactic acid, 25%		X	90/32
Lactic acid, conc.	90/32	90/32	90/32
Magnesium chloride	300/149	90/32	X
Malic acid	X		
Manganese chloride	X		
Methyl chloride	90/32	X	210/99
Methyl ethyl ketone	80/27	60/16	210/99
Methyl isobutyl ketone	90/32		210/99
Muriatic acid	X		
Nitric acid, 5%	X	X	X
Nitric acid, 20%	X	X	X
Nitric acid, 70%	X	X	X
Nitric acid, anhydrous	X	X	X
Nitrous acid, conc.	80/27	X	
Oleum		X	X
Perchloric acid, 10%			X
Perchloric acid, 70%			X
Phenol	X	X	570/299
Phosphoric acid, 50–80%	X	X	X
Picric acid	X	X	X
Potassium bromide, 30%	80/27		
Salicylic acid	90/32		210/99
Silver bromide, 10%	X		
Sodium carbonate	120/49	60/16	90/32
Sodium chloride, to 30%	210/99	60/16	210/99
Sodium hydroxide, 10%	210/99	60/16	210/99
Sodium hydroxide, 50%	X	X	X
Sodium hydroxide, conc.	X	X	X
Sodium hypochlorite, 20%	X	X	80/27
Sodium hypochlorite, conc.	X	X	X
Sodium sulfide, to 50%	X	X	X
Stannic chloride	X	X	X
Stannous chloride	X	X	X
Sulfuric acid, 10%	X	X	200/93
Sulfuric acid, 50%	X	X	X

TABLE 18.8 Continued

Chemical	Maximum temp. (°F/°C)		
	Copper	Aluminum bronze	Red brass
Sulfuric acid, 70%	X	X	X
Sulfuric acid, 90%	X	X	X
Sulfuric acid, 98%	X	X	X
Sulfuric acid, 100%	X	X	X
Sulfuric acid, fuming	X	X	X
Sulfurous acid	X	X	90/32
Toulene	210/99	90/32	210/99
Trichloroacetic acid	80/27	X	80/27
Zinc chloride	X	X	X

The chemicals listed are in the pure state or in a saturated solution unless otherwise indicated. Compatibility is shown to the maximum allowable temperature for which data are available. Incompatibility is shown by an X. A blank space indicates that data are unavailable. When compatible, corrosion rate is <20 mpy.
Source: Ref. 6.

III. COPPER-ZINC ALLOYS (BRASSES)

The principal alloying ingredient of the brasses is zinc. Other alloying ingredients include lead, tin, and aluminum. Machinability is improved by the addition of lead, which does not improve corrosion resistance. Strength and dealloying resistance is increased by the addition of approximately 1% tin. The protective surface film is stabilized by the addition of aluminum. Refer to Table 18.11 for the chemical composition of selected brasses.

Brass alloys of copper can be subject to dealloying (dezincification); a type of corrosion in which the brass dissolves as an alloy and the copper constituent redeposits from solution onto the surface of the brass as a metal in a porous form. The zinc constituent may be deposited in place as an insoluble compound or carried away from the brass as a soluble salt. The corrosion can take place uniformly or locally. The latter, called plug-type dealloying, is prone to take place in alkaline, neutral, or slightly acid environments. The former, called layer-type dealloying, is more apt to occur in acid environments.

Brass alloys containing less than 15% zinc are resistant to dealloying. Alloying elements such as tin and arsenic inhibit this type of corrosion.

Brass alloys containing in excess of 15% zinc are subject to stress corrosion cracking. Moist ammonia in the presence of air will cause stress

TABLE 18.9 High Copper Alloys: Maximum[a] Composition (%)

UNS no.	Cu	Fe	Ni	Co	Be	Pb	P	Zn	Sn	Si	Al
C17000	Balance	b	b	b	1.6–1.79	—	—	—	—	0.20	0.20
C17200	Balance	b	b	b	1.8–2.00	—	—	—	—	0.20	0.20
C17300	Balance	b	b	b	1.8–2.00	0.20–0.60	—	—	—	0.20	0.20
C18000[c]	Balance	0.10	2.5	—	—	—	—	—	0.4	0.7	—
C19200	96.7 min.	0.8–1.2	—	—	—	—	0.01–0.04	—	—	—	—
C19400	97.0 min.	2.1–2.6	—	—	—	0.03	0.015–0.15	0.05–0.20	—	—	—

[a]Unless shown as a range or minimum.
[b]M + Co: 0.20 min.; Ni + Fe + Co: 0.60 max.
[c]Also available in cast form as copper alloy UNS C81540.
Source: Ref. 3.

TABLE 18.10   Mechanical and Physical Properties of Wrought High Copper Alloys

| | Alloy | | |
Property	C17200 cold worked and aged	C18100 cold worked and aged	C19400 strip H04
Modulus of elasticity $\times$ 10^6 (psi)	18.5	18.2	17.5
Yield strength 0.5% ext. 0.2% offset (ksi)	195	66	55
Elongation in 2 in. (%)	5	10	7
Rockwell hardness	C-44	—	B-73
Field strength 10^8 cycles (ksi)	55	—	21
Shear strength (ksi)	115	—	—
Density (lb/in.^{3})	0.298	0.319	0.322
Coefficient of thermal expansion at 70–572°F (10^{-6}/°F)	9.9	10.7	9.2
Thermal conductivity (Btu/ft^2/ft/hr/°F)	62	187	150
Specific heat	0.10	0.094	0.092
Electrical conductivity (% IACS)	22	80	65

TABLE 18.11   Copper-Zinc Alloys: Maximum[a] Composition (%)

UNS no.	Cu	Pb	Fe	Zn	Sn	P	Al	Other
C27000	63.0–68.5	0.10	0.07	Balance	—	—	—	—
C28000	59.0–63.0	0.30	0.07	Balance	—	—	—	—
C44300	70.0–73.0	0.07	0.06	Balance	0.8–1.2	—	—	0.02–0.10 As
C44400	70.0–73.0	0.07	0.06	Balance	0.8–1.2	—	—	0.20–0.10 Sb
C44500	70.0–73.0	0.07	0.06	Balance	0.8–1.2	0.02–0.10	—	—
C46400	59.0–62.0	0.20	0.10	Balance	0.50–1.0	—	—	—
C46500	59.0–62.0	0.20	0.10	Balance	0.50–1.0	—	—	0.02–0.10 As
C46600	59.0–62.0	0.20	0.10	Balance	0.50–1.0	—	—	0.02–0.10 Sb
C46700	59.0–62.0	0.20	0.10	Balance	0.50–1.0	0.02–0.10	—	—
C68700	76.0–79.0	0.07	0.02	Balance	—	—	1.8–2.5	0.02–0.10 As

[a]Unless shown as a range.
Source: Ref. 3.

corrosion cracking of copper alloys. As long as other factors are present the amount of ammonia present need not be great. Relative resistance of the various copper alloys to stress corrosion cracking is as follows:

> *Low resistance*
> Brasses containing >15% zinc
> Brasses containing >15% zinc and small amounts of lead, tin, or
>   aluminum
> *Intermediate resistance*
> Brasses containing <15% zinc
> Aluminum bronzes
> Nickel-silvers
> Phosphor bronzes
> *Good resistance*
> Silicon bronzes
> Phosphorized copper
> *High resistance*
> Commercially pure copper
> Cupronickles

Residual stresses remaining in the metal as the result of cold forming can also lead to stress corrosion cracking. It is possible to remove these stresses by heating the metal to a high enough temperature so that the metal recrystalizes. A stress relieving anneal at a lower temperature can also be provided that will not substantially change the mechanical properties.

As stated previously alloys containing less than 15% zinc resist dealloying. These same alloys are generally more corrosion resistant than high zinc–bearing alloys. They are also more resistant to stress corrosion cracking than the high zinc alloys. Red brass (C23000) is a typical alloy in this group containing 15% zinc. Mechanical and physical properties are given in Table 18.12. The corrosion resistance of red brass is basically the same as that of copper. Refer to Table 18.8. Alloy C23000 is commonly used for piping. Table 18.13 shows the mechanical and physical properties of red brass pipe, and Table 18.14 gives the allowable design stress.

The high zinc brasses, such as C27000, C28000, C44300, and C46400 are more resistant to sulfides than the low zinc brasses. Admiralty brasses (C44300 through C44500) and naval brasses (C46400 through C46700) find use in petroleum refinery operations because of their resistance to dry hydrogen sulfide and sulfides in general. Their dealloying resistance is due to the presence of tin. In addition minor amounts (usually less than 0.10%) of arsenic, antimony, or phosphorus are added to further increase the dealloying resistance. These alloys also find application as marine hardware and pro-

TABLE 18.12 Mechanical and Physical Properties of Wrought Copper-Zinc Brass Alloys

Property	Alloy		
	C23000 rod H04	C26000 strip H02	C26800 rod H07
Modulus of elasticity × 10⁶ (psi)	43	16	—
Yield strength 0.5% ext. (ksi)	52	52	40
Elongation in 2 in. (%)	10	25	45
Rockwell hardness	B-65	B-70	B-55
Field strength 10⁸ cycles (ksi)	—	22	—
Shear strength (ksi)	43	42	36
Density (lb/in.³)	0.316	0.308	0.306
Coefficient of thermal expansion at 70–572°F (10⁻⁶/°F)	10.4	11.1	11.3
Thermal conductivity (Btu/ft²/ft/hr/°F)	92	70	67
Specific heat	0.09	0.09	0.09
Electrical conductivity (% IACS)	37	28	27

peller shafts and for structural use. Table 18.15 shows the mechanical and physical properties of three of these alloys.

The leaded brasses (C31200 through C38500 have improved machinability as a result of the addition of lead. Table 18.16 gives the mechanical and physical properties of these alloys.

IV. COPPER-TIN ALLOYS

These are known as phosphor bronzes or tin bronzes and are probably the oldest alloys known, having been the bronzes of the Bronze Age. Even today artifacts produced during the Bronze Age are still in existence. Items such as statues, vases, bells, and swords have survived hundreds of years of exposure to a wide variety of environments, testifying to the corrosion resistance of these materials. Table 18.17 lists the compositions of these alloys.

In general the tin bronzes are noted for their high strength. The mechanical and physical properties are shown in Table 18.18. Those containing more than 5% tin are especially resistant to impingement attack. Their main application is in water service for such items as valves, valve components, pump casings, and similar items. Because of their corrosion resistance in stagnant waters they also find wide application as components in fire protection systems.

TABLE 18.13 Mechanical and Physical Properties of Red Brass Pipe

Modulus of elasticity × 10^6 (psi)	17
Tensile strength × 10^3 (psi)	40
Yield strength 0.2% offset × 10^3 (psi)	15
Elongation in 2 in. (%)	50
Brinell hardness	50
Density (lb/in.^{3})	0.316
Specific gravity	8.75
Specific heat (Btu/lb °F)	0.09
Thermal conductivity at 32–212°F (Btu/ft^2/hr/°F/in.)	1100
Coefficient of thermal expansion at 31–212°F × 10^{-6} (in./in. °F)	9.8

V. COPPER-ALUMINUM ALLOYS

These alloys are commonly referred to as aluminum bronzes. They are available in both wrought and cast form. The ability of copper to withstand the corrosive effects of salt and brackish water is well known. Copper artifacts recovered from sunken ships have been identifiable and in many cases usable after hundreds of years under the sea. During the early 1900s aluminum was added to copper as an alloying ingredient. It was originally added to give strength to copper while maintaining corrosion resistance of the base metal. As it developed the aluminum bronzes were more resistant to direct chemical attack because aluminum oxide plus copper oxide was formed. The two oxides are complementary and often give the alloy superior corrosion resistance. Table 18.19 lists the alloys generally used for corrosion resistance.

Aluminum bronzes have progressed from simple copper-aluminum alloys to more complex alloys with the addition of iron, nickel, silicon, manganese, tin, and other elements.

TABLE 18.14 Allowable Design Stress for Red Brass Pipe

Operating temp. (°F/°C)	Allowable design stress (psi)
≥300°F/149°C	8000
350/177	6000
400/204	3000
450/232	2000

TABLE 18.15 Mechanical and Physical Properties of Wrought Copper-Zinc-Tin Brass Alloys

Property	Alloy		
	C42500 strip H08	C46400 rod H02	C48200 rod H02
Modulus of elasticity $\times$ 10^6 (psi)	16	15	15
Yield strength 0.5% ext. (ksi)	75	53	53
Elongation in 2 in. (%)	4	20	15
Rockwell hardness	B-92	B-82	B-82
Field strength 10^8 cycles (ksi)	—	—	—
Shear strength (ksi)	—	44	41
Density (lb/in.^{3})	0.317	0.304	0.303
Coefficient of thermal expansion at 70–572°F (10^{-6}/°F)	11.2	11.8	11.8
Thermal conductivity (Btu/ft^2/ft/hr/°F)	64	67	67
Specific heat (Btu/lb°F)	0.09	0.09	0.09
Electrical conductivity (% IACS)	28	26	26

TABLE 18.16 Mechanical and Physical Properties of Wrought Copper-Zinc-Lead Brasses

Property	Alloy		
	C34500 rod H02	C36000 rod H02	C37700 extruded rod
Modulus of elasticity $\times$ 10^6 (psi)	—	15	—
Yield strength 0.5% ext. (ksi)	58	45	20
Elongation in 2 in. (%)	25	25	45
Rockwell hardness	B-80	B-75	F-78
Field strength 10^8 cycles (ksi)	—	—	—
Shear strength (ksi)	42	38	30
Density (lb/in.^{3})	0.307	0.307	0.305
Coefficient of thermal expansion at 70–572°F (10^{-6}/°F)	11.3	11.4	11.5
Thermal conductivity (Btu/ft^2/ft/hr/°F)	67	67	69
Specific heat (Btu/lb°F)	0.09	0.09	0.09
Electrical conductivity (% IACS)	26	26	27

TABLE 18.17 Wrought Copper-Tin Alloys: Maximum[a] Composition (%)

UNS no.	Cu	Pb	Fe	Sn	Zn	P
C51000	Balance	0.05	0.10	4.2–5.8	0.30	0.03–0.35
C51100	Balance	0.05	0.10	3.5–4.9	0.30	0.03–0.35
C52100	Balance	0.05	0.10	7.0–9.0	0.20	0.03–0.35
C52400	Balance	0.05	0.10	9.0–11.0	0.20	0.03–0.35
C54400	Balance	3.5–4.5	0.10	3.5–4.5	1.5–4.5	0.01–0.50

[a]Unless shown as a range.
Source: Ref. 3.

The aluminum content is primarily responsible for the microstructure of these bronzes. A single-phase alpha structure is present in alloys containing approximately 8% aluminum. As the aluminum content is increased other constituents appear in the microstructure. They may be beta, alpha-gamma eutectoid, or a combination, depending on the aluminum content, the cooling rate from solidification, and the heat treatment. An iron-rich intermetallic compound is generally formed when iron is added, while the addition of nickel above 2% tends to introduce an additional kappa phase.

TABLE 18.18 Mechanical and Physical Properties of Wrought Copper-Tin Bronze Alloys

	Alloy	
Property	C50500 strip H04	C51100 strip H089
Modulus of elasticity $\times 10^6$ (psi)	17	16
Yield strength 0.5% ext. (ksi)	50	80
Elongation in 2 in. (%)	8	3
Rockwell hardness	B-75	B-93
Field strength 10^8 cycles (ksi)	—	—
Shear strength (ksi)	—	—
Density (lb/in.^{3})	0.321	0.320
Coefficient of thermal expansion at 70–572°F (10^{-6}/°F)	90	99
Thermal conductivity (Btu/ft^2/ft/hr/°F)	50	48.4
Specific heat (Btu/lb°F)	0.09	0.09
Electrical conductivity (% IACS)	48	20

TABLE 18.19 Wrought Copper-Aluminum Alloys: Maximum[a] Composition (%)

UNS no.	Cu	Al	Fe	Ni	Mn	Si	Sn	Zn	Other
C60800	92.5–94.8	5.0–6.5	0.010	—	—	—	—	—	0.02–0.25 As, 0.10 Pb
C61000	90.0–93.0	6.0–8.5	0.50	—	—	0.10	—	0.20	0.02 Pb
C61300	88.6–92.0	6.0–8.5	2.0–3.0	0.15	0.10	0.10	0.20–0.50	0.05	0.01Pb
C61400	88.0–92.5	6.0–8.0	1.5–3.5	—	0.10	—	—	0.20	0.01 Pb
C61500	89.0–90.5	7.7–8.3	—	1.8–2.2	—	—	—	—	0.015 Pb
C61800	86.9–91.0	8.5–11.0	0.5–1.5	—	—	0.10	—	0.02	0.02 Pb
C62300	82.2–89.5	8.5–11.0	2.0–4.0	1.0	0.50	0.25	0.60	—	—
C63000	78.0–85.0	9.0–11.0	2.0–4.0	4.0–5.5	1.5	0.25	0.20	0.30	—
C63200	75.9–84.5	8.5–9.5	3.0–5.0	4.0–5.5	3.5	0.10	—	—	0.02 Pb

[a]Unless shown as a range.
Source: Ref. 3.

The mechanical and physical properties of typical aluminum bronzes are shown in Table 18.20.

Aluminum bronzes are resistant to nonoxidizing mineral acids such as phosphoric and sulfuric. The presence of an oxidizing agent controls their resistance.

These alloys are also resistant to many organic acids such as acetic, citric, formic, and lactic. The possibility of copper pickup by the finished product may limit their use. Such a pickup may discolor the product even though it is very, very low.

Dealloying is rarely seen in all alpha,, single-phase alloys such as UNS C60800, C61300, or C61400. When dealloying does occur it is in conditions of low pH and high temperature.

Alloy C61300 is used to fabricate vessels to handle acetic acid because its good corrosion resistance, strength, and heat conductivity make it a good choice for acetic acid processing. Alkalies such as sodium and potassium hydroxides can also be handled by aluminum bronze alloys.

Aluminum bronzes are also used as condenser tube sheets in both fossil fuel and nuclear power plants to handle fresh, brackish, and seawaters for cooling, particularly alloys C61300, C61400, and C63000.

VI. COPPER-NICKEL ALLOYS

The copper-nickels are single-phase solution alloys, with nickel as the principal alloying ingredient. The alloys most important for corrosion resistance are those containing 10 and 30% nickel. Table 18.21 lists these wrought alloys. Iron, manganese, silicon, and niobium may be added. Iron improves the impingement resistance of these alloys if it is in solid solution. Iron present in small microprecipitates can be detrimental to corrosion resistance. In order to aid weldability niobium is added.

Of the several commercial copper-nickel alloys available, alloy C70600 offers the best combination of properties for marine application and has the broadest application in seawater service. Mechanical and physical properties are shown in Table 18.22. Alloy 706 has been used aboard ships for seawater distribution and shipboard fire protection. It is also used in many desalting plants. Exposed to seawater, alloy 706 forms a thin but tightly adhering oxide film on its surface. To the extent that this film forms, copper-nickel does in fact "corrode" in marine environments. However, the copper-nickel oxide film is firmly bonded to the underlying metal and is nearly insoluble in seawater. It therefore protects the alloy against further attack once it is formed. Initial corrosion rates may be in the range of from less than 1.0 to about 2.5 mils per year. In the absence of turbulence, as would be the case in a properly designed piping system, the copper-nickels

TABLE 18.20 Mechanical and Physical Properties of Wrought Copper-Aluminum Alloys

	Alloy	
Property	C61300 rod H04	C63000 rod H02
Modulus of elasticity $\times 10^6$ (psi)	17	17
Yield strength 0.5% ext. (ksi)	48	60
Elongation in 2 in. (%)	35	15
Rockwell hardness	B-80	B-96
Field strength 10^8 cycles (ksi)	—	38
Shear strength (ksi)	40	—
Density (lb/in.^{3})	0.287	0.274
Coefficient of thermal expansion at 70–572°F (10^{-6}/°F)	9.0	9.0
Thermal conductivity (Btu/ft^2/ft/hr/°F)	32	21
Specific heat (Btu/lb°F)	0.09	0.09
Electrical conductivity (% IACS)	12	7

corrosion rate will decrease with time, eventually dropping as low as 0.05 mpy after several years of service.

Another important advantage of the oxide film developed on alloy 706 is that it is an extremely poor medium for the adherence and growth of marine life forms. Algae and bromades, the two most common forms of marine biofouling simply will not grow on alloy 706. Alloy 706 piping therefore remains clean and smooth, neither corroding appreciably nor becoming encrusted with growth.

TABLE 18.21 Chemical Composition of Wrought Cupronickels (%)

UNS no.	Cu	Ni	Fe	Mn	Other
C70600	Balance	9.0–11.0	1.0–1.8	1.0 max.	Pb 0.05 max., Zn 1.0 max.
C71500	Balance	29.0–33.0	0.40–0.7	1.0 max.	Pb 0.05 max., Zn 1.0 max.
C71900	Balance	29.0–32.0	0.25 max.	0.5–1.0	Cr 2.6–3.2, Zr 0.08–0.2, Ti 0.02–0.08

TABLE 18.22 Mechanical and Physical Properties of Wrought
Copper-Nickel Alloys

	Alloy	
Property	C70600 tube	C71500 rod H02
Modulus of elasticity $\times 10^6$ (psi)	18	22
Yield strength 0.5% ext. (ksi)	57	70
Elongation in 2 in. (%)	16	15
Rockwell hardness	B-72	B-80
Field strength 10^8 cycles (ksi)	—	—
Shear strength (ksi)	—	42
Density (lb/in.^{3})	0.323	0.323
Coefficient of thermal expansion at 70–572°F (10^{-6}/°F)	19.5	9.0
Thermal conductivity (Btu/ft^2/ft/hr/°F)	26	17
Specific heat (Btu/lb°F)	0.09	0.09
Electrical conductivity (% IACS)	9	4

Flow velocities play an important role in seawater corrosion. Velocity-dependent erosion-corrosion is often associated with turbulence and entrained particles. Recommended velocities for continuous service are as follows:

Pipe size (in.)	Velocity (ft/s)
≤3	5
4–8 with short radius bends	6.5
≥4–8 with long radius bends	11
≥8	14

Table 18.23 shows the mechanical and physical properties of alloy 706 pipe.

Alloy C71500 finds use in many of the same applications as alloy C70600. The mechanical and physical properties are shown in Table 18.22. Sulfides as low as 0.007 ppm in seawater can induce pitting in both alloys, and both alloys are highly susceptible to accelerated corrosion as the sulfide concentration exceeds 0.01 ppm.

TABLE 18.23 Mechanical and Physical Properties of 90-10 Copper-Nickel Alloy 706 Pipe

Modulus of elasticity × 10^6 (psi)	18
Tensile strength × 10^3 (psi)	
$4\frac{1}{2}$ in. O.D.	40
$5\frac{1}{2}$ in. O.D.	38
Yield strength at 0.5% extension under load × 10^3 (psi)	
$4\frac{1}{2}$ in. O.D.	15
$5\frac{1}{2}$ in. O.D.	13
Elongation in 2 in. (%)	25
Density (lb/in.^{3})	0.323
Specific heat (Btu/lb °F)	0.09
Thermal conductivity (Btu/hr ft/ft^2 °F)	26
Coefficient of thermal expansion at 68–572°F × 10^{-6} (in./in. °F)	9.5

Alloy C71900 is a cupronickel to which chromium has been added. It was developed for naval use. The chrome addition strengthens the alloy by spinodal decomposition. This increases the yield strength from 20.5 ksi for C71500 to 45 ksi for C71900. This alloy also has improved resistance to impingement. There is, however, some sacrifice in general corrosion resistance, pitting, and crevice corrosion under stagnant or low-velocity conditions.

The copper-nickels are highly resistant to stress corrosion cracking. Of all the copper alloys they are the most resistant to stress corrosion cracking in ammonia and ammoniacal environments. Although not used in these environments because of cost they are resistant to some nonoxidizing acids, alkalies, neutral salts, and organics.

VII. COPPER-BERYLLIUM ALLOYS

These alloys are also known as beryllium bronze. They are the strongest of the copper-based alloys, able to be heat treated to exceptionally high strength levels—on the order of 155 ksi. Only 2 wt% beryllium is required to achieve this strength level. Applications include springs, electrical switch contacts, diaphragms, and fasteners in the cold-worked state.

VIII. CAST COPPER ALLOYS

The UNS designations for cast copper alloys consist of numbers C80000 through C99999. Temper designations, which define metallurgical conditions, heat treatment, and/or casting method, further describe the alloy. A summary of the temper numbers will be found in Table 18.24, while the complete listing can be found in ASTM B601. The various alloy categories and applications are shown in Table 18.25.

As with other metals the composition of the cast copper alloys varies from that of the wrought alloys. Copper casting possess some advantages over wrought copper in that the casting process permits greater latitude in alloying because hot- and cold-working properties are not important. This is particularly true relative to the use of lead as an alloying ingredient. The chemical composition of the more common copper alloys are given in Table 18.26. Commercially pure copper alloys are not commonly cast.

Copper alloys are not normally selected because of their corrosion resistance alone, but rather for that characteristic plus one or more other properties. In many applications conductivity may be the deciding factor.

A. Mechanical and Physical Properties

Table 18.27 presents the mechanical and physical properties of cast copper alloys. As with the wrought alloys, the heat-treated copper-beryllium alloys have the highest strengths, followed by the aluminum bronzes. Since the copper-beryllium alloys are relatively expensive they have been limited to special applications, such as aircraft parts. It should be noted that beryllium poses a health hazard; therefore OSHA regulations should be consulted before processing and selecting copper-beryllium alloys.

The aluminum bronzes have a wide range of attractive properties including wear resistance, high hardness, good castability, and excellent corrosion resistance. Heat treating of aluminum bronzes will increase their strength. Manganese bronzes have properties similar to the aluminium

TABLE 18.24 Temper Designations of Cast Copper Alloys

Designation	Description
010	Cast and annealed (homogenized)
011	As-cast and precipitation heat-treated
M01 through M07	As-cast; includes casting method
TQ00	Quench hardened
TQ30	Quenched and tempered
TQ50	Quench hardened and temper-annealed
TX00	Spinodal-hardened
TB00	Solution heat-treated
TF00	Solution heat-treated and precipitation hardened

TABLE 18.25 Applications of Cast Copper Alloys

Coppers (C80100–C81200)	Electrical connectors, water-cooled apparatus, oxidation resistant applications.
High copper alloys (C81400–C82800)	Electromechanical hardware; high-strength beryllium-copper alloys are used in heavy duty electromechanical equipment, resistance welding machine components, inlet guide vanes, golf club heads, and components of undersea cable repeater housing.
Brasses (C83000–C87900)	Red brasses: water valves, pipe fittings, plumbing hardware. Leaded yellow brasses: gears, machine components, bolts, nuts. Silicon bronze/ brasses: bearings, gears, intricately shaped pump and valve components, plumbing goods.
Bronzes (C90200–C95900)	Unleaded tin bronzes: bearings, pump impellers, valve fittings, piston rings. Leaded tin bronzes: most popular bearing alloy, corrosion resistant valves and fittings, pressure retaining parts. Nickel-tin bronzes: bearings, pistons, nozzles, feed mechanisms.
Copper-nickels (C96200–C96900)	Pump components, impellers, valves, pipes, marine products, offshore platforms, desalination plants.
Copper-nickel-zinc (C97300–C97800)	Ornamental, architectural, and decorative trim, low-pressure valves and fittings, food, dairy, and beverage industries.
Copper-lead (C98200–C98840)	Special purpose bearings; alloys have relatively low strength and poor impact properties and generally require reinforcement.

Source: Ref. 1.

bronzes and are used in many of the same applications. Machinability of the bronzes is improved by the addition of lead. By adding lead the corrosion resistance is reduced.

The bronze alloys do not contain zinc or nickel as the main alloying ingredient. The primary cast bronze alloys are copper-tin (tin bronze), copper-aluminum (aluminum bronze), and copper-silicon (silicon bronze).

Brass is a copper base alloy containing zinc as the main alloying ingredient. These are the most common casting alloys and have a favorable combination of strength, corrosion resistance, and cost. They are good general purpose alloys.

TABLE 18.26 Chemical Composition of Cast Copper Alloys (wt%)

UNS no.	ASTM spec	Cu	Zn	Sn	Pb	Mn	Al	Fe	Si	Ni	Cb
C83600	B584	85	5	5	5	—	—	—	—	—	—
C85200	B584	72	24	1	3	—	—	—	—	—	—
C85800	B176	61	36	1	1	—	—	—	—	—	—
C86200	B584	63	27	—	—	3	4	3	—	—	—
C86300	B584	61	21	—	—	3	6	3	—	—	—
C87200	B584	89	5	1	—	1.5	1.5	2.5	3	—	—
C87300	B584	95	—	—	—	1	—	—	3	—	—
C87800	B176	87	14	—	—	—	—	—	4	—	—
C90300	B584	88	4	8	—	—	—	—	—	—	—
C90500	B584	88	2	10	—	—	—	—	—	—	—
C92200	B61	85	4	6	1.5	—	—	—	—	—	—
C92300	B584	83	3	7	7	—	—	—	—	—	—
C95200	B184	88	—	—	—	—	9	3	—	—	—
C95400	B148	85	—	—	—	—	11	4	—	—	—
C95500	B148	81	—	—	—	—	11	4	—	4	—
C95800	B148	81	—	—	—	—	9	4	—	4	—
C96200	B369	87.5	—	—	—	0.9	—	1.5	0.1	10	—
C96400	B369	68	—	—	—	—	—	1	—	30	1

Cast copper alloys can be joined in a fashion similar to the wrought alloys. Table 18.28 summarizes the joining processes for copper alloy castings.

B. Corrosion Resistance

The brasses are the most useful of the copper alloys. They find application in seawater with the higher strength, higher hardness materials used under high-velocity and turbulent conditions. In general brass has less corrosion resistance in aqueous solutions than the other copper alloys, although red brass is superior to copper for handling hard water. The addition of zinc does improve the resistance to sulfur compounds, but decreases the resistance to season cracking in ammonia. Refer to the section on dezincification under wrought copper alloys. The brasses also find application in boric acid, neutral salts (such as magnesium chloride and barium chloride), organics (such as ethylene glycol and formaldehyde), and organic acids.

The next major group of copper alloys are the bronzes, which from a corrosion standpoint are very similar to the brasses. Copper-aluminum (aluminum bronze), copper-silicon (silicon bronze), and copper-tin (tin bronze)

TABLE 18.27 Mechanical and Physical Properties of Cast Copper Alloys

UNS no.	Tensile strength (ksi)	Yield strength (ksi)	Brinell hardness	Electrical conductivity (% IACS)	Thermal conductivity (W/mK)	Elongation (%)
C83600	30	14	60	15	72	20
C85200	35	12	—	—	—	25
C85800	55	30	—	—	—	—
C86200	90	45	—	—	—	18
C86300	110	60	—	—	—	12
C86500	65	25	100	22	85.5	—
C87200	45	18	—	—	—	20
C87300	45	18	—	—	—	—
C87800	83	50	—	—	—	—
C90300	40	18	70	12	74.8	20
C90500	40	18	—	—	—	20
C92200	34	16	65	14	69.6	—
C93200	30	14	65	12	58.8	—
C95200	65	25	131	—	—	35
C95400	75	30	170	13	58.7	14
C95500	90	40	187	—	—	12
C95800	85	35	159	7	36	24
C96200	45	25	—	11	45.2	20
C96400	60	32	140	5	28.5	20

TABLE 18.28 Joining Processes for Copper Casting Alloys

UNS. no.	Alloy type	GTAW/GMAW[a]	Brazing	Soldering
C81300–C82200	High copper	Fair	Good	Good
C82400–C82800	Cu-Be	Fair	Fair	Fair
C83800–C85200	Leaded brasses	Not recommended	Good	Excellent
C85300	70-30 brass	Good	Excellent	Excellent
C85400–C85800	Yellow brass	Not recommended	Good	Good
C87200–C87600	Silicon brass	Fair	Fair	Not recommended
C90200–C91600	Tin bronze	Fair	Good	Excellent
C92300–C94500	Leaded tin bronze	Not recommended	Fair	Excellent
C95200–C95700	Aluminum bronze	Good/excellent	Good	Good
C96200–C96400	Copper-nickel	Fair/good	Excellent	Excellent
C97300–C97800	Nickel silvers	Not recommended	Excellent	Excellent

Note: Oxyacetylene welding is generally not recommended. Exceptions are some yellow brasses, C87200 silicon bronze, C87600 silicon brass, and tin bronze (fair).
[a]GTAW, gas tungsten arc welding; GMAW, gas metal arc welding.

are the main cast bronze alloys. The addition of aluminum to the bronzes improves resistance to high-temperature oxidation, increases the tensile properties, and provides excellent resistance to impingement corrosion. They are resistant to many nonoxidizing acids. Oxidizing acids and metallic salts will cause attack. Alloys having more than 8% aluminum should be heat treated since it improves corrosion resistance and toughness. Aluminum bronzes are susceptible to stress corrosion cracking in moist ammonia.

Silicon bronze has approximately the same corrosion resistance as copper, but better mechanical properties and superior weldability. The corrosion rates are affected less by oxygen and carbon dioxide contents than with other copper alloys. Silicon bronzes can handle cold dilute hydrochloric acid, cold and hot dilute sulfuric acid, and cold concentrated sulfuric acid. They have better resistance to stress corrosion cracking than the common brasses. In the presence of high-pressure steam silicon bronze is susceptible to embrittlement.

Tin bronze is less susceptible to stress corrosion cracking than brass, but has less resistance to corrosion by sulfur compounds. The addition of 8–10% tin provides good resistance to impingement attack. Tin bronze has good resistance to flowing seawater and some nonoxidizing acids (except hydrochloric acid).

The final group of copper alloys are the copper-nickel (cupronickels) alloys. They exhibit the best resistance to corrosion, impingement, and stress corrosion cracking of all the copper alloys. They are among the best alloys for seawater service and are immune to season cracking. Dilute hydrochloric, phosphoric, and sulfuric acids can be handled. They are almost as resistant as Monel to caustic soda.

REFERENCES

1. GT Murray. Handbook of Materials Selections for Engineering Applications. New York: Marcel Dekker, 1997.
2. PA Schweitzer. Encyclopedia of Corrosion Technology. New York: Marcel Dekker, 1998.
3. JM Cieslewicz. Copper and copper alloys. In: PA Schweitzer, ed. Corrosion and Corrosion Protection Handbook, 2nd ed. New York: Marcel Dekker, 1988.
4. PA Schweitzer. Corrosion of copper and copper alloys. In: PA Schweitzer, ed. Corrosion Engineering Handbook. New York: Marcel Dekker, 1996.
5. PA Schweitzer. Corrosion Resistant Piping Systems. New York: Marcel Dekker, 1994.
6. PA Schweitzer. Corrosion Resistance Tables, 4th ed., Vols. 1–3. New York: Marcel Dekker, 1995.

7. GT Murray. Introduction to Engineering Materials. New York: Marcel Dekker, 1993.
8. JL Gossett. Corrosion resistance of cast alloys. In: PA Schweitzer, ed. Corrosion Engineering Handbook. New York: Marcel Dekker, 1996.
9. GW George, PA Breig. Cast alloys. In: PA Schweitzer, ed. Corrosion and Corrosion Engineering Handbook. New York: Marcel Dekker, 1988.

19

Aluminum and Aluminum Alloys

Aluminum is one of the most prevalent metallic elements in the solid portion of the earth's crust, comprising approximately 8%. It is always present in a combined form, usually a hydrated oxide, of which bauxite is the primary ore. Metallic aluminum is very active thermodynamically and seeks to return to the natural oxidized state through the process of corrosion.

Aluminum is second to iron in terms of production and consumption and is the most important metal of commerce in the United States. As the resources of other metals are depleted further the relative position of aluminum will increase. Aluminum can be extracted from many resources. Even though such extraction may be more expensive than from bauxite, it will still be economically feasible. The increased cost of extraction will be offset substantially by increased recycling of aluminum, which saves 95% of the energy required for extraction, and by any new smelting processes which require less energy.

Approximately 18 million tons of aluminum are produced each year by the electrolytic reduction of aluminum oxide. Beverage cans alone consume approximately 1.5 million tons, of which 75% is composed of recycled metal. The energy required for recycling is about 5% of that required for electrolytic reduction of aluminum oxide.

Aluminum as a structural metal has the advantages of light weight (2.7 g/m^3 density compared to 7.83 g/m^3 for iron), corrosion resistance, ease of fabrication, and appearance. In strength-to-density ratio (i.e., the specific strength) aluminum alloys have approximately three times that of structural steel and are roughly equivalent to the strength of steels used in gears, shafts, and axles.

Aluminum alloys are produced in the usual flat rolled sheet and plate, in tube, rod, wire, and bar forms, as well as castings, forgings, and stampings.

In addition to being low in cost and having a high strength-to-weight ratio, they are readily fabricated and joined by most of the commonly used methods. The aluminum alloys possess a high resistance to corrosion by most atmospheres and waters, many chemicals, and other materials. Their salts are nontoxic—allowing applications with beverages, foods, and pharmaceuticals, are white or colorless—permitting applications with chemical and other materials without discoloration, and are not damaging to the ecology. Other desirable properties of aluminum and its alloys include a high electrical conductivity, high thermal conductivity, high reflectivity, and noncatalytic action. They are also nonmagnetic.

I. CLASSIFICATIONS AND DESIGNATIONS

Wrought aluminum and aluminum alloys are classified based on their major alloying element via a four-digit numbering system, as shown in Table 19.1. These alloy numbers and their respective tempers are covered by the American National Standards Institute (ANSI) standard H35.1. In the 1XXX group the second digit indicates the purity of the aluminum used to manufacture this particular grade. The zero in the 10XX grade indicates that the aluminum is essentially of commercial purity, while second digits of 1 through 9 indicate special control of one or more individual impurity elements. In the 2XXX through 7XXX alloy groups the second digit indicates an alloy modification. If the second digit is zero, the alloy is the original alloy, while numbers 1 through 9 are assigned consecutively as the original alloy becomes modified. The last two digits serve only to identify the different alloys in the group and have no numerical significance.

TABLE 19.1 Wrought Aluminum and Aluminum Alloy Designation

Series designation	Alloying materials
1XXX	99.9% min. Al
2XXX	Al-Cu, Al-Cu-Mg, Al-Cu-Mg-Li, Al-Cu-Mg-Si
3XXX	Al-Mn, Al-Mn-Mg
4XXX	Al-Si
5XXX	Al-Mg, Al-Mg-Mn
6XXX	Al-Mg-Si, Al-Mg-Si-Mn, Al-Mg-Si, Cu
7XXX	Al-Zn, Al-Zn-Mg, Al-Zn-Mg-Mn, Al-Zn-Mg-Cu

The classification shown in Table 19.1 is based on the major alloying ingredient as shown below:

Series designation	Major alloying ingredient
1XXX	Aluminum ≥99.0%
2XXX	Copper
3XXX	Manganese
4XXX	Silicon
5XXX	Magnesium
6XXX	Magnesium and silicon
7XXX	Zinc
8XXX	Other elements
9XXX	Unused series

The Unified Numbering System (UNS), as used for other metals, is also used for aluminum. A comparison for selected aluminum alloys between the Aluminum Association designation and the UNS designation is shown in Table 19.2.

II. TEMPER DESIGNATIONS

The Aluminum Association has adopted a temper designation system that is used for all product forms, excluding ingots. Aluminum alloys are hardened and strengthened by either deformation at room temperature, referred to as strain hardening, and designated by the letter H, or by an aging heat treatment designated by the letter T. When a wrought alloy is annealed to attain its softest condition the letter O is used in the temper designation. If the product has been shaped without any attempt to control the amount of hardening the letter F (as-fabricated) is used for the temper designation. The letter W is used for the solution-treated (as-quenched) condition. The strain-hardened and heat-treated conditions are further subdivided according to the degree of strain hardening and the type of heat treating (called aging).

III. STRAIN HARDENED SUBDIVISIONS

A. H1X—Strain Hardened Only

The second digit indicates the degree of strain hardening. Full strain hardening is represented by H18. This is produced at about 75% reduction in area at room temperature. A product strained to about half maximum strain hardening is designated by H14, which is attained by approximately 35% reduction in area at room temperatures.

TABLE 19.2 Aluminum Association
Designation System and UNS
Equivalencies for Aluminum Alloys

Aluminum Association designation	UNS designation
1050	A91050
1080	A91080
1100	A91100
1200	A91200
2014	A92014
2024	A92024
3003	A93003
3004	A93004
4043	A94043
4047	A94047
5005	A95005
5052	A95052
5464	A95464
6061	A96061
6063	A96063
6463	A96463
7005	A97005
7050	A97050
7075	A97075

B.   H2X—Strain Hardened and Partially Annealed

The second digit indicates tempers ranging from quarter hard to full hard obtained by partial annealing of a room worked material. Typical tempers are H22, H24, H26, and H28. The highest number represents the highest hardness and strength, as in the H1X series.

C.   H3X—Strain Hardened and Stabilized

Alloys whose mechanical properties have been stabilized by a low-temperature heat treatment or as a result of heat introduced during fabrication are included in the H3X series. This designation applies only to those alloys that, unless stabilized, gradually age soften at room temperature. As before a second digit is used to indicate the degree of strain hardness remaining after stabilization. Typical designations are H32, H34, H36, and H38, again the highest number relating to the highest hardness and strength.

IV. HEAT TREATED SUBDIVISIONS

Temper designations are shown as follows:

T1   Cooled from an elevated process temperature and aged (natural aging) at ambient temperature to a substantially stable condition.

T2   Cooled from an elevated-temperature forming process, cold worked and naturally aged to a substantially stable condition.

T3   Solution heat treated, cold worked, and naturally aged to a substantially stable condition. Solution heat treatment involves heating the product to an elevated temperature, usually in the 932–1022°F (500–550°C) range. At this temperature the alloy becomes a single-phase solid solution which is cooled rapidly (quenched) so that a second phase does not form. This is designated as the W condition, which is an unstable condition. Consequently properties for this condition will not be found in the handbooks.

T4   Solution heat treated and naturally aged to a substantially stable condition.

T5   Cooled from an elevated-temperature forming process and artificially aged (holding for a specific time a temperature above the ambient temperature).

T6   Solution heat treated and artificially aged. The aging process is usually maintained until maximum or near maximum strength is achieved.

T7   Solution heat treated and overaged. Products that have been aged beyond the point of maximum strength to provide some special characteristics such as improved corrosion resistance or ease of forming are included in this designation.

T8   Solution heat treated, artificially aged, and cold worked.

T10   Cooled from an elevated-temperature forming process, cold worked, and artificially aged.

V. CHEMICAL COMPOSITION

The word aluminum can be misleading since it is used for both the pure metal and for the alloys. Practically all commercial products are composed of aluminum alloys.

For aluminum to be considered as "unalloyed" it must have a minimum content of 99.00% of aluminum. Most unalloyed specifications range from 99.00 to 99.75% minimum aluminum.

To assist in the smelting process elements such as bismuth and titanium are added, while chromium, manganese, and zirconium are added for grain control during solidification of large ingots. Elements such as copper, magnesium, nickel, and zinc are added to impart properties such as strength, formability, stability at elevated temperature, etc. Some unintentional impurities are present coming from trace elements contained in the ore, from pickup from ceramic furnace linings, or from the use of scrap metal in recycling.

There are three types of composition listings in use. First there is the nominal, or target, composition of the alloy. This is used in discussing the generic types of alloys, their uses, etc. Second are the alloy limits registered with the Aluminum Association which are the specification limits against which alloys are produced. In these limits intentional alloying elements are defined with allowable ranges. The usual impurity elements are listed as the maximum amount allowed. Rare trace elements are grouped into an "each other" category. Each trace element cannot exceed a specified "each" amount and the total of all trace elements cannot exceed the slightly higher "total" amount. The third listing consists of the elements actually present as found in an analyzed sample. Examples of all three types of these compositions for unalloyed aluminum (1160), a heat-treatable alloy (2024), and a non–heat-treatable alloy (3004) are shown in Table 19.3.

When the melt is analyzed the intentioned elements must be within the prescribed ranges, but not necessarily near the midpoint if the range is wide. For example, as shown in Table 19.3 copper in alloy 2024 can be skewed to a high content (sample 2) to improve strength or low content to improve toughness (sample 3). The producer will target for the nominal composition when the allowable range is only about 0.5 percentage point or less.

Specified impurity elements must be at or below the maximum limit. Individual nonspecified impurities should be less than the "0.05% each" level with the total less than 0.15%. It should be recognized that, as shown in Table 19.3, some impurity elements will be present, but usually well below the allowed limit. Every sample will not contain all of the impurity elements, but some amount of iron and silicon are usually present except in ultrarefined pure aluminum. Certain alloys are produced in several levels of purity, with the less pure levels being less expensive and the higher purity levels improving some property. For example, when the iron and silicon levels are both less than 0.10%, toughness is improved.

It is important to know that other metallic elements are present and necessary for desired properties. Many elements combine with one another and with aluminum to produce intermetallic compounds that are either soluble or insoluble in the aluminum matrix. The presence of second-phase particles is normal and they can be seen and identified by metallographic examination.

TABLE 19.3 Comparison of Composition Listings of Aluminum and Aluminum Alloys

Alloy	Si	Fe	Cu	Mn	Mg	Cr	Ni	Zn	Ti	Others Each	Others Total
Nominal (target) chemical compositon of wrought alloys (%)[a]											
1160	(99.6% min. Al; all other elements 0.040%)										
2024	—	—	4.4	0.6	1.5	—	—	—	—		
3004	—	—	1.2	1.0	—	—	—	—	—		
Registered chemical composition limits of wrought alloys (%)[b]											
1160	0.25	0.35	0.05	0.03	0.03	—	—	0.05	0.03	0.03	0.15
2024	0.50	0.40	3.8–4.9	0.30–0.9	1.2–1.8	0.10	—	0.25	0.05	0.05	0.15
3004	0.30	0.70	0.25	1.0–1.5	0.8–1.3	—	—	0.25	0.05	0.05	0.15
Analysis of aluminum samples alloying elements present (%)[c]											
1160 sample 1	0.080	0.100	0.00	0.00	0.00	0.00	0.00	0.000	0.020		
2024 sample 2	0.25	0.32	4.77	0.61	1.77	0.00	0.00	0.025	0.03		
2024 sample 3	0.10	0.12	4.40	0.55	1.45	0.00	0.00	0.00	0.02		
3004 sample 4	0.030	0.42	0.00	1.25	1.10	0.00	0.00	0.00	0.03		

[a]Aluminum and normal impurities constitute remainder.
[b]Alloying elements shown as a required range; impurity elements are the maximum tolerable. Aluminum and trace impurities constitute remainder.
[c]Remainder is aluminum.

The nominal chemical composition of representative aluminum wrought alloys are given in Table 19.4.

There are two types of wrought alloys: non–heat treatable of the 1XXX, 3XXX, 4XXX, and 6XXX series and heat treatable of the 2XXX, 6XXX, and 7XXX series. In the non–heat-treatable type strengthening is achieved by strain hardening, which may be augmented by solid solution and dispersion hardening. Strengthening is produced in the heat-treatable type by (1) a solution heat treatment at 800–1050°F (400–565°C) to dissolve soluble alloying elements, (2) quenching to retain them in solid solution, and (3) a precipitation or aging treatment, either naturally at ambient temperature or more commonly artificially at 240–380°F (115–195°C), to pre-

TABLE 19.4 Nominal Chemical Compositions of Representative Aluminum Wrought Alloys

Alloy	\ Percent of alloying elements								
	Si	Cu	Mn	Mg	Cr	Zn	Ti	V	Zr
Non–heat-treatable alloys									
1060	99.60% min. Al								
1100	99.00% min. Al								
1350	99.50% min. Al								
3003		0.12	1.2						
3004			1.2	1.0					
5052				2.5	0.25				
5454			0.8	2.7	0.12				
5456			0.8	5.1	0.12				
5083			0.7	4.4	0.15				
5086			0.45	4.0	0.15				
7072[a]						1.0			
Heat-treatable alloys									
2014	0.8	4.4	0.8	0.50					
2219		6.3	0.30				0.06	0.10	0.18
2024		4.4	0.6	1.5					
6061	0.6	0.28		1.0	0.20				
6063	0.4			0.7					
7005			0.45	1.4	0.13	4.5	0.04		0.14
7050		2.3		2.2		6.2			
7075		1.6		2.5	0.23	5.6			

[a]Cladding for Alclad products

cipitate these elements in an optimum size and distribution. Designations for these treatments have been discussed previously.

A high resistance to general corrosion is exhibited by all of the non–heat-treatable alloys. Because of this, selection is usually based on other factors. Alloys of the 1XXX series have relatively low strength. Alloys of the 3XXX series have the same desirable properties as those of the 1XXX series but with higher strength. Magnesium added to some alloys in this series provides additional strength, but the amount is low enough that the alloys still behave more like those with manganese alone than like the stronger Al-Mg alloys of the 5XXX series. Alloys of the 4XXX series are low strength alloys used for brazing and welding products and for a cladding in architectural products.

The strongest non–heat-treatable alloys are those of the 5XXX series, and in most products they are more economical than alloys of the 1XXX and 3XXX series in terms of strength per unit cost. Magnesium is one of the most soluble elements in aluminum and, when dissolved at an elevated temperature, it is largely retained in solution at lower temperatures, even though its equilibrium solubility is greatly exceeded. It produces considerable solution hardening. Additional strength is produced by strain hardening.

Alloys of the 5XXX series have the same high resistance to general corrosion as the other non–heat-treatable alloys in most environments. In addition they exhibit a better resistance in slightly alkaline solutions than that of any other aluminum alloy. These alloys are widely used because of their high as-welded strength when welded with a compatible 5XXX series filler wire.

Of the heat-treatable alloys those of the 6XXX series exhibit a high resistance to general corrosion, equal to or approaching that of the non–heat-treatable alloys. A high resistance to corrosion is also exhibited by alloys of the 7XXX series that do not contain copper as an alloying ingredient. All other heat-treatable alloys have a lower resistance to general corrosion.

VI. MECHANICAL AND PHYSICAL PROPERTIES

Pure aluminum can only be strengthened by solid solution strengthening up to the solid solubility limit of the strengthening element. Heat-treatable alloys are strengthened by solution heat treatment and by precipitation heat treatment.

A. Non–Heat-Treatable Alloys

If an element has appreciable solubility (a few percent) in solid aluminum at relatively low temperatures (about 212°F/100°C), then its addition to aluminum strains the lattice, thereby increasing strength. Elements having an appreciable effect are magnesium, copper, silicon, and silver, in that order. Magnesium is the element used most commercially, and additions of up to 3 wt% are added without any effect on corrosion resistance. An example is shown in Table 19.5. The O temper strength increases from 70 to 195 MPa as the percent of magnesium is increased from 0 to 2.5%. In a similar manner the H34 temper increased from 100 to 260 MPa, and the H38 temper increased from 130 to 290 MPa. The effect of solid solution strengthening by the addition of magnesium can be seen by comparing the figures within columns in Table 19.5.

As long as the alloying element remains in solution its addition has little effect on corrosion resistance. If an excess of the alloying element is added, it will precipitate out as a separate phase, which usually has an adverse effect on corrosion, particularly if the second phase is segregated to specific regions rather than being randomly distributed. Alloys containing 3.5–5.5% magnesium can pose a problem when used for long periods at temperatures in the range of 176–347°F (80–175°C). Exposure at higher temperatures is less of a problem because the precipitate agglomerates into fewer particles, reducing the adverse effect.

All alloys can be cold worked to increase strength. This decreases properties such as elongation, forming, and toughness, but has little effect on corrosion resistance. Table 19.6 shows typical tensile properties of representative non–heat-treatable aluminum wrought alloys in various tempers.

B. Heat-Treatable Alloys

Solution heat treatment involves a eutectic alloy in which the solubility of the alloying element is much greater at high temperatures (698–1012°F/ 370–545°C) than at room or slightly elevated temperatures. When held at an elevated temperature above the solvus temperature the alloying element will dissolve. Rapid cooling (quenching) to room temperature retains much of the dissolved alloying element in a supersaturated solid solution. The actual amount retained depends on the degree of supersaturation, the mobility of the alloying element, and the rate of cooling. The first known heat-treatable alloy systems were the Al-Cu alloys, which are still extensively used. In order to retain all of the copper in solid solution a cooling rate greater than 1040°F (560°C) per second is required. This is only attainable in thin sheet or small parts, such as rivets, although cooling rates of 338°F (170°C) per second or faster will develop substantially improved strengths. As the copper precipitates out of solid solution, it forms intermetallic particles which tend to form along grain boundaries, resulting in susceptibility to intergranular corrosion. In the as-quenched condition, the temper desig-

TABLE 19.5 Examples of Strengthening Aluminum Alloy Sheet by Solid Solution Hardening Through Increased Magnesium Content or Cold Working by Rolling at Temperature of 121°F (50°C)[a]

Alloy	Nominal magnesium (wt%)	O temper ultimate tensile strength (MPa)	H34 temper ultimate tensile strength (MPa)	H38 temper ultimate tensile strength (MPa)
1100	0.0	70	100	130
5005	0.8	125	160	200
5050	1.4	145	190	220
5252	2.5	145	260	240

[a]The O temper receives no cold reduction, the H34 temper about 38% cold reduction, and the H38 temper about 75% cold reduction.
Source: Ref. 3.

nation is T4 when no subsequent cold work is performed and T3X for subsequent cold work. The second digit indicates the degree of cold work.

Precipitation heat treatment, or artificial aging, greatly improves mechanical properties, and if the amount of aging is sufficient, corrosion resistance is improved by eliminating the susceptibility for localized intergranular corrosion. Artificial aging is accomplished by the as-quenched alloy at an elevated temperature, usually in the range of 212–392°F (100–200°C), for a sufficient time to permit fine precipitates to form. These fine precipitates form not only at the grain boundary, but randomly on lattice vacancies throughout the grains.

The artificial aging time is inversely proportional to the temperature and can vary greatly with individual alloys. The aging process for most 2XXX alloys can be accelerated by introducing small amounts of cold work prior to the aging operation. This cold work increases the local strain and dislocations around the insoluble intermetallic Al-Mn particles. Table 19.7 illustrates how alloy 2024 (nominal 4.4% copper) can be strengthened from the fully annealed O temper by (1) solution heat treatment (T42), (2) the addition of cold work (T31), (3) artificial aging (T63), or (4) cold work plus artificial aging (T81 and T86).

Other heat-treatable alloys are the 6XXX aluminum-silicon and the 7XXX aluminum-zinc systems. Cold work prior to aging has only a small effect on the strengths of these alloys.

In most cases the 6XXX alloys are used in the artificially aged, peak strength, T6 temper. Since these alloys have superior formability in the T4 temper they are frequently supplied in this temper. After forming, the finished product is artificially aged.

TABLE 19.6 Typical Tensile Properties of Representative
Non–Heat-Treatable Aluminum Wrought Alloys in Various Tempers[a]

| Alloy and temper | Strength (MPa) | | Elongation (%) | |
	Ultimate	Yield	In 50 mm[b]	In 5D[c]
1060-O	70	30	43	
-H12	85	75	16	
-H14	100	90	12	
-H16	115	105	8	
-H18	130	125	6	
1100-O	90	35	35	42
-H14	125	125	9	18
-H18	165	150	5	13
3003-O	110	40	30	37
-H14	150	145	8	14
-H18	200	185	4	9
3004-O	180	70	20	22
-H34	240	200	9	10
-H38	285	250	5	5
5052-O	195	90	25	27
-H34	260	215	10	12
-H38	290	255	7	7
5454-O	250	115	22	
-H32	275	205	10	
-H34	305	240	10	
-H111	260	180	14	
-H112	250	125	18	
5456-O	310	160		22
-H111	325	230		16
-H112	310	165		20
-H116, H321	350	255		14
5083-O	290	145		20
-H116, H321	315	230		14
5086-O	260	115	22	
-H116, H32	290	205	12	
-H34	325	255	10	
-H112	270	130	14	

[a]Averages for various sizes, product forms, and methods of manufacture; not to be specified as engineering requirements or used for design purposes.
[b]1.60-mm-thick specimen.
[c]12.5-mm-diameter specimen.

TABLE 19.7   Examples of Strengthening Aluminum 2024 Alloy Sheet
by Solution Heat Treatment (SHT) to the T42 and T31 Tempers and
SHT Plus Precipitation Heat Treatment to the T62, T81, and
T86 Temper[a]

Mode	Temper (MPa)					
	O	T42	T31	T62	T81	T86
UTS	220	425	440	435	460	480
TYS	95	260	290	345	440	440

[a]UTS, ultimate tensile strength; TYS, ultimate 2% yield strength.
Note: The T31 tempers are stretched 1−3% cold work and the T86 temper is
stretched 6−10% cold work prior to aging. The O temper is used as a baseline
representing no thermal nor mechanical strengthening.

The 6XXX alloys are the easiest of the heat-treatable alloys to fabricate. Because of their moderately high strength, good machinability, and
high resistance to corrosion in natural environments alloys 6061-T6 and
6063-T3 are used for most general applications. Alloy 6061-T6 provides
higher tensile properties, on the order of 60−70 MPa (8−10 ksi), than alloy
6063-T6, but the latter has a greater resistance to pitting corrosion.

The 7XXX alloys are not used in the as-quenched temper since they
are not metallurgically stable and gradually increase in strength by natural
aging at room temperature. Alloys 7075 and 7050 have shown a continual
change for more than 20 years. Since other characteristics are also changed
by this natural aging process 7XXX products are normally artificially aged
to a stable temper within a month or two after quenching. A minimum 4-
day room temperature interval between quenching and artificial aging is
usually specified to guarantee attaining a more consistent strength in the
final temper.

Many 7XXX-T6 products can be susceptible to exfoliation or stress
corrosion cracking if stressed in the T or ST directions. Consequently alloys
such as 7075, 7050, 7150, and 7055 are generally aged beyond maximum
strength to a variety of T7 type tempers (e.g., T73, T74, T76, and T77) to
provide highest strength attainable together with a specified degree of corrosion resistance. Table 19.8 shows the typical tensile properties of representative heat-treatable alloys in various tempers.

The aluminum alloys most commonly used for piping systems are
alloys 1060, 3003, 5052, 6001, and 6063. Of these alloy 6063 is the most
widely used since it has good mechanical properties at reasonable cost. Table

TABLE 19.8 Typical Tensile Properties of Representative Heat-Treatable Aluminum Wrought Alloys in Various Tempers[a]

Alloy and temper	Strength (MPa)		Elongation (%)	
	Ultimate	Yield	In 50 mm[b]	In 5D[c]
2014-O	185	95		16
-T4, T451	425	290		18
-T6, T651	485	415		11
2219-O	170	75	18	
-T37	395	315	11	
-T87	475	395	10	
2024-O	185	75	20	20
-T4, T351	470	325	20	17
-T851	480	450	6	
-T86	515	490	6	7
6061-O	125	55	25	27
-T4, T451	240	145	22	22
-T6, T651	310	275	12	15
6063-O	90	50		
-T5	185	145	12	
-T6	240	215	12	
-T83	255	240	9	
7005-O	195	85		20
-T63, T6351	370	315		11
7050-T76, T7651	540	485		10
-T736, T73651	510	455		10
7075-O	230	105	17	14
-T6, T651	570	505	11	9
-T76, T7651	535	470		10
-T73, T7351	500	435		11

[a]Averages for various sizes, product forms, and methods of manufacture; not to be specified as engineering requirements or used for design purposes.
[b]1.60-mm-thick specimen.
[c]12.5-mm-diameter specimen.

19.9 lists the mechanical and physical properties of these alloys, while Table 19.10 and Table 19.11 list the allowable design stress values for welded and seamless construction, respectively.

   Table 19.12 shows the physical properties of selected wrought aluminum alloys.

TABLE 19.9 Mechanical and Physical Properties of Aluminum Alloys

	Alloy			
Property	3003-3	5052-0	6061-T6	6063-T6
Modulus of elasticity $\times 10^6$ (psi)	10	10.2	10	10
Tensile strength $\times 10^3$ (psi)	17	41	45	35
Yield strength 0.2% offset $\times 10^3$ (psi)	8	36	40	31
Elongation in 2 in. (%)	40	25	12	12
Density (lb/in.^{3})	.099	.097		.098
Specific gravity	2.73	2.68		2.70
Specific heat (Btu/hr °F)	0.23	0.23		
Thermal conductivity (Btu/hr/ft^2/°F/in.)	1070	960	900	1090
Coefficient of thermal expansion $\times 10^{-6}$ (in./°F/in.)				
at −58−68°F	12		12.1	12.1
at 68−212°F	12.9	13.2	13.0	13.0
at 68−392°F	13.5		13.5	13.6
at 68−572°F	13.9		14.1	14.2

TABLE 19.10 Allowable Design Stress for Aluminum Alloys, Welded Construction

Operating temp. (°F/°C)	Allowable stress (psi)				
	1060	3003	5052	6061	6063
100/38	1650	3350	6250	6000	4250
150/66	1650	3150	6250	5900	4200
200/93	1600	2900	6200	5700	4000
250/121	1450	2700	6000	5400	3800
300/149	1250	2400	5400	5000	3600
350/177	1200	2100	4650	4200	2750
400/204	1050	1800	3500	3200	1900

TABLE 19.11 Allowable Design Stress for
Aluminum Alloys, Seamless Construction

Operating temp. (°F/°C)	Allowable stress (psi)	
	3003-0	6061-T-6
−452/−269	3350	12650
100/38	3350	12650
150/66	3350	12650
200/93	3350	12650
250/121	3250	12250
300/149	2400	10500
350/177	2100	
400/204	1800	

TABLE 19.12 Typical Physical Properties of Selected Wrought
Aluminum Alloys

Alloy	Density (g/cm³)	Electrical conductivity (% IACS)	Thermal conductivity at 25°C (W/mK)	Coefficient of thermal expansion (10⁻⁶/°C)
1100-0	2.70	61	222	24
2014-T6	2.80	34	154	22
2024-T4	2.70	30	121	23
3003-H14	2.73	41	159	23
3004-H38	2.72	42	163	24
4032-T6	2.68	35	138	19
5005-0	2.69	52	200	24
5050-H34	2.69	50	193	24
5052-H34	2.68	30	138	24
5054-H32	2.67	33	130	23
6061-T6	2.69	43	165	24
6063-T6	2.70	53	200	20
7075-T6	2.81	33	130	24

Source: Ref. 4.

VII. CORROSION OF ALUMINUM

The resistance of aluminum to corrosion is dependent upon the passivity of a protective oxide film. The thermodynamic conditions under which this film forms in aqueous solutions are expressed by the potential–pH diagram according to Pourbaix. (Refer to Fig. 19.1). Note from the diagram that aluminum is passive only in the pH range of 4 to 9. The limits of passivity depend on the form of oxide present, the temperature, and the low dissolution of aluminum that must be assumed for inertness. (Theoretically this value cannot be zero for any metal.) At a pH of about 5 the various forms of aluminum oxide all exhibit a minimum solubility.

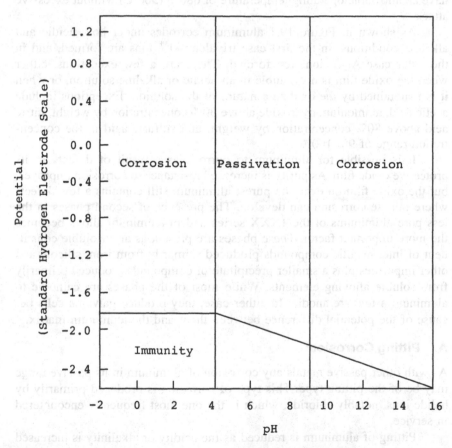

FIGURE 19.1 Potential–pH diagram according to Pourbaix for aluminum at 77°F (25°C) with an oxide film of hydrargillite. (From Ref. 5.)

When the protective oxide film is formed in water and atmospheres at ambient temperatures, it is only a few nanometers thick and structureless. Thicker fims are formed at higher temperatures. These may consist of a thin structureless barrier layer next to the aluminum and a thicker crystalline layer next to the barrier layer. Highly protective films of boehmite (aluminum oxide hydroxide, (A100H) are formed in water near its boiling point, particularly if it is made slightly alkaline. In water or steam at still higher temperatures, thicker, more protective films are formed.

A protective film in water or steam no longer develops starting at a temperature of about 445°F (230°C), and the reaction progresses rapidly until eventually all the aluminum exposed to this media is converted to oxide. Special alloys containing iron and nickel retard this reaction. These alloys have an allowable operating temperature of 680°F (360°C) without excessive attack.

As shown in Figure 19.1 aluminum corrodes under both acidic and alkaline conditions. In the first case trivalent Al^{3+} ions are formed, and in the latter case Al_2O_3 ions are formed. There are a few exceptions, either when the oxide film is not soluble in an acidic or alkaline solution, or when it is maintained by the oxidizing nature of the solution. Exceptions include acetic acid, ammonium hydroxide above 30% concentration by weight, nitric acid above 80% concentration by weight, and sulfuric acid in the concentration range of 98–100%.

It is possible for aluminum to corrode as a result of defects in its protective oxide film. As purity is increased resistance to corrosion improves, but the oxide film on even the purest aluminum still contains a few defects where minute corrosion can develop. The presence of second phases in the less pure aluminums of the 1XXX series and in aluminum alloys becomes the more important factor. These phases are present as an insoluble constituent of intermetallic compounds produced primarily from iron, silicon, and other impurities plus a smaller precipitate of compounds produced primarily from soluble alloying elements. While most of the phases are cathodic to aluminum a few are anodic. In either case, they produce galvanic cells because of the potential difference between them and the aluminum matrix.

A. Pitting Corrosion

As with other passive metals any corrosion of aluminum in its passive range may be of the pitting type. This type of corrosion is produced primarily by halide ions, notably chloride, which is the one most frequently encountered in service.

Pitting of aluminum is reduced as the acidity or alkalinity is increased beyond the passive range of aluminum, at which point the corrosion attack becomes more nearly uniform.

Pits that are almost invisible to the naked eye will develop in polluted outdoor atmospheres. Their growth is relatively rapid during the first few years of exposure, but it eventually stops and they seldom exceed 200 μm. These pits have no effect on the mechanical strength of the structure, but the bright appearance of the surface is gradually replaced by a gray patina of corrosion products. If soot is present, it will become absorbed by the corrosion products and the patina will become dark. Exterior areas exposed to rain will generally age uniformly, but areas sheltered from the washing action of the rain will corrode and produce an uneven gray discoloration. By regularly washing these sections this condition can be prevented.

B. Galvanic Relations

The galvanic series of aluminum alloys and other metals representative of their electrochemical behavior in seawater and in most natural waters and atmospheres is shown in Table 19.13. The effect of alloying elements in determining the position of aluminum alloys in the series is shown in Fig. 19.2. These elements, primarily copper and zinc, affect electrode potential only when they are in solid solution.

As can be seen in Table 19.13 aluminum and its alloys become the anode in galvanic cells with most metals and corrode sacrificially to protect them. Only magnesium and zinc are more anodic and corrode to protect aluminum. Neither aluminum nor cadmium corrode sacrificially in a galvanic cell because they have nearly the same electrode potential.

The degree to which aluminum is polarized in a galvanic cell will determine the degree to which aluminum corrodes when coupled to a more cathodic metal. Contact with copper and its alloys should be avoided because of the low degree of polarization of these metals. Aluminum may be used in contact with stainless steel and chromium in atmospheric and other mild environments with only a slight increase of corrosion. In these environments the two metals polarize highly; therefore the additional corrosion current impressed into the aluminum with them in the galvanic cell is small.

When in contact with other metals the ratio of exposed aluminum to that of the more cathodic metal should be kept as high as possible. This reduces the current density on the aluminum. In order to minimize corrosion, paints and other coatings may be applied to both the aluminum and cathodic metal or to the cathodic metal alone, but never applied to only the aluminum since it is very difficult to apply and maintain free of defects.

Cathodic metals in nonhalide salt solutions usually corrode aluminum to a lesser degree than in solutions of halide salts. This is because the aluminum is less likely to be polarized to its pitting potential. Galvanic corrosion is reduced in any solution when the cathodic reactant is removed.

Table 19.13 Electrode Potentials of Representative
Aluminum Alloys and Other Metals[a]

Aluminum alloy or other metal[b]	Potential (V)
Chromium	+0.18 to −0.40
Nickel	−0.07
Silver	−0.08
Stainless steel (300 series)	−0.09
Copper	−0.20
Tin	−0.49
Lead	−0.55
Mild carbon steel	−0.58
2219-T3, T4	−0.64[c]
2024-T3, T4	−0.69[c]
295.O-T4 (SC or PM)	−0.70
295.O-T6 (SC or PM)	−0.71
2014-T6, 355.O-T4 (SC or PM)	−0.78
355.O-T6 (SC or PM)	−0.79
2219-T6, 6061-T4	−0.80
2024-T6	−0.81
2219-T8, 2024-T8, 356.O-T6 (SC or PM), 443.O-F (PM), cadmium	−0.82
1100, 3003, 6061-T6, 6063-T6, 7075-T6,[c] 443.O-F (SC)	−0.83
1060, 1350, 3004, 7050-T73,[c] 7075-T73[c]	−0.84
5052, 5086	−0.85
5454	−0.86
5456, 5083	−0.87
7072	−0.96
Zinc	−1.10
Magnesium	−1.73

[a]Measured in an aqueous solution of 53 g of NaCl and 3 g of H_2O_2
per liter at 25°C; 0.1 N calomel reference electrode.
[b]The potential of an aluminum alloy is the same in all tempers wherever the temper is not designated.
[c]The potential varies ±0.01 to 0.02 V with quenching rate.

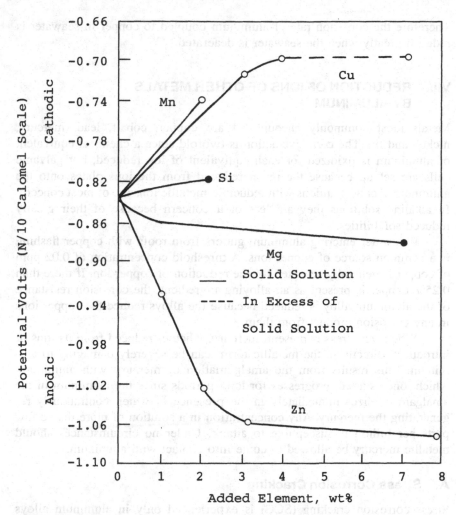

FIGURE 19.2 Effect of alloying elements on the electrode potential of aluminum. (From Ref. 5.)

Therefore the corrosion rate of aluminum coupled to copper in seawater is reduced greatly when the seawater is deaerated.

VIII. REDUCTION OF IONS OF OTHER METALS BY ALUMINUM

Metals most commonly encountered are copper, cobalt, lead, mercury, nickel, and tin. The corrosive action is twofold since a chemical equivalent of aluminum is oxidized for each equivalent of ion reduced, but galvanic cells are set up because the metal reduced from the ions plates onto the aluminum. Acidic solutions with reducible metallic ions are of most concern. In alkaline solutions they are less of a concern because of their greatly reduced solubilities.

Rainwater entering aluminum gutters from roofs with copper flashing is a common source of copper ions. A threshold concentration of 0.02 ppm of copper is generally accepted for the reduction of copper ion. If more than 0.25% copper is present as an alloying ingredient, the corrosion resistance of the aluminum alloy is reduced because the alloys reduce the copper ions in any corrosion product from them.

Whenever stress is present, mercury, whether reduced from its ions or introduced directly in the metallic form, can be severely damaging to aluminum. This results from the amalgamation of mercury with aluminum, which, once started, progresses for long periods since the aluminum in the amalgam oxidizes immediately in the presence of water, continuously regenerating the mercury. Any concentration in a solution of more than a few parts per billion is susceptible to attack. Under no circumstances should metallic mercury be allowed to come into contact with aluminum.

A. Stress Corrosion Cracking

Stress corrosion cracking (SCC) is experienced only in aluminum alloys having appreciable amounts of copper, magnesium, silicon, and zinc as alloying elements. The cracking is normally intergranular and may be produced whenever alloying ingredients precipitate along grain boundaries, depleting the regions adjacent to them of these ingredients. Metallurgical treatment of these alloys can improve or prevent stress corrosion cracking in aluminum alloys. The process of stress corrosion cracking can be retarded greatly, if not completely eliminated, by cathodic protection.

Stress corrosion cracking of an aluminum alloy in a susceptible temper is determined by the magnitude and duration of a tensile stress acting on its surface. Resistance to SCC is highest for stressing parallel to the longitudinal direction of grains and lowest for stressing across the minimum thickness of grains. Therefore, in wrought alloys having an elongated grain structure, and in products thick enough for stressing in all directions, resistance to SCC in the short transverse direction may be the controlling factor in applying these alloys.

For SCC to take place water or water vapor must be present; otherwise cracking will not occur. The presence of halides will accelerate cracking further.

Sufficient amounts of alloying elements are present in wrought alloys of the 2XXX, 5XXX, 6XXX, and 7XXX series that make them subject to SCC. Special treatment can cause SCC in the 6XXX series, but cracking has never been experienced in commercial alloys Tempers have been developed to provide a very high resistance to stress corrosion cracking in the other three alloy series.

B. Exfoliation Corrosion

Exfoliation corrosion is a leafing or delamination of the product. Wrought aluminum products, in certain tempers, are subject to this type of corrosion. Alloys of the 2XXX, 5XXX, and 7XXX series are the most prone to this type of corrosion. Both exfoliation corrosion and stress corrosion cracking in alloys of this series are associated with decomposition of solid solution selectively along boundaries. Consequently, metallurgical treatment that improves resistance to SCC also improves resistance to exfoliation corrosion; however, resistance to the latter is usually achieved first.

Exfoliation corrosion is infrequent and less severe in wrought alloys of the non–heat-treatable type.

C. Weathering

Aluminum alloys, except those containing copper as a major alloying element, have a high resistance to weathering in most atmospheres. After an initial period of exposure the depth of attack decreases to a low rate. The loss in strength also decreases in the same manner after the initial period, but not to as low a rate.

This "self-limiting" characteristic of corrosive attack during weathering also occurs with aluminum alloys in many other environments.

D. Waters

Wrought alloys of the 1XXX, 3XXX, and 5XXX series exhibit excellent resistance to high purity water. When first exposed a slight reaction takes place, producing a protective oxide film on the alloys within a few days, after which pickup of aluminum by water becomes negligible. The presence of carbon dioxide or oxygen dissolved in the water does not appreciably affect the corrosion resistance of these alloys; neither is the corrosion resistance affected by the chemicals added to the water to minimize the corrosion of steel because of the presence of these gasses.

These same alloys are also resistant to many natural waters, their resistance being greater in neutral or slightly alkaline waters and less in acidic waters.

Resistance to corrosion by seawater is also high. General corrosion is minimal. Corrosion of these alloys in seawater is primarily of the pitting type. The rates of pitting usually range from 3–6 μm/year during the first year and from 0.8–1.5 μm/year averaged over a 10 year period. The lower rate for the longer period indicates the tendency of older pits to become inactive.

Alloys of the 5XXX series have the highest resistance to seawater and are widely used for marine applications.

E. General Corrosion Resistance

All of the non–heat-treatable alloys have a high degree of corrosion resistance. These alloys, which do not contain copper as a major alloying ingredient, have a high resistance to corrosion by many chemicals. They are compatible with dry salts of most inorganic chemicals and, within their passive range of pH 4–9 in aqueous solutions, with most halide salts, under conditions at which most alloys are polarized to their pitting potentials. In most other solutions where conditions are less likely to occur that will polarize the alloys to these potentials, pitting is not a problem.

Aluminum alloys are not compatible with most inorganic acids, bases, and salts with pH outside the passive range of 4–9.

Aluminum alloys are resistant to a wide variety of organic compounds including most aldehydes, esters, ethers, hydrocarbons, ketones, mercaptans, other sulfur-containing compounds, and nitro compounds. They are also resistant to most organic acids, alcohols, and phenols, except when these compounds are nearly dry and near their boiling points. Carbon tetrachloride also exhibits this behavior.

Aluminum alloys are most resistant to organic compounds halogenated with chlorine, bromine, and iodine. They are also resistant to highly polymerized compounds.

It should be noted that the compatibility of aluminum alloys with mixtures of organic compounds cannot always be predicted from their compatibility with each of the compounds. For example, some aluminum alloys are corroded severely in mixtures of carbon tetrachloride and methyl alcohol, even though they are resistant to each compound alone. Caution should be exercised in using data for pure organic compounds to predict performance of the alloys with commercial grades that may contain contaminents. Ions of halides and reducible metals, commonly copper and chloride, frequently have been found to be the cause of excessive corrosion of aluminum alloys in commercial grades of organic chemicals that would not have been predicted from their resistance to pure compounds.

Regardless of environment pure aluminum has the greatest corrosion resistance, followed by the non–heat-treatable alloys, and finally the heat-treatable alloys. The two most frequently used alloys are 3003 and 3004. The 3XXX series of alloys are not susceptible to the more drastic forms of localized corrosion. The principal type of corrosion encountered is pitting corrosion. With a low copper content of <0.05% the 3003 and 3004 alloys are almost as resistant as pure aluminum.

Large quantities of aluminum are used for household cooking utensils and for the commercial handling and processing of foods. Aluminum and aluminum alloys such as foil, foil laminated to plastics, and cans are used for the packaging of foods and beverages. For most applications, lacquers and plastic laminated coatings are applied to the alloys. Because of the long periods of exposure only the smallest amount of corrosion can be tolerated.

Refer to Table 19.14 for the compatibility of aluminum with selected corrodents. Reference 2 provides a more detailed listing.

F. Alclad Products

Alclad products are used where perforation of a product cannot be tolerated. These products consist of a core alloy and a more anodic cladding alloy, usually representing 10% or less of the total thickness, which is metallurgically bonded to one or both surfaces of the core alloy. Any corrosion which takes place progresses only to the cladding–core interface, then spreads laterally as a result of the cathodic protection provided by the core of the cladding.

IX. CAST ALUMINUM

There is no single commercial designation system for aluminum castings. The most widely used system is that of the Aluminum Association. It consists of a four-digit numbering system incorporating a decimal point to separate the third and fourth digits. The first digit identifies the alloy group, as listed in Table 19.15.

Aluminum castings are of two types: heat treatable, corresponding to the same type of wrought alloys where strengthening is produced by dissolution of soluble alloying elements and their subsequent precipitation, and

TABLE 19.14   Compatibility of Aluminum Alloys with Selected Corrodents[a]

Chemical	Maximum temp. °F	Maximum temp. °C	Chemical	Maximum temp. °F	Maximum temp. °C
Acetaldehyde	360	182	Barium sulfide	X	
Acetamide	340	171	Benzaldehyde	120	49
Acetic acid, 10%	110	43	Benzene	210	99
Acetic acid, 50%	130	54	Benzene sulfonic acid, 10%	X	
Acetic acid, 80%	90	32	Benzoic acid, 10%	400	204
Acetic acid, glacial	210	99	Benzyl alcohol	110	43
Acetic anhydride	350	177	Benzyl chloride	X	
Acetone	500	260	Borax	X	
Acetyl chloride	X		Boric acid	100	38
Acrylonitrile	210	99	Bromine gas, dry	60	16
Adipic acid	210	99	Bromine gas, moist	X	
Allyl alcohol	150	66	Bromine liquid	210	99
Allyl chloride	X		Butadiene	110	43
Alum	110	43	Butyl acetate	110	43
Aluminum acetate	60	16	Butyl alcohol	210	99
Aluminum chloride, aqueous	X		n-Butylamine	90	32
Aluminum chloride, dry	60	16	Butyl phthalate	X	
Aluminum fluoride	120	49	Butyric acid	180	82
Aluminum hydroxide	80	27	Calcium bisulfite	X	
Aluminum nitrate	110	43	Calcium carbonate	X	
Aluminum sulfate	X		Calcium chlorate	140	60
Ammonia gas	X		Calcium chloride, 20%	100	38
Ammonium carbonate	350	177	Calcium hydroxide, 10%	X	
Ammonium chloride, 10%	X		Calcium hydroxide, sat.	X	
Ammonium chloride, 50%	X		Calcium hypochlorite	X	
Ammonium chloride, sat.	X		Calcium nitrate	170	77
Ammonium fluoride, 10%	X		Calcium oxide	90	32
Ammonium fluoride, 25%	X		Calcium sulfate[b]	210	99
Ammonium hydroxide, 25%	350	177	Caprylic acid	300	149
Ammonium hydroxide, sat.	350	177	Carbon bisulfide	210	99
Ammonium nitrate	350	177	Carbon dioxide, dry	570	299
Ammonium persulfate	350	177	Carbon dioxide, wet	170	77
Ammonium phosphate	X		Carbon disulfide	210	99
Ammonium sulfate, 10–40%	X		Carbon monoxide	570	299
Ammonium sulfide	170	77	Carbon tetrachloride	X	
Ammonium sulfite	X		Carbonic acid	80	27
Amyl acetate	350	177	Cellosolve	210	99
Amyl alcohol	170	77	Chloracetic acid, 50% water	X	
Amyl chloride	90	32	Chloracetic acid	X	
Aniline[b]	350	177	Chlorine gas, dry	210	99
Antimony trichloride	X		Chlorine gas, wet	X	
Aqua regia, 3:1	X		Chlorobenzene	150	66
Barium carbonate	X		Chloroform, dry	170	77
Barium chloride, 30%	180	82	Chlorosulfonic acid, dry	170	77
Barium hydroxide	X		Chromic acid, 10%	200	93
Barium sulfate	210	99	Chromic acid, 50%	100	38

TABLE 19.14 Continued

Chemical	Maximum temp. °F	Maximum temp. °C	Chemical	Maximum temp. °F	Maximum temp. °C
Chromyl chloride	210	99	Muriatic acid	X	
Citric acid, 15%	210	99	Nitric acid, 5%	X	
Citric acid, conc.	70	21	Nitric acid, 20%	X	
Copper acetate	X		Nitric acid, 70%	X	
Copper carbonate, 10%	X		Nitric acid, anhydrous	90	32
Copper chloride	X		Nitrous acid, conc.	X	
Copper cyanide	X		Oleum	100	38
Copper sulfate	X		Perchloric acid, 10%	X	
Cresol	150	66	Perchloric acid, 70%	X	
Cupric chloride, 5%	X		Phenol	210	99
Cyclohexane	180	81	Phosphoric acid, 50–80%	X	
Cyclohexanol	X		Picric acid	210	99
Dichloroethane (ethylene dichloride)	110	43	Potassium bromide, 30%[b]	80	27
Ethylene glycol	100	38	Salicylic acid	130	54
Ferric chloride	X		Silver bromide, 10%	X	
Ferric chloride, 50% in water	X		Sodium carbonate	X	
Ferric nitrate, 10–50%	X		Sodium chloride	X	
Ferrous chloride	X		Sodium hydroxide, 10%	X	
Fluorine gas, dry	470	243	Sodium hydroxide, 50%	X	
Fluorine gas, moist	X		Sodium hydroxide, conc.	X	
Hydrobromic acid, dilute	X		Sodium hypochlorite, 20%	80	27
Hydrobromic acid, 20%	X		Sodium hypochlorite, conc.	X	
Hydrobromic acid, 50%	X		Sodium sulfide, to 50%	X	
Hydrochloric acid, 20%	X		Stannic chloride	X	
Hydrochloric acid, 38%	X		Stannous chloride, dry	X	
Hydrocyanic acid, 10%	100	38	Sulfuric acid, 10%	X	
Hydrofluoric acid, 30%	X		Sulfuric acid, 50%	X	
Hydrofluoric acid, 70%	X		Sulfuric acid, 70%	X	
Hydrofluoric acid, 100%	X		Sulfuric acid, 90%	X	
Hypochlorous acid	X		Sulfuric acid, 98%	X	
Iodine solution, 10%	X		Sulfuric acid, 100%	X	
Ketones, general	100	38	Sulfuric acid, fuming	90	32
Lactic acid, 25%	80	27	Sulfurous acid	370	188
Lactic acid, conc.[c]	100	38	Thionyl chloride	X	
Magnesium chloride	X		Toluene	210	99
Malic acid	210	99	Trichloroacetic acid	X	
Methyl chloride	X		White liquor	100	38
Methyl ethyl ketone	150	66	Zinc chloride	X	
Methyl isobutyl ketone	150	66			

[a]The chemicals listed are in the pure state or in a saturated solution unless otherwise indicated. Compatibility is shown to the maximum allowable temperature for which data are available. Incompatibility is shown by an X. When compatible, the corrosion rate is <20 mpy.
[b]Material subject to pitting.
[c]Material subject to intergranular corrosion.
Source: Ref. 2.

TABLE 19.15   Designation of Aluminum Castings

Series	Alloy system
1XXX.X	99.9% minimum aluminum
2XXX.X	Aluminum plus copper
3XXX.X	Aluminum plus silicon plus magnesium
	Aluminum plus silicon plus copper
	Aluminum plus silicon plus copper plus magnesium
4XXX.X	Aluminum plus silicon
5XXX.X	Aluminum plus magnesium
6XXX.X	Currently unused
7XXX.X	Aluminum plus zinc
8XXX.X	Aluminum plus tin
9XXX.X	Currently unused

non–heat treatable, in which strengthening is produced primarily by constituents of insoluble or undissolved alloying elements. Tempers of heat-treatable casting alloys are designated by an F. Alloys of the heat-treatable type are usually thermally treated subsequent to their casting, but for a few in which a considerable amount of alloying elements are retained in solution during casting they may not be thermally treated after casting; thus they may be used in both the F and fully strengthened T tempers.

The 1XXX.X series is assigned to pure aluminum. Besides ingot, the only major commercial use of pure aluminum castings is electrical conductor parts such as collector rings and bus bars. Because of their low strength these alloys are usually cast with integral steel stiffeners.

The 2XXX.X series of the aluminum + copper alloys were the first type of casting alloys used commercially and are still used. They provide medium to high strength but are difficult to cast. These alloys are the least corrosion resistant and can be susceptible to SCC in the maximum strength of T6 temper.

The 3XXX.X alloys provide the best combination of strength and corrosion resistance. They are produced in both as-cast (F) tempers and heat-treated tempers T5 through T7.

The 4XXX.X castings are the most prevalent because of their superior casting characteristics. They provide reasonably good corrosion resistance but low to medium strength.

The 5XXX.X castings provide the highest resistance to corrosion and good machinability and weldability. However, they have low to medium strength and are difficult to cast, being limited to sand castings or simple permanent mold shapes.

The 7XXX.X castings find limited applications. They are difficult to cast and are limited to simple shapes. They have medium to good resistance to corrosion and high melting points.

The 8XXX.X castings were designed for bearings and bushings in internal combustion engines. Required properties are the ability to carry high compressive loads and good fatigue resistance.

Nominal chemical compositions of representative aluminum alloys are shown in Table 19.16.

A. Mechanical and Physical Properties

The properties of cast products will vary considerably depending on whether the alloy will be used in the as-cast or the cast and heat-treated condition. The mechanical and physical properties for selected alloys are listed in Table 19.17.

B. Casting Processes

Aluminum castings can be produced by all casting processes, of which die, permanent mold, and sand casting account for the greatest production. The selection of a casting alloy must take into consideration the casting characteristics as well as the properties.

TABLE 19.16 Nominal Chemical Compositions of Representative Aluminum Casting Alloys

Alloy	Alloying elements (%)				
	Si	Cu	Mg	Ni	Zn
Alloys not normally heat treated					
360.0	9.5		0.5		
380.0	8.5	3.5			
443.0	5.3				
514.0			4.0		
710.0		0.5	0.7		6.5
Alloys normaly heat treated					
295.0	0.8	4.5			
336.0	12.0	1.0	1.0	2.5	
355.0	5.0	1.3	0.5		
356.0	7.0		0.3		
357.0	7.0		0.5		

TABLE 19.17 Typical Properties of Selected Aluminum Casting Alloys

Alloy no.	Yield strength (ksi)	Elongation in 50 mm (%)	Electrical conductivity (% IACS)	Thermal conductivity at 25°C (W/mk)	Coefficient of thermal expression at 20–300°C (10^{-6}/°C)
208.0-F (s)	14	2.5	31	121	23.9
295.0-T4 (s)	16	8.5	35	138	24.8
336.0-T5 (p)	28	1.0	—	—	—
355.0-T6 (s)	24.6	3.0	—	—	—
-T6 (p)	35	4.0	—	—	—
-T61 (s)	36	3.0	—	—	—
-T62 (p)	52	1.5	—	—	—
356.0-T6 (s)	24	3.5	34	151	23.4
-T6 (p)	27	5.0	—	—	—
-T7 (s)	30	2.0	—	—	—
-T7 (p)	24	6.0	—	—	—
357.0-T6 (s)	43	2.0	—	—	—
-T6 (p)	43	5.0	—	—	—
-T7 (s)	37	3.0	—	—	—
-T7 (p)	30	5.0	—	—	—
360.0-F (d)	25	3.0	37	142	22.9
380.0-F (d)	24	3.0	27	96	22.5
390.0-F (d)	35	1.0	25	134	18.5
-T5 (d)	39	1.0	24	134	18.0
443.0-F (d)	16	9.0	37	142	23.8
-F (s)	8	8.0	37	142	24.1
413.0-F (d)	21	2.5	39	121	22.5
514.0-T (s)	12	9.0	35	146	25.9
518.0-F (d)	27	8.0	24	96	26.1
520.0-T4 (s)	48	16.0	21	88	25.0
710.0-F (s)	25	3.0	—	—	—
713.0-T5 (s)	30	4.0	35	140	24.1

s, sand mold; d, die; p, permanent mold.

In castings, as well as in wrought products, copper is the alloying element most deleterious for general corrosion resistance. Alloys such as 356.0, 513.0, and 414.0 that do not contain copper as an alloying ingredient have a high resistance to general corrosion comparable to that of non–heat-treatable wrought alloys. Corrosion resistance decreases as the copper content of the alloy increases.

C. Corrosion Resistance

In general the corrosion resistance of a cast aluminum alloy is equivalent to that of the comparable wrought aluminum alloy.

REFERENCES

1. GT Murray. Introduction to Engineering Materials. New York: Marcel Dekker, 1993.
2. PA Schweitzer. Corrosion Resistance Tables, 4th ed., Vols. 1–3. New York: Marcel Dekker, 1995.
3. BW Lifka. Corrosion of aluminum and aluminum alloys. In: PA Schweitzer, ed. Corrosion Engineering Handbook. New York: Marcel Dekker, 1996.
4. GT Murray. Handbook of Materials Selection for Engineering Applications. New York: Marcel Dekker, 1997.
5. EH Hollingsworth, HY Hunsicher. Aluminum alloys. In: PA Schweitzer, ed. Corrosion and Corrosion Engineering Handbook, 2nd ed. New York: Marcel Dekker, 1988.
6. PA Schweitzer. Corrosion Resistant Piping Systems. New York: Marcel Dekker, 1998.
7. PA Schweitzer. Encyclopedia of Corrosion Technology. New York: Marcel Dekker, 1998.
8. JL Gossett. Corrosion resistance of cast alloys. In: PA Schweitzer, ed. Corrosion Engineering Handbook. New York: Marcel Dekker, 1996.
9. GW George, PA Breig. Cast alloys. In: PA Schweitzer, ed. Corrosion and Corrosion Engineering Handbook, 2nd ed. New York: Marcel Dekker, 1988.

20

Zinc and Zinc Alloys

Zinc, as a pure metal, finds relatively few applications basically because of its poor mechanical properties. It is relatively weak. Unalloyed zinc used for roofing up to about the 1960s was soft and had a tendency to creep. If the bays of fully supported roofing were too wide, they would eventually sag in the middle.

The single largest use of zinc is in the application of zinc coatings (galvanizing) to permit the most efficient use of steel and to conserve energy.

I. CORROSION OF ZINC

Depending upon the nature of the environment zinc has the ability to form a protective layer made up of basic carbonates, oxides, or hydrated sulfates. Once the protective layers have formed, corrosion proceeds at a greatly reduced rate. Consideration of the corrosion of zinc is primarily related to the slow general dissolution from the surface. Even with a considerable moisture content air is only slightly corrosive to zinc. Below 390°F (200°C), the film grows very slowly and is very adherent. Zinc-coated steel behaves similarly to pure zinc.

The pH of the environment governs the formation and maintenance of the protective film. Within the pH range of 6 to 12.5 the corrosion rate is low. Corrosive attack is most severe at pH values below 6 and above 12.5.

Uniform corrosion rates of zinc are not appreciably affected by the purity of the zinc. However, the addition of some alloying elements can increase the corrosion resistance of zinc.

A. White Rust (Wet Storage Stain)

White rust is a form of general corrosion which is not protective. It is more properly called wet storage stain because it occurs in storage where water is present, but only a limited supply of oxygen and carbon dioxide is available. Wet stain formation will be accelerated by the presence of chlorides and sulfates.

White rust is a white, crumbly, and porous coating consisting of $2ZnCO_3 \cdot 3Zn(OH)_2$ together with ZnO and voluminous β-$Zn(OH)_2$. The surface underneath the white products is often dark gray.

This coating is found particularly on newly galvanized bright surfaces, particularly in crevices between closely packed sheets whose surfaces have come into contact with condensate or rain water and the moisture cannot dry up quickly. If the zinc surfaces have already formed a protective film prior to storage, the chances are that no attack will take place.

Short-term protection against wet storage staining can be provided by chromating or phosphating. Painting after galvanizing will also provide protection.

Materials stored outdoors should be arranged so that all surfaces are well ventilated and that water can easily run off of the surfaces. If possible, new zinc surfaces should not be allowed to come into contact with rain or condensate water during transit or storage. This is the best way of preventing wet storage stain. Figure 20.1 illustrates the stacking of galvanized parts out of doors.

B. Bimetallic Corrosion

The ratio of the area of metals in contact, the duration of wetness, and the conductivity of the electrolyte will determine the severity of corrosive attack. Seawater, which is a highly conducting solution, will produce a more severe bimetallic corrosion than most fresh waters, which generally have a lower conductivity. A film of moisture condensed from the air or rain water can dissolve contaminants and produce conditions conducive to bimetallic corrosion. (See Ref. 4.)

Bimetallic corrosion is less severe under atmospheric exposure than under immersed conditions. In the former, attack will occur only when the surface is wet, which depends on several factors such as the effectiveness of drainage, presence or retention of moisture in crevices, and the speed of evaporation.

Under normal circumstances galvanized steel surfaces may safely be in contact with types 304 and 316F stainless steel, most aluminum alloys, chrome steel (>12% Cr), and tin, provided the area ratio of zinc to metal is

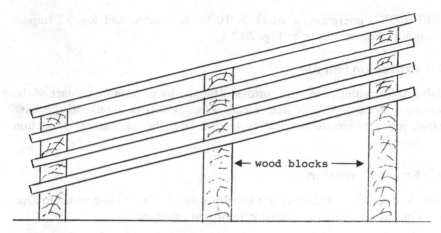

FIGURE 20.1 Stacking of galvanized parts out of doors.

1:1 or lower and oxide layers are present on both aluminum alloys and the two stainless steels.

Prevention of bimetallic corrosion can be accomplished by preventing the flow of the corrosion currents between the dissimilar metals in contact. This can be by either insulating the dissimilar metals from each other (breaking the metallic path) or by preventing the formation of a continuous bridge of conductive solution between the two metals (breaking the electrolytic path).

If electrical bonding is not required, the first method may be possible by providing insulation when under immersed conditions. For example, a zinc-coated steel bolt and nut may be fitted with an insulating bushing and washers where it passes through a steel surface that cannot be coated.

The second method may be accomplished by the application of paint or plastic coatings to the immersed parts of the metal. If it is not practical to coat both metals, it is preferable to coat the more noble metal, not the zinc.

C. Intergranular Corrosion

If pure zinc-aluminum alloys are exposed to temperatures in excess of 160°F (70°C) under wet or damp conditions, intergranular corrosion may take place. The use of these alloys should be restricted to temperatures below 160°F (70°C) and impurities controlled to specific limits of 0.006% each for lead and cadmium and to 0.003% for tin.

Impact strength can decrease as a result on intergranular corrosion as well as by aging. At 140°F (60°C), in high humidity, the loss is minimal. At

203°F (95°C) intergranular attack is 10 times greater, and loss of impact strength increases. (Refer to Fig. 20.2.)

D. Corrosion Fatigue

Galvanized coatings can stop corrosion fatigue by preventing contact of the corrosive substances with base metal. Zinc, which is anodic to the base metal, provides electrochemical protection after the mechanical protection has ceased.

E. Stress Corrosion

Zinc or zinc-coated steels are not usually subjected to stress corrosion. Zinc can also prevent stress corrosion cracking in other metals.

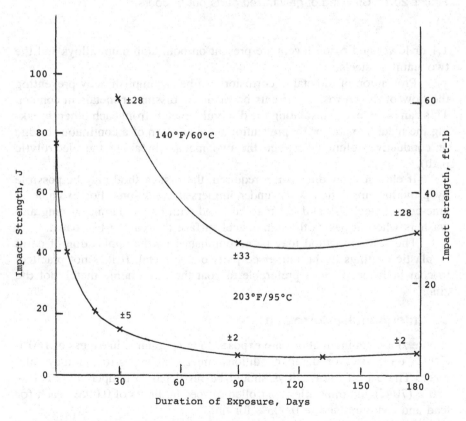

FIGURE 20.2 Effect of intergranular corrosion of zinc-aluminum alloys on impact strength.

II. ZINC COATINGS

Approximately half of the world's production of zinc is used to protect steel from rust. The reasons for this wide application are

1. Prices are relatively low.
2. Due to large reserves, an ample supply of zinc is available.
3. There is a great flexibility in application procedures resulting in many different qualities with well-controlled layer thicknesses.
4. Steel provides good cathodic protection.
5. Many special alloy systems have been developed with improved corrosion protection properties.

A. Principle of Protection

Zinc coatings protect the substrate by means of cathodic control. Cathodic overpotential of the surface is increased by the coating, which makes the corrosion potential more negative than that of the substrate. The coating layer acts as a sacrificial anode for iron and steel substrates when the substrates are exposed to the atmosphere. The coating layer provides cathodic protection for the substrate by galvanic action. Zinc is therefore considered a sacrificial metal.

The electrical conductivity of the electrolyte, the temperature, and the surface condition determine the galvanic action of the coating. An increase in the cathodic overpotential is responsible for the corrosion resistance of the coating layer. Figure 20.3 shows the principles of cathodic control protection by a sacrificial metal coating. The corrosion rate, i_{corr}, of zinc-coated iron becomes lower than that of uncoated iron since the cathodic overpotential of the surface is increased by zinc coating, and the exchange current density of dissolved oxygen, i_{oc}, on zinc is lower than that on iron. If a small part of iron is exposed to the atmosphere, the electrode potential of the exposed iron, E_{corr}, is equal to the corrosion potential of zinc since the exposed iron is polarized cathodically by the surrounding zinc, so that little corrosion occurs on the exposed iron. Zinc ions dissolved predominately from the zinc coating form the surrounding barrier of corrosion products at the defect, thereby protecting the exposed iron.

Sacrificial metal coatings protect iron and steel by two or three protective abilities such as

1. Original barrier action of coating layer
2. Secondary barrier action of corrosion product layer
3. Galvanic action of coating layer

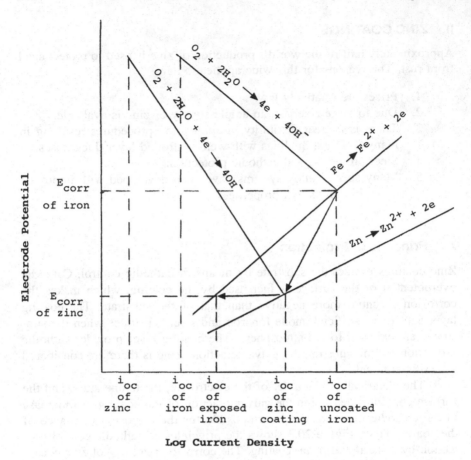

FIGURE 20.3 Cathodic control protection.

The surface oxide film and the electrochemical properties based on the metallography of the coating material provide the original barrier action. The original barriers of zinc and zinc alloy coatings result from electrochemical properties based on the structure of the coating layer.

Nonuniformity of the surface condition generally induces the formation of a corrosion cell. Such nonuniformity results from detects in the surface oxide film, localized distribution of elements, and the difference in crystal face or phase. These nonuniformities of surface cause the potential difference between portions of the surface, thereby promoting the formation of a corrosion cell.

Many corrosion cells are formed on the surface, accelerating the corrosion rate as a sacrificial metal and its alloy-coated materials are exposed in the natural atmosphere. During this time corrosion products are gradually formed and converted to a stable layer after a few months of exposure. Once the stable layer has been formed the corrosion rate becomes constant. This secondary barrier of corrosion protection regenerates continuously over a long period of time. In most cases the service life of a sacrificial metal coating depends on the secondary barrier action of the corrosion product layer.

Zinc metal coatings are characterized by their galvanic action. Exposure of the base metal as a result of mechanical damage polarizes the base metal cathodically to the potential of the coating layer, as shown in Fig. 20.3, so that little corrosion takes place on the exposed base metal. A galvanic couple is formed between the exposed part of the base metal and the surrounding coating metal. Since zinc is more negative in electrochemical potential than iron or steel, the zinc acts as an anode and the exposed base metal behaves as a cathode. Consequently the dissolution of the zinc layer around the defect is accelerated and the exposed part of base metal is protected against corrosion. Figure 20.4 shows a schematic illustration of the galvanic action of a zinc coating.

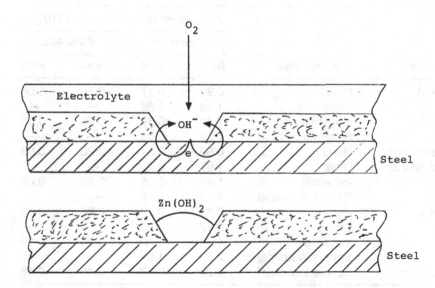

FIGURE 20.4 Schematic illustration of galvanic action of a zinc metallic coating.

The loss of metal coating resulting from corrosion determines the service life of the coating. The degree of loss is dependent on the time of wetness on the metal surface and the type of concentration of pollutants in the atmosphere. Table 20.1 shows the average corrosion losses of zinc and 55% Al-Zn coatings in various locations and atmospheres. The losses were calculated from the mean values of time of wetness and the average corrosion rate during wet duration. The time of wetness of walls is 40% that of roofs. Coating alloys and coating thicknesses can be decided from Table 20.1 since the corrosion losses of zinc and Al-Zn alloy are proportional to exposure time.

As can be seen from the table, G90 sheet, which has a 1 mil zinc coating, cannot be used for a roof having a durability of 10 years in any atmosphere except in a rural area. Were this sheet to be used in an urban, marine, or industrial atmosphere it would have to be painted for protection.

The ability to select a particular alloy or to specify a particular thickness of coating depends on the environment to which it will be exposed, the service life required, and the type of coating process used. Zinc coatings can be applied in many ways.

TABLE 20.1 Average Corrosion Losses of Zinc and 55% Al-Zn Coatings for 10 Years

		Average corrosion loss (mil[a]/10 yr)			
		Zinc		55% Al-Zn	
Location	Atmosphere	Roof	Wall	Roof	Wall
Inland	Rural	0.92	0.17	0.15	0.06
	Urban	1.48	0.59	—	—
	Industrial	1.40	0.56	0.25	0.06
	Severe industrial	1.59	0.64	—	—
Inland shore of lake or marsh	Rural	0.59	0.24	0.20	0.08
	Urban	1.97	0.79	—	—
	Industrial	1.40	0.56	0.20	0.08
	Severe industrial	2.12	0.85	—	—
Coast	Rural	0.74	0.23	0.25	0.10
	Urban	2.47	0.99	—	—
	Industrial	1.75	0.70	0.25	0.10
	Severe industrial	2.65	1.06	—	—
Seashore	Severe industrial	2.06	0.82	0.46	0.18

[a]25.4 μm

Source: Ref. 3.

III. ZINC COATINGS

The six most commonly used zinc coating procedures are

1. Hot dipping
2. Zinc electroplating
3. Mechanical coating
4. Sheradizing
5. Thermally sprayed coatings
6. Zinc dust painting

A. Hot Dipping

This is a process in which cleaned steel is immersed in molten zinc or zinc alloy, and a reaction takes place to form a metallurgically bonded coating.

The coating is integral with the steel because the formation process produces zinc-iron alloy layers overcoated with zinc. Continuity and uniformity are good since any discontinuities are readily visible as "black spots."

Coating thicknesses can be varied approximately 50–125 μm on tube and products. Thicker coatings up to 250 μm can be obtained by grit-blasting before galvanizing. Sheet and wire normally receive thicknesses of 10–30 μm.

Conventional coatings that are applied to finished articles are not formable. The alloy layer is abrasion resistant but brittle on bending. Special coatings with little or no alloy layer are readily formed (e.g., on sheet) and resistance welded.

A chromate conversion coating over the zinc coating prevents wet storage stain, while phosphate coatings provide a good base (on a new sheet) for paints. Weathered coatings are often painted after 10–30 years for longer service.

Hot dip galvanizing is the most important zinc coating process. All mild steel and cast iron can be coated by this process. The thickness and structure of the coating will depend on the alloying elements. Approximately half of the steel that is coated is in the form of sheet, approximately one-quarter is fabricated work, while the remainder is tube or wire. Metallurgically the processes used for tube and fabricated work are similar, while the process used for sheet has small additions to the zinc, which reduce the quantity of iron-zinc alloy in the coating, which provides flexibility.

B. Zinc Electroplating

This process is sometimes mistakenly referred to as electrogalvanizing. In this process zinc salt solutions are used in the electrolytic deposition of a layer of zinc on a cleaned steel surface.

This process provides good adhesion, comparable with other electro-plated coatings. The coating is uniform within the limitations of "throwing power" of the bath. Pores are not a problem as exposed steel is protected by the adjacent zinc.

Coating thicknesses can be varied at will but are usually 2.5–15 μm. Thicker layers are possible but are not generally economical.

Electroplated steel has excellent formability and can be spot welded. Small components are usually finished before being plated.

Chromate conversion coatings are used to prevent wet storage stain, while phosphate conversion coatings are used as a base for paint.

This process is normally used for simple, fairly small components. It is suitable for barrel plating or for continuous sheet and wire. No heating is used in this process except for hydrogen embrittlement relief on high strength steels.

Electroplated steel is very ductile and consequently this process is widely used for the continuous plating of strip and wire, where severe deformation may be required.

The coating on steel from this process gives a bright and smooth finish. It is used for decorative effect to protect delicate objects where rough or uneven finishes cannot be tolerated (e.g., instrument parts). It is also used for articles that cannot withstand the pretreatment or temperatures required in other coating processes.

It was previously mentioned that the term "electrogalvanizing" is sometimes used to describe this process. This is misleading since the chief characteristic of galvanizing is the formation of a metallurgical bond at the zinc–iron interface. This does not occur in electroplating.

C. Mechanical Coating

This process involves the agitating of suitably prepared parts to be coated, with a mixture of nonmetallic impactors (e.g., glass beads), zinc powder, a chemical promoter, and water. All types of steel can be coated. However, this process is less suitable for parts heavier than 0.5 pound (250 g) because the tumbling process reduces coating thickness at the edges.

The adhesion is good compared to electroplated coatings. Thicknesses can be varied as desired from 5 μm to more than 70 μm. However, the coating is not alloyed with the steel, nor does it have the hard, abrasion-resistant iron-zinc alloy layers of galvanized or sheradized coatings. Conversion coatings can be applied.

D. Sheradizing

The articles to be coated are tumbled in a barrel containing zinc dust at a temperature just below the melting point of zinc, usually around 716°F

(380°C). In the case of spring steels the temperature used is somewhat lower. By means of a diffusion process the zinc bonds to the steel forming a hard, even coating of zinc-iron compounds. The coating is dull gray in color and can readily be painted if necessary.

The finish is continuous and very uniform, even on threaded and irregular parts. This is a very useful finish for nuts and bolts, which, with proper allowance for thickness of coats, can be sheradized after manufacture and used without retapping the threads.

The thickness of coating can be controlled. Usually a thickness of 30 μm is used for outdoor applications and 15 μm for indoor.

E. Thermally Sprayed Coatings

In this process droplets of semimolten zinc are sprayed from a special gun that is fed with either wire or powder onto a grit-blasted surface. The semimolten droplets coalesce with some zinc oxide present at each interface between droplets. Electrical continuity is maintained both throughout the coating and with the iron substrate so that full cathodic protection can be obtained since the zinc oxide forms only a small percentage of the coating.

The sprayed coating contains voids (typically 10–20% by volume) between coalesced particles. These voids have little effect on the corrosion protection since they soon fill up with zinc corrosion products and are thereafter impermeable. However, the use of a sealer to fill the voids improves appearance in service and adds to life expectancy, but more important it provides a better surface for subsequent application of paints.

There are no size or shape limitations regarding the use of this process.

IV. CORROSION OF ZINC COATINGS

In general zinc coatings corrode in a similar manner as solid zinc. However, there are some differences. For example, the iron-zinc alloy present in most galvanized coatings has a higher corrosion resistance than solid zinc in neutral and acid solutions. At points where the zinc coating is defective, the bare steel is cathodically protected under most conditions.

The corrosion of zinc coatings in air is an approximate straight-line relationship between weight loss and time. Since the protective film on zinc increases with time in rural and marine atmospheres of some types, under these conditions the life of the zinc may increase more than in proportion to thickness. However, this does not always happen.

Zinc coatings are used primarily to protect ferrous parts against atmospheric corrosion. These coatings have good resistance to abrasion by solid pollutants in the atmosphere. General points to consider are:

1. Corrosion increases with time of wetness.
2. The corrosion rate increases with an increase in the amount of sulfur compounds in the atmosphere.
3. Chlorides and nitrogen oxides usually have a lessor effect but are often very significant in combination with sulfates.

Zinc coatings resist atmospheric corrosion by forming protective films consisting of basic salts, notably carbonate. The most widely accepted formula is $3Zn(OH)_2 \cdot 2ZnCO_3$. Environmental conditions that prevent the formation of such films, or conditions that lend to the formation of soluble films, may cause rapid attack on the zinc.

The duration and frequency of moisture contact is one such factor. Another factor is the rate of drying because a thin film of moisture with high oxygen concentration promotes reaction. For normal exposure conditions the films dry quite rapidly. It is only in sheltered areas that drying times are slow, so that the attack on zinc is accelerated significantly.

The effect of atmospheric humidity on the corrosion of a zinc coating is related to the conditions that may cause condensation of moisture on the metal surface and to the frequency and duration of the moisture contact. If the air temperature drops below the dew point, moisture will be deposited. The thickness of the piece, its surface roughness, and its cleanliness also influence the amount of dew deposited. Lowering the temperature of a metal surface below the air temperature in a humid atmosphere will cause moisture to condense on the metal. If the water evaporates quickly, corrosion is usually not severe and a protective film is formed on the surface. If water from rain or snow remains in contact with zinc when access to air is restricted and the humidity is high, the resulting corrosion can appear to be severe (wet storage stain) since the formation of a protective basic zinc carbonate is prevented.

In areas having atmospheric pollutants, particularly sulfur oxides and other acid-forming pollutants, the time of wetness becomes of secondary importance. These pollutants can also make rain more acidic. However, in less corrosive areas time of wetness assumes a greater proportional significance.

In the atmospheric corrosion of zinc, the most important atmospheric contaminant to be considered is sulfur dioxide. At relative humidities of about 70% or above, it usually controls the corrosion rate.

Sulfur oxides and other corrosive species react with the zinc coating in two ways: dry deposition and wet deposition. Sulfur dioxide can deposit on a dry surface of galvanized steel panels until a monolayer of SO_2 is formed. In either case the sulfur dioxide that deposits on the surface of the zinc forms a sulfurous or other strong acid, which reacts with the film of zinc oxide, hydroxide, or basic carbonate to form zinc sulfate. The conversion of sulfur dioxide to sulfur-based acids may be catalized by nitrogen compounds in the air (NO_x compounds). This factor may affect corrosion rates in practice. The acids partially destroy the film of corrosion products, which will then reform from the underlying metal, thereby causing continuous corrosion by an amount equivalent to the film dissolved, hence the amount of SO_2 absorbed.

Chloride compounds have less effect than sulfur compounds in determining the corrosion rate of zinc. Chloride is most harmful when combined with acidity due to sulfur gases. This is prevalent on the coast in highly industrial areas.

Atmospheric chlorides will lead to the corrosion of zinc, but to a lesser degree than the corrosion of steel, except in brackish water and flowing seawater. Any salt deposit should be removed by washing. The salt content of the atmosphere will usually decrease rapidly further away from the coast. Corrosion also decreases with distance from the coast, but the change is more gradual and erratic because chloride is not the primary pollutant affecting zinc corrosion. Chloride is most harmful when combined with acidity resulting from sulfur gases.

Other pollutants also have an effect on the corrosion of galvanized surfaces. Deposits of soot or dust can be detrimental because they have the potential to increase the risk of condensation onto the surface and hold more water in position. This is prevalent on upward-facing surfaces. Soot (carbon) absorbs large quantities of sulfur, which are released by rain water.

In rural areas overmanuring of agricultural land tends to increase the ammonia content of the air. The presence of normal atmospheric quantities of ammonia does not accelerate zinc corrosion, and petroleum plants where ammonium salts are present show no appreciable attack on galvanized steel. However, ammonia will react with atmospheric sulfur oxides, producing ammonium sulfate, which accelerates paint film corrosion as well as zinc corrosion. When ammonia reacts with NO_x^- compounds in the atmosphere, ammonium nitrite and nitrate are produced. Both compounds increase the rate of zinc corrosion, but less than SO_2 or SO_3.

Because of the Mears effect (wire corrodes faster per unit of area than more massive materials) galvanized wire corrodes some 10–80% faster than galvanized sheet. However, the life of rope made from galvanized steel wires is greater than the life of the individual wire. This is explained by the fact that the parts of the wire that lie on the outside are corroded more rapidly, and when the zinc film is penetrated in these regions, the uncorroded zinc inside the rope provides cathodic protection for the outer regions.

Galvanized steel also finds application in the handling of various media. Table 20.2 lists the compatibility of galvanized steel with selected corrodents.

V.   ZINC ALLOYS

Small alloy additions are made to zinc to improve grain size, give work hardening, and improve properties such as creep resistance and corrosion resistance. There are a number of proprietary compositions available containing additions of copper, manganese, magnesium, aluminum, chromium, and titanium.

A.   Zinc–5% Aluminum Hot Dip Coatings

This zinc alloy coating is known as Galfan. Galfan coatings have a corrosion resistance up to three times that of galvanized steel. The main difference between these two coatings lies in the degree of cathodic protection they afford. This increase in corrosion protection is evident in both a relatively mild urban industrial atmosphere and in a marine atmosphere, as can be seen in Table 20.3. The latter is particularly significant because unlike galvanizing the corrosion rate appears to be slowing after about 4 years, and conventional galvanized steel would show rust in 5 years (see Fig. 20.5). The slower rate of corrosion also means that the zinc–5% aluminum coatings provide full cathodic protection to cut edges over a longer period. Refer to Table 20.4.

Because Galfan can be formed with much smaller cracks than can be obtained in conventional galvanized coatings, it provides bulges. This reduced cracking means that less zinc is exposed to the environment, which increases the relative performance factor compared with galvanized steel.

B.   Zinc–55% Aluminum Hot Dip Coatings

These coatings are known as Galvanlume and consist of zinc–55% aluminum–1.5% silicon. This alloy is sold under such trade names as Zaluite, Aluzene, Alugalva, Alyafort, Aluzink, and Zincalume. Galvalume exhibits superior corrosion resistance over galvanized coatings in rural, industrial, marine, and severe marine environments. However, this alloy has limited cathodic protection and less resistance to some alkaline conditions and is subject to weathering discoloration and wet storage staining. The latter two disadvantages can be overcome by chromate passivation, which also improves its atmospheric corrosion resistance.

Initially, a relatively high corrosion loss is observed for Galvalume sheet as the zinc-rich portion of the coating corrodes and provides sacrificial

TABLE 20.2   Compatibility of Galvanized Steel with Selected Corrodents

Acetic acid	U	Diamylamine	G
Acetone	G	Dibutyl amine	G
Acetonitrile	G	Dibutyl phthalate	G
Acrylonitrile	G	Dichloroether	G
Acrylic latex	U	Diethylene glycol	G
Aluminum chloride, 26%	U	Dipropylene gylcol	G
Aluminum hydroxide	U	Ethanol	G
Aluminum nitride	U	Ethyl acetate	G
Ammonia, dry vapor	U	Ethyl acrylate	G
Ammonium acetate solution	U	Ethyl amite, 69%	G
Ammonium bisulfate	U	n-Ethyl butylamine	G
Ammonium bromide	U	2-Ethyl butyric acid	G
Ammonium carbonate	U	Ethyl ether	G
Ammonium chloride, 10%	U	Ethyl hexanol	G
Ammonium dichloride	U	Fluorine, dry, pure	G
Ammonium hydroxide	U	Formaldehyde	G
Ammonium hydroxide vapor	U	Fruit juices	S
Ammonium molybdate	G	Hexalene glycol	G
Ammonium nitrate	U	Hexanol	G
Argon	G	Hexylamine	G
Barium nitrate, solution	S	Hydrochloric acid	U
Barium sulfate, solution	S	Hydrogen peroxide	S
Beeswax	U	Iodine gas	U
Borax	S	Isohexanol	G
Bromine, moist	U	Isooctane	G
2-Butanol	G	Isopropyl ether	G
Butyl acetate	G	Lead sulfate	U
Butyl chloride	G	Lead sulfite	S
Butyl ether	G	Magnesium carbonate	S
Butyl phenol	G	Magnesium chloride, 42.5%	U
Cadmium chloride solution	U	Magnesium fluoride	G
Cadmium nitrate	U	Magnesium hydroxide, sat.	S
Cadmium sulfate	U	Magnesium sulfate, 2%	S
Calcium hydroxide, sat.	U	Magnesium sulfate, 10%	U
Calcium hydroxide, 20%	S	Methyl amine alcohol	G
Calcium sulfate, sat.	U	Methyl ethyl ketone	G
Cellosolve acetate	G	Methyl isobutyl ketone	G
Chloric acid, 20%	U	Methyl propyl ketone	G
Chlorine, dry	G	Nickel ammonium sulfate	U
Chlorine, water	U	Nickel chloride	U
Chromium chloride	U	Nickel sulfate	S
Chromium sulfate, solution	U	Nitric acid	U
Copper sulfate, solution	U	Nitrogen, dry, pure	G
Decyl acrylate	G	Nonylphenol	G

TABLE 20.2 Continued

Oxygen, dry, pure	G	Silver bromide	U
Oxygen, moist	U	Silver chloride, pure, dry	S
Paraldehyde	G	Silver chloride, moist, wet	U
Perchloric acid, solution	S	Silver nitrate, solution	U
Permanganate, solution	S	Sodium acetate	S
Peroxide, pure, dry	S	Sodium aluminum sulfate	U
Peroxide, moist	U	Sodium bicarbonate, sat.	U
Phosphoric acid, 0–33%	G	Sodium bisulfate	U
Polyvinylacetate latex	U	Sodium carbonate, solution	U
Potassium bichromate, 14%	G	Sodium chloride, solution	U
Potassium bichromate, 20%	S	Sodium hydroxide, solution	U
Potassium carbonate, 10%	U	Sodium nitrate, solution	U
Potassium carbonate, 50%	U	Sodium sulfate, solution	U
Potassium chloride, solution	U	Sodium sulfide	U
Potassium disulfate	S	Sodium sulfite	U
Potassium fluoride, 5–20%	G	Styrene, monomeric	G
Potassium hydroxide	U	Styrene oxide	G
Potassium nitrate, 5–10%	S	Triethylene glycol	G
Potassium peroxide	U	1,1,2-Trichloroethane	G
Potassium persulfate, 10%	U	1,2,3-Trichloropropane	G
Propyl acetate	G	Vinyl acetate	G
Propylene glycol	G	Vinyl butyl ether	G
Propionaldehyde	G	Vinyl ethyl ether	G
Propionic acid	U	Water, potable, hard	G

G = suitable application; S = borderline application; U = not suitable application.

TABLE 20.3 Five-Year Outdoor Exposure Results of Galfan Coating

	Thickness loss (μm)		Ratio of improvement
Atmosphere	Galvanized	Galfan	
Industrial	15.0	5.2	2.9
Severe marine	>20.0	9.5	>2.1
Marine	12.5	7.5	1.7
Rural	10.5	3.0	3.5

Source: Ref. 1.

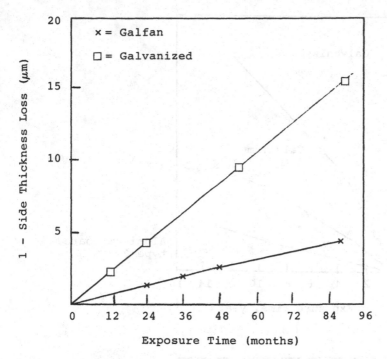

FIGURE 20.5 Seven-year exposure of Galfan and galvanized steel in a severe marine atmosphere.

TABLE 20.4 Comparison of Cathodic Protection for Galvanized and Galfan Coatings

Environment	Amount of bare edges exposed after 3 years (coating recession from edge) (mm)	
	Galvanized	Galfan
Severe marine	1.6	0.1
Marine	0.5	0.06
Industrial	0.5	0.05
Rural	0.1	0.0

Source: Ref. 1.

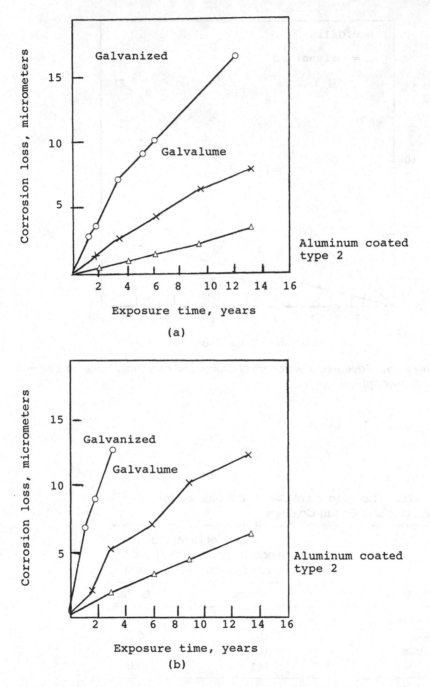

FIGURE 20.6 Thirteen-year exposure of Galvalume in marine and industrial atmospheres. (From Ref. 3.)

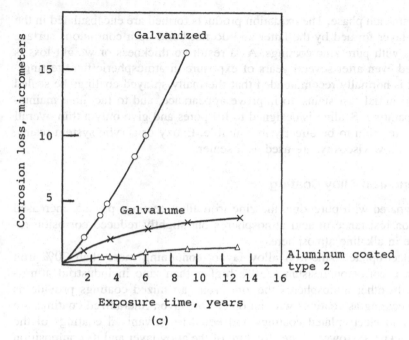

FIGURE 20.6 Continued

protection at cut edges. This takes place in all environments. After approximately 3 years, the corrosion–time curves take on more gradual slopes, reflecting a change from active, zinc-like behavior to passive aluminum-like behavior as the interdentric regions fill with corrosion products. It has been predicted that Galvalume sheets should outlast galvanized sheet of equivalent thickness by at least two to four times over a wide range of environments. Figures comparing the corrosion performance of galvanized sheet and Galvalume sheet are shown in Fig. 20.6.

Galvalume sheets provide excellent cut-edge protection in very aggressive conditions, where the surface does not remain too passive. However, it does not offer as good protection on the thicker sheets in mild rural conditions, where zinc–5% aluminum coatings provide good general corrosion resistance, and when sheared edges are exposed or localized damage to the coating occurs during fabrication or service, the galvanic protection is retained for a longer period.

C. Zinc–15% Aluminum Thermal Spray

Zinc–15% aluminum coatings are available as thermally sprayed coatings. These coatings have a two-phase structure consisting of a zinc-rich and an aluminum-rich phase. The oxidation products formed are encapsulated in the porous layer formed by the latter and do not build up a continuous surface layer as with pure zinc coatings. As a result no thickness or weight loss is observed even after several years of exposure in atmospheric field testing.

It is normally recommended that thermally sprayed coatings be sealed to avoid initial rust stains, to improve appearance, and to facilitate maintenance painting. Sealing is designed to fill pores and give only a thin overall coating, too thin to be directly measurable. Epoxy or acrylic system resins, having a low viscosity, are used as a sealer.

D. Zinc-Iron Alloy Coating

As compared with pure zinc, the zinc-iron alloy coatings provide increased corrosion resistance in acid atmospheres but slightly reduced corrosion resistance in alkaline atmospheres.

Electroplated zinc-iron alloy layers containing more than 20% iron provide a corrosion resistance 30% higher than zinc in industrial atmospheres. In other atmospheres the zinc-iron galvanized coatings provide as good a coating as coatings with an outer zinc layer. Sheradized coatings are superior to electroplated coatings and equal to galvanized coatings of the same thickness. However, the structure of the alloy layer and its composition affect the corrosion resistance.

If the zinc layer of a galvanized coating has weathered, or the zinc-iron alloy layer forms the top layer after galvanizing, brown areas may form. Brown staining can occur on sheradized or hot dip galvanized coatings in atmospheric corrosion through the oxidation of iron from the zinc-iron alloy layers or from the substrate. Such staining is usually a dull brown, rather than the bright red-brown of uncontrolled rust. Usually there is a substantial intact galvanized layer underneath, leaving the life of the coating unchanged. Unless the esthetic appearance is undesirable, no action need be taken.

VI. CAST ZINC

For over 60 years zinc alloy castings have been produced. They offer the advantage that they can be cast to near net or net shape and compete with aluminum, magnesium, bronze, and iron casting alloys.

Cast zinc finds applications mainly for small die castings stressed at low levels. There are two basic groups of zinc casting alloys. The first group contains about 4% by weight of aluminum as the primary alloying ingredient and are known as Zamak 3, 5, and 7. They are used for pressure die casting. The second group comprises the ZA (zinc-aluminum) series of which ZA-8, ZA-12, and ZA-27 are the most popular. The numbers indicate the nominal weight percent of the major alloying element, aluminum. Application of these zinc casting alloys are shown in Table 20.5.

The Zamak family of castings are resistant to atmospheric corrosion and behave like pure zinc. Higher aluminum content will increase the general corrosion of zinc. All of the ZA family of alloys (8–29% aluminum) corrode less than 99.99% zinc. The ZA alloys perform better than pure zinc in acid-contaminated industrial plants.

Zinc alloys possess excellent machinability. Most of the zinc alloys can be machined by the common procedures. The use of cutting fluids is recommended.

TABLE 20.5 Applications of Zinc Casting Alloys

Alloy	Applications
Zamak 3 Al 4, Cu 0.25, Fe 0.1 UNS Z33521	Most widely used zinc casting alloy. Has excellent plating and finishing characteristics. Locks, screwdrivers, seatbelt buckles, faucets, carburetors, gears, fan housings, computer parts.
Zamak 5 Al 4, Cu 1, Fe 0.1 UNS Z35530	Has greater hardness and strength than Zamak 3. Good creep resistance. Automotive parts, household appliances and fixtures, office and building hardware.
Zamak 7 Al 4, Cu 0.25 UNS Z33522	A high purity form of Zamak 3, but with better ductility and fluidity and more expensive. Similar uses to Zamak 3, especially those requiring more extreme forming operations.
ZA 8 AL 8.4, Cu 1, Fe 0.1 UNS Z35630	For pressure die castings, sand and permanent mold. First choice when considering die casting of ZA family. Agricultural equipment, electrical fittings, hand tools, plated parts.
ZA 12 Al 11, Cu 0.8, UNS Z35630	Preferred for permanent mold casting. Also popular for sand castings. Used for higher strength applications. Journal bearings, agricultural equipment, pump components, plated parts.
ZA 27 Al 26, Cu 2, Fe 0.1 UNS Z35840	Components requiring optimum strength and hardness and light weight. Excellent bearing properties and dumping characteristics. Engine mounts and drive train, general hardware, ratchet wrench, winch components, sprockets, gear housings. Not normally electroplated. Die castings must be by cold chamber method.

Source: Ref. 2.

TABLE 20.6 Typical Properties of Zinc Casting Alloys

Alloy	Yield strength 0.2% offset (ksi)	Elongation in 50 mm (%)	Modulus of elasticity ×10^6 (psi)	Brinell hardness	Impact (ft-lb)	Fatigue strength ×10^3 (psi)	Electrical conductivity (% IACS)	Thermal conductivity (W/mK)	Coefficient of thermal expansion (10^{-6}/°F)
Zamak 3[a]	32	10	12.4	82	43	6.9	27	113.0	15.2
Zamak 5[a]	33	7	12.4	91	48	8.2	26	108.9	15.2
Zamak 7[a]	—	13	12.4	80	43	6.8	27	113.0	15.2
ZA 8[b]	29	1.7	12.4	85	15	—	27.7	114.7	12.9
ZA 8[c]	30	1.3	12.4	87	—	7.5	27.7	114.7	12.9
ZA 8[a]	42	8	12.4	103	31	15	27.7	114.7	12.9
ZA 12[b]	31	1.5	12.0	94	19	—	28.3	116.1	13.4
ZA 12[c]	34	2.2	12.0	—	—	—	28.3	116.1	13.4
ZA 12[a]	46	5	12.0	100	21	17	28.3	116.1	13.4
ZA 27[b]	54	4.6	11.3	113	35	25	29.7	125.5	14.4
ZA 27[c]	55	2.5	11.3	114	—	—	29.7	125.5	14.4
ZA 27[a]	54	2.5	11.3	119	9	21	29.7	125.5	14.4

[a]Die cast.
[b]Sand cast.
[c]Permanent mold.

Zinc alloys may be joined by the common mechanical methods including threaded fasteners, crimping, riveting, staking, and force fits. They may also be joined by welding and soldering operations. Welding is best accomplished by using the arc-shielded inert gas process.

The mechanical properties of zinc casting alloys are shown in Table 20.6.

REFERENCES

1. FC Porter. Corrosion Resistance of Zinc and Zinc Alloys. New York: Marcel Dekker, 1994.
2. GT Murray. Handbook of materials Selection for Engineering Applications. New York: Marcel Dekker, 1997.
3. PA Schweitzer. Corrosion Resistant Linings and Coatings. New York: Marcel Dekker, 2001.
4. PA Schweitzer. Atmospheric Degradation and Corrosion Control. New York: Marcel Dekker, 1999.
5. JL Gossett. Corrosion resistance of cast alloys. In: PA Schweitzer, ed. Corrosion Engineering Handbook. New York: Marcel Dekker, 1996.

21

Titanium

Titanium is the ninth most abundant element on earth and the fourth most abundant metal. It is more plentiful than chromium, copper, or nickel, which are commonly employed as alloys to resist corrosion. Titanium and its alloys are noted for their high strength-to-weight ratios and excellent corrosion resistance. Although the needs of the aerospace industry for better strength-to-weight ratio structured materials was recognized little use was made of titanium until the commercialization of the Kroll process which made titanium sponge available in about 1950.

Although having the advantages of being highly corrosion resistant in oxidizing environments, a low density (specific gravity 4.5, approximately 60% that of steel) and a high tensile strength (60,000 psi) its widespread use has been limited somewhat by cost. However, as consumption has increased and new technologies have been developed to reduce the high cost, usage has increased and will probably continue to increase further. At the present time it is competitive with nickel base alloys. Thinner sections, coupled with decreased maintenance requirements and longer life expectancy in many applications, permit titanium equipment installations to be cost effective despite a higher initial cost. Increasing usage has been found in automotive applications, chemical processing equipment, pulp and paper industry, marine vehicles, medical prostheses, and sporting goods. Applications for some of the more popular alloys are shown in Table 21.1.

The titanium alloys, unlike other nonferrous alloys, are not separated into wrought and cast categories. Most of the widely used casting alloys are based on the traditional wrought compositions.

Metallurgists have separated titanium alloys into categories according to the phases present, namely,

TABLE 21.1   Applications of Titanium Alloys

Alloy	Applications
ASTM grade 12	
UNS R53400	Chemical process industries. Used in hot brines, heat exchangers, and chlorine cells.
ASTM grade 9	
UNS R56320	Chemical processing and handling equipment. Has high degree of immunity to attack by most mineral acids and chlorides, boiling seawater, and organic compounds.
UNS R54210	Cryogenic applications.
UNS R56210	Marine vehicle hulls. Has high fracture toughness.
TI 6211 and	Aerospace industry, medical prostheses, marine
UNS R56400	equipment, chemical pumps, high performance automotive components.
ASTM grade 5	
UNS R56400	Golf club heads, auto parts, working tools.
UNS R58030	Aircraft fasteners, springs, orthodontic applicances.
UNS R54810	Airframe and turbines.

1. Commercially pure or modified titanium
2. Alpha and near-alpha alloys
3. Alpha-beta alloys
4. Beta alloys

I. TITANIUM ALLOYS

These alloys have strengths comparable with alloy steels, while the weight is only 60% that of steel. In addition the corrosion resistance of titanium alloys is superior to aluminum and stainless steels under most conditions. Titanium's low magnetic permeability is also notable.

The chemical composition of unalloyed titanium grades and titanium alloys are covered by ASTM specifications. Table 21.2 lists the compositions of representative grades. These alloys are all available in various product forms covered by ASTM specifications as shown in Table 21.3.

ASTM Grades 1, 2, 3, and 4 cover unalloyed titanium. Grade 2 is most often used for corrosion resistance. Grade 1 possesses better ductility but lower strength; grades 3 and 4 possess higher strength.

TABLE 21.2 Chemical Composition of Titanium Alloys

Element	Ti-50A[a] (ASTM grade 2)	Ti-6Al-4V[a] (ASTM grade 5)	Ti-Pd (ASTM grade 7)	Ti-Code 12[a] (ASTM grade 12)
Nitrogen, max.	0.03	0.05	0.03	0.03
Carbon, max.	0.10	0.10	0.10	0.08
Hydrogen, max.	0.015	0.015	0.015	0.015
Iron, max.	0.30	0.40	0.30	0.30
Oxygen, max.	0.25	0.20	0.25	0.25
Aluminum	—	5.5–6.75	—	—
Vanadium	—	3.5–4.5	—	—
Palladium	—	—	0.12–0.25	—
Molybdenum	—	—	—	0.2–0.4
Nickel	—	—	—	0.6–0.9
Titanium	Remainder	Remainder	Remainder	Remainder

[a]Timet designation.

TABLE 21.3   ASTM Titanium Specifications

ASTM B 265-76	Titanium and titanium alloy strip, sheet, and plate
ASTM B 337-76	Seamless and welded titanium and titanium alloy pipe
ASTM B 338-76	Seamless and welded titanium and titanium alloy tubes for condensers and heat exchangers
ASTM B 348-76	Titanium and titanium alloy bars and billets
ASTM B 363-76	Seamless and welded unalloyed titanium welding fittings
ASTM B 367-69	(1974) Titanium and titanium alloy castings
ASTM B 381-76	Titanium and titanium alloy forgings

Grade 7 alloy, compared with unalloyed titanium, possesses an improved corrosion resistance. This alloy, as grade 11, is used for improved
formability.

Grade 12 is a lower cost alternative for grades 7 and 11 and is suitable
for some applications. The palladium of alloys 7 and 11 has been replaced
with 0.8% nickel and 0.3% molybdenum.

Grade 5 is an alloy having high strength and toughness and is a general
purpose alloy finding numerous applications in the aerospace industry. Its
corrosion resistance is inferior to the unalloyed grades.

The general properties of titanium alloys are shown in Table 21.4.

A.   Physical Properties

As previously stated, titanium is a light metal with a density slightly over
half that of iron- or copper-based alloys. Its modulus of elasticity is also
only approximately half that of steel. Titanium has a low expansion coefficient and a relatively high electrical resistance. The specific heat and thermal
conductivity of titanium are similar to those of stainless steel. The physical
properties are shown in Table 21.5.

When designing or fabricating process equipment it is important that
the modulus of elasticity and coefficient of expansion be carefully considered.

B.   Mechanical Properties

The mechanical properties of annealed titanium alloys are shown in Table
21.6. As will be noted strength and ductility of titanium alloys are comparable to those of other corrosion resistant alloys. As will be seen in Table
21.7 the allowable stress values from the ASME Pressure Vessel Code are
a function of temperature. Alloy grade 5 is not covered by the ASME Code.

TABLE 21.4 General Properties of Titanium Alloys

ASTM grade	UNS no.	Properties
1 (CP)	R50250	Ductility, lower strength
2 (CP)	R50400	Good balance of moderate strength and ductility
3 (CP)	R50550	Moderate strength
7 and 11	R52400 R52250	Improved resistance to reducing acids and superior crevice corrosion resistance
16	—	Resistance similar to grade 7, but at lower cost
12	R53400	Reasonable strength and improved crevice corrosion resistance; lower cost
9	R56320	Medium strength and superior pressure code design allowances
18	—	Same as grade 9 but with improved resistance to reducing acids and crevice corrosion
5	R56400	High strength and toughness

CP = chemically pure.

TABLE 21.5 Physical Properties of Titanium

Property	Ti-50A (grade 2)	Ti-6Al-4V (grade 5)	Ti-Pd (grade 7)	Ti-code 12 (grade 12)
Modulus of elasticity, tension $\times 10^6$ (psi)	14.9	16.5	14.9	15.0
Modulus of elasticity, torsion $\times 10^6$ (psi)	6.5	6.1	6.5	6.2
Density (lb/in.^{3})	0.163	0.160	0.163	0.164
Specific heat at 75°F (BTU/lb °F)	0.125	0.135	0.125	0.130
Thermal conductivity at 75°F (BTU/ft^2 hr°F in.)	114	50	114	132
Coefficient of expansion at 32–600°F $\times 10^{-6}$ (in./in.°F)	5.1	5.1	5.1	5.4
Electrical resistivity at 75°F ($\mu\Omega$-cm)	56	171	56.7	52

TABLE 21.6 Minimum Mechanical Properties of Annealed Titanium Alloys

Alloy	Tensile strength (ksi)	Yield strength 0.2% (ksi)	Elongation in 2 in. (%)
Ti-50A (grade 2)	50	40	20
Ti-6Al-4V (grade 5)	130	120	10
Ti-Pd (grade 7)	50	40	20
Ti-code 12 (grade 12)	70	50	18

Titanium and its alloys maintain excellent strength at low temperatures, as can be seen from Fig. 21.1.

Titanium exhibits excellent fatigue properties, as can be seen in Table 21.8. Titanium and its alloys exhibit a high fatigue strength/tensile strength ratio at ambient temperatures. It is in the range of 0.5 to 0.6.

The allowable design stress for welded and annealed titanium pipe grade 2 is as follows:

Temperature (°F/°C)	Maximum allowable stress (psi)
100/38	10,600
150/66	10,200
200/93	9,300
250/121	8,400
300/149	7,700
350/177	7,100
400/204	6,500
450/232	6,100
500/260	5,600
550/288	5,300
600/316	4,800

II. TYPES OF CORROSION

As with other metals titanium owes its corrosion resistance to a protective oxide film. This film is relatively thin but very stable. It is attacked by only a few substances, notable of which is hydrofluoric acid.

Titanium's strong affinity for oxygen makes it capable of healing film ruptures almost instantly in any environment having oxygen and a trace of

TABLE 21.7 Design Stresses for Titanium Plate[a]

For metal temperatures not exceeding (°F)	Allowable stress values (ksi)				
	Ti-50A (grade 2)	Titanium (grade 3)	Ti-Pd (grade 7)	Ti-3Al-2.5V (grade 9)[c]	Ti-0.3MO-0.8Ni (grade 12)[b]
100	12.5	16.3	12.5	24.3	17.5
200	10.9		10.9		16.4
300	9.0	11.7	9.0	20.0	14.2
400	7.7		7.7		12.5
500	6.6		6.6		11.4
600	5.7	6.0	5.7	16.6	

[a] ASME Section VIII, Division 1—Pressure Vessels.
[b] Case BC78-326.
[c] Estimated values. Grade 9 in the process of being qualified.

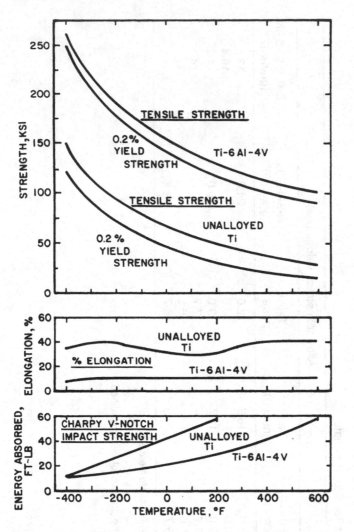

FIGURE 21.1 Effect of temperature on strength, ductility, and toughness on grade 2 and grade 5 titanium alloys.

moisture present. The protective film may not be formed in anhydrous conditions devoid of oxygen. Consequently these applications should be avoided.

A. Erosion Corrosion

Erosion corrosion is the removal of a protective film as a result of being swept away above a critical water velocity. When this occurs accelerated

TABLE 21.8 Fatigue Strength of Titanium

Alloy	Tensile strength (ksi)	Fatigue strength, 10^8 cycles (ksi)	FS/TS ratio
Unalloyed Ti	52.6	32.7	0.62
Unalloyed Ti	92.2	47.1	0.51
Ti-6A1-4V	140.7	85.5	0.60
Ti-6A1-4V	163.4	92.8	0.56

corrosion attack takes place. For some metals the critical water velocity is as low as 2–4 ft/s. In seawater titanium has a critical velocity of 90 ft/s.

B. General Corrosion

General corrosion of titanium is encountered in hot reducing acid solutions. In such environments when oxidizing agents and certain multivalent metal ions are present the titanium may be passivated. Many sulfuric and hydrochloric acid streams contain enough impurities in the form of ferric ions, cupric ions, etc., to passivate titanium, permitting its use.

Table 21.9 provides the compatibility of titanium with selected corrodents.

C. Galvanic Corrosion

For most environments titanium will be the cathodic member of any galvanic couple. As a result the coupling of titanium with dissimilar metals usually does not accelerate the corrosion of titanium. The exception is in reducing environments where titanium does not passivate. Under these conditions titanium has a potential similar to aluminum and will corrode when coupled to more noble metals.

When coupled with a dissimilar metal titanium may accelerate the corrosion of the other member of the couple. As a result hydrogen may be evolved on the surface of the titanium, resulting in the formation of a surface hydride film which is stable and which will not cause any problems. However, if the temperature exceeds 170°F (77°C), hydriding may cause embrittlement.

If it is necessary to have titanium in contact with a dissimilar metal the critical part should be constructed of titanium since the titanium will not usually be attacked.

TABLE 21.9 Compatibility of Titanium, Zirconium, and Tantalum with Selected Corrodents[a]

Chemical	Maximum temp. (°F/°C)		
	Titanium	Zirconium	Tantalum
Acetaldehyde	300/104	250/121	90/32
Acetamide			
Acetic acid, 10%	260/127	220/104	302/150
Acetic acid, 50%	260/127	230/110	302/150
Acetic acid, 80%	260/127	230/110	302/150
Acetic acid, glacial	260/127	230/110	302/150
Acetic anhydride	280/138	250/121	302/150
Acetone	190/88	190/88	302/150
Acetyl chloride		80/27	80/27
Acrylic acid			
Acrylonitrile	210/99	210/93	210/99
Adipic acid	450/232		210/99
Allyl alcohol	200/93	200/93	300/149
Allyl chloride		200/93	
Alum	200/93	210/99	90/32
Aluminum acetate			
Aluminum chloride, aqueous	10%	40%	302/150
	310/154	200/93	
Aluminum chloride, dry		37%	
	200/93	210/99	302/150
Aluminum fluoride	80/27	X	X
Aluminum hydroxide	190/88	200/93	100/38
Aluminum nitrate	200/93		80/27
Aluminum oxychloride			
Aluminum sulfate	210/99	210/99	302/150
Ammonia gas		100/38	
Ammonium bifluoride			
Ammonium carbonate	200/93		200/93
Ammonium chloride, 10%	210/99	210/99	302/150
Ammonium chloride, 50%	190/88	220/104	302/150
Ammonium chloride, sat.	203/93		302/150
Ammonium fluoride, 10%	90/32	X	X
Ammonium fluoride, 25%	80/27	X	X
Ammonium hydroxide, 25%	80/27	210/99	302/150
Ammonium hydroxide, sat.	210/99	210/99	302/150
Ammonium nitrate	210/99	210/99	210/99
Ammonium persulfate	80/27	220/104	90/32
Ammonium phosphate, 10%	210/99	210/99	302/150
Ammonium sulfate, 10–40%	210/99	210/99	302/150
Ammonium sulfide			90/32

TABLE 21.9 Continued

Chemical	Maximum temp. (°F/°C)		
	Titanium	Zirconium	Tantalum
Ammonium sulfite			210/99
Amyl acetate	210/99	210/99	302/150
Amyl alcohol	200/93	200/93	320/160
Amyl chloride		210/99	302/150
Aniline	210/99	210/99	210/99
Antimony trichloride	110/43		210/99
Aqua regia, 3:1	80/27	X	302/150
Barium carbonate	80/27	210/99	90/32
Barium chloride, 25%	210/99	210/99	210/99
Barium hydroxide	210/99	200/93	302/150
Barium sulfate	210/99	210/99	210/99
Barium sulfide	90/32	90/32	90.32
Benzaldehyde	100/38	210/99	210/99
Benzene	230/110	230/110	230/110
Benzene sulfonic acid, 10%		210/99	210/99
Benzoic acid	400/204	400/204	210/99
Benzyl alcohol	210/99	210/99	210/99
Benzyl chloride			230/110
Borax	190/88		X
Boric acid	210/99	210/99	300/149
Bromine gas, dry	X	X	302/150
Bromine gas, moist	190/88	60/16	302/150
Bromine, liquid	X	60/16	570/299
Butadiene			80.27
Butyl acetate	210/99	210/99	80/27
Butyl alcohol	200/93	200/93	80/27
n-Butylamine	210/99		
Butyl phthalate	210/99	210/99	210/99
Butyric acid	210/99	210/99	302/150
Calcium bisulfide			
Calcium bisulfite	210/99	90/32	80/27
Calcium carbonate	230/110	230/110	230/110
Calcium chlorate	140/60		210/99
Calcium chloride	310/154	210/99	302/150
Calcium hydroxide, 10%	210/99	210/99	302/150
Calcium hydroxide, sat.	230/110	210/99	302/150
Calcium hypochlorite	200/93	200/93	302/150
Calcium nitrate	210/99		80/27
Calcium oxide			
Calcium sulfate	210/99	210/99	210/99
Caprylic acid	210/99	210/99	300/149

TABLE 21.9   Continued

Chemical	Maximum temp. (°F/°C)		
	Titanium	Zirconium	Tantalum
Carbon bisulfide	210/99		210/99
Carbon dioxide, dry	90/32	410/210	310/154
Carbon dioxide, wet	80/27		300/149
Carbon disulfide	210/99		210/99
Carbon monoxide	300/149		
Carbon tetrachloride			302/150
Carbonic acid	210/99	210/99	300/149
Cellosolve	210/99	210/99	210/99
Chloracetic acid, 50% water	210/99	210/99	210/99
Chloracetic acid	210/99	210/99	302/150
Chlorine gas, dry	X	90/32	460/238
Chlorine gas, wet	390/199	X	570/299
Chlorine, liquid		X	300/149
Chlorobenzene	200/93	200/93	300/149
Chloroform	210/99	210/99	210/99
Chlorosulfonic acid	210/99		210/99
Chromic acid, 10%	210/99	210/99	302/150
Chromic acid, 50%	210/99	210/99	302/150
Chromyl chloride	60/16		210/99
Citric acid, 15%	210/99	210/99	302/150
Citric acid, conc.	180/82	180/82	302/150
Copper acetate		200/93	300/149
Copper carbonate	80/27		300/149
Copper chloride	200/93	X	300/149
Copper cyanide	90/32	X	300/149
Copper sulfate	210/99	210/99	300/149
Cresol	210/99		
Cupric chloride, 5%	210/99	X	300/149
Cupric chloride, 50%	210/99	190/88	90/32
Cyclohexane			
Cyclohexanol			
Dichloroacetic acid	280/138	350/177	260/127
Dichloroethane			
Ethylene glycol	210/99	210/99	90/32
Ferric chloride	300/149	X	302/150
Ferric chloride, 50% in water	210/99	X	302/150
Ferric nitrate, 10–50%	90/32		210/99
Ferrous chloride	210/99	210/99	210/99
Ferrous nitrate			
Fluorine gas, dry	X	X	X
Fluorine gas, moist	X	X	X

TABLE 21.9 Continued

Chemical	Maximum temp. (°F/°C)		
	Titanium	Zirconium	Tantalum
Hydrobromic acid, dilute	90/32	80/27	302/150
Hydrobromic acid, 20%	200/93	X	302/150
Hydrobromic acid, 50%	200/93	X	302/150
Hydrochloric acid, 20%	X	300/149	302/150
Hydrochloric acid, 38%	X	140/60	302/150
Hydrocyanic acid, 10%			
Hydrofluoric acid, 30%	X	X	X
Hydrofluoric acid, 70%	X	X	X
Hydrofluoric acid, 100%	X	X	X
Hypochlorous acid	100/38		302/150
Iodine solution, 10%	90/32		
Ketones, general	90/32		
Lactic acid, 25%	210/99	300/149	302/150
Lactic acid, conc.	300/149	300/149	300/149
Magnesium chloride	300/149		302/150
Malic acid	210/99	210/99	210/99
Manganese chloride, 5–20%	210/99	210/99	210/99
Methyl chloride	210/99		210/99
Methyl ethyl ketone	210/99	210/99	210/99
Methyl isobutyl ketone	200/93	200/93	210/99
Muriatic acid	X		302/150
Nitric acid, 5%	360/182	500/260	302/150
Nitric acid, 20%	400/204	500/260	302/150
Nitric acid, 70%	390/199	500/260	302/150
Nitric acid, anhydrous	210/99	90/32	302/150
Nitrous acid, conc.			300/149
Oleum			X
Perchloric acid, 10%	X		302/150
Perchloric acid, 70%	X	210/99	302/150
Phenol	90/32	210/99	302/150
Phosphoric acid, 50–80%	X	180/82	302/150
Picric acid	90/32		200/93
Potassium bromide, 30%	200/93	200/93	90/32
Salicylic acid	90.32		210/99
Silver bromide, 10%			90/32
Sodium carbonate	210/99	210/99	210/99
Sodium chloride	210/99	250/151	302/150
Sodium hydroxide, 10%	210/99	210/99	X
Sodium hydroxide, 50%	200/93	200/93	X
Sodium hydroxide, conc.	200/93	210/99	X
Sodium hypochlorite, 20%	200/93	100/38	302/150

TABLE 21.9 Continued

Chemical	Maximum temp. (°F/°C)		
	Titanium	Zirconium	Tantalum
Sodium hypochlorite, conc.			302/150
Sodium sulfide, to 10%	210/99	X	210/99
Stannic chloride, 20%	210/99	210/99	300/149
Stannous chloride	90.32		210/99
Sulfuric acid, 10%	X	300/149	302/150
Sulfuric acid, 50%	X	300/149	302/150
Sulfuric acid, 70%	X	210/99	302/150
Sulfuric acid, 90%			302/150
Sulfuric acid, 98%	X	X	302/150
Sulfuric acid, 100%	X	X	300/149
Sulfuric acid, fuming	X		X
Sulfurous acid	170/77	370/188	300/149
Thionyl chloride			300.149
Toluene	210/99	80/27	300/149
Trichloroacetic acid	X	X	300/149
White liquor		250/121	
Zinc chloride			210/99

[a]The chemicals listed are in the pure state or in a saturated solution unless otherwise indicated. Compatibility is shown to the maximum allowable temperature for which data are available. Incompatibility is shown by an X. A blank space indicates that data are unavailable. When compatible, corrosion rate is <20 mpy.
Source: Ref. 3.

D. Hydrogen Embrittlement

In most cases the oxide film on titanium acts as an effective barrier to penetration by hydrogen. However, hydrogen absorption has been observed in alkaline solutions at temperatures above the boiling point. Acidic conditions that cause the oxide films to be unstable may also result in embrittlement under conditions in which hydrogen is generated on titanium surfaces. It appears that embrittlement only occurs if the temperature is sufficiently high, i.e., above 170°F (77°C), permitting hydrogen to diffuse into the titanium. Otherwise surface hydride films on the surface are not detrimental.

Gaseous hydrogen has no embrittlement effects on titanium. The presence of as little as 2% moisture effectively prevents the absorption of molecular hydrogen up to a temperature as high as 600°F (315°C). However, this may reduce the ability of titanium to resist erosion, resulting in higher corrosion rates.

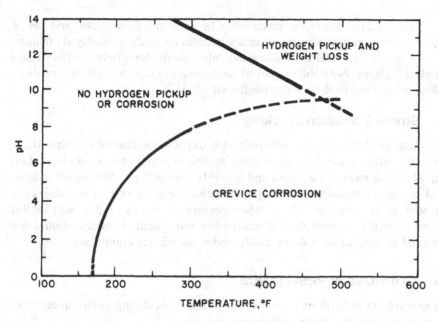

FIGURE 21.2 Effect of temperature and pH on crevice corrosion of grade 2 unalloyed titanium in saturated brine. (From Ref. 4.)

E. Crevice Corrosion

Crevice corrosion of titanium is most often observed in bromide, iodide, and sulfate solutions. Dissolved oxygen or other oxidizing agents present in the solution are depleted in the restricted volume of solution in the crevice. As a result the potential of the metal in the crevice becomes more negative than the metal exposed to the bulk solution. Thus an electrolytic cell is formed with the metal in the crevice acting as the anode and the metal outside the crevice acting as the cathode. Consequently the resulting current will cause metal to dissolve at the anode. Titanium chlorides formed in the crevice are unstable and tend to hydrolyze, forming small amounts of hydrochloric acid. This reaction is very slow at first, but in the restricted volume of the crevice it can reduce the pH of the solution to a value as low as 1. This reduces the potential still further until corrosion becomes severe.

The crevice corrosion resistance can be improved by alloying titanium with elements such as nickel, molybdenum, or palladium. Consequently, grade 12 and the titanium-palladium alloys are more resistant to crevice corrosion than unalloyed titanium.

Figure 21.2 shows the relationship between the temperature and pH of saturated brine at which corrosive attack initiates on grade 2 unalloyed titanium.

Figures 21.3 and 21.4 show the relationship for grade 7 (Ti-Pd) and grade 12 alloys. Note the improved resistance to crevice corrosion of these alloys compared to that of the unalloyed grade 2.

F.   Stress Corrosion Cracking

Titanium grades 1 and 2 (unalloyed) with oxygen contents of less than 0.2% are susceptible to cracking in absolute methanol and higher alcohols, certain liquid metals such as cadmium and possibly mercury, red-fuming nitric acid, and nitrogen tetraoxide. Cracking tendencies are increased when halides are present in the alcohols. Water, when present in excess of 2%, will inhibit stress cracking in alcohols and red-fuming nitric acid. Titanium should not be used in contact with these media under anhydrous conditions.

III.   CORROSION RESISTANCE

In general titanium offers excellent resistance in oxidizing environments and poor resistance in reducing environments.

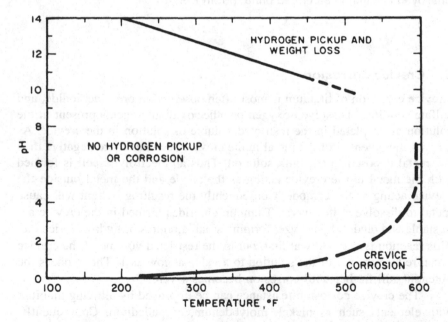

FIGURE 21.3 Effect of temperature and pH on crevice corrosion of grade 7 titanium alloy in saturated NaCl brine. (From Ref. 4.)

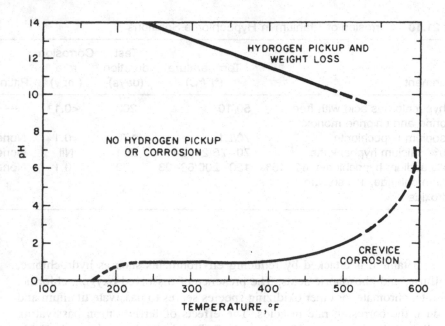

FIGURE 21.4 Effect of temperature and pH on crevice corrosion of grade 12 titanium alloy in saturated NaCl brine. (From Ref. 4.)

It has excellent resistance to moist chlorine gas, chlorinated brines, and hypochlorites (refer to Table 21.10). Titanium is attacked rapidly by dry chlorine gas. If the moisture is low enough, the titanium can ignite and burn. Approximately 1% water is required under static conditions at room temperature. At 392°F (200°C) approximately 1.5% water is required.

Titanium is immune to corrosion in all natural waters, including highly polluted seawater, at temperatures up to the boiling point. Titanium has replaced copper-based alloys that were corroding in the presence of sulfides as well as stainless steels that were suffering from pitting and stress corrosion cracking caused by chlorides.

Titanium is also resistant to steam up to temperatures as high as 600°F (315°C) and pressures up to 2000 psi. It is as resistant to erosion of wet steam as that of the 18Cr-8Ni stainless steels.

Titanium offers excellent resistance to oxidizing acids such as nitric and chromic acids. However, it is not recommended for use in red-fuming nitric acid, particularly if the water content is below 1.5% and the nitrogen dioxide content is above 2.5%. Pyrophoric reactions have occurred in this environment.

TABLE 21.10 Corrosion of Titanium in Hypochlorite Solutions

Environment	Temperature (°F/°C)	Test duration (days)	Corrosion rate (mpy)	Pitting
17% hypochlorous acid with free chlorine and chlorine monoxide	50/10	203	<0.1	—
16% sodium hypochlorite	70/21	170	<0.1	None
18–20% calcium hypochlorite	70–75/21–24	204	Nil	None
1.5–4% sodium hypochlorite, 12–15% sodium chloride, 1% sodium hydroxide	150–200/66–93	72	0.1	None

Titanium is attacked by reducing environments such as hydrochloric, sulfuric, and phosphoric acids. The presence of dissolved oxygen, chlorine, nitrate, chromate, or other oxidizing species serves to passivate titanium and reduce the corrosion rate in acids. The effects of ferric ion on passivating unalloyed titanium grade 12 and grade 7 (Ti-Pd alloy) are shown in Table 21.11, while Table 21.12 shows a similar comparison in sulfuric acid. Table 21.13 shows a similar effect of ferric ion in passivating titanium against corrosion by a 10 wt% hydrochloric acid solution.

Titanium generally shows good corrosion resistance to organic compounds. Many organic compounds are absorbed on titanium surfaces and act as inhibitors. When the temperature is high in an anhydrous environment and the organic compound dissociates, hydrogen embrittlement of the titanium is a consideration.

TABLE 21.11 Effect of Ferric Chloride Inhibitor on Corrosion in Boiling HCl

HCl wt%	Alloy	Corrosion rate (mpy)	
		No FeCl$_3$	2 g/L FeCl$_3$
3.32	Ti-50A	740	0.2
3.32	Ti-Code 12	606	1.0
3.32	Ti-Pd	3	0.1
4.15	Ti-50A	843	0.4
4.15	Ti-Code 12	1083	2.3
4.15	Ti-Pd	6	0.3

TABLE 21.12   Corrosion Rates of Titanium Alloys in Boiling H_2SO_4, mpy

H_2SO_4 wt%	Ti-Pd	Ti-Code 12	Unalloyed, Ti-50A
0.54	0.1	0.6	252
1.08	0.2	35.4	440
1.62	2.1	578	713
2.16	3.8	759	945
2.70	14.9	1331	1197
5.40	29.8	2410	2047

Titanium is also quite resistant to oxidizing organic acids. Only a few organic acids are known to attack titanium; these are hot nonaerated formic acid, hot oxalic acid, concentrated trichloroacetic acid, and solutions of sulfamic acid.

Titanium is resistant to acetic acid, terephthalic acid, and adipic acids. It also exhibits good resistance to citric, tartaric, carbolic, stearic, lactic, and tannic acids.

IV.   FABRICATION

The properties of titanium must be kept in mind during the design stage. For example, low modulus of elasticity means greater springback on forming. Shrinkage and distortion of welds may be greater than is generally experienced with other metals. This, combined with low strength at elevated temperatures, may require more careful attention to alignment during welding. Lower ductility might require more generous radii on bending or preheating before forming.

Special attention must be paid to the welding of titanium. Molten titanium absorbs oxygen, nitrogen, and hydrogen readily on exposure to air

TABLE 21.13   Effect of Inhibitors on Acid Corrosion of Titanium

Environment	Temperature (°F)	Corrosion rate (mpy)
10% HCl	Boiling	>2360
10% HCl + 16 g/L $FeCl_3$	Boiling	8
20% H_2SO_4	212	>2360
20% H_2SO_4 + 2.5 g/L $CuSO_4$	212	>2

or moisture. To a lesser degree, solid titanium will also absorb these elements, at least down to temperatures in the range of 600 to 800°F (315 to 425°C). Inert gas shielding during welding is imperative down to temperatures in the range of 600 to 800°F (315 to 425°C).

Good techniques for inert gas shielding of welds, which involve the use of trailing shields as well as back shields, are necessary. These techniques, combined with careful procedures to exclude all moisture, dirt, and other foreign matter from the weld area, assure good welds.

REFERENCES

1. PA Schweitzer. Encyclopedia of Corrosion Technology. New York: Marcel Dekker, 1998.
2. PA Schweitzer. Corrosion Resistant Piping Systems. New York: Marcel Dekker, 1994.
3. PA Schweitzer. Corrosion Resistance Tables, 4th ed. Vols. 1–3. New York: Marcel Dekker, 1995.
4. LC Covington, PA Schweitzer. Titanium. In: PA Schweitzer, ed. Corrosion and Corrosion Protection Handbook, 2nd ed. New York: Marcel Dekker, 1989.
5. GT Murray. Handbook of Materials Selection. New York: Marcel Dekker, 1997.

22

Zirconium and Zirconium Alloys

Zirconium was initially identified in 1789 by the German chemist M. H. Klaproth, but the metal itself was not isolated until 1824. J. J. Berzelius produced a brittle impure metal powder. It was not until 100 years later, in 1924, that the iodide decomposition process was developed to produce a pure, ductile metal. Although the "iodide crystal bar" process is slow and expensive it is still used to purify titanium, zirconium, and hafnium.

With the advent of nuclear reactors there was a need for a suitable structural metal having good high-temperature corrosion resistance, resistance to irradiation damage, and transparency to thermal neutrons needed for nuclear reaction.

During the 1940s zirconium, along with other metals, was studied as to its suitability for nuclear applications. Zirconium was developed for these applications because it is essentially transparent to neutrons. However, hafnium, which is naturally present in the zirconium, must be removed since it absorbs neutrons at 500–600 times the rate of zirconium. For nuclear applications the hafnium is removed; for other applications zirconium alloys will usually contain approximately 4% hafnium by weight since the natural occurring minerals contain 1 to 4% hafnium.

In 1947 W. J. Kroll, who had previously developed a production process for titanium by the reduction of titanium tetrachloride with magnesium in an inert atmosphere, developed a similar process for zirconium. However, the cost was still high.

By the mid-1950s several nuclear submarines had been authorized by the U.S. Congress, and nuclear power plants were on the horizon. Development work was continued to reduce the cost of zirconium.

In 1958 zirconium became available outside of the U.S. Navy programs. Zirconium was beginning to be used by the chemical process industry because of its broad range of corrosion resistance. Applications were also found as surgical tools and instruments because of zirconium's corrosion resistance and biocompatibility. Zirconium is also used as an alloying element for iron-, copper-, magnesium-, aluminum-, molybdenum-, and titanium-based alloys.

At times zirconium is referred to as a rare metal, which is incorrect. Of the chemical elements occurring in the earth's crust, zirconium is ranked 19th in abundance. It is more abundant than nickel, copper, zinc, chromium, lead, and cobalt. The most important source for zirconium is zircon ($ZnO_2 \cdot SiO_2$), which is in the form of beach sand found in several regions around the world.

I. GENERAL INFORMATION

Zirconium and its alloys can be classified into two major categories: nuclear and non-nuclear. The major difference between these two categories is in the hafnium content. Nuclear grades of zirconium alloys are essentially free of hafnium (<100 ppm). Non-nuclear grades of zirconium may contain as much as 4.5% hafnium, which has an enormous effect on zirconium's nuclear properties but little effect on its mechanical and chemical properties. The commercially available grades of zirconium alloys are shown in Table 22.1.

The majority of the nuclear grade material is produced as tubing, which is used for nuclear fuel rod claddings, guide tubes, pressure tubes, and ferrule spacer grids. Sheets and plates are used for spacer grids, water channels, and channel boxes for nuclear fuel bundles.

Non-nuclear zirconium applications make use of ingots, forgings, pipes, tubes, plates, sheet, foils, bars, wires, and castings to construct highly corrosion resistant equipment. Included are heat exchangers, condensers, reactors, columns, piping systems, agitators, evaporators, tanks, pumps, valves, and packing.

II. PHYSICAL PROPERTIES

Zirconium is a lustrous, grayish white, ductile metal. The physical properties are shown in Table 22.2. There are three of these properties worth mentioning:

1. Zirconium's density is considerably lower than those of iron- and nickel-based alloys.

TABLE 22.1 Commercially Available Grades of Zirconium Alloys

Alloy designation (UNS no.)	Zr + Hf, min.	Hf, max.	Composition (%)							
			Sn	Nb	Fe	Cr	Ni	Fe + Cr	Fe + Cr + Ni	O, max.
Nuclear grades										
Zircaloy-2 (R60802)	—	0.010	1.20–1.70	—	0.07–0.20	0.05–0.15	0.03–0.08	—	0.18–0.38	—
Zircaloy-4 (R60804)	—	0.010	1.20–1.70	—	0.18–0.24	0.07–0.13	—	0.28–0.37	—	—
Zr–2.5Nb (R60901)	—	0.010	—	2.40–2.80	—	—	—	—	—	—
Chemical grades										
Zr 702 (R60702)	99.2	4.5	—	—	—	—	—	0.2 max.	—	0.16
Zr 704 (R60704)	97.5	4.5	1.0–2.0	—	—	—	—	0.2–0.4	—	0.18
Zr 705 (R60705)	95.5	4.5	—	2.0–3.0	—	—	—	0.2 max.	—	0.18
Zr 706 (R60706)	95.5	4.5	—	2.0–3.0	—	—	—	0.2 max.	—	0.16

TABLE 22.2 Typical Physical and Mechanical Properties of Zirconium Grades 702 and 705

Properties	Zr 702	Zr 705
Physical		
Atomic number	40	—
Atomic weight	91.22	—
Atomic radius		
Å (zero charge)	1.60–1.62	—
Å (+4 charge)	0.80–0.90	—
Density		
at 20°C (g/cm³)	6.510	6.640
lb/in.³	0.235	0.240
Crystal structure		
α phase	Hexagonal close-packed (below 865°C)	
β phase	Body-centered cubic (above 865°C)	Body-centered cubic (above 854°C)
α + β phase		Hexagonal close-packed + body-centered cubic (below 854°C)
Melting point	3365°F (1852°C)	3344°F (1840°C)
Boiling point	7910°F (4377°C)	7916°F (4380°C)
Coefficient of thermal expansion per °C at 73°F (25°C)	5.89×10^{-6}	6.3×10^{-6}
Thermal conductivity at 300–800 K		
Btu ft/hr ft² °F	13	10
W/m-K	22	17.1

Specific heat [Btu/lb/°F(32–212°F)]	0.068	0.067
Vapor pressure (mmHg)		
at 3632°F (2000°C)	0.01	—
at 6512°F (3600°C)	900.0	—
Electrical resistivity at 68°F (20°C) (μ cm)	39.7	55.0
Temperature coefficient of resistivity per °C at 68°F (20°C)	0.0044	—
Latent heat of fusion (cal/g)	60.4	—
Latent heat of vaporization (cal/g)	1550	—
Mechanical		
Modulus of elasticity		
10^6 psi	14.4	14.0
GPa	99	97
Shear modulus		
10^6 psi	5.25	5.0
GPa	36	34
Poisson's ratio at ambient temperature	0.35	0.33

2. Zirconium has a low coefficient of expansion. This is an advantage for equipment requiring a close tolerance. Its coefficient of thermal expansion is approximately two-thirds that of titanium, about one-third that of type 316 stainless steel, and approximately one-half that of Monel.
3. The thermal conductivity of zirconium is high, more than 30% greater than those of stainless alloys.

All of these properties make zirconium very fabricable for constructing compact, efficient equipment.

III. MECHANICAL PROPERTIES

All zirconium ores contain a few percent hafnium, whose chemical and metallurgical properties are similar to those of zirconium. However, their nuclear properties are vastly different. While hafnium is a neutron absorber, zirconium is not. Consequently for nuclear applications the hafnium must be removed, but not for non-nuclear applications. For this reason there are nuclear and non-nuclear grades of zirconium.

With the exception of the thermal neutron cross-section, the presence of hafnium in the zirconium does not affect the mechanical properties. As a result the counterparts of nuclear and non-nuclear grades of zirconium are interchangeable in mechanical properties. However, specification requirements for nuclear materials are more extensive than those for non-nuclear materials.

Table 22.3 gives the minimum ASTM requirements for room temperature mechanical properties of non-nuclear zirconium. It will be noted that Zr 705 is the stronger of the alloys, while Zr 706 has been developed for

TABLE 22.3 Non-nuclear Minimum ASTM Requirements for the Room-Temperature Mechanical Properties of Zirconium Alloys

Alloy	Minimum tensile strength (MPa)	Minimum yield strength 0.2% offset (MPa)	Minimum elongation (%)	Bend test radius[a]
Zr 702	380	207	16	5 T
Zr 704	414	240	14	5 T
Zr 705	552	380	16	3 T
Zr 706	510	345	20	2.5 T

[a]Bend tests are not applicable to material more than 4.75 mm thick. T is the thickness of the bend test specimen.

severe forming applications. Table 22.4 gives the ASME mechanical requirements of Zr 702 and Zr 705 used for infrared pressure vessels.

Table 22.5 lists the fatigue limits of selected zirconium alloys.

Tensile properties versus temperature is shown in Fig. 22.1 for Zr 702, Zr 704, and Zr 705. The minimum creep rate versus stress curves for Zr 702 and Zr 705 are shown in Fig. 22.2. The mechanical and physical properties of zirconium pipe grades 702 and 705 are shown in Table 22.6, and the maximum allowable stress values are shown in Table 22.7.

IV. CORROSION PROTECTION

In spite of the reactive nature of zirconium metal, the zirconium oxide (ZrO_2) film which forms on the surface is among the most insoluble compounds in a broad range of chemicals. Excellent corrosion protection is provided in most media. When mechanically destroyed the oxide film will regenerate itself in many environments. When placing zirconium in a corrosive medium there is no need to thicken this oxide film.

For mechanical reasons it is desirable to preoxidize zirconium, which provides a much improved performance against sliding forces; although it can be damaged by striking action. Other advantages are also realized.

Several methods are available to produce the oxide film. They include anodizing, autoclaving in hot water or steam, formation in air, and formation in molten salts.

A. Anodizing

A very thin film is formed by anodizing (<0.5 μm). As the thickness of the film grows, the color changes. Although the film formed is attractive, it does not have the adhesion of thermally produced films and has very limited ability to protect the metal from mechanical damage.

B. Autoclave Film Formation

The nuclear reaction industry uses this method. These films, in addition to providing a slower corrosion rate, reduce the rate of hydrogen absorption.

C. Film Formation in Air or Oxygen

This is the most common method used in the chemical process industry. The film is formed during the final stress relief of a component in air at 1022°F (550°C) for 0.5 to 4 hr. It will range in color from straw yellow to an iridescent blue or purple to a powdery tan or light gray. These colors are not indications of metal contamination. This treatment does not cause sig-

TABLE 22.4 ASME Mechanical Requirements for Zr 702 and Zr 705 Used for Unfired Pressure Vessels

Material form and condition	ASME specification no.	Alloy grade	Tensile strength (MPa)	Minimum yield strength (MPa)	Maximum allowable stress in tension for metal temperature not exceeding °C (MPa)						
					40	95	150	205	260	315	370
Flat-rolled products	SB 551	702	359	207	90	76	64	48	42	41	33
		705	552	379	138	115	98	86	78	72	62
Seamless tubing	SB 523	702	359	207	90	76	64	48	42	41	33
		705	552	379	138	115	98	86	78	72	62
Welded tubing[a]	SB 523	702	359	207	77	65	55	41	36	35	28
		705	552	379	117	97	83	73	66	59	52
Forgings	SB 493	702	359	207	90	76	64	48	42	41	33
		705	552	379	138	115	98	86	78	72	62
Bar	SB 550	702	359	207	90	76	64	48	42	41	33
		705	552	379	138	115	98	86	78	72	62

[a]85% joint efficiency was used to determine the allowable stress value for welded tube. Filler material shall not be used in the manufacture of welded tube.

TABLE 22.5 The 10^7 Fatigue Limits for Zirconium Alloys

Alloy	Fatigue limit (MPa)	
	Smooth	Notched
Iodide Zr	145	55
Zircaloys or Zr 705 (annealed 2 hr at 732°C)	283	55
Zr–2.5% Nb (aged 4 hr at 566°C)	290	55

nificant penetration of oxygen into the metal, but it does form an oxide layer that is diffusion bonded to the base metal.

D. Film Formation in Molten Salts

In this process, developed and patented by TWC, zirconium subjects are treated in a fused sodium cyanide containing 1–3% sodium carbonate or, in a eutectic mixture of sodium and potassium chlorides, with 5% sodium carbonate. Treatment is carried out at temperatures ranging from 1112–1472°F (600–800°C) for several hours. A thick, protective, strongly cohesive oxide

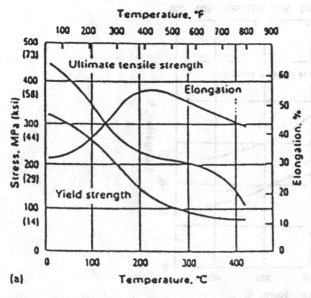

FIGURE 22.1 Tensile properties versus temperature curves for zirconium alloys: (a) Zn 702; (b) Zn 704; (c) Zn 705. (From Ref. 1.)

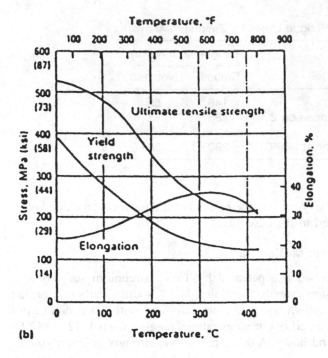

(b)

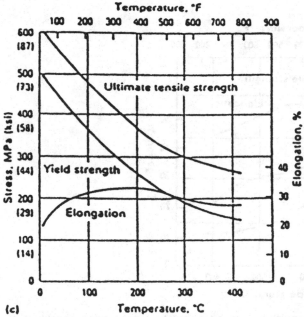

(c)

FIGURE 22.1   Continued

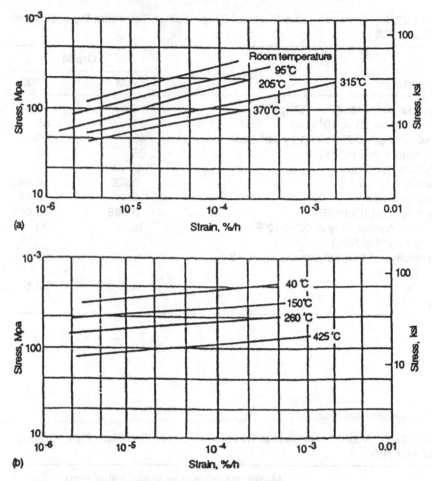

FIGURE 22.2 Maximum creep rate versus stress curves for zirconium alloys: (a) Zn 702; (b) Zn 705. (From Ref. 1.)

film ranging from 20 to 30 μm is formed. This film has improved resistance to abrasion and galling over thick films produced by other methods.

E. Electrochemical Protection

Zirconium performs well in most reducing environments as a result of its ability to take oxygen from water to form stable passive films. Most passive metals and alloys require the presence of an oxidizing agent such as oxygen in order to form a protective oxide film. Zirconium's corrosion problems

TABLE 22.6   Mechanical and Physical Properties of Zirconium Pipe, Grades
702 and 705

	Grade	
Property	702	705
Modulus of elasticity $\times\ 10^6$ (psi)	14.4	14.0
Tensile strength $\times\ 10^3$ (psi)	65	85
Yield strength 0.2% offset by 10^3 (psi)	45	65
Elongation in 2 in. (%)	25	20
Rockwell hardness	B-90	
Density (lb/in.^{3})	0.235	0.240
Specific gravity	6.51	6.64
Specific heat (Btu/lb°F)	0.068	0.067
Thermal conductivity at 32–212°F (Btu/ft^2/hr/°F/in.)	95	95
Coefficient of thermal expansion $\times\ 10^{-6}$ (in./in.°F)		
at 60°F	3.2	3.5
at 60–750 °F	3.7	

TABLE 22.7   Maximum Allowable Stress Values for Zirconium Pipe, Grades
702 and 705

Operating temperature (°F/°C)	Maximum allowable stress value (psi)			
	Grade 702		Grade 705	
	Seamless	Welded[a]	Seamless	Welded[a]
100/38	13000	11100	20000	17000
200/93	11000	9400	16600	14100
300/149	9300	7900	14200	12000
400/204	7000	6000	12500	10600
500/260	6100	5200	11300	9600
600/316	6000	5100	10400	8500
700/371	4800	4100	9900	7600

[a]8.5% joint efficiency has been used in determining the allowable stress values for
welded pipe.

TABLE 22.8 Corrosion Rate of Zirconium in 500 ppm Fe^{3+} Solution After 32 Days

Environment	Acidity	Temperature (°C)	Penetration rate (mpy)	
			Unprotected	Protected
10% HCl	3 N	60	7.1	<0.1
		120	51	<0.1
Spent acid (15% Cl)	5 N	65	36	<0.1
		80	36	<0.1
20% HCl	6 N	60	3.6	<0.1
		107	59	<0.1

Source: Ref. 1.

can be controlled by converting the corrosive condition to a more reducing condition.

By impressing a potential that is arbitrarily 50–100 mV below its corrosion potential, zirconium becomes corrosion resistant in oxidizing chloride solutions. Tables 22.8 and 22.9 demonstrate the benefits of electrochemical protection in controlling pitting and SCC. Pitting penetration in oxidizing chloride solutions is considerably higher than general corrosion rates,

TABLE 22.9 Time to Failure of Welded Zirconium U Bends in 500 ppm Fe^{3+} Solution After 32 Days

Environment	Acidity	Temperature (°C)	Time to failure (days)	
			Unprotected	Protected
10% HCl	3 N	60	<0.1	NF
		120	<0.1	NF
Spent acid (15% Cl)	5 N	65	<0.3	NF
20% HCl	6 N	60	NF	NF
		107	<0.1	NF
28% HCl	9 N	60	2	NF
		94	<0.1	NF
32% HCl	10 N	53	1	32
		77	<0.1	20
37% HCl	12 N	30	0.3	NF
		53	1	NF

NF, no failure.
Source: Ref. 1.

which may be low on unprotected zirconium. Electrochemical protection eliminates this local attack.

As can be seen from Table 22.9 unprotected welded zirconium U bends cracked in all but one case shortly after exposure. Protected U bends resisted cracking for the 32-day test period in all but one acid concentration. From these tests it is obvious that electrochemical protection provides an improvement to the corrosion properties of zirconium in oxidizing solutions.

V. FORMS OF CORROSION

The more common forms of corrosion to which zirconium is susceptable, other than general (uniform) corrosion, are covered in this section.

A. Pitting

Zirconium will pit in acidic chloride solutions because its pitting potential is greater than its corrosion potential. The presence of oxidizing ions, such as ferric and cupric ions, in acidic chloride solutions may increase the corrosion potential to exceed the pitting potential. Therefore pitting may occur. However, zirconium does not pit in most other halide solutions. Under certain conditions nitrate and sulfate ions can inhibit the pitting.

One of the critical factors in pitting is surface condition. A metal with a homogeneous surface is less likely to pit and less likely to be vulnerable to other forms of localized corrosion. A common method used to homogenize a metal's surface is by pickling. Results of tests show that pickled zirconium may perform well in boiling 10% $FeCl_3$ and even ClO_2, while zirconium with a normal surface finish is unsuitable for handling these solutions.

B. Stress Corrosion Cracking

Zirconium and its alloys resist SCC in many media, such as NaCl, $MgCl_2$, NaOH, and H_2S, which cause SCC on common metals and alloys. However, zirconium is susceptible to SCC in environments such as $FeCl_3$, $CuCl_2$, halide or halide-containing methanol, concentrated HNO_3 64–69% H_2SO_4, and liquid mercury or cesium.

Stress corrosion cracking of zirconium can be prevented by

1. Avoiding high sustained tensile stresses
2. Modifying the environment, e.g., changing pH concentration or adding an inhibitor
3. Maintaining a high-quality surface film (one low in impurities, defects, and mechanical damage)

4. Applying electrochemical techniques
5. Shot peening
6. Achieving a crystallographic texture with the hexagonal basal planes perpendicular to the cracking path

C. Fretting Corrosion

When the protective oxide coating of zirconium is damaged or removed, fretting may occur. It takes place when vibration contact is made at the interface of tight-fitting, highly bonded surfaces. If the vibration cannot be removed mechanically the addition of a heavy oxide coating on the zirconium may eliminate the problem. This coating reduces friction and prevents the removal of the passive film.

D. Galvanic Corrosion

The protective oxide film that forms on zirconium causes zirconium to assume a noble potential similar to that of silver. It is possible for zirconium to become activated and corrode at vulnerable areas when in contact with a noble metal. Vulnerable areas include areas with damaged oxide films and grain boundaries.

Other less noble metals will corrode when in contact with zirconium when its oxide film is intact.

E. Crevice Corrosion

Zirconium is among the most resistant of all of the corrosion resistant metals to crevice corrosion. However, it is not completely immune to crevice corrosion in a broad sense. For example, crevice corrosion will occur when a dilute sulfuric acid solution is allowed to concentrate within a crevice.

F. Corrosion Resistance

Zirconium is a highly corrosion resistant metal. It reacts with oxygen at ambient temperatures and below to form an adherent, protective oxide film on its surface. Said film is self-healing and protects the base metal from chemical and mechanical attack at temperatures as high as 662°F (350°C). In a few media, such as hydrofluoric acid, concentrated sulfuric acid, and oxidizing chloride solutions, it is difficult to form this protective film. Therefore zirconium cannot be used in these media without protective measures previously discussed. Refer to Table 21.9 for the compatibility of zirconium with selected corrodents. The corrosion resistance of all zirconium alloys is similar.

Zirconium has excellent corrosion resistance to seawater, brackish water, and polluted water. It is insensitive to changes such as chloride concentration, temperature, pH, crevice formation, flow velocity, and sulfur-containing organisms.

Zirconium also resists attack by all halogen acids except hydrofluoric, which will attack zirconium at all concentrations. One of the most impressive corrosion resistant properties is its resistance to hydrochloric acid at all concentrations, even above boiling. The isocorrosion diagram for zirconium is shown in Fig. 22.3.

Nitric acid poses no problem for zirconium. It can handle 9% HNO_3 below the boiling point and 70% HNO_3 up to 482°F (250°C) with corrosion rates of less than 5 mpy. Refer to Fig. 22.4.

The nature of sulfuric acid is complicated. Dilute solutions are reducing in nature. At or above 65%, sulfuric acid solutions become increasingly oxidizing. Referring to Fig. 22.5, it will be noted that zirconium resists attack by H_2SO_4 at all concentrations up to 70% and at temperatures to boiling and above. In the 70–80% range of concentration the corrosion resistance of

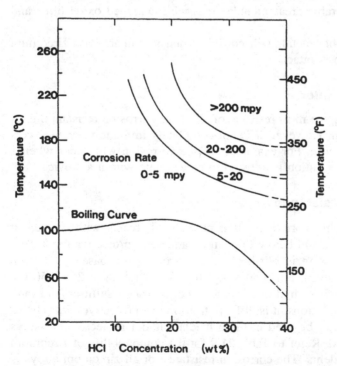

FIGURE 22.3   Isocorrosion diagram for zirconium in HCl. (From Ref. 1.)

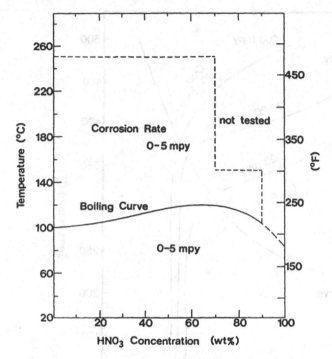

FIGURE 22.4   Isocorrosion diagram for zirconium in HNO_3. (From Ref. 1.)

zirconium depends strongly on temperature. In higher concentrations, the corrosion rate of zirconium increases rapidly as the concentration increases. The presence of chlorides in H_2SO_4 has little effect on the corrosion resistance of zirconium unless oxidizing agents are also present.

Zirconium resists attack in phosphoric acid at concentrations up to 55% and temperatures exceeding the boiling point. Above 55% concentration the corrosion rate may increase greatly with increasing temperature. Zirconium performs ideally handling dilute acid at elevated temperatures. If the phosphoric acid contains more than a trace of fluoride ions, zirconium may be attacked.

Zirconium is resistant to most alkalies including, sodium hydroxide, potassium hydroxide, calcium hydroxide, and ammonium hydroxide.

Most salt solutions, including halogen, nitrate, carbonate, and sulfate, will not attack zirconium. Corrosion rates are usually very low up to the boiling point. The exceptions are strong oxidizing chloride salts such as $FeCl_3$ and $CuCl_2$. In these media the corrosion resistance of zirconium is

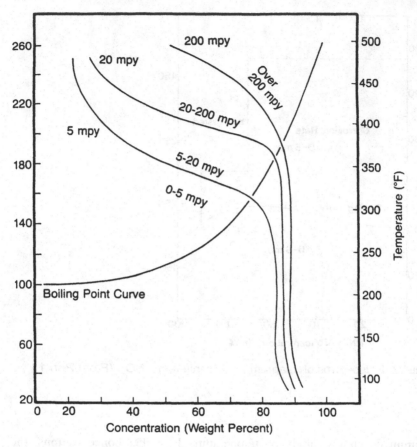

FIGURE 22.5 Isocorrosion diagram for zirconium in H_2SO_4. (From Ref. 1.)

dependent on the surface conditions. When the zirconium has a good surface finish it becomes quite resistant to pitting.

Zirconium possesses excellent resistance in most organic solutions. Corrosion is experienced when halogens are present and there is a lack of water. For example, if water is added to alcohol solutions with halide impurities zirconium's susceptibility to SCC will be suppressed. Table 22.10 provides the corrosion rate of zirconium in selected organic solutions.

VI. WELDING OF ZIRCONIUM

Welding of zirconium requires proper shielding of the weld puddle and hot bead from the air. The shielding gas must be highly pure argon, helium, or

TABLE 22.10 Corrosion Rates for Zirconium in Organic Solutions

Environment	Concentration (wt%)	Temperature (°C)	Corrosion rate (mpy)
Acetic acid	5–99.5	35 to boiling	<0.07
Acetic anhydride	99.5	Boiling	0.03
Aniline hydrochloride	5, 20	35–100	<0.01
Chloroacetic acid	100	Boiling	<0.01
Citric acid	10–50	35–100	<0.2
Dichloroacetic acid	100	Boiling	<20
Formic acid	10–90	35 to boiling	<0.2
Lactic acid	10–85	35 to boiling	<0.1
Oxalic acid	0.5–25	35–100	<0.5
Tartaric acid	10–50	35–100	<0.05
Tannic acid	25	35–100	<0.1
Trichloroacetic acid	100	Boiling	>50
Urea reactor	58% urea, 17% NH_3, 15% CO_2, 10% H_2O	193	<0.1

a mixture of the two. Absorption of oxygen and nitrogen by the metal results in weld embrittlement. Cleanliness is another important requirement to prevent weld embrittlement.

The most common method used to weld zirconium is the gas tungsten arc welding technique (GTAW). Other welding methods include metal arc gas welding (MAGW), plasma arc welding, electron beam welding, spot welding, friction welding, and resistance welding.

VII. CAST ZIRCONIUM

When cast, zirconium must be melted and poured into graphite molds in a vacuum. Zirconium alloys can be machined by conventional methods following three rules:

1. Slow speeds
2. Heavy feeds
3. A flood coolant system using a water-soluble oil lubricant

Zirconium has a tendency to gall and work harden; therefore a higher than normal clearance angle on tools is needed to penetrate the previously work-

TABLE 22.11   Cast Zirconium Alloys

ASTM specification and grade	Total residual elements, max.	Hf, max	Sn	Cb	Minimum strength, (ksi/MPa)		Maximum hardness (HB)
					Tensile	Yield	
ASTM B752 Grade 702C	0.4	4.5	0.3 max	—	55/380	40/276	210
ASTM B752 Grade 704C	0.4	4.5	0.3 max	—	60/413	40/276	235
ASTM B752 Grade 705C	0.4	4.5	1–2	2–3	70/483	50/345	235

hardened surface and cut a clean coarse chip. Cemented carbide and high speed tools can be used.

Three grades of cast zirconium are covered by ASTM B752 and are shown in Table 22.11.

REFERENCES

1. T-L Yau. Corrosion of zirconium. In: PA Schweitzer, Corrosion Engineering Handbook. New York: Marcel Dekker, 1996.
2. PA Schweitzer. Corrosion Resistant Piping Systems. New York: Marcel Dekker, 1994.
3. GT Murray. Handbook of Materials Selection for Engineering Applications. New York: Marcel Dekker, 1997.
4. PA Schweitzer. Corrosion Resistance Tables, 4th ed., Vols. 1–3. New York: Marcel Dekker, 1995.

23

Tantalum and Tantalum Alloys

Tantalum is not a new material. Its first commercial use at the turn of the century was as filaments in light bulbs. Later, when it became apparent that tantalum was practically inert to attack by most acids, applications in the laboratory and in the chemical and medical industries were developed. The rise of the electronics industry accelerated the development of many new applications.

Much of this growth can be attributed to a broader range of tantalum powders and mill products available from the producers; high melting point, ability to form a dielectric oxide film, and chemical inertness. With these applications, new reduction, melting, and fabrication techniques have led to higher purities, higher reliabilities, and improved yields to finished products.

In 1802 the Swedish chemist Anders Ekelberg discovered a new element in some mineral samples from Sweden and Finland. He called the new element tantalum and it was given the chemical symbol Ta.

Tantalum has a number of especially unique characteristics. It is difficult to dissolve, it melts at a high temperature, it is very heavy, it is very uncommon, and it is usually associated with a very similar element, niobium.

There are 92 naturally occurring elements which make up the earth's crust. These elements are not present in equal amounts—two elements, oxygen and silicon, make up about 75% of the crust. Eight elements, oxygen, silicon, aluminum, iron, calcium, sodium, potassium, and magnesium, make up 96.5% of the earth's crust. The other 88 elements make up only 3.5%, including tantalum which represents 0.0002% of the crust.

If the tantalum were equally distributed in the rocks of the earth it would be uneconomical to recover. Fortunately, tantalum is concentrated in a few unusual types of rocks in quantities sufficient to allow economical mining procedures. Most of the crushed rocks are composed of those three most abundant elements, oxygen, silicon, and aluminum. These three elements combine in a number of ways to form the aluminum silicate minerals. Many of the other elements incorporate themselves into these aluminum silicate minerals, but tantalum, because of some of its peculiar properties, does not. It combines with a few other elements to form special tantalum-rich minerals. The most important tantalum minerals are tantalite, microlite, and wodginite and are found in rock formations knows as pegamites.

Pegamites are coarse-grained rocks formed when molten rock material was cooled slowly. They range in size from 1 in. to many feet in diameter. Found in the pegamites are many rare elements such as tantalum, niobium, tin, lithium, and beryllium.

The only operating mine in North America is located at Bernic Lake in Manitoba, Canada. The other important mine in the Americas is found in Brazil. In the humid tropics the rocks weather and rot to great depths. Many times the rocks in which tantalum minerals were formed have completely weathered and have been carried away by running water. The heavy tantalum minerals tend to be concentrated in deposits called placers. These can be panned or washed with machinery, much as gold was recovered during the Gold Rush.

Because tin and tantalum are often found together, tantalum is a by-product of the tin industry. Because most of the tantalum deposits are small, hard to find, and very expensive to mine, the result is a high priced ore and a correspondingly high priced metal.

I. TANTALUM MANUFACTURE

The first production route for tantalum was by powder metallurgy. Tantalum powder, produced by one of several reduction techniques, is pressed into suitably sized bars and then sintered in a vacuum at a temperature in excess of 3800°F (2100°C). When completed the pressed and sintered bars are ready for processing into mill shapes. Forging, rolling, swaging, and drawing of tantalum is performed at room temperature on standard metal working equipment with relatively few modifications.

The powder metallurgy route, although still in use and adequate for many applications, has two major limitations: (1) the size of the bar capable of being pressed and sintered to a uniform density limits the size of the finished shape available, and (2) the amount of residual interstitial impurities such as oxygen, carbon, and nitrogen remaining after sintering adversely affects weldability.

The use of vacuum melting overcomes these limitations. Either melting technique is capable of producing ingots that are big enough and high enough in purity to meet most requirements of product size and specification adequately, providing that starting materials are selected with care.

The greatest volume of tantalum is supplied as powder for the manufacture of solid electrolytic capacitors.

Because it is necessary to distinguish between capacitor grade power and melting grade powder, manufacturers of electronic components tend to use the term "capacitor grade" when ordering forms such as wire, foil, and sheet to identify end use and desired characteristics.

The term capacitor grade means that the material should have the ability to form an oxide film of certain characteristics. Capacitor grade in itself does not mean an inherently higher purity, cleaner surface, or different type of tantalum. It does mean that the material should be tested using carefully standardized procedures for electrical properties. If certain objective standards, such as formation voltage and leakage current are not available against which to test the material, the use of the phrase capacitor grade is not definitive.

Metallurgical grade could be simply defined as noncapacitor grade.

II. MECHANICAL PROPERTIES

The room-temperature mechanical properties of tantalum are dependent on chemical purity, the amount of reduction in cross-sectional area, and temperatures of final annealing. Annealing time apparently is not critical. Close control over the many factors that affect mechanical properties are mandatory to ensure reproducible mechanical behavior. Typical mechanical properties of tantalum are shown in Table 23.1.

Tantalum can be strengthened by cold work, with a resulting loss in ductility. Because certain residual impurities have pronounced effects on

TABLE 23.1 Typical Mechanical Properties: Annealed Tantalum Sheet

Thickness	Yield strength 0.2% (psi)	Ultimate tensile strength (psi)	Elongation (%)	Rockwell hardness	
				15T	B
0.005 deep draw	29,000	41,000	22	—	—
0.005 regular	44,000	55,000	18	—	—
0.010 regular	40,000	52,000	32	—	—
0.030 regular	35,000	45,000	40	75	—
0.060 regular	35,000	45,000	42	78	48

ductility levels and metallurgical behavior, the purpose of most consolidation techniques is to make the material as pure as possible. Cold working methods are used almost without exception to preclude the possibility of embrittlement by exposure to oxygen, carbon, nitrogen, and hydrogen at even moderate temperatures. Temperatures in excess of 800°F (425°C) should be avoided.

There are basically three structures that can be ordered: (1) unannealed, (2) stress relieved, and (3) annealed.

In the unannealed condition the structure will be typically wrought fibrous. Yield and tensile strength will be increased with a corresponding decrease in elongation, as shown in Table 23.2. The amount of work hardening will be dependent on the amount of cold reduction since the last anneal. The rate of work hardening is rapid for the first 30% of reduction. The rate then diminishes so that there will be no appreciable strengthening until reductions of over 90% are taken. There is actually no limit to the amount of cold work that the metal can take; there are only equipment limitations or mechanical limitations, such as poor shape control in rolling or excessive thining when forming, which dictate periodic heat treatment in vacuum to soften the metal.

Unannealed tantalum may be preferred for machinability. Corrosion behavior is not affected nor is the susceptibility to interstitial contamination changed. Unalloyed sheet 0.030 in. thick and less can make 1 in. thickness bends, but the annealed condition is preferred when bending because the metal is not as stiff or springy.

Stress relieving at 1850°F (1010°C) in vacuum reduces yield and tensile strength and raises elongation levels. These properties will be interme-

TABLE 23.2   Typical Mechanical Properties: Tantalum Sheet with Increasing Cold Work

Cold work (%)	Yield strength 0.2% (psi)	Ultimate tensile strength (psi)	Elongation (%)	Hardness (VHN)
30	70,600	74,200	18	189
50	82,200	86,000	9	192
80	100,500	109,200	4	235
90	117,800	123,400	2	239
95	127,000	135,500	1	265
98	—	135,500	1	280

diate between annealed and unannealed. Stress relieving has been used more as a matter of expedience than design. Fabricators need some ductility to allow them to roll tubes into tube sheets. Initially the only tubular heat-treating vacuum furnaces were limited to 1850°F (1010°C) maximum. This limitation dictated the use of stress relieving. Since newer furnaces allowing full annealing are now available stress relieving is falling into disuse.

Tantalum specified in the annealed condition is in its softest most ductile condition. The usual objective of the procedure is to choose an annealing temperature that will result in complete recrystallization but avoid excessive grain growth. This temperature is approximately 2150 ± 75°F (1175 ± 32°C). The temperature for recrystallation is, however, considerably affected by the purity and amount of cold work before annealing.

When the amount of reduction is limited, complete recrystallization is very difficult unless the annealing temperature is substantially increased. Purity is at the core of the problem. Unless sufficient work (about 75% reduction) is put into the material, tantalum does not have the impurities present to act as nucleation sites for grain growth unless an inordinate amount of energy in the form of annealing heat is added. This will result in recrystallization, but to a very large grain size. The larger the cross-section, the more severe is this problem. Recrystallization to a finer grain size becomes more readily attainable. The tensile test is normally used to determine the state of anneal. ASTM specifications for annealed tantalum are shown in Table 23.3.

III. PHYSICAL PROPERTIES

Tantalum, the 73rd element in the periodic table, is a member of the group V elements. It is typically one of the "reactive metals." Many of its prominent features resemble those of titanium and zirconium, and it is reasonably compatible and comparable with those metals as well as with molybdenum and tungsten. The physical properties of tantalum are shown in Table 23.4.

The thermal and electrical conductivities of tantalum are high relative to many of the other base metals and are considerably better than widely used alloys such as Hastelloys, stainless steels, and Monel.

Tantalum also has a usefully high value of elastic modulus, a desirable property for provision of structural rigidity of engineering components. Tantalum's value of elastic modulus is equivalent to those of nickels and steels and considerably more than those of brasses, aluminums, titanium, glasses, and graphite. In addition this value does not decrease significantly when tantalum is heated to higher temperatures.

TABLE 23.3 ASTM Specification Limits for Tensile Properties: Test Procedure ASTM E8-61T

	Ultimate psi		Maximum yield (psi)	Minimum elongation (%)	Maximum hardness
	Minimum	Maximum			
Cold worked	75,000			2	
Stress relieved					
any section >0.021 in.	55,000			10	
any section <0.021 in.	55,000			7.5	
Annealed					
any section >0.021 in.		55,000	45,000	25	80 + 5T
any section <0.021 in. to 0.005 in. min.		55,000	45,000	15	

TABLE 23.4 Physical Properties of Tantalum

Atomic weight	180.9
Density	16.6 g/cm^3 (0.601 lb/in.^{3})
Melting point	5432°F (2996°C)
Vapor pressure at 1727°C	9.525 × 10^{-11} mmHg
Linear coefficient of expansion	1135K; 5.76 × 10^{-6}/°C
	1641K; 9.53 × 10^{-6}/°C
	2030K; 12.9 × 10^{-6}/°C
	2495K; 16.7 × 10^{-6}/°C
Thermal conductivity	20°C; 0.130 cal/cm-s °C
	100°C; 0.131 cal/cm-s °C
	1430°C; 0.174 cal/cm-s °C
	1630°C; 0.186 cal/cm-s °C
	1830°C; 0.198 cal/cm-s °C
Specific heat	100°C; 0.03364 cal/g
Electrical conductivity	13.9% IACS
Electrical resistivity	−73°C; 9.0 $\mu\Omega$/cm
	75°C; 12.4 $\mu\Omega$/cm
	127°C; 18.0 $\mu\Omega$/cm
	1000°C; 54.0 $\mu\Omega$/cm
	1500°C; 71.0 $\mu\Omega$/cm
	2000°C; 87.0 $\mu\Omega$/cm

IV. TANTALUM-BASED ALLOYS

There are certain advantages to the use of tantalum-based alloys:

1. Alloying with a less expensive material reduces the cost of the material while still retaining essentially all of the corrosion resistant properties.
2. The use of light material will reduce the overall weight.
3. Depending upon the alloying ingredient, the physical strength of tantalum may be improved.

A. Tantalum-Tungsten Alloys

The tantalum-tungsten alloys are probably the most common. The addition of 2 to 3% tungsten will raise the strength of the tantalum by 30 to 50%. By also adding 0.15% niobium, a marked increase in the corrosion resistance to concentrated sulfuric acid at 392°F (200°C) is noted. In the lower temperature ranges, 347°F (175°C) or less, the resistance of the alloy is equal to that of pure tantalum.

When the tungsten concentration is increased to 18% or higher, the alloys exhibit essentially no corrosion rate in 20% hydrofluoric acid. This is a definite advantage over pure tantalum.

B. Tantalum-Titanium Alloys

The tantalum-titanium alloys are receiving a great deal of study because this series of alloys shows considerable promise of providing a less expensive, lower weight alloy having a corrosion resistance almost comparable with that of tantalum. Tantalum-titanium alloys show excellent resistance in nitric acid at 374°F (190°C) and at the boiling point.

C. Tantalum-Molybdenum Alloys

When these alloys are exposed to concentrated sulfuric acid and concentrated hydrochloric acid they are extremely resistant, and the properties of tantalum are retained, as long as the tantalum concentration is higher than 50%.

V. CORROSION RESISTANCE

Tantalum forms a thin, impervious, passive layer of tantalum oxide on exposure to oxidizing or slightly anodic conditions, even at a temperature as low as 77°F (25°C). Chemicals or conditions which attack tantalum, such as hydrofluoric acid, are those which penetrate or dissolve this oxide film, in the case of fluoride ion by forming the complex TaF_5^{2-}. Once the oxide layer is lost the metal loses its corrosion resistance dramatically.

When in contact with most other metals tantalum becomes cathodic. In galvanic couples in which tantalum becomes the cathode, nascent hydrogen forms and is absorbed by the tantalum, causing hydrogen embrittlement. Caution must be taken to electrically isolate tantalum from other metals or otherwise protect it from becoming cathodic.

Tantalum is inert to practically all organic and inorganic compounds at temperatures under 302°F (150°C). The only exceptions to this are hydrofluoric acid (HF) and fuming sulfuric acid. At temperatures under 302°F (150°C) it is inert to all concentrations of hydrochloric acid, to all concentrations of nitric acid (including fuming), to 98% sulfuric acid, to 85% phosphoric acid, and to aqua regia (refer to Table 23.5).

Corrosion is first noticed at about 375°F (190°C) for 70% nitric acid, at about 345°F (175°C) for 98% sulfuric acid, and at about 355°F (180°C) for 85% phosphoric acid (refer to Fig. 23.1).

Hydrofluoric acid, anhydrous HF, or any acid medium containing fluoride ion will rapidly attack the metal. One exception to fluoride attack appears to be in chromium plating baths. Hot oxalic acid is the only organic acid known to attack tantalum. The corrosion rates of tantalum in various acid media are given in Table 23.6.

Referring to Fig. 23.1 it will be seen that tantalum shows excellent resistance to reagent grade phosphoric acid at all concentrations below 85% and temperatures under 374°F (190°C). However, if the acid contains more than a few parts per million of fluoride, as is frequently the case with commercial acid, corrosion of tantalum may take place. Corrosion tests should be run to verify the suitability under these conditions.

Figure 23.2 indicates the corrosion resistance of tantalum to hydrochloric acid over the concentration range of 0–37% and temperature to 374°F (190°C).

Figure 23.3 indicates the corrosion resistance of tantalum to nitric acid in all concentrations and at all temperatures to boiling. The presence of chlorides in the acid does not reduce its corrosion resistance.

Fused sodium and potassium hydroxides and pyrosulfates dissolve tantalum. It is attacked by concentrated alkaline solutions at room temperature; it is fairly resistant to dilute solutions. Tantalum's resistance to oxidation by various gases is very good at low temperatures, but it reacts rapidly at high temperatures. Only HF and SO_3 attack the metal under 212°F (100°C); most gases begin to react with it at 570 to 750°F (300 to 400°C). As the temperature and concentration of such gases as oxygen, nitrogen, chlorine, hydrogen chloride, and ammonia are increased, oxidation becomes more rapid; the usual temperature for rapid failure is 930–1200°F (500–700°C). The

TABLE 23.5   Materials to Which Tantalum Is Completely Inert,
Up to at Least 302°F (150°C)

Acetic acid	Hydrogen
Acetic anhydride	Hydrogen chloride
Acetone	Hydrogen iodide
Acids, mineral (except HF)	Hydrogen peroxide
Acid salts	Hydrogen sulfide
Air	Iodine
Alcohols	Hypochlorous acid
Aluminum chloride	Lactic acid
Aluminum sulfate	Magnesium chloride
Amines	Magnesium sulfate
Ammonium chloride	Mercury salts
Ammonium hydroxide	Methyl sulfuric acid
Ammonium phosphate	Milk
Ammonium sulfate	Mineral oils
Amyl acetate	Motor fuels
Amyl chloride	Nitric acid, industrial fuming
Aqua regia	Nitric oxides
Barium hydroxide	Nitrogen
Body fluids	Nitrosyl chloride
Bromine, wet or dry	Nitrous oxides
Butyric acid	Organic chlorides
Calcium bisulfate	Oxalic acid
Calcium chloride	Oxygen
Calcium hydroxide	Perchloric acid
Calcium hypochlorite	Petroleum products
Carbon tetrachloride	Phenols
Carbonic acid	Phosphoric acid, <4 ppm F
Carbon dioxide	Phosphorus
Chloric acid	Phosphorus chlorides
Chlorinated hydrocarbons	Phosphorus oxychloride
Chlorine oxides	Phthalic anhydride
Chlorine, water and brine	Potassium chloride
Chlorine, wet or dry	Potassium dichromate
Chloroacetic acid	Potassium iodide, iodine
Chrome-plating solutions	Potassium nitrate
Chromic acid	Refrigerants
Citric acid	Silver nitrate
Cleaning solutions	Sodium bisulfate, aqueous
Copper salts	Sodium bromide
Ethyl sulfate	Sodium chlorate
Ethylene dibromide	Sodium chloride
Fatty acids	Sodium hypochlorite
Ferric chloride	Sodium nitrate
Ferrous sulfate	Sodium sulfate
Foodstuffs	Sodium sulfite
Formaldehyde	Sugar
Formic acid	Sulfamic acid
Fruit products	Sulfur
Hydriodic acid	Sulfur dioxide
Hydrobromic acid	Sulfuric acid, under 98%
Hydrochloric acid	Water

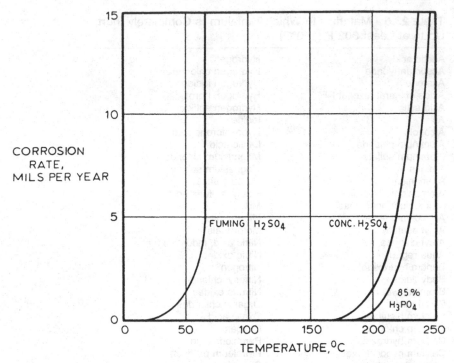

FIGURE 23.1 Corrosion rates of tantalum in fuming sulfuric acid, concentrated sulfuric acid, and 85% phosphoric acid. (From Ref. 2.)

conditions under which tantalum is attacked are noted in Table 23.7. Refer to Table 21.9 for the compatibility of tantalum with selected corrodents.

VI. WELDING

Tantalum is one of the reactive metals. Like titanium, zirconium, and niobium, tantalum will react with contaminants when exposed to them at temperatures as low as 600°F (315°C). It must be remembered that contamination is a time–temperature relationship. Seconds at a high temperature, as in spot welding, or some minutes at 600°F (315°C) when extruding may not be harmful. But there is danger everywhere when heat is present. Once exposed to any atmosphere except inert gases under closely controlled conditions, the metal will, in all probability, become embrittled. Only hydrogen contamination can be removed by outgassing by heating to 1200°F (650°C) and cooling in a vacuum.

TABLE 23.6 Corrosion Rates of Tantalum in Selected Media

| Medium | Temperature | | Corrosion rate |
	(°C)	(°F)	(mpy)
Acetic acid	100	212	Nil
AlCl₃ (10% solution)	100	212	Nil
NH₄Cl (10% solution)	100	212	Nil
HCl 20%	21	70	Nil
	100	212	Nil
Conc.	21	70	Nil
HNO₃ 20%	100	212	Nil
70%	100	212	Nil
65%	170	338	1
H₃PO₄, 85%	25	76	Nil
	100	212	Nil
H₂SO₄			
10%	25	76	Nil
40%	25	76	Nil
98%	25	76	Nil
98%	50	122	Nil
98%	100	212	Nil
98%	200	392	3
98%	250	482	Rapid
H₂SO₄ fuming (15% SO₃)	23	73	0.5
	70	158	Rapid
Aqua regia	25	78	Nil
Chlorine, wet	75	167	Nil
H₂O			
Cl₂ sat.	25	76	Nil
Sea	25	76	Nil
Oxalic acid	21	70	Nil
	96	205	0.1
NaOH 5%	21	70	Nil
	100	212	0.7
10%	100	212	1
40%	80	176	Rapid
HF, 40%	25	76	Rapid

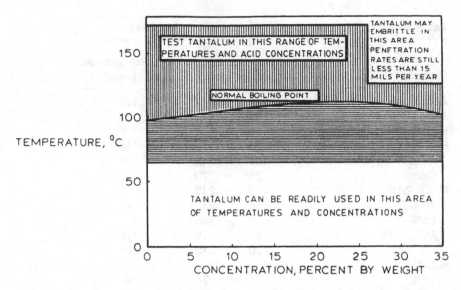

FIGURE 23.2   Corrosion resistance of tantalum in hydrochloric acid at various concentrations. (From Ref. 2.)

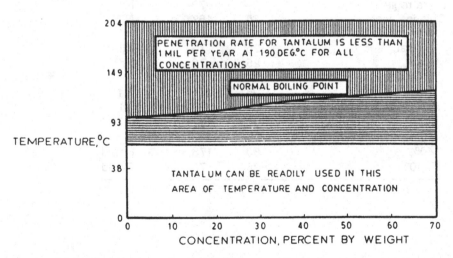

FIGURE 23.3   Corrosion resistance of tantalum in nitric acid at various concentrations and temperatures. (From Ref. 2.)

TABLE 23.7 Temperatures at Which Various Media Attack Tantalum

Medium	State	Remarks
Air	Gas	At temperatures over 572°F (300°C)
Alkaline solutions	Aqueous	At pH > 9, moderate temperature, some corrosion
Ammonia	Gas	Pits at high temperature and pressures
Bromine	Gas	At temperatures over 572°F (300°C)
Chlorine, wet	Gas	At temperatures over 482°F (250°C)
Fluorides, acid media	Aqueous	All temperatures and concentrations
Fluorine	Gas	At all temperatures
HBr, 25%	Aqueous	Begins to corrode at temperatures over 374°F (190°C)
Hydrocarbons	Gas	React at temperatures around 2732°F (1500°C)
HCl 25%	Aqueous	Begins to corrode at temperatures over 374°F (190°C)
HF	Aqueous	Corrodes at all temperatures and pressures
Hydrogen	Gas	Causes embrittlement, especially at temperatures over 752°F (400°C)
HBr	Gas	At temperatures over 752°F (400°C)
HCl	Gas	At temperatures over 662°F (350°C)
HF	Gas	At all temperatures
Iodine	Gas	At temperatures over 572°F (300°C)
Nitrogen	Gas	At temperatures over 572°F (300°C)
Oxalic acid, sat. solution	Aqueous	At temperatures of about 212°F (100°C)
Oxygen	Gas	At temperatures over 662°F (350°C)
H_3PO_4, 85%	Aqueous	Corrodes at temperatures over 356°F (180°C), at higher temperatures for lower concentrations
Potassium carbonate	Aqueous	Corrodes at moderate temperatures depending on concentration
Sodium carbonate	Aqueous	Corrodes at moderate temperatures depending on concentration
NaOH, 10%	Aqueous	Corrodes at about 212°F (100°C)
NaOH	Molten	Dissolves metal rapidly over 608°F (320°C)
Sodium pyrosulfate	Molten	Dissolves metal rapidly over 752°F (400°C)
H_2SO_4, 98%	Aqueous	Begins to corrode at temperatures over 347°F (175°C); lower concentrations begin to corrode at higher temperatures
H_2SO_4 (oleum) (over 98% H_2SO_4)	Fuming	Corrodes at all temperatures
Sulfuric trioxide	Gas	At all temperatures
Water	Aqueous	Corrodes at pH > 9, reacts at high temperatures

Any fusion weld must be performed in an atmosphere free of contamination. This means not only the weld puddle, but all hot metal must be protected. Four different procedures are used to achieve this protection. Although one method may be preferable to another, each has its own limitations.

A.   Electron Beam Welding

This method requires highly sophisticated expensive equipment. Electron beam welding produces good quality welds with narrow weld zones and good penetration. It is useful for intricate, hard-to-reach welds, particularly fillets and tees of different cross-sections. Equipment tooling and setup are all expensive.

B.   Flow-Purged Chamber

A flow purged chamber is used when work is too large to fit any available chamber and the joints are too complex to permit open-air welding. The enclosure is constructed of polyethylene sheet and masking tape. Argon flowing through the "bag" displaces or mixes with entrapped air to a level at which welding can be performed. Argon must be allowed to flow until the work is cooled.

C.   Dry Box, or Vacuum-Purge Chamber

The dry box provides the best inert gas atmosphere possible. The parts to be welded are placed in the box, the chamber sealed, and a vacuum is pulled on the box. The box is then backfitted with high-purity tank argon to a slight positive pressure. Welding is done from the outside by inserting hands into rubber gloves through the sides of the chamber.

The major limitations are the size of the dry box, which limits the size of the work that can be handled, and the skill of the operator.

D.   Open-Air Welding

Only relatively simple joints allowing adequate shielding are possible. Protection must be given to the arc, to the heated metal in front, to the sides, and to the cooling metal behind and underneath the weld bead. The protective atmosphere is provided by using a gentle but adequate gas flow from the maximum-sized cup feasible consistent with good visibility. A blanket of inert gas must be supplied by properly constructed trailing shields, which provide a gas flow until its metal is cooled below the critical temperature. Backup shielding on the underside from the weld bead must also be used to give protection until the metal is cooled.

Absolute cleanliness is essential for good tantalum welding using any of the foregoing techniques. A generally accepted procedure to remove the naturally occurring oxide film before welding is to use an emery cloth to slightly abrade the surfaces to be joined. Then etch with a nitric-hydrofluoric acid solution. Finally, rinse before welding with a solvent such as acetone to be sure that all of the grease is removed. Handle the work only with clean, lint-free nylon gloves.

REFERENCES

1. GT Murray. Handbook of Materials Selection for Engineering Applications. New York: Marcel Dekker, 1997.
2. JB Lambert. Corrosion of tantalum. In: PA Schweitzer, ed. Corrosion Engineering Handbook. New York: Marcel Dekker, 1996.
3. PA Schweitzer. Corrosion and Corrosion Protection Handbook, 2nd ed. New York: Marcel Dekker, 1988.
4. PA Schweitzer. Encyclopedia of Corrosion Technology. New York: Marcel Dekker, 1998.
5. PA Schweitzer. Corrosion Resistance Tables, 4th ed., Vols. 1–3. New York: Marcel Dekker, 1995.
6. PA Schweitzer. Corrosion Resistant Piping Systems. New York: Marcel Dekker, 1994.

24

Niobium (Columbium) and Niobium Alloys

The discoveries of niobium and tantalum were almost simultaneous; however, the similarity of their chemical properties caused great confusion of the early scientists, who tried to establish their separate identities. The confusion was compounded by some scientists using two different names for the same discovery.

In 1801 an English chemist C. Hatchett found a new element. Since he found the element in a black stone discovered near Connecticut he named it columbium after the country of its origin, Columbia, a synonym for America.

A Swedish chemist, A. G. Ekeberg, discovered tantalum only one year later in 1802. He gave it the name tantalum because of the tantalizing difficulty he had in dissolving the oxide of the new metal in acids.

It was not until 1865 that Mangnac separated niobium and tantalum by using the differences in the solubilities of their double fluorides of potassium. In 1905 Dr. W. von Bolten introduced both tantalum and niobium to industry.

The dual nomenclature of niobium and columbium caused confusion and controversy. Niobium was preferred in Europe, and columbium was preferred in the United States. Finally, at the Fifteenth International Union of Chemistry Congress in Amsterdam in 1949 the name niobium was chosen as the recognized international name.

Niobium is a soft, ductile metal which can be cold worked over 90% before annealing becomes necessary. The metal is somewhat similar to stainless steel in appearance. It has a moderate density of 8.57 gm/cc compared to the majority of the high melting point metals, being less than molybdenum at 10.2 gm/cc and only half that of tantalum at 16.6 gm/cc. The physical

TABLE 24.1 Physical Properties of Niobium

Melting point (°C)	2468
Boiling point (°C)	4927
Density (g/cm³)	8.57
Thermal neutron absorption cross-section (barns)	1.1
Electronegativity (Pauling's units)	1.6
Thermal conductivity (s cm°C)	
at 0°C J	0.523
at 1600°C J	0.691
Coefficient of thermal expansion at 20°C ($\times 10^{-6}$/°C)	7.1
Electrical resistivity ($\mu\Omega$)	15
Temperature coefficient $\times 10^{-3}$ (°C)	3.95
Volume electrical conductivity (% IACS)	13.3
Specific heat (J/g)	
at 15°C	0.268
at 1227°C	0.320
Heat capacity (J/mol °C)	
at 0°C	24.9
at 1200°C	29.7
at 2700°C	33.5

properties of niobium are shown in Table 24.1, while mechanical properties are shown in Table 24.2.

Niobium finds many applications as an alloy in a wide variety of end uses, such as beams and girders in buildings and offshore drilling towers, special industrial machinery, oil and gas pipelines, railroad equipment, and automobiles. It is also used as an additive in superalloys for jet and turbine engines.

Niobium is used extensively in aerospace equipment and missiles because of its relative light weight and because it can maintain its strength at elevated temperatures.

TABLE 24.2 Mechanical Properties of Niobium

Modulus of elasticity $\times 10^6$ (kg/cm²)	1.05
Poisson's ratio	0.38
Hardness (VHN)	60–100
Resistance to thermal shock	Good
Workability ductile-to-brittle transition (°C)	−150
Stress relieving temperature (°C)	800

Niobium is available in the form of sheet, foil, rod, wire, and tubing.

I. CORROSION RESISTANCE

A readily formed adherent passive oxide film is responsible for niobium's corrosion resistance. Its corrosion properties resemble those of tantalum, but it is slightly less resistant in aggressive media such as hot concentrated mineral acids.

Niobium is susceptible to hydrogen embrittlement if cathodically polarized by either galvanic coupling or by impressed potential.

Except for hydrofluoric acid, niobium is resistant to most organic and mineral acids at all concentrations below 212°F (100°C). This includes hydrochloric, hydriodic, hydrobromic, nitric, sulfuric, and phosphoric acids. It is especially resistant under oxidizing conditions such as concentrated sulfuric acid and ferric chloride or cupric chloride solutions.

Niobium experiences corrosion rates of less than 1 mpy in ambient aqueous alkaline solutions. Even though the corrosion rate may not seem excessive at higher temperatures, niobium is embrittled even at low concentrations. Niobium is also embrittled in salts such as sodium and potassium carbonates and phosphates that hydrolyze to form alkaline solutions.

As long as a salt solution does not hydrolyze to form an alkali, niobium has excellent corrosion resistance. It is resistant to chloride solutions even with oxidizing agents present. It does not corrode in 10% ferric chloride at room temperature, and it is resistant to attack in seawater.

Niobium is inert in most common gasses (bromine, chlorine, nitrogen, hydrogen, oxygen, carbon dioxide, carbon monoxide, and sulfur dioxide, wet or dry) at 212°F (100°C). However, at higher temperatures niobium will be attacked, in some cases catastrophically.

Niobium is resistant to attack in many liquid metals to relatively high temperatures as illustrated in the following table:

Liquid metal	Maximum temp. (°F/°C)
Bismuth	950/510
Gallium	952/500
Lead	1562/850
Lithium	1832/1000
Mercury	1112/600
Sodium	1832/1000
Potassium	1832/1000
Uranium	2552/1400
Zinc	842/450

Niobium's resistance may be reduced by the presence of excessive amounts of gas impurities.

In general niobium is less expensive than tantalum but possesses similar corrosion resistant properties and is often considered as an alternative for tantalum. Refer to Table 24.3 for the compatibility of niobium with selected corrodents.

II. FABRICATION

Niobium has excellent cold working properties. It can be easily forged, rolled, or swaged directly from the ingot at room temperature. Annealing is required after the area has been reduced by about 90%. Heat treatment at 2192°F (1200°C) for 1 hr causes complete recrystallization of material cold worked over 50%. The annealing process must be performed in an inert gas or in a high vacuum at pressures below 1×10^{-4} Torr. Because of the difficulty ensuring the purity of an inert gas, vacuum is preferred.

The sheet metal can be formed easily by general sheet metal working techniques. The low rate of work hardening eases these operations by reducing springback.

III. MACHINING

The usual machining techniques may be used with niobium, but special attention should be paid to tool angles and lubrication due to the tendency of niobium to gall. It machines in a manner similar to lead or soft copper. Tooling recommendations are as follows:

Approach angle	15 to 20°
Side rake	30 to 35°
Side and end clearance	5°
Plan relief angle	15 to 20°
Nose radius	0.020 to 0.030 in.
Cutting speed	60 to 80 ft/min with HSS
	250 to 300 ft/min with carbide
Feed, roughing	0.008 to 0.12 in./revolution
Feed, finishing	0.005 in. max./revolution
Depth of cut	0.030 to 0.125 in.

IV. DRILLING

Standard high speed drills may be used, but the peripheral lands wear badly and care must be exercised to ensure that the drill has not worn undersize.

TABLE 24.3   Compatibility of Niobium with Selected Corrodents

Corrosion	Concentration wt%	Temperature (°F/°C)	Corrosion rate (mpy)
Acetic acid	5–99.7	Boiling	Nil
Aluminum chloride	25	Boiling	0.2
Aluminum potassium sulfate	10	Boiling	Nil
Aluminum sulfate	25	Boiling	Nil
Ammonium chloride	40	Boiling	10
Ammonium hydroxide		RT	Nil
Bromine, liquid		68/20	Nil
Bromine, vapor		68/20	1.0
Calcium chloride	70	Boiling	Nil
Citric acid	10	Boiling	1.0
Copper nitrate	40	Boiling	Nil
Copper sulfate	40	219/104	1.0
Ferric chloride	10	RT, boiling	Nil
Formaldehyde	37	Boiling	0.1
Formic acid	10	Boiling	Nil
Formic acid	50	Boiling	1.0
Hydrochloric acid	1	Boiling	Nil
Hydrochloric acid (aerated)	15	140/60	Nil
Hydrochloric acid (aerated)	15	212/100	1.0
Hydrochloric acid (aerated)	30	95/35	1.0
Hydrochloric acid (aerated)	30	140/60	2.0
Hydrochloric acid (aerated)	30	212/100	5.0
Hydrochloric acid	37	RT	1.0
Hydrochloric acid	37	140/60	1.0
Lactic acid	10–85	Boiling	1.0
Magnesium chloride	47	Boiling	1.0
Mercuric chloride	sat.	Boiling	0.1
Nickel chloride	30	Boiling	Nil
Nickel nitrate	40	219/104	1.0
Nitric acid	50	176/80	5.0
Nitric acid	65	RT	Nil
Nitric acid	65	Boiling	1.0
Nitric acid	70	482/250	1.0
Oxalic acid	10	Boiling	5.0
Peroxide	30	RT	1.0
Peroxide	30	Boiling	2.0
Phosphoric acid	50	86/30	Nil
Phosphoric acid	50	194/90	5.0
Phosphoric acid	60	Boiling	20
Phosphoric acid	85	86/30	Nil

TABLE 24.3 Continued

Corrosion	Concentration wt%	Temperature (°F/°C)	Corrosion rate (mpy)
Phosphoric acid	85	RT	0.1
Phosphoric acid	85	190/88	2.0
Phosphoric acid	85	212/100	5.0
Phosphoric acid	85	311/155	150
Phosphoric acid	85	Boiling	150
Potassium carbonate	1–10	RT	1.0
Potassium carbonate	10–20	208/98	Embrittle
Potassium hydroxide	5–40	RT	Embrittle
Potassium hydroxide	1–5	208/98	Embrittle
Potassium phospate	10	RT	1.0
Seawater, natural	10	Boiling	Nil
Sodium bisulfate	40	Boiling	5.0
Sodium carbonate	10	RT	1.0
Sodium carbonate	10	Boiling	20
Sodium chloride, sat., pH 1		Boiling	1.0
Sodium hydroxide	1–40	RT	5.0
Sodium hydroxide	1–10	208/98	Embrittle
Sulfuric acid	5–40	RT	Nil
Sulfuric acid	25	212/100	5.0
Sulfuric acid	98	RT	Embrittle
Sulfuric acid	10	Boiling	50
Sulfuric acid	40	Boiling	20
Sulfuric acid	60	Boiling	50
Sulfuric acid	60	194/90	2.0
Sulfuric acid	65	307/153	100
Sulfuric acid	70	332/167	200
Tartaric acid	20	RT, boiling	Nil
Trichloroacetic acid	50	Boiling	Nil
Trichloroethylene	90	Boiling	Nil
Zinc chloride	40–70	Boiling	Nil
Zirconium chloride	70	Boiling	Nil
Zirconium chloride	88	Boiling	Nil

RT, room temperature.

TABLE 24.4 Mechanical and Physical Properties of Alloy WC-103

Density (lb/in.^{3})	0.320
Melting point (°F/°C)	4280 ± 90/2350 ± 50
Thermal expansion × 10^{-6} (cm/cm °C^{-1})	8.73 ± 0.09
Specific heat (Btu/°F/lb)	0.832
Modulus of elasticity × 10^6 (psi)	
at Room temperature	13.1
at 2200°F	9.3

V. WELDING

TIG welding is recommended for niobium using the same setups and practices as used for titanium, zirconium, and tantalum. Refer to those chapters for welding details.

VI. NIOBIUM-TITANIUM ALLOYS

This alloy is fabricated in all forms although it is generally used in multifilamentry cables. The alloys are manufactured in both grade 1 and grade 2 types. Grade 2 material has a higher allowable tantalum content which has no effect on the superconducting properties. A rivet grade niobium–55% titanium alloy is also available, which finds application in the aircraft industry.

VII. WC-103 ALLOY

This is a niobium–10 hafnium–1 titanium alloy. Application is primarily in aerospace programs because of its weight savings over other materials and ability to withstand high stress levels and high temperatures up to 2700°F (1482°C). Mechanical and physical properties are shown in Table 24.4.

TABLE 24.5 Physical and Mechanical Properties of WC-1Zr Alloy

Density (lb/in.^{3})	0.31
Melting point (°F/°C)	4365 ± 15/2410 ± 10
Thermal conductivity at 250°C (Btu/hr ft^2/°F/ft)	24.2
Specific heat at 70°F (Btu/°F lb)	0.065
Modulus of elasticity at room temperature × 10^6 (psi)	10
Charpy impact at 32°F (0°C) (ft/lb)	100

TABLE 24.6 Chemical Composition of Niobium Alloys

	Composition (wt%)					
Alloy	Hf	Ti	W	Y	Zr	Nb
WC-103	10	1				Balance
WC-129Y	10		10	0.1		Balance
WC-1Zr					1	Balance
WC-752			10		2.5	Balance

WC-103 alloy can be easily fabricated and worked. TIG is the best welding method. The same precautions must be observed as with pure niobium.

VIII. WC-1 Zr ALLOY

The creep strength of pure niobium is greatly improved by the addition of 1% zirconium. This is a medium strength alloy which is less expensive than the higher strength alloys such as WC-103. It is used in applications where a high temperature material is required with low loads, such as a load-free thermal shield. Physical and mechanical properties are shown in Table 24.5.

IX. GENERAL ALLOY INFORMATION

In addition to the two niobium alloys discussed, there are other high strength and medium strength alloys available. The chemical compositions are given in Table 24.6, while Table 24.7 gives the tensile properties.

TABLE 24.7 Tensile Properties of Niobium Alloys

Alloy	Ultimate tensile strength $\times 10^3$ (psi)	Yield strength 0.2% offset $\times 10^3$ (psi)	Elongation in 1 in. (%)
Nb	25	11	35
WC-103	54	38	20
WC-129Y	80	60	20
WC-1Zr	35	15	20
WC-752	75	55	20

25

Magnesium Alloys

When strength-to-weight ratio is an important consideration magnesium alloys compete with aluminum alloys. Magnesium has a density of 1.74 g/cm^3, which is 36% less than that of aluminum. However, aluminum is less expensive and has a greater corrosion resistance. The oxide film formed on magnesium provides only limited protection, unlike the adherent protective oxide film on aluminum.

Magnesium alloys have the best strength-to-weight ratio for the commonly die cast metals, in general a better machinability, and often a higher production rate. The primary application of magnesium alloys are for die cast products.

Magnesium alloys are designated by a series of letters and numbers. The first two letters indicate the two principal alloying elements; these are followed by two numbers that state the weight percentages of each element. The next letter in the sequence denotes the alloy developed; the letter C, e.g., indicates the third alloy of the series. AZ91C describes the third alloy standardized that contains nominally 9% aluminum and 1% zinc. Heat treatments are designated in a manner similar to that used for the aluminum alloys, i.e., H10, slightly strain hardened; H23 to H26, strain hardened and partially annealed; and T6 solution heat treated and artificially aged. Some letters used are different from the chemical symbols, e.g., E, rare earths, H, thorium, K, zirconium, and W, yttrium. Table 25.1 lists the chemical composition of selected magnesium alloys.

I. PROPERTIES

Magnesium alloys have high electrical and thermal conductivities, high impact resistance, good damping capacity, and low inertia. The physical prop-

TABLE 25.1 Chemical Composition of Selected Magnesium Alloys

UNS no.	ASTM no.	Composition (wt%)						
		Mn	Zn	Cu	Zr	Al	RE	Y
M16710	ZC71		6.5	1.2				
	ZW3		3.0		6.0			
M11312	AZ31		1			13		
M18410	WE54A						2	5
M16631	ZC6356		6	3				
M18430	WE43A						3	4
	AZM		1			6		
M11918T6		0.3	0.7			9		
M10602F		0.2				6		
M10100F		0.3				10		
M11810T4		0.3	1			8		

TABLE 25.2 Mechanical Properties of Selected Magnesium Alloys

UNS no.	Condition	Tensile yield strength (ksi)	Elongation in 50 mm (%)	Brinell hardness	Fatigue strength $\times 10^8$ cys unnotched (ksi)
M16710	Extruded	29	5	—	—
M16710	Heat treated	43	3	—	—
M11312	Extruded	23	10	50–65	—
M18410	Cast, heat treated	30	4	80–90	14.8
M16631	Cast, heat treated	23	4	55–65	14.5
M18430	Cast, heat treated	28	7	75–95	13.1
M11918	T6 temper	16	—	—	—
M10602	F temper	19	—	—	—
M10100	F temper	69	—	—	—
M11810	T4 temper	69	—	—	—

erties do not vary appreciably with composition. For screening purposes the following values may be used: tensile modulus, 45 GPa; thermal conductivity 30 W/m K; thermal coefficient of expansion, $26 \times 10^{-6}/°C$.

The Mg-Al and Mg-Zn alloys are precipitation hardenable. The Mg-Al alloys have strength on the order of 50 ksi. Mechanical properties of selected magnesium alloys are shown in Table 25.2.

II. CORROSION RESISTANCE

Magnesium resists corrosion in fresh water, hydrofluoric acid, pure chromic acid, fatty acids, dilute alkalies, aliphatic and aromatic hydrocarbons, pure halogenated organic compounds, dry fluorinated hydrocarbons, and ethylene glycol solutions. Ambient temperature dry gases, such as chlorine, iodine, bromine, and fluorine, do not attack magnesium. In coastal atmospheres, the high-purity alloys such as M11918 offer better corrosion resistance than steel or aluminum.

Magnesium is rapidly attacked by seawater, many salt solutions, most mineral acids, methanol and ethanol, most wet gases, and halogenated organic compounds when wet or hot.

Since magnesium is anodic to most metals it is very often used as a sacrificial anode in cathodic protection systems.

26

Lead and Lead Alloys

Lead is a weak metal being unable to support its own weight; therefore alloys have been developed to improve its physical and mechanical properties.

Chemical lead is lead with traces of copper and silver left in from the original ore. It is not economical to recover the copper and silver. The copper content is believed to improve the general corrosion resistance and to add stiffness.

Antimonial lead (also called hard lead) is an alloy containing from 2 to 6% antimony to improve the mechanical properties. It is used in places where greater strength is needed. Hard lead can be used in services up to 200°F (93°C). Above this temperature strength and corrosion resistance are reduced. Cast lead-antimony alloys containing 6 to 14% antimony have a tensile strength of 7000 to 8000 psi with elongation decreasing from 24 to 10%. The lead-antimony alloys in the range of 2 to 8% antimony are susceptible to heat treatment which increases their strength; however, this treatment is rarely employed.

Tellurium lead is a lead alloy containing a fraction of a percent of tellurium. This alloy has better resistance to fatigue failure caused by vibration because of its ability to work harden under strain.

There are also a number of proprietary alloys to which copper and other elements have been added to improve corrosion resistance and creep resistance.

I. CORROSION RESISTANCE

The corrosion resistance of lead is due primarily to the protective film formed by the insolubility of some of its corrosion products.

TABLE 26.1   Compatibility of Lead with Selected Corrodents[a]

Corrodent	Temperature (°F/°C)	Corrodent	Temperature (°F/°C)
Acetic acid	U	Hydrofluoric acid, to 50%	100/38
Acetic anhydride	80/27	Hydrofluoric acid, 70%	U
Acetone	190/88	Hydrogen peroxide	U
Acetone, 50% water	212/100	Hydrogen sulfide, wet	U
Acetophenone	140/60	Hypochorous acid	U
Allyl alcohol	220/104	Jet fuel, JP-4	170/77
Allyl chloride	U	Kerosene	170/77
Aluminum chloride	U	Lactic acid	U
Ammonium nitrate	U	Lead acetate	U
Arsenic acid	U	Lead sulfate	150/66
Barium hydroxide	U	Magnesium chloride	U
Barium sulfide	U	Magnesium hydroxide	U
Boric acid	130/54	Magnesium sulfate	150/66
Butyric acid	U	Mercuric chloride	U
Calcium bisulfite	U	Methyl alcohol	150/66
Calcium chloride	U	Methyl ethyl ketone	150/66
Calcium hydroxide	U	Methyl isobutyl ketone	150/66
Calcium hypochlorite	U	Monochlorobenzene	U
Carbon bisulfide	170/77	Nickel nitrate	212/100
Cardon bioxide, dry	170/77	Nickel sulfate	212/100
Carbon dioxide, wet	180/82	Nitric acid	U
Carbonic acid	U	Oleic acid	U
Chlorobenzene	150/66	Oleum	80/27
Chloroform	140/60	Oxalic acid	U
Chromic acid, 10–50%	212/100	Phenol	90/32
Citric acid	U	Phosphoric acid, to 85%	150/66
Copper sulfate	140/60	Picric acid	U
Cresylic acid	U	Potassium carbonate	U
Dichloroethane	150/66	Potassium cyanide	U
Ethyl acetate	212/100	Potassium dichromate 30%	130/54
Ethyl chloride	150/66	Potassium hydroxide	U
Ferric chloride	U	Potassium nitrate	80/27
Ferrous chloride	U	Potassium permanganate	U
Fluorine gas, dry	200/93	Potassium sulfate, 10%	80/27
Formic acid, 10–85%	U	Propane	80/27
Hydrobromic acid	U	Pyridine	100/38
Hydrochloric acid	U	Salicylic acid	100/38
Hydrocyanic acid	U	Silver nitrate	U

TABLE 26.1   Continued

Corrodent	Temperature (°F/°C)	Corrodent	Temperature (°F/°C)
Sodium bicarbonate	80/27	Sodium perborate	U
Sodium bisulfate	90/32	Stannic chloride	U
Sodium bisulfite	90/32	Stannous chloride	U
Sodium carbonate	U	Stearic acid	U
Sodium chloride, to 30%	212/100	Sulfite liquors	100/38
Sodium cyanide	U	Sulfur dioxide, dry	180/82
Sodium hydroxide, to 50%	U	Sulfur dioxide, wet	160/71
Sodium hydroxide, 70%	120/49	Sulfuric acid, to 50%	212/100
Sodium hypochlorite	U	Sulfuric acid, 60–70%	180/82
Sodium nitrate	U	Sulfuric acid, 80–100%	100/38

The chemicals listed are in the pure state or in a saturated solution unless otherwise indicated. Compatibility is shown to the maximum allowable temperature for which data are available. Incompatibility is shown by U. When compatible, corrosion rate is less than 20 mpy.
Source: Ref. 1.

Lead is resistant to atmospheric exposures, particularly atmospheres in which a protective $PbSO_4$ film forms.

Being amphotoric, lead is corroded by alkalies at moderate or high rates depending on aeration, temperature, and concentration. In caustics, lead is limited to concentrations of 10% maximum up to 195°F (90°C). It will resist cold, strong amines but is attacked by dilute aqueous amine solutions.

Lead will be attacked by hydrochloric and nitric acids as well as organic acids if they are dilute or if they contain oxidizing agents. It is resistant to sulfuric, sulfurous, chromic, and phosphoric acids and cold hydrofluoric acid.

Lead will be attacked by soft aggressive waters, but is resistant to most natural waters. Because of the toxicity of lead salts, lead should not be used to handle potable (drinking) water.

Lead is initially anodic to more highly alloyed materials, but due to a film of insoluble corrosion products on its surface it may become cathodic in time. There have been instances where alloy 20 valves have undergone accelerated attack in lead piping systems handling sulfuric acid services. In such applications it is necessary to electrically isolate the valves from the piping to prevent galvanic action.

Because of the toxicity problems associated with lead burning (joining process), applications of lead have been greatly reduced in modern practice.

Refer to Table 26.1 for the compatibility of lead with selected corrodents.

REFERENCE

1. PA Schweitzer. Corrosion Resistance Tables, 4th ed., Vols. 1–3. New York: Marcel Dekker, 1995.

Index